Concepts in

Biotechnology

Concepts in Biotechnology

Revised edition

Editors

D Balasubramanian

C F A Bryce

K Dharmalingam

J Green

Kunthala Jayaraman

COSTED–IBN

Universities Press

Universities Press (India) Private Limited

Registered Office
3-6-747/1/A & 3-6-754/1, Himayatnagar,
Hyderabad 500 029 (A.P.), India
e-mail: info@universitiespress.com

Distributed by
Orient Blackswan Private Limited

Registered Office
3-6-752, Himayatnagar, Hyderabad 500 029 (A.P.), India

Other Offices
Bangalore / Bhopal / Bhubaneshwar / Chennai / Ernakulam
Guwahati / Hyderabad / Jaipur / Kolkata / Lucknow / Mumba
New Delhi / Patna

First published 1996
Reprinted 1998, 2001
Revised edition 2004
Reprinted 2005, 2009

ISBN: 978 81 7371 483 2

Typeset by
OSDATA, Hyderabad 500 029

Printed in India at
Graphica Printers
Hyderabad 500 013

Published by
Universities Press (India) Private Limited
3-6-747/1/A & 3-6-754/1, Himayatnagar
Hyderabad 500 029 (A.P), India

To

Prof R R Daniel
for his remarkable ability to foster
international cooperation in a
discipline far removed from his own,
and for his patience, insight and genuine interest in
the publication of this book

Contents

About ICSU and COSTED–IBN

The International Council of Scientific Unions (ICSU) was created in 1931 to promote international scientific activity in the different branches of science and their applications for the benefit of humanity. Since its creation, it has vigorously pursued a policy of non-discrimination, affirming the rights and freedom of scientists throughout the world to engage in international scientific activity without regard to such factors as citizenship, religion, creed, political stance, ethnic origin, race, colour, language, age or sex. ICSU is a non-governmental organisation with 92 National Scientific Members, 23 Disciplinary Scientific Unions and 29 Scientific Associates. ICSU also creates interdisciplinary bodies and joint initiatives which include new activities and research of interest to its member bodies. One such joint initiative is the Committee on Science and Technology in Developing Countries and the International Biosciences Networks (COSTED–IBN); it is sponsored by UNESCO.

COSTED was set up as a special Scientific Committee of ICSU in 1966 and the IBN in 1979. Since the main objective of both these bodies is to promote science and technology in developing countries, they were merged as COSTED–IBN in 1993. COSTED–IBN has its headquarters in Madras, India. The Central Secretariat receives its operational expenses from the Government of India, through the Indian National Science Academy. It has 19 National Scientific Members and more than 30 Corresponding Members. There are Regional Secretariats located in Latin America, Africa, the Arab region and Asia. Further information about COSTED–IBN and its activities can be obtained from The Scientific Secretary, COSTED–IBN, 24 Gandhi Mandap Road, Madras 600 025, India.

Preface to the Second Edition

This is the second edition of the book, the first having come out eight years ago. When we started the project, encouraged and ably supported by COSTED-IBN (the Committee on Science and Technology in Developing Countries and the Intenational Biosciences Networks), the subject had but a handful of textbooks that could be used effectively for the undergraduate and first year Master's level curriculum. It thus filled a felt need, particularly in the colleges and universities of the countries of the South, the main focus of COSTED-IBN. We were gratified by the response that the book received across national borders. Universities Press had to run several reprints of the book. One of the features of the book that students found attractive was the boxed items, namely the assignments, self-assessment exercises, and the responses to assignments. In this edition, therefore, we have decided to keep these and expand them a bit as well. After all, we believe that *Concepts in Biotechnology* should be a basic pedagogic textbook that introduces the subject to the starting student, and yet kindles and teases his/her interest a little.

Over these eight years, biotechnology has evolved and developed in rapid strides. Yet the basic concepts and paradigms have not. Thus, we have retained the original thrust of presentation of topics, and the approach to them. Experience in the classrooms, and response from students and teachers of several developing countries have prompted us not to change the basic structure of the book, and also not to add too many topics to a book of this nature, which is meant as an introduction to the area. We have, however, added two chapters, namely biotechnology of foods and beverages, and biotechnology of the environment, based on suggestions from students and faculty from these areas of biotechnology.

Biotechnology is a field that also obeys the IT principle – the moment a book is published, it is obsolete. Thus, a third and revised edition will have to come out within a few years. We hope the present second edition serves the purpose of today's curricula.

D Balasubramanian
February, 2004

Foreword

The Committee on Science and Technology in Developing Countries and the International Biosciences Networks (COSTED–IBN) is a non-profit, non-governmental, interdisciplinary body under the aegis of the International Council of Scientific Unions (ICSU), co-sponsored by UNESCO. The broad mandate of COSTED–IBN is to devise mechanisms for linking science and technology to development in developing countries. It undertakes activities that reflect on specific needs of developing countries for scientific and technological development. Consequently, the thrust areas for COSTED–IBN activities include, *inter alia,* emerging scientific fields of special concern to developing countries.

The promise that biotechnology holds for developing countries is well recognised. An organised approach aimed at the applications of biotechnology in a variety of sectors would prove beneficial for national development. Biotechnology is an important tool that can be applied to various economic sectors of special interest to developing countries, such as the production of food crops, livestock management, human health care, chemical industries, and environmental management. The majority of developing countries lack, however, the financial resources, the infrastructure for basic and applied research and the necessary qualified manpower to exploit such technologies. They must therefore utilise the rapid advances in biotechnology with the optimum use of their modest financial resources, relatively abundant and cheap human resources and the existing indigenous biotechnological knowledge. In doing so, each developing country should evolve its own strategy for reaping the benefits of biotechnology by adapting the indigenous know-how and national research capabilities to developmental problems. Education and training form an essential part of this strategy and enable the building up of the right kind of qualified scientific and technical personnel who can develop innovative solutions to local needs and problems. It is seen that in a majority of developing countries, the curricular programmes in biological sciences are quite often highly compartmentalised and sometimes outdated. Very few of these countries offer regular courses in biotechnology at the graduate and postgraduate levels.

COSTED–IBN became aware of the urgent need for developing a suitable curriculum, possibly in the form of a model textbook aimed at providing undergraduate and postgraduate students in developing countries with a strong base in this emerging, interdisciplinary area. While there are a good number of textbooks on biotechnology in the market, these are very often produced by experts and institutions in developed countries, where the teaching methods, the environment for research and development, infrastructural facilities, problems and needs are to a large extent quite different from those existing in developing countries. Moreover, the prices of these books are not within the reach of the average student in developing countries.

Against this background, COSTED–IBN undertook a project on curriculum development in biotechnology and this textbook is a part of that endeavour. The contents of the textbook are based on a model curriculum drawn up by the editorial team after due consultations with a group of experts constituting the Advisory Committee for the project and taking note of the existing syllabi in biological sciences in some of the developing countries.

The chapters of the book have been contributed by distinguished experts in the field from all over the world; special efforts were made to identify and invite a number of experts from developing countries also to contribute to the textbook. The textbook covers the fundamental principles and concepts which form the basis for the subject and illustrates their applications in selected areas such as health care, agriculture, animal systems, bioprocess technologies and the environment. The textbook also provides in its last chapter, an insight into the impact of biotechnology applications on international competition, trade, societies in developing countries, their economy, way of life and social structure.

COSTED–IBN are to be congratulated on this initiative and on its successful completion. The sponsorship of the project by the group of Biological Unions of ICSU, UNESCO and the ICSU Union of Microbiological Societies (IUMS) is gratefully acknowledged.

November 1995

J.C.I. DOOGE
President, ICSU

Acknowledgements

This textbook is the outcome of a COSTED–IBN project on curriculum development in biotechnology for developing countries. The project involved the development of a model curriculum in biotechnology for undergraduate study, especially aimed at students in developing countries, and the publication of a textbook based on this model curriculum. We hope that this book will be widely used not only in institutions in the developing world but also elsewhere and thereby contribute to improving the teaching of this very important and crucial area which holds great promise for economic development.

COSTED–IBN gratefully acknowledges the generous financial support for this activity from the Bio-Unions of the International Council of Scientific Unions (ICSU), the United Nations Educational, Scientific, and Cultural Organisation (UNESCO) and the International Union of Microbiological Societies (IUMS). The book owes its uniqueness in appeal to the truly international authorship of its various chapters. We are indebted to all the authors who have contributed to the book. COSTED–IBN would like to place on record its deep gratitude to the editorial team for their careful editing and integration of the chapters into the final form. Professor Bryce and his colleagues at Napier University, Edinburgh, U.K. deserve a hearty compliment for enhancing the content and appeal of this textbook through the addition of assignments and self assessment exercises. We are also grateful to them for their help in preparing the final manuscript. The excellent production of the book by Universities Press, an associate of Orient Longman Limited, is deeply appreciated.

Readers may communicate their suggestions for improvement of this textbook to COSTED–IBN and the publishers.

R.R. DANIEL AND VEENA RAVICHANDRAN
Project Coordinators, COSTED–IBN

How to use this book

This book has been designed to present an overview of biotechnology. As biotechnology is concerned with the applications of living cells or their components to the benefit of humankind, the early chapters review the structure and function of biological molecules and living cells, and the way in which cellular structure and function is controlled by the genetic makeup of the cell. These chapters will be of particular benefit for readers without a strong background in cell biology, but it would be of benefit to all readers to review them before reading further into the book. The diverse technologies associated with applications of living systems are then described, together with their various applications. Finally the impact of biotechnology on the environment, and ethical and social implications of this technology, are discussed. The chapters should thus be studied in the order in which they are presented.

Each chapter starts with a chapter outline. Before studying the text examine the contents and familiarise yourself with the general content of the chapter.

The text is designed to be worked through rather than simply read, so it would be helpful if you have a pen/pencil and paper at hand when studying the text. As you read through the text you will encounter assignments which are designed to lead you into discovering facts for yourself using previous knowledge. When you come to an assignment in the text you should read the assignment and make sure what it is you are being asked to do, then put the book down and think of the correct response. Preferably put your response down on paper. Only when you have done this should you compare your response with the model response given at the end of the chapter. As you read through the text you will also encounter self assessment exercises, designed to reinforce important points. Some of these exercises are designed to test your memory of what you have read, but the majority are designed to test your understanding of what you have read. When you encounter a self assessment exercise, think of the correct response and preferably write it down. Only when you have done this should you compare your response with the correct response given at the end of the chapter.

Because this book has been designed to present an overview of biotechnology, it is thereby limited in scope, and is introductory in nature. To direct readers towards appropriate further study of areas of interest, a further reading list is provided at the end of each chapter. The title of the article/book generally gives a good indication of the area covered.

Because the book is limited in content, it has not always been possible to define all scientific and technical terms used in the text. Important scientific/technical terms not defined in the text are thus listed and defined in a glossary at the end of the book.

1 From cell biology to biotechnology

The Office of Technology Assessment of the United States Congress has accepted a wide definition of biotechnology as "any technique that uses living organisms, or substances from those organisms, to make or modify a product, to improve plants or animals, or to develop microorganisms for specific uses". With this broad definition, one readily sees that biotechnology is not a modern practice, but has been practised for centuries. In the area of agriculture, farmers have been crossing plants to produce hybrids and varieties that are improvements over existing ones. Indeed, the Austrian monk Gregor Mendel, who started the field of genetics over a hundred years ago, did that by crossing pea plants. Enthusiastic gardeners have, over the years, generated over four hundred varieties of the rose plant — using the methods of grafting and selecting. Veterinarians and livestock breeders have done similarly with animals. In the field of health, Edward Jenner of Britain realised as early as 1820 that prior infection protected one against the recurrence of smallpox; using this, he vaccinated a dairymaid with previously killed pox virus and provided her with defence against the dreaded disease. The great French scientist Louis Pasteur, who died one hundred years ago, showed how invisible germs spoil milk, wine and cheese, and how the simple act of boiling inactivates them — and in this way developed the important technique which goes by his name — pasteurisation. He also rescued the wine industry of France from disastrous losses by the introduction of proper strains of yeast, proper conditions of aeration, temperatures and storage and other methods in what today is known as downstream processing.

Biotechnology has been intuitively and empirically used in the kitchen for quite some time. Marination, caramelisation, food preservation using naturally occurring substances, pickling, fermentation, tenderising meat using papaya extracts, flavour enhancement using chemicals, gelatinisation, use of skins, bladders and collodion membranes for perm-selective separation — these are some examples of the practice of food biotechnology.

Medicine came of age only in the twentieth century. Yet, it owes its present-day sophistication to centuries of a whole host of experiments — many of which were crude, tentative and even bizarre, but several of which were curiously successful; later research provided the molecular and technological rationale behind some of these success stories. Some instances are the use of saliva to control bleeding and prevent further infection; liver extracts as health builders for convalescents, the herbal products of India, SouthEast Asia, China and Korea as antidepressants, anti-hypertensives, and speciality medicines. The success of the Chinese practice of acupuncture has been understood only in the last decades as arising due to the modulation of the levels of hormones, neurochemicals, agonists and antagonists in the body.

The above historical account provides the background or the perspective against which biotechnology may be viewed. Biotechnology, over the years, has involved using biological systems at any level of complexity or organisation for useful purposes. These range from the level of molecules, molecular assemblies (such as liposomes, emulsions, soaps), cells and their organelles (such as cell membranes, mitochondria, chloroplasts, protoplasts), tissues and organs, to whole organisms themselves. Table 1.1 illustrates how biotechnology can be practised using molecules and materials at various levels of biological organisation.

Table 1.1 Practising biotechnology at various levels

System	Examples	Utilisation
At the level of small molecules	Steroids, lipids, sugars vitamins, coenzymes	Drugs, packaging or encapsulating materials, nutrition, food, cleaning and detergency
Macromolecular level	Proteins, nucleic acids, polysaccharides	Catalysis, energy sources, copying and reproduction
Organelle level	Membranes, cell extracts, chloroplasts, mitochondria	Separation, energy transduction, *in vitro* biochemistry
Cells with walls removed	Protoplasts	Hybrid cells and cell fusion
Cells	All types of microbial, plant and animal cells	Biochemicals, genetic engineering, various purposes of biotechnology
Tissues	From plants and animals for use in medicine and surgery	Biochemicals
Organs and organisms	Plants, fruitflies, nematodes, frogs, mice, rats and rabbits	Chemicals, biochemicals, toxins, immunochemicals, drug testing, transgenics

Historically, biotechnology started out from the whole organism level. The farmer or the gardener crossed plants to produce hybrids or grafted two plants to produce newer varieties. When the new variety bred true, multiplication of an individual to a mass level became possible. A single experiment could thus be repeated on a mass scale thereby becoming a technology. Veterinarians and animal breeders kept careful pedigree records of farm animals and of racehorses. Closely watching the genetic quality of the offspring, they were able to assess which male and female animals would be the best to produce offspring.

When biology came of age, it became increasingly reductionist and enabled us to understand many such macroscopic features in terms of constituent tissues and cells of plants and animals. Agriculture, horticulture, veterinary practice and animal husbandry came to be practised at a level lower than the individual organism. At the cellular level it became understood that there are special types of animal cells called stem cells. These are the initiators or the starting point of the process of cell differentiation. Many cells in the course of their life commit themselves to differentiation and change their morphology, structure and, therefore, function. For example, a myoblast differentiates to produce the long fibrous muscle cells; the cells of the eye lens epithelium change their morphology from a compact globular shape into long fibrous entities that go to make the bulk body of the eye lens. A stem cell, on the other hand, has the ability to divide without limit and to give rise to a progeny which can be terminally differentiated; the stem cell is not at the end of the pathway of differentiation and when it divides, the daughter cell still has a choice of remaining a stem cell like its parent or to go irreversibly towards terminal differentiation (see Fig. 1.1).

This is the reason why a stem cell is referred to as pluripotent and can generate a variety of cells that go to make various tissues and organs. The pluripotent haematopoietic stem cell that is produced in the bone marrow has the ability to become an erythrocyte (red blood cell), a neutrophil, a basophil, a lymphocyte, a monocyte, a macrophage, a megakaryocyte or a mast cell (see Fig. 1. 2).

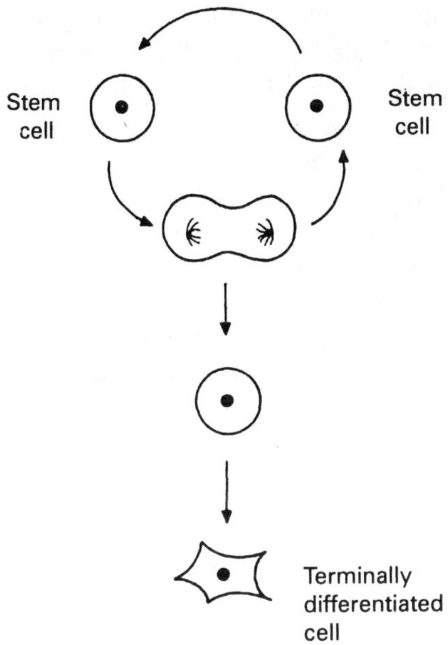

Fig. 1.1 *Differentiation by stem cells.*

Understanding the process and triggering the selective pathway to the production of any one of these would be of advantage in biotechnology. Thus one goes from the organism or the organ level to the cell level in the practice of biotechnology.

In the case of plant cells, differentiation is not necessarily a terminal process such as happens in most animal cells. It is possible to make a plant cell either divide or switch its development. The process of wound healing in plants exemplifies this. When a plant tissue is wounded, cortical cells proliferate around the wound and heal the wound through a mass of undifferentiated cells referred to as callus. It was discovered about 15 years ago that callus cells can be dissociated to form a suspension of single cells, and that this suspension can be maintained in cell culture. When such a culture is nurtured with appropriate nutrients and plant hormones the cells can be induced to grow and produce the whole plant itself (see Fig. 1.3). This discovery has led to the development of plant biotechnology, particularly the micropropagation of plants through callus culture or what is referred to as tissue culture. These examples highlight the growing importance of biotechnology in agriculture and horticulture.

Microbiology is a rich source for biotechnology. Microbes are small, they do not need elaborate inputs for their growth, they multiply rapidly, often with a generation time of less than thirty minutes, they produce mutants naturally and, most importantly from the human health point of view, produce secondary metabolites that are used as antibiotics. Growing microbes, selecting out strains of particular interest as producers of drugs and fine chemicals or for other uses, growing them in large batch cultures and harvesting them is an important area of biotechnology that is practised by the pharmaceutical and chemical industries. It has been possible to introduce mutations randomly in microbial populations and to select mutants of interest. This finding has played an important role in microbial technology.

However, the most important advance from the biotechnological point of view came from those in the molecular genetics of microbes about 20 years ago. It became possible to isolate plasmids which are autonomous, non-chromosomal, cyclic DNA molecules found in the cytoplasm of microbial cells. It further became possible to selectively cut and open the circular DNA that constitutes the plasmid through the use of specific enzymes called restriction endonucleases. Soon it also became possible to introduce DNA sequences into linearised (cut) plasmid DNA and reseal (ligate) the circle. This "cut and paste" method has allowed the introduction of foreign DNA into the plasmids. The discovery of

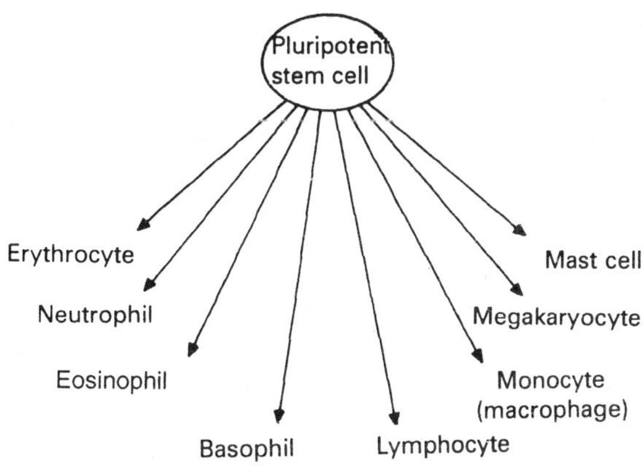

Fig. 1.2 *Differentiation by haematopoietic stem cells.*

Fig. 1.3 *Differentiation by isolated plant cells.* Top left: callus tissue. Bottom left: shoots arising from callus tissue. Right: complete plantlet formed by root induction.

restriction enzymes for cutting DNA at specific spots and ligases that covalently join DNA molecules through phosphodiester bonds were revolutionary steps in molecular engineering. It became possible to gently remove the plasmids from microbial cells and to reintroduce them at will (see Fig. 1.4). This gave birth to the method that is now known as recombinant DNA technology or genetic engineering. Any DNA can be stitched on to any other, and thus it became possible to express mammalian proteins in microbial systems. The advantage of rapid multiplication of microbes elevates the method from the laboratory scale to the industrial scale.

Introduction of foreign genes into organisms is not limited to microbes, but is also possible in plants and animals. This creates opportunity for transgenic technology. This is also called "add a gene and thus a trait". The traits could be any — resistance to pests, increased production of nutritive agents such as protein, novel molecules that were not present in the original organisms, and so on. In essence, the earlier methods of trying a variety of experiments by 'hit and miss' methods on the whole organism or the whole plant or animal are now replaced by this transgenic technology.

Proteins and nucleic acids are polymers that are made of a small number of constituent monomers. The back-bone chemical bond that connects the monomers in each case is a simple one that repeats along the chain — the peptide bond in proteins and the phosphodiester bond in nucleic acids. This simplicity allows for automated stepwise assembly line chemical synthesis of these molecules in the laboratory. Automated peptide synthesisers and gene assemblers are available on the market. These have helped in the design of peptide drugs through what has come to be known as the combinatorial chemistry approach, and the large scale amplification of DNA chains through the "polymerase chain reaction". While the synthesis of peptides is an "open" or "free" affair, that of DNA is generally a template-assisted one, or a copying process. This lets us vary a base or a monomer sequence in the chain at will, and thus to do what is known as "site-directed mutagenesis", or do gene mutations in the laboratory in a specified manner. Since proteins are synthesised in the cell from the relevant gene sequences, this method enables the synthesis and study of protein chains with predetermined changes in the amino acid sequence.

Lastly, there is the application of cell biology to the technology of production of antibodies and vaccines, and for the generation of diagnostic kits. The immune system of mammals can be challenged with any "foreign substance", or antigen, to produce antibodies — molecules that fight the antigen by combining with it and removing it from the system. However, all normal cells have a limited life span and this limits the time of effectiveness of immunity. The only cells that are "immortal" are several types of cancer cells that proliferate endlessly. Recently it has become possible to exploit this situation by combining the antibody-producing immune cells (lymphocytes) with a certain class of cancer cells (myeloma) to give an immortal cell line or "hybridoma" that produce the antibody in a cloned fashion *ad infinitum*. This hybridoma technology has caused a revolution in disease detection and treatment.

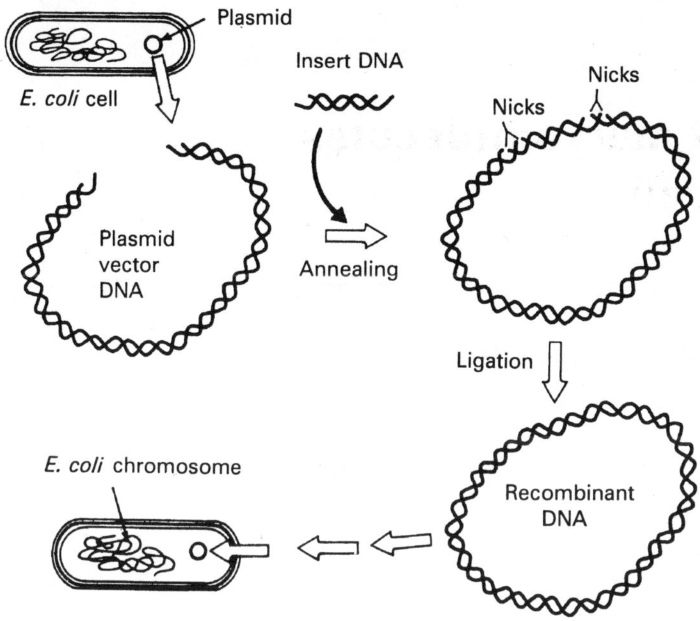

Fig. 1.4 *Genetic engineering.*
Insertion of foreign genes into a bacterium.

One thus sees the spread of biotechnology in diverse areas such as agriculture, drug design, animal health, human health care, industrial chemistry, in environmental protection and in biodiversity. In this book we shall look at several facets of such applications. The running theme would be largely at three levels — the cell level, the gene level and the molecular level.

The introduction of biotechnology to produce new molecules, new plants and new animals also poses challenges of the type that had not been faced before. Some of these are technological and some of these are ethical. If a new product is made through the introduction of a gene from another organism, is it appropriate to patent such a product or process? Patenting had so far dealt with inventions of the kind that had not been encountered in the natural world before. Yet what one encounters in biotechnology is the shuffling of naturally occurring genes from one cell to another. How should one view this — as an invention? If so, could one patent animals and even humans? Likewise, what we understand today in reproductive biology lets us know beforehand whether the yet-to-be-born baby is male or female. It might even be possible to select the sex of the baby ahead of time. What would be the ethical implications of this? We look at some of these issues briefly in this book.

In writing this book we have assumed only a high school knowledge of chemistry and physics. The level of biology assumed is a little higher but, as far as possible, we have attempted to keep each chapter as a read-alone, self-contained one. The reader would find, however, that a study of the ensuing three chapters to be helpful for the later ones.

2 Interplay of macromolecules in a living cell

Chapter outline

2.1 Biological molecules

Cells contain molecules of great variety; these molecules may be classified according to some main structural characteristics. In the first place, there are small molecules, composed of a few carbon atoms, which can then combine among themselves or with other kinds of molecules, organic or inorganic, increasing the possibilities of variation and complexity.

The main classes of structural and functional molecules in cells are carbohydrates (or sugars), fatty acids, amino acids and nucleotides. These molecules may be linked together to form larger molecules referred to as polymers.

Assignment 2.1

Give the terms, each beginning with 'poly', that describe a polymer of (i) sugar molecules, (ii) amino acid molecules, (iii) nucleotides.

A large polymeric molecule, or more than one polymer associated into a large molecule, can be referred to as a macromolecule. A polymer composed of repeating units of the same molecule is referred to as a homopolymer; for example starch is a homopolymer composed of linked glucose molecules. A heteropolymer is composed of two or more different molecules; for example proteins are complex polymers composed of a variety of amino acids.

2.1.1 *Carbohydrates and polysaccharides*

Sugars are molecules composed of carbon, hydrogen and oxygen, which participate in the proportion $(CH_2O)n$, and for this reason they are called carbohydrates (hydrates of carbon). The structure of glucose is shown in Fig. 2.1.

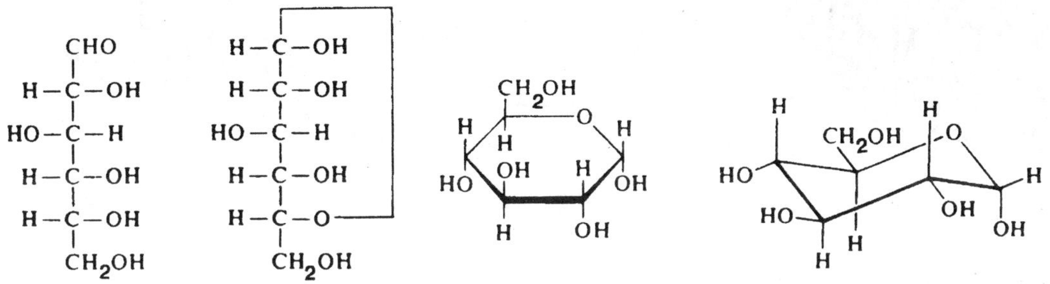

Fig. 2.1 *Representations of the α-D-glucose molecule.*

Units of glucose are capable of polymerising in numbers up to 5000 or more, to form large polysaccharide molecules, such as cellulose, glycogen and starch. Glycogen is the most important form of glucose storage in animal cells. Starch is produced by plants in grains, and it is a major source of food for humans. Starch thus represents the direct storage form of solar energy, which can be used by animals. Cellulose is perhaps the most abundant molecule on earth, as it is the main constituent of plant fibre and wood.

Besides glucose and other simple sugars (or monosaccharides), derivatives of sugars such as uronic acids, amino sugars, or acetylated sugars, may also form polymers. These polysaccharides play different roles in organisms: some of them are lubricants, such as those present in joints and mucous membranes; others are distinctive molecules of cells or microorganisms, responsible for their identification (this is the case, for instance, in the polysaccharides which distinguish the blood cells of groups of individuals from those of others).

2.1.2 *Lipids*

Lipids are composed of a relatively small number of fatty acid derivatives. Esters of glycerol with fatty acids are known as neutral fats, and are an important storage material of plants and animals, having a very high energy content. Other esters, such as lanolin or beeswax, are esters of fatty acids with high molecular weight alcohols. Phospholipids are somewhat more complex than neutral fats, since in their composition, besides glycerol and fatty acids, phosphoric acid and several organic bases (especially choline, ethanolamine or serine) or alcohols (inositol) participate.

Although lipids are not macromolecules, they conglomerate to form the basic structure of membranes of cells by virtue of the structure they acquire when present in water. As Fig. 2.2 shows, the hydrocarbon parts of phospholipids are repelled by water (they are hydrophobic), forcing them to associate in the form of a double layer. The polar portion, formed by phosphoric acid, the nitrogen base and glycerol, constitute a polar "head", which may be in contact with water.

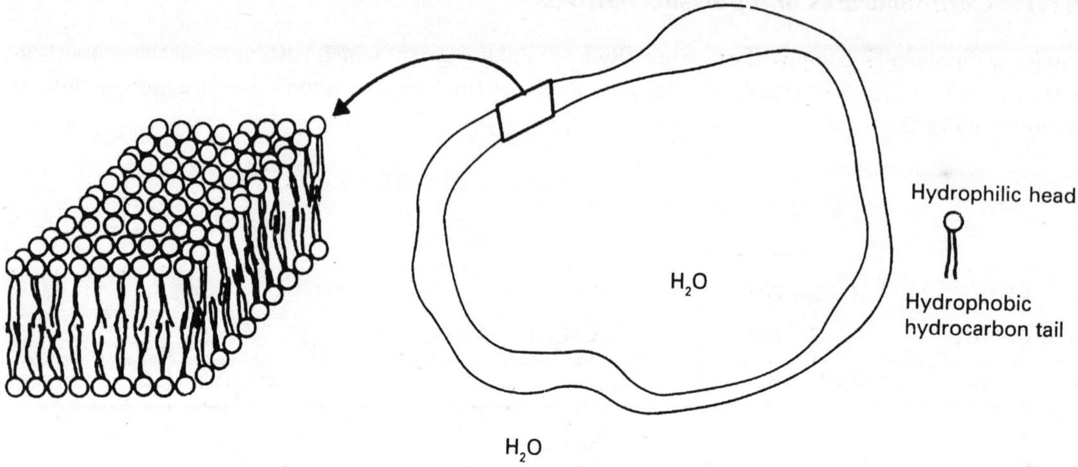

Fig. 2.2 *The structure of phospholipids in water.*

> ***Assignment 2.2***
>
> What advantages are there to having a cell and (for eukaryotic cells) its internal compartments enclosed by a phospholipid membrane?

The phospholipid bilayer is the basic structure of cell membranes, on which other components are included to provide the functional characteristics of the membranes, such as transport, communication, and hormonal responses.

2.1.3 Proteins

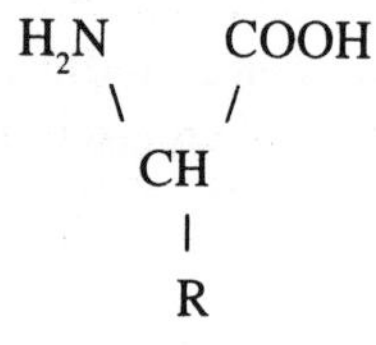

Fig. 2.3 *General structure of amino acids.*

Proteins are important macromolecules and are formed by units, called amino acids, which have the general structure shown in Fig. 2.3.

Peptide bonds form between the carboxyl groups of one amino acid and the amino groups of another to form a polypeptide, as shown in Fig. 2.4.

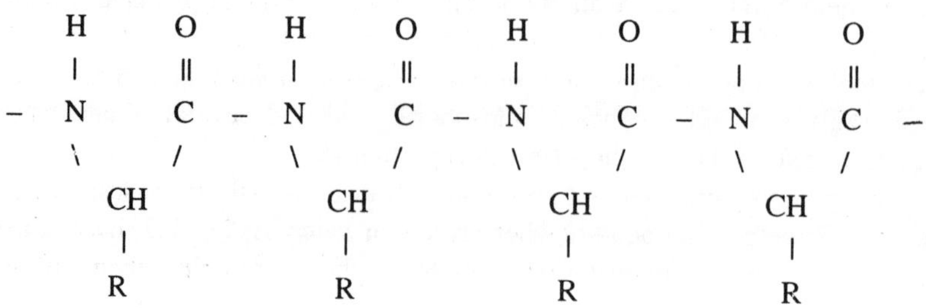

Fig. 2.4 *Structure of a polypeptide.*

The primary structure of a protein relates to the sequence of the amino acids in the polypeptide. The secondary structure of a protein relates to essentially regular, repeating relationships which amino acids have with their near neighbours in the linear polypeptide, principally forming α-helices and β-structures. The tertiary structure of the protein arises from folding of the polypeptide chain upon itself, to give a three-dimensional structure. An example of this is given in Fig. 2.5.

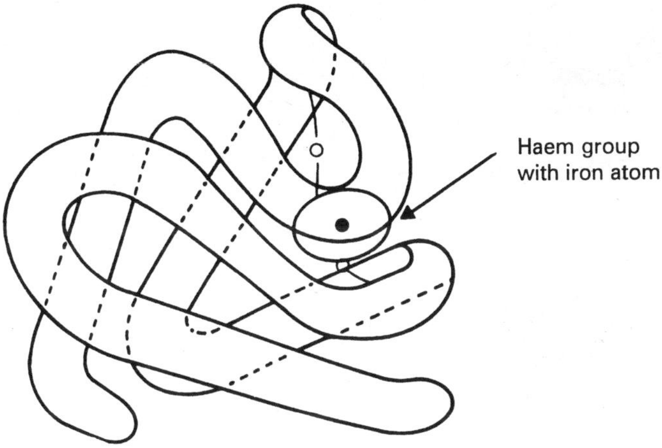

Fig. 2.5 *The tertiary structure of haemoglobin.*

2.1.4 *Nucleic acids*

Nucleic acids are macromolecules formed from repeating units called nucleotides (Fig.2.6). A nucleotide consists of a purine or pyrimidine base, linked to a sugar-phosphate. In deoxyribonucleic acid (DNA), a purine (adenine[A] or guanine [G]) or a pyrimidine (thymine[T] or cytosine[C]) is linked to deoxyribose sugar-phosphate. In ribonucleic acid (RNA), a purine (adenine or guanine) or a pyrimidine (uracil [U] or cytosine [C]) is linked to a ribose sugar-phosphate. In nucleic acids the purines or pyrimidines are referred to as bases, and a base attached to a sugar (without the phosphate groups) is referred to as a nucleoside.

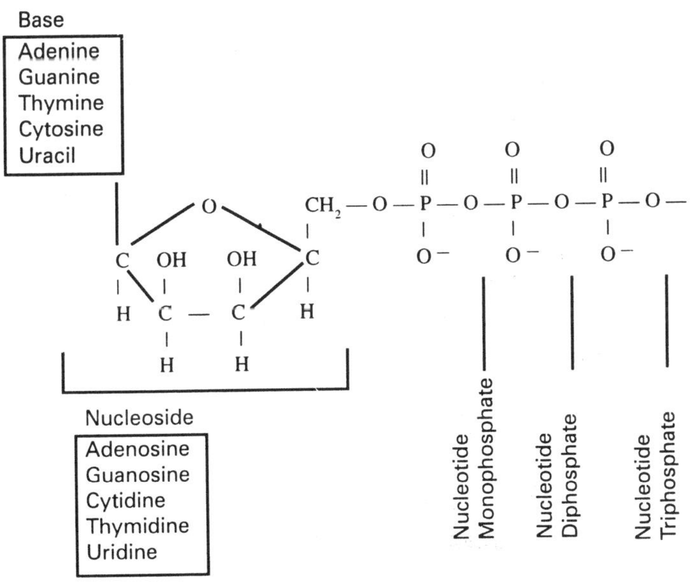

Fig. 2.6 *Structure of nucleotides.*
In this figure the sugar molecule is a deoxyribose sugar — for ribonucleosides and ribonucleotides this would be replaced with a ribose sugar molecule.

Nucleotides are polymerised by the reaction between the

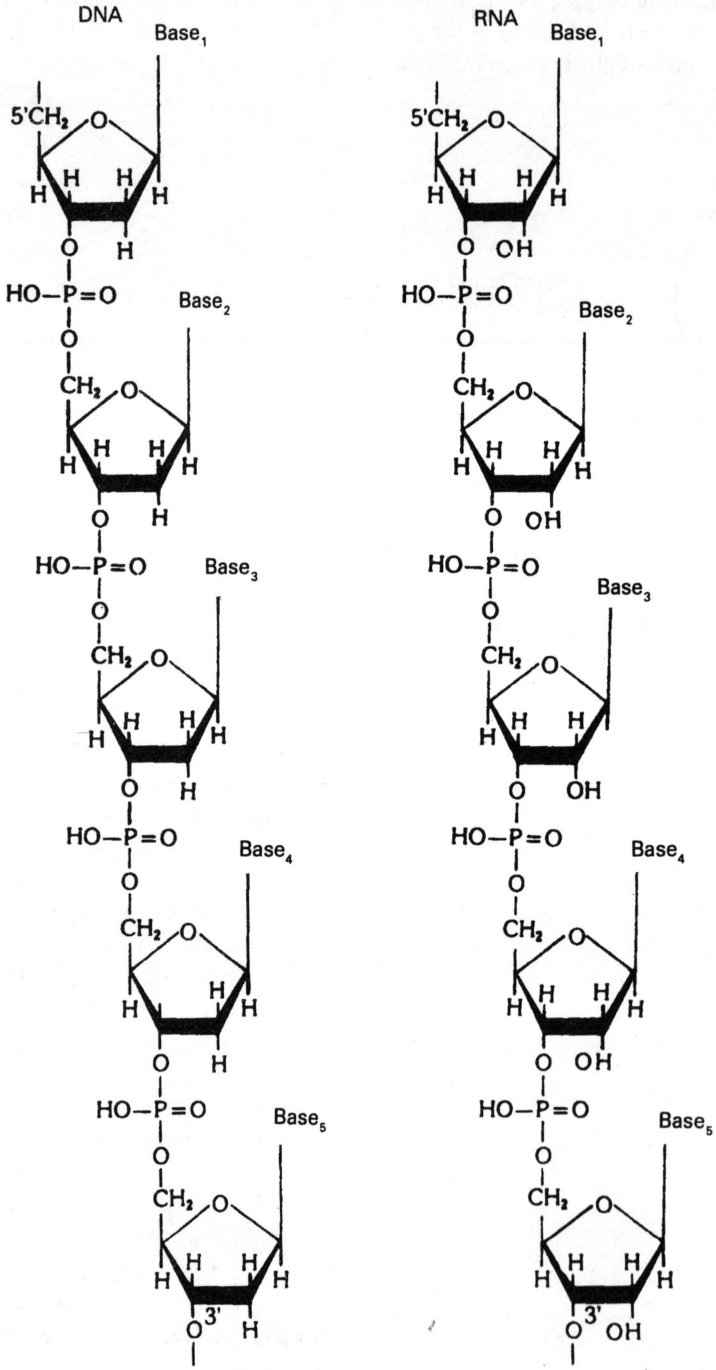

Fig. 2.7 *Structure of DNA and RNA polynucleotides.*

phosphate group of one nucleotide with the sugar moiety (part) of another, to produce a long polymer called a polynucleotide (Fig. 2.7). The DNA macromolecule in living cells is normally present in the form of two nucleotide polymers closely associated with each other to form a twisted spiral (a double helix). The two strands (polymers) in the double helix are held together by hydrogen bonding between adjacent bases in the helix (Fig. 2.8). Thus in DNA, across the double helix structure, bonding between the two strands is such that an A base in one strand bonds with a T base in the other strand (and *vice versa*) and a C base in one strand bonds with a G base in the other strand (and *vice versa*). The two strands in a DNA molecule thus have a complementary arrangement of bases.

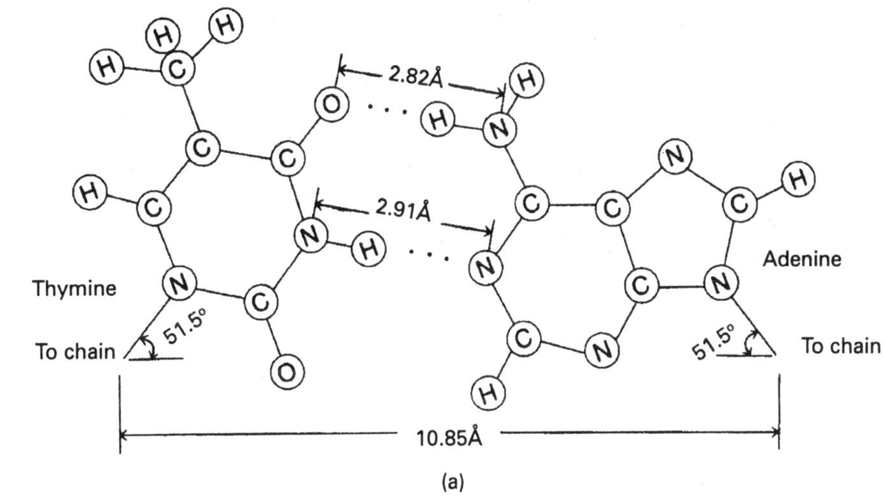

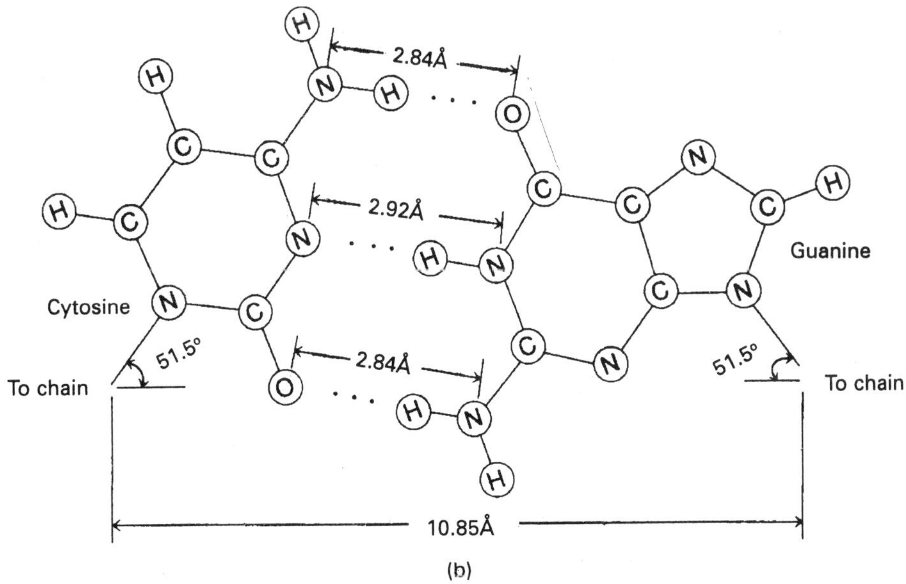

Fig. 2.8 *Hydrogen bonding in the DNA double helix.*

Assignment 2.3

The sequence of nucleotides in one strand of DNA in a double helix is

- A T T G C A T C -

What will the complementary base sequence in the other strand in the double helix be?

DNA comprises the major genetic information in living cells and organisms. The genetic message (or code) is determined by the sequence of bases in the DNA polynucleotide (see Chapter 4). Bearing in mind that multicellular organisms usually are derived from a single cell, and that for all organisms it is vital to pass down from generation to generation an exact copy of the genetic information contained in DNA, it stands to reason that a mechanism exists for the exact replication of DNA. The complementary nature of the two strands of the DNA molecule and the sequential mode of replication of the two strands in the replication process ensure that this is the case. This is described in more detail in Chapter 4. It is important to remember that although DNA may contain the genetic information of a cell, its replication and transmission to daughter cells (the cells formed by cell replication) depends on the existence of other cellular elements (molecules, macromolecules and cellular structures). Viruses are a good example of this, being composed basically of nucleic acid (DNA or RNA) surrounded by a protective coat; their nucleic acid can only be replicated once the virus has infected a host cell with its nucleic acid and 'hijacked' the synthetic machinery of the infected cell.

Nucleic acids, having their phosphate groups dissociated, acquire negative charge, and associate frequently with proteins. In eukaryotic cells, DNA associates with histones, proteins with a strong positive charge. Other proteins participate to organise DNA in the form of complex structures called chromosomes, among which the different genes are distributed. Ribosomes, responsible for protein synthesis with the participation of other RNAs, contain ribosomal RNA (rRNA) which is synthesised in the nucleolus as a precursor called pre-rRNA. Once formed, pre-rRNA immediately associates with proteins. Pre-rRNA is further processed, and finally a structure, the ribosome, composed of two subunits is formed. As well as rRNA, ribosomes contain proteins, several of which are enzymes essential for the synthesis of proteins.

When cells are growing and dividing, they go through a series of stages or phases. The first or G1 (first gap) phase is one of growth and preparation for cell division, for which duplication of DNA will be required. Duplication of DNA takes place during a rather short time during the cell cycle; the so called S (synthetic) phase. After a second gap phase called G2, the cell divides during the M (mitotic) phase . In this way, an equal set of DNA molecules can be transferred to each of the daughter cells.

2.2 Relationship between nucleic acids and proteins; protein synthesis

2.2.1 Protein synthesis

Besides transmission of information from one generation of cells to another, it is necessary to use the genetic information in DNA in the direction of cellular function. To this purpose, an intricate system exists in cells to synthesise proteins, as outlined below.

(1) A molecule of RNA, called mRNA (messenger RNA), is synthesised from DNA, according to a very simple rule, similar to that of DNA duplication. A complementary correspondence exists between the nucleotides of DNA and those of the newly synthesised molecule of RNA; thus from adenine in the DNA, a uracil appears in mRNA (not a thymine as in DNA); from cytosine in DNA, a guanine appears in mRNA; from a thymine in DNA, an adenine appears in mRNA; from

guanine in DNA, a cytosine appears in mRNA. In this form, mRNA carries the message contained in DNA (which is situated in the nucleus in eukaryotic cells) to the cytoplasm of the cell. This message is code-written in "three letter words"; that is, different combinations of bases in the DNA in groups of three (termed codons) determine how each amino acid is placed after another in the protein formed from the DNA template. The process of synthesising mRNA from DNA is referred to as transcription, as the base sequence (or codon sequence) in the DNA template is transcribed or 'rewritten' into a complementary sequence on the mRNA synthesised from the template.

> ***Assignment 2.4***
>
> What will be the complementary sequence transcribed in mRNA from the DNA template sequence A T T G C A T C?

(2) Complex structures, ribosomes, are present in the cytoplasm of the cell, to carry out the "translation" of the message brought by mRNA into a polypeptide chain. Ribosomes are composed of ribosomal RNA, enzymes and other proteins required to carry out the whole process.
(3) A series of molecules of RNA, called transfer RNA (tRNA) are capable of each recognising on the one hand a specific amino acid, and on the other a specific codon in the mRNA.
(4) Several enzymes, one for each amino acid, exist in the cytoplasm, and they couple each to a specific tRNA and "activate", that is, provide each amino acid with the energy required to attach to another, to form a chain.
(5) The process of translation is shown in Fig. 2.9; the first step is the binding of mRNA to a ribosome.

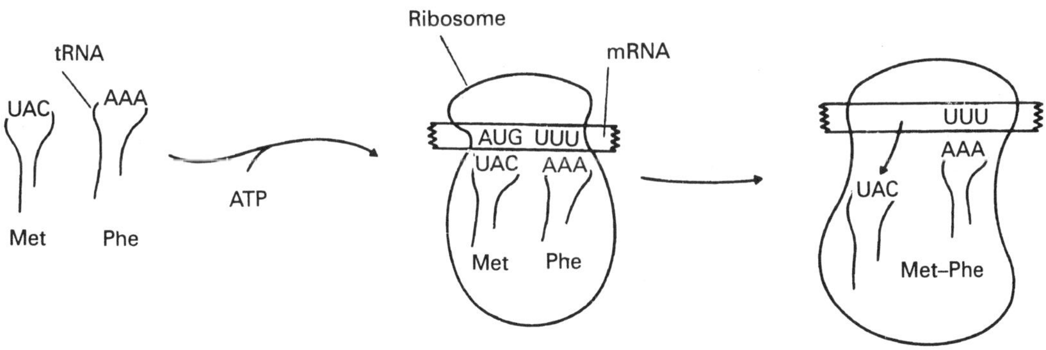

Fig. 2.9 *The basic steps of translation.*

As each specific type of tRNA binds its corresponding amino acid, to form aminoacyl tRNA, it finds its way to the ribosome where mRNA has bound. The process of combining the amino acid and its corresponding tRNA has the effect of providing each amino acid with a means of identifying its place in the "message" carried by mRNA.

Thus, the mRNA has a message written in the sequence of its triplets of bases (codons). When attached to the ribosome, the mRNA is sequentially 'read' and each codon in it is recognised by a specific tRNA carrying a specific amino acid. Amino acids are thus sequentially added to the growing polypeptide, according to the sequence of codons in the mRNA (which itself reflects the sequence of codons in the DNA from which the mRNA was transcribed).

The specificity of the process is determined by the fact that each type of tRNA has in a loop in its structure a triplet of bases, termed an anticodon, which is complementary to the codon. For example, the AUG codon on mRNA is recognised by tRNA with the CAU anticodon (expressed in this way because of the antiparallel nature of the binding), which carries the amino acid methionine. Thus the codon AUG in mRNA is translated into methionine in the growing polypeptide. If the next codon along the mRNA is UUU, this is recognised by tRNA with the AAA anticodon which carries the amino acid phenylalanine. Thus the next amino acid added to the polypeptide is phenylalanine. Ribosomal enzymes catalyse the peptide bond formation between the amino acids.

The process of synthesis of a polypeptide from an mRNA template is called translation, as the sequence of codons (message) is translated or "converted" into a sequence of amino acids.

2.2.2 Protein structure

As already described, proteins are synthesised as amino acid chains, which attain great complexity. The first level of organisation of amino acids within polypeptides is simply the order in which the amino acids are arranged; this is called the primary structure. The second level of organisation is provided by the interactions of amino acids with their neighbours, and they may organise as helices or as sheets; this organisation is known as the secondary structure. A third level of organisation is produced by the attractions or repulsions existing between amino acids in the polypeptide chains of proteins, due to the presence of charges, the formation of hydrogen bonds, van der Waals interactions, or the repulsion by water of hydrophobic residues of the amino acids. The structural characteristics of proteins resulting from this kind of interactions are known as the tertiary structure of proteins. Finally, some proteins are constituted by more than one polypeptide chain, giving rise to further interactions between the several chains that constitute them; this level of organisation is known as the quaternary structure of proteins. The different and multiple interactions of the amino acid chains leading to the formation of the final three-dimensional structure are essential for the activity of the protein.

Proteins are usually large molecules and within cells they may be structural (building blocks) or functional (with enzymatic or transport functions). For functional proteins, their effective action results from interactions within only a small portion of the molecule, which for transport proteins or enzymes is called the active site. The interaction of an enzyme with its substrate, for instance, is that of one molecule (the enzyme), frequently several thousand times larger than the other (the substrate). However, although the function itself takes place in this very small site, even slight changes in the structure of sites or amino acid sequences located far from the active site may produce great changes or even the complete loss of the enzyme function by altering the conformation of the molecule.

Self assessment exercise 2.1

Which of the following could correctly be referred to as a heteropolymer?

(i) A polysaccharide containing glucosamine and glucose.
(ii) A polypeptide containing six different amino acids.
(iii) A strand of DNA.
(iv) mRNA.

Self assessment exercise 2.2

The following list gives stages in the synthesis of a protein from a DNA template in a eukaryotic cell. List the stages in the correct sequence (the order in which the stages occur).

(i) mRNA attaches to a ribosome.
(ii) The amino acid is added to the growing polypeptide by formation of a peptide bond.
(iii) mRNA migrates from the nucleus to the cytoplasm.
(iv) An aminoacyl tRNA binds to its specific codon on mRNA.
(v) mRNA is transcribed from its DNA template.

2.3 Proteins as enzymes, and metabolic pathways

Proteins are responsible for complex functions; probably the best known is that of enzymes, which have the role of accelerating the chemical reactions of cells. It is important to point out that enzymes do not take part in the chemical reaction themselves, but only accelerate greatly (catalyse) changes that would otherwise take place extremely slowly.

Enzymes are important in metabolism; chemical processes in which substrates are converted into energy and cell products through a series of reactions. The main difference between enzymes and inorganic or man made catalysts is that enzymes are finely regulated by many factors, starting with the concentration of the substrates on which they act. Enzymes are also regulated by the internal conditions of the cell, or environmental factors such as pH, temperature and ionic strength. The presence of surrounding molecules may be important, as is the case of membrane proteins, for which the nature of surrounding phospholipids is of fundamental importance.

Enzymes act with different velocities according to their intrinsic properties, derived essentially from their structure. These intrinsic properties also respond to environmental factors such as pH, temperature and ionic strength. Many enzymes seem to be designed specially to function in special environments. For instance, pepsin has the capacity to work at very low pH, since it has to act in the stomach, where pH is close to 1. Enzymes of thermophilic bacteria are capable of working optimally at very high temperatures, while animal or other bacterial enzymes are inactivated at temperatures above 40°C.

One difference between enzymes and inorganic (or, in general, artificial) catalysts, is the fact that velocity of reaction is dependent on substrate concentration; this is due to the fact that enzymes (E) act by reacting with the substrate (S) to produce the product (P) according to the following reaction:

$$E + S \rightarrow ES \rightarrow E + P$$

The theoretical treatment of this reaction produces the equation of Michaelis, which is the following:

$$v = \frac{V_m \, [S]}{K_m + [S]}$$

in which, v is the actual velocity of the enzyme reaction; V_m is the maximal velocity that can be reached; [S] is the substrate concentration, and K_m is the substrate concentration which results in $v = 0.5\, V_m$.

> ***Assignment 2.5***
>
> At a given concentration of substrate, what effect would variation in the value of K_m have on the velocity of the reaction?

The two constants, V_m and K_m can be obtained under defined conditions, and allow the characterisation of the catalytic capacity of an enzyme. Of course, these values vary depending on the temperature, pH and other characteristics of the medium in which enzymes act.

Metabolic pathways consist of series of enzymes functioning after one another, in such a way that the product of the first is the substrate for the second, the product of the second is the substrate for the third, and so on. Enzymes catalyse steps in a metabolic pathway and the slowest step may become the limiting reaction in the series. Enzymes are actually interconnected in a network of reactions in which a single molecule may be the substrate of several enzymes, or enzymes may act on substrates coming from different pathways.

Enzymes, as proteins, are synthesised by the mechanisms already described, but are also degraded by enzymes called proteases. Some enzymes are found in cells in constant amounts, which depend on the relation between their synthesis and degradation rates. Others are regulatable by virtue of the capacity that cells have to regulate their rate of production by induction or repression mechanisms.

There are other mechanisms of metabolic regulation in cells, depending on the presence of enzymes whose activity is sensitive to inhibition from other substances or metabolites in the cells, allowing such metabolites to modulate the rates at which metabolism progresses in some of the multienzyme systems.

A given pathway may exist in different types of cells or organisms, with rather similar characteristics. Glycolysis, for instance, appears in more or less the same form in brain, muscle or adipose tissue cells. However, in some organisms, variations exist which, besides allowing physiological adaptation for organisms, have allowed industrial use for the production of particular substances. An example has existed for centuries, with the use of yeast for fermentation and the production of wine and alcoholic beverages, taking advantage of the variation of glycolysis in this organism, which produces ethanol and CO_2 instead of lactate or pyruvate. This same pathway can be modified also, by altering the function of its enzymes, or modifying its products, to produce glycerol. Industrial products produced from microbial cells in culture, and the ways in which metabolic pathways can be manipulated to maximise product yields are described in Chapter 8.

2.4 Proteins as transport systems

There are proteins principally located in cell membranes which, under the same general principles of enzymes, act on small or large molecules, but instead of transforming them, simply catalyse their movement from one side of a membrane to the other. These are the transport proteins, of which there are different kinds with different mechanisms of action.

Membranes, constituted as a basic double layer of phospholipids, are impermeable to most molecules, unless they are small and hydrophobic. The effective boundary produced by membranes, however, requires a large variety of transport proteins to allow the movements of molecules, both in and out of the cell or of a particular cellular compartment. The movement of a molecule through a membrane may occur according to its electrochemical potential. Glucose, for instance, may move into many cells depending on its concentration difference between the exterior and the interior of the cell. Ions, however, may move against their concentration gradient, provided a membrane electric potential difference exists, which assures its accumulation in one side of the membrane. In this form, cells are capable of concentrating molecules or ions, using as the driving force the electric potential difference at both sides of the membrane.

The general function of transport proteins allows them to be classified as primary or secondary, depending on the connection they have with the energy sources in the cells. Primary transporters or "pumps" are those which can directly use energy (which may be ATP, reducing power or light) to "pump" ions into or out of the cells or compartments, mostly against their concentration gradients. These membrane proteins are both the basis of energy transformations and also may establish the conditions required to move other substances into or out of the cell. One example is that of bacteriorhodopsin of *Rhodospirillum rubrum*, a marine bacterium. This molecule can capture light energy and use it to move protons (hydrogen ions: H^+) to the outside of the membrane, generating both a membrane potential and a pH difference between the inside and the outside of the bacterial cell (Fig. 2.10). There is another protein, called ATPase, which can do the same by using the energy of ATP while splitting it to ADP and inorganic phosphate. ATP (adenosine triphosphate) contains "energy-rich" phosphate bonds. Upon release of a phosphate group from ATP, ADP (adenosine diphosphate) is formed, and the bond energy is released and made available to drive cellular processes which require an energy input. ATPase can, however, function in the reverse sense; if a proton gradient exists, when protons go through it, it can catalyse the reverse reaction, that is, the synthesis of ATP. *Rhodospirillum rubrum,* in this way, possesses a simple mechanism to transform light energy into ATP; bacteriorhodopsin pumps protons to the outside of the cell, and when they come back through the ATPase (which is also called, more properly, ATP synthetase), they provide the energy required for the synthesis of ATP.

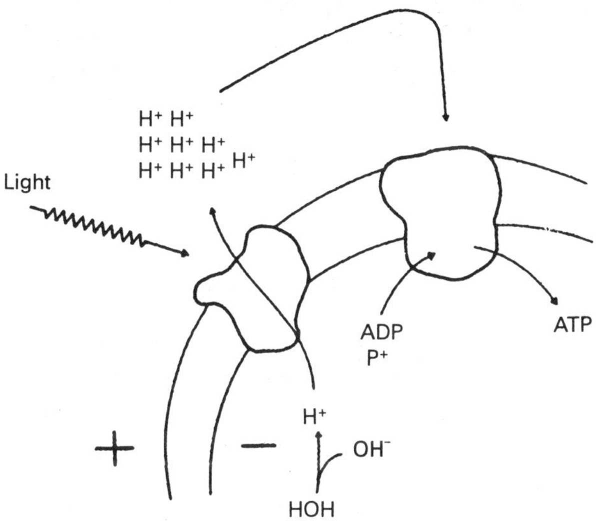

Fig. 2.10 *The coupling of bacteriorhodopsin and proton "pumping" to ATP synthesis.*

Mitochondria and bacterial cells have in their membranes a system called the electron transport chain, and some of its components have the capacity to pump protons outward. The membranes also have an ATP synthetase which can utilise the proton gradient and synthesise ATP from ADP and inorganic phosphate.

Animal cells have a membrane enzyme, the Na^+–K^+–ATPase, which uses the energy in ATP to pump potassium ions inwards and sodium ions outwards in such a way that it produces the normally high internal potassium and external sodium concentrations that cells require to function. For each ATP molecule hydrolysed, three potassium ions are pumped out of the cell and two sodium ions in, and a negative potential is generated on the inside of the membrane.

Plant cells and fungi also have an ATPase capable of pumping protons to the outside of cells. This enzyme, which is different from the bacterial or mitochondrial ATPase, generates both a proton gradient, higher in the outside, and a negative potential on the inside of the membrane. The membrane potential difference of the cells and organelles, usually negative inside, produces a driving force that can be used by other carrier proteins called secondary transporters to transport ions to the inside.

Another kind of membrane transport protein is able to allow the entrance of neutral molecules by coupling the movement of an ion (which can be driven inside the cells or organelles) to that of another molecule which is moved inside and can be concentrated against its concentration gradient. This mechanism receives the name of symport, or cotransport. Another of these secondary transport mechanisms can move other ions into cells or organelles by a mechanism called antiport, or countertransport.

Finally, cells have in their membranes selective pores, called channels, which allow the passage of ions or other molecules, driven by the combination of their concentration gradient and membrane potential.

There are also more complicated transport systems, in which membrane proteins recognise molecules in the outside of the cell, and trigger a series of profound changes. These changes result, not in the simple passage of the molecule into the cell, but in its internalisation, by endocytosis (a process by which the membrane engulfs the external particle and forms a vesicle which separates from the membrane, migrates to the lysosomes, and fuses with them, usually to degrade the particle that has been taken up). This is, for instance, the mechanism by which cholesterol is taken up by the cells from plasma, in which it is carried by lipoproteins. This is also the mechanism by which some viruses enter animal cells; they have evolved to exploit this specialised system to enter the cells.

Self assessment exercise 2.3

Choose the correct completion to the following statement.

Rhodospirillum rubrum transforms energy from light into chemical energy available for cellular processes by:

(i) using the energy captured by bacteriorhodopsin to pump protons to the outside of the membrane; as protons flow back across the membrane their energy is used to convert ATP to ADP.

(ii) using the energy captured by bacteriorhodopsin to pump protons to the outside of the membrane; as protons flow back across the membrane their energy is used to convert ADP to ATP.

(iii) using the energy captured by bacteriorhodopsin to activate ATPase which pumps protons to the outside of the cell; as protons flow back across the membrane their energy is used to convert ADP to ATP.

2.5 Concluding remarks

Biological molecules, by virtue of their particular structures, associate in remarkable ways, generating comparatively simple unicellular organisms to highly complex muticellular organisms, such as human beings. The genetic blueprint of each cell is determined by the sequence of base molecules in DNA, and this determines the structure of proteins (enzymes) which synthesise the biological molecules which form that cell.

2.6 Suggested further reading

1. BIOTOL Series (1992) — *The fabric of cells*
   — *Biomolecules: analysis and properties.*
   Butterworth-Heinemann, Oxford.
2. Garrett, R.H. and Grisham, C.M. (1995). *Biochemistry.* Saunders College Publishing, Fort Worth, Texas.
3. Zubay, G.L., Parson, W.W. and Vance, D.E. (1995). *Principles of biochemistry.* W.C. Brown Publishing, Dubuque, Iowa.

2.7 Responses to assignments

2.1 (i) A polymer of sugar molecules can be described as a polysaccharide.
(ii) A polymer of amino acids can be described as a polypeptide.
(iii) A polymer of nucleotides can be described as a polynucleotide.

2.2 The membrane is essentially hydrophobic. Thus the only molecules passing easily through such a membrane will be hydrophobic (and, in addition, relatively small). The majority of molecules which a cell needs to uptake or excrete in order to metabolise and grow are hydrophilic, and will thus not pass easily through such a membrane. If specific mechanisms can be developed to transport such hydrophilic molecules across the membrane (see Section 2.4) then the cell can control what goes in and out of the cell or its compartments, and can thus optimise the internal environment.

2.3 Remember, complementary base pairing across the helix is A–T and G–C. Thus, in the double helix the pairing would be

A	T	T	G	C	A	T	C
|	|	|	|	|	|	|	|
T	A	A	C	G	T	A	G

The complementary sequence is thus T A A C G T A G. Strictly speaking the correct complementary sequence should read G A T G C A A T since it is a convention that such a sequence is written with a polarity 5' → 3'. Because the two complementary sequences are related in an antiparallel sense, this explains the apparent reversal of the sequence. This is explained in Chapter 4.

2.4 Remember from the bases A, T, G and C in DNA, the bases U, A, C and G, respectively appear in mRNA. Thus the mRNA transcribed would be

A	T	T	G	C	A	T	C	(DNA template)
U	A	A	C	G	U	A	G	(mRNA template)

2.5 If the value of K_m is comparatively small in relation to the substrate concentration, then $\frac{[S]}{K_m + [S]}$ is approximately unity (~ 1.0) and so $v \sim V_m$ (the reaction velocity is at maximum). If K_m is comparatively large in relation to the substrate concentration, then $\frac{[S]}{K_m + [S]} < 1.0$, and $v < V_m$ (the reaction velocity will be below maximum).

2.8 Responses to self assessment exercises

2.1 All the polymers listed can correctly be referred to as heteropolymers, as they are composed of repeating molecules of different types.

2.2 The correct sequence is as follows.
(i) mRNA is transcribed from a DNA template.
(ii) mRNA migrates from the nucleus to the cytoplasm.
(iii) mRNA attaches to a ribosome.
(iv) An aminoacyl tRNA binds to its specific codon on mRNA.
(v) The amino acid is added to the polypeptide by peptide bond formation.

2.3 The correct response is (ii).
Answer (i) is incorrect, as the high energy molecule formed by the return of protons across the membrane is ATP, not ADP.
Answer (iii) is incorrect. As it happens, even if an organism used an ATPase system to pump protons out of the membrane, and used the protons returning across the membrane to regenerate ATP, this would not generate energy overall (ATP is used to pump protons out of the cell in using the ATPase system).

3 Structural and functional dynamics of the cell

Chapter outline

3.1 Introduction

The cell is the elementary structural and functional unit of organisation of all living organisms. The simplest organisms, such as bacteria, are made of single cells. The most complex organisms such as mammals are made of millions of cells which are structurally and functionally differentiated in order to play different complementary roles in the life of the whole organism. Some organisms can live either as single cells or as organised multicellular systems, depending on the metabolic conditions imposed on them by the environment. Information needed for the determination of the structure and the physiology of the cell is stored in the nucleotide sequence of the deoxyribonucleic acid (DNA) which is packed in the chromosome(s), where it is associated with specific proteins ensuring its maintenance and expression. In the case of unicellular organisms a distinction can be made between prokaryotes, such as bacteria, the cells of which are devoid of a nucleus and whose chromosome (called a nucleoid) is in direct contact with the cytoplasm, and eukaryotes, such as yeast, in which the chromosomes are packed in the nucleus, separated from the cytoplasm by a specifically permeable nuclear membrane.

In multicellular organisms, the process of cell differentiation leads to an elaborate functional organisation into tissues and organs. In some cases this differentiation results in the formation of very specialised types of cells, for example the red blood cell which is devoid of a nucleus. These differentiation processes result from programmed cascades of variations in the pattern of gene expression. Several types of mechanisms are involved in these phenomena, such as pleiotropic interactions of some regulatory proteins with DNA, and the establishment of gradients of some metabolites before cell division.

Biochemical studies of cells provide fundamental information about the mechanisms by which living organisms reproduce, adapt, differentiate and evolve. These studies are also important for the development of biotechnologies for the health and comfort of human kind. The aim of this chapter is to emphasise some aspects of cell organisation and biochemistry which can be useful for the development of biotechnologies.

3.2 Evolution of the cell

The cellular organisation that is observed today in prokaryotes and eukaryotes, actually reflects the different steps through which "living matter" evolved and gradually organised during evolution from the primitive elements of the universe. This evolution can be described as the progressive formation of cells, and their differentiation and association supported by progressively developing auto-reproductive chemical processes. Cell-like structures developed during the early steps of the appearance of life on earth. It is now well established experimentally that, under the reducing conditions which probably prevailed on earth about 4.5 billion years ago, temperature, ultraviolet irradiation and electric discharges could have promoted the chemical synthesis of amino acids, nucleotides, lipids and sugars, as well as the oligomérisation of these basic materials. Already in these early developments, specific pairing of nucleotides in nucleic acids probably played a role in providing a crude mechanism to keep information about favourable structures. It is thought that due to spontaneous molecular interactions, these polymers of different nature associated to form some vesicles, insoluble in the water of original ponds and lakes, and semi-permeable to chemicals. Within these vesicles, self-organisation of the spontaneously synthesised macromolecules led to the formation of pro-organelles. The stability of these vesicles is maximum around an optimal size, a phenomenon which results from the difference between the variations of internal volume and external surface, leading these vesicles to spontaneously divide like cells, a purely physical phenomenon which results from variation of surface tension. Further evolution led to the association in these vesicles of "working molecules" (proteins, catalysts) and "informative molecules" (nucleic acids), giving rise to self-dividing precytes (Fig. 3.1). A precyte is defined as "a cell-like structure arising through self-organisation of completely or partially abiotically formed constituents"

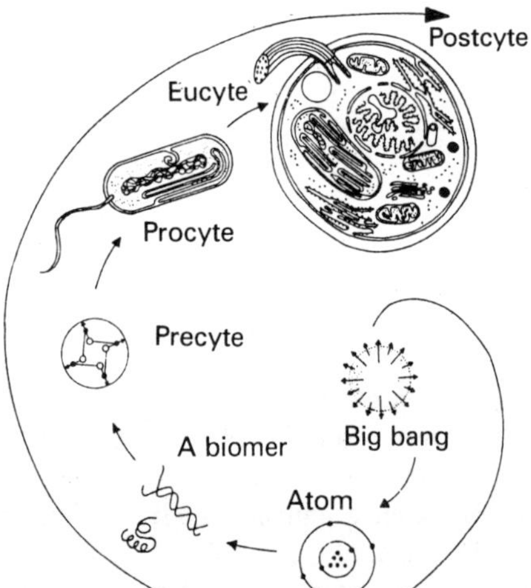

Fig. 3.1 *Schematic representation of the emergence and evolution of the cell (from Schwemmler, 1984).*

> ***Assignment 3.1***
>
> What advantages and disadvantages would arise from the localisation of synthetic processes within water-insoluble semi-permeable vesicles?

It is the evolution of precytes into procytes (prokaryotes) which corresponds to the real appearance of living organisms as we define them. An important step in the subsequent appearance of eucytes (Fig. 3.1) was the separation in these organisms of DNA from the rest of the cell; confined in the nucleus, packed and associated with specific proteins (including histones), and divided into several chromosomes. The passage from procytes to eucytes was probably not a single event, but rather occurred in different organisms at different times. Eucytes are considerably more differentiated than procytes and they show subcellular compartmentation into organelles which play specific physiological roles, such as oxido-reductive metabolism in mitochondria and photosynthesis in chloroplasts. Figure 3.2 illustrates the subcellular organisation of a flagellated unicellular alga. A cytoskeleton, made of actin and tubulin maintains the stability of this organisation and allows for the dynamics of the cell.

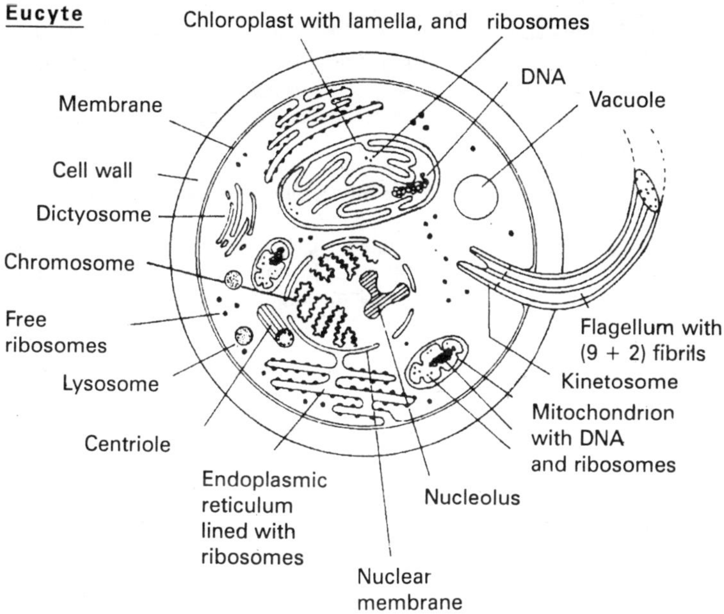

Fig. 3.2 *Schematic organisation of a chloroplastic flagellated cell (from Schwemmler, 1984).*

Eucytes organised themselves into multicellular organisms in which cells differentiate to form specialised tissues which play specific roles. Some species still exist which are illustrations of this evolution; they live either as single cells or as organised multicellular organisms, depending on the nutritive value of the environment. This is the case with myxobacteria such as *Dictyostelium discoideum*, the cells of which can associate and differentiate to form the different parts (foot plates, stalk and body) of fruiting bodies in which some of the cells are transformed into spores upon starvation (Fig. 3.3).

Unicellular procytes are generally immortal since they reproduce by simple cell division (scissiparity). It is an interesting phenomenon that when eucytes associated and differentiated to form tissues, the cells became mortal, similar to multicellular organisms made of highly differentiated cells and which reproduce sexually. Thus, in nature, evolution with differentiation has death as its price.

Differentiated cells can reach a further step of evolution to so-called postcytes (Fig. 3.1), in which two different organisms associate to live in symbiosis, for their mutual benefit. This phenomenon is considered to lead to further differentiation of the two partners in order to adjust to each other in a better way. In the case of highly differentiated postcytes, the partners are no longer viable when they are separated. A

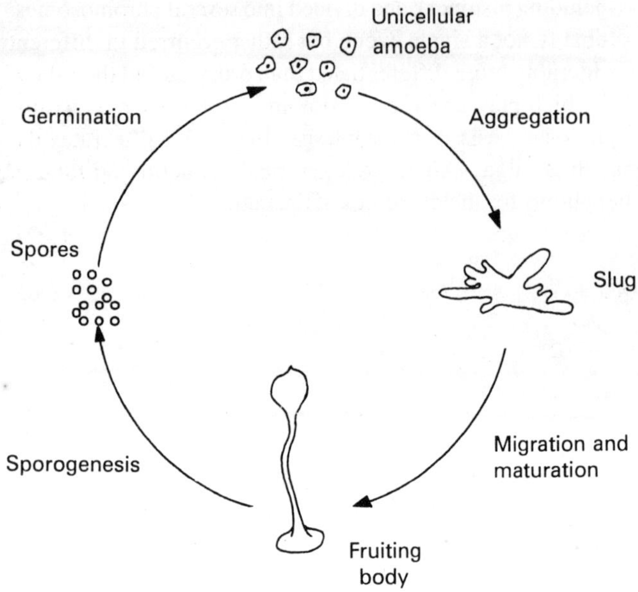

Fig. 3.3 *The life cycle of Dictyostelium discoideum.*
Under favourable conditions the unicellular organisms develop and divide. Upon starvation, the cells aggregate into a slug-like mass which migrates and settles to differentiate into a fruiting body, which, after maturation, liberates spores. Upon return of favourable conditions, these spores germinate to give unicellular forms.

well-known example of symbiosis is that of leguminous plants whose roots are infected by *Rhizobium*, a nitrogen assimilating bacterium which induces the formation of nodules in the roots of the plants it inhabits, providing nitrogen derivatives for the plants. A remarkable example of such symbiosis was described in 1977 when rich biotypes were discovered in the depth of oceans, around hot hydrothermal volcanic vents. The strangest species living there, *Riftia pachyptila*, is a one-meter long worm which resides in a self-excreted tube made of biopolymers (Fig. 3.4).

This animal does not possess a digestive tract and has only a simple blood circulatory system which transports blood from an external "panache", allowing the animal to absorb oxygen, carbon dioxide and sulphur from the volcanic waters in which it lives. These elemental nutrients are brought to the main part of the body (trophosome) which is a mass of irrigated cells. These cells are filled with cells of a bacterium that is chemolithoautotrophic, obtaining energy from the oxidation of sulphur, and using this energy to synthesise the carbon metabolites needed for itself and its host *Riftia* from carbon dioxide.

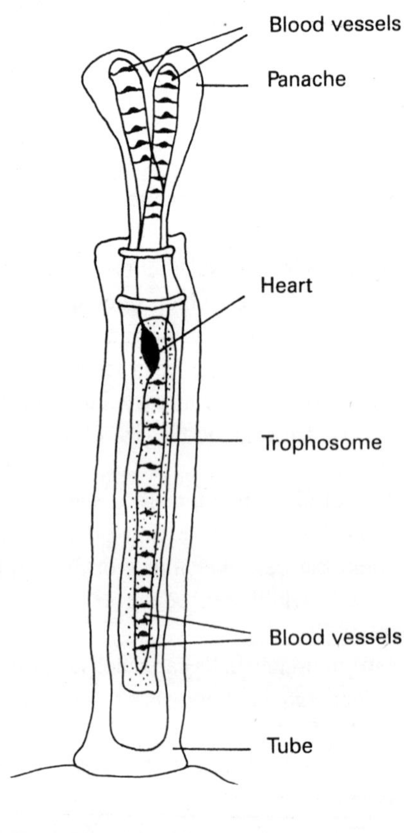

Fig. 3.4 *Schematic anatomy and organisation of Riftia pachyptila.*
This differentiated worm lives inside a self-excreted tube. The main part of its body (trophosome) is a blood-irrigated mass of cells. These cells are filled with cells of a chemolithoautotrophic bacterium. A simple heart sends blood to the panache where sulphide, carbon dioxide and oxygen are absorbed and transported to the bacteria which obtain their energy from the oxidation of sulphide. They use this energy to synthesise the metabolites which are necessary for their own metabolism and that of *Riftia*.

Self assessment exercise 3.1

Match each item in Column A with its correct definition in Column B.

Column A	Column B
(i) Postcyte	(a) A cell with internal compartmentation into organelles, including a nucleus.
(ii) Eucyte	(b) A cell which has differentiated in order to live in symbiosis with another cell.
(iii) Precyte	(c) A sac-like structure formed by the spontaneous association of abiotically-formed water-insoluble polymers.
(iv) Procyte	(d) A cell-like structure arising by self-organisation of abiotically-formed polymers.
	(e) A simple cell with little internal compartmentation and without a compartmentalised nucleus.

3.3 Cell differentiation and enzyme components

Cell differentiation to form highly specialised tissues is accompanied by modifications of cellular metabolism, a phenomenon linked to variations of enzymatic component of the cells. Such an adaptation is well illustrated by the case of the pyrimidine pathway. This pathway is responsible for the biosynthesis of the pyrimidine nucleotide triphosphates, precursors of nucleic acids.

All living cells, from bacteria to man, rely on two metabolic pathways for the biosynthesis of their pyrimidine nucleotides :

(i) the *de novo* pathway in which pyrimidine bases are synthesised entirely from simple precursors and combined to ribose phosphate;

(ii) the salvage pathway, in which the nucleotides are synthesised simply using the bases or nucleosides arising from the degradation of nucleic acids or from the cells' environment.

Depending on the cell type, or on some steps of cellular differentiation, the relative contribution of these two metabolic pathways can vary. For instance, it was found that the level of aspartate transcarbamylase (ATCase), the second enzyme of the *de novo* pathway is very high in rapidly dividing cells. A high level of activity of this enzyme is now considered to be a marker of cell activation or proliferation. In many instances of highly differentiated cells which do not proliferate, the genes coding for the *de novo* pathway enzymes are not expressed, and these cells rely only on the salvage pathway. In contrast, proliferative cells contain high levels of the enzymes of the *de novo* pathway. This is observed, for instance, in hepatic cells; upon liver regeneration after partial hepatectomy, ATCase activity of the cells increases by more than two orders of magnitude. Similarly, ATCase activity increases significantly during the development of tumours (in which cells grow rapidly). It was shown that there is a virtually linear relationship between aspartate transcarbamylase activity and the frequency of cell division in the cases of leukaemia, hepatoma and tumours of the central nervous system. In the human adult, neurons and glial cells (which do not divide) do not show detectable ATCase activity, and these cells rely entirely on the salvage pathway for the biosynthesis of their pyrimidine nucleotides. When tumours of these tissues develop, ATCase reappears in the tumour cells and its activity can increase by more than three orders of magnitude.

Such a phenomenon is also observed in the course of normal physiological processes of cell differentiation. In *Plasmodium falciparum*, the malaria parasite, ATCase activity disappears during the phase of development when this parasite is located in erythrocytes (red blood cells). During the cell cycle of the alga *Chlorella pyronoidosa* the cellular level of ATCase parallels the pattern of DNA synthesis In a similar way, when the bacterium *Bacillus subtilis* enters into the process of sporulation, ATCase is rapidly inactivated and degraded.

3.4 Cellular organisation of metabolic pathways; multienzyme complexes

The increasing complexity of the cell which goes with evolution is accompanied by the emergence of a more complex organisation of cellular metabolism. Cell life can be considered to be the result of integration of thousands of chemical reactions in microorganisms, tens of thousands in higher organisms. Each of these single reactions is catalysed by a specific enzyme. Metabolic pathways, series of sequential reactions, lead to the biosynthesis of molecules which are necessary for the development of the cell (amino acids for the biosynthesis of proteins, nucleotides for the biosynthesis of nucleic acids, etc.).

Evolution is accompanied by the development of these metabolic pathways into an increasingly integrated organisation. In microorganisms, the enzymes which catalyse the different sequential reactions are generally independent of each other and the metabolites must diffuse from the catalytic site of the enzyme where they are synthesised, to the catalytic site of the enzyme catalysing the next step, where they will be used as substrates. However, in higher organisms the enzymes of a given pathway tend to organise into multifunctional proteins or multienzyme complexes.

Assignment 3.2

What advantages and disadvantages would arise from the formation of multienzyme complexes containing enzymes involved in the sequential steps in a metabolic pathway?

Again the pyrimidine pathway provides an interesting example of such increasing complexity. The first reaction of this pathway consists of the production of carbamyl phosphate from glutamine, ATP and bicarbonate, catalysed by the enzyme carbamyl phosphate synthetase. In fact, two proteins are involved in the catalysis of this reaction, and in the bacterium *E. coli* these two proteins are non-covalently associated. Carbamyl phosphate synthetase (CPSase) of this microorganism is made of two subunits (Fig. 3.5). The small subunit is a glutamine amino transferase (GATase) which picks up the NH_2-group of the guanido group of glutamine and transfers it to the catalytic site of the CPSase subunit.

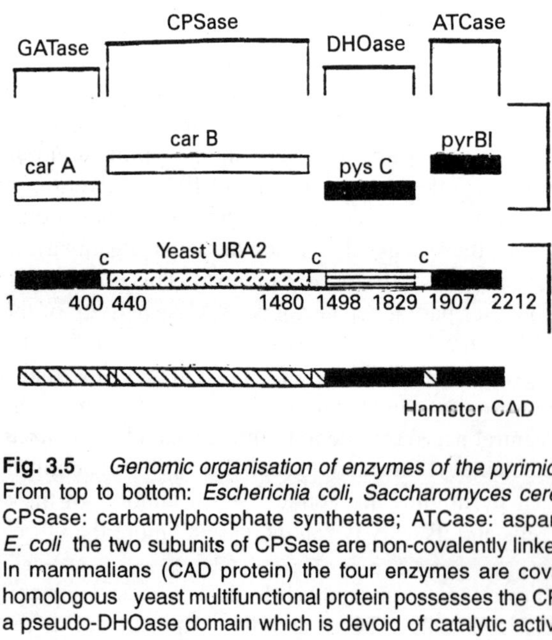

Fig. 3.5 *Genomic organisation of enzymes of the pyrimidine biosynthetic pathway.*
From top to bottom: *Escherichia coli, Saccharomyces cerevisiae, hamster.* GATase: glutamine aminotransferase; CPSase: carbamylphosphate synthetase; ATCase: aspartate transcarbamylase; DHOase: dihydro-orotase. In *E. coli* the two subunits of CPSase are non-covalently linked and ATCase and DHOase are independent enzymes. In mammalians (CAD protein) the four enzymes are covalently linked, forming a multifunctional protein. The homologous yeast multifunctional protein possesses the CPSase (including the GATase) and ATCase activities plus a pseudo-DHOase domain which is devoid of catalytic activity.

In the second reaction catalysed by aspartate transcarbamylase (ATCase), carbamyl phosphate is used to carbamylate the alpha amino group of aspartate to give carbamyl aspartate. This compound is then cyclised into dihydroorotate during the third reaction catalysed by dihydroorotase (DHOase). In mammals these reactions are catalysed by a multifunctional polypeptide chain (CAD protein). A comparison of the primary structure of this multifunctional chain with that of the corresponding but independent enzymes present in *E. coli* shows that in mammals the amino acid sequences which are homologous to the bacterial enzymes are separated by "linkers", short sequences which are not present in the corresponding *E. coli* enzymes. This observation suggests that during evolution these enzymes became covalently associated as a result of gene fusion. This conclusion is supported by the fact that the different functional domains of the multienzyme proteins can be separated by proteolysis limited to the linkers without loss of their catalytic properties. The organisation of the same enzymes in the yeast *Saccharomyces cerevisiae* shows an interesting intermediate state of organisation (Fig. 3.5). In this microorganism the multifunctional protein contains only the GATase, CPSase and ATCase activities. It was at first surprising to observe that the multifunctional polypeptide chain which catalyses these three reactions in yeast has about the same molecular weight as the mammalian CAD protein. This observation was explained by the fact that the yeast multifunctional protein contains an amino acid sequence which is homologous to DHOase but which does not have enzymatic activity. *Saccharomyces cerevisiae* possesses another gene for DHOase which is located on another chromosome. It is interesting to note that yeast pseudo-DHOase shows a higher homology with mammalian DHOase than with yeast DHOase. These observations are certainly of importance regarding the evolution of the pyrimidine nucleotides metabolic pathway, although their exact significance is still obscure. Also, in terms of allosteric regulation, the yeast multifunctional protein shows properties which are intermediate between those of the *E. coli* enzymes and the mammalian CAD protein (Fig. 3.6).

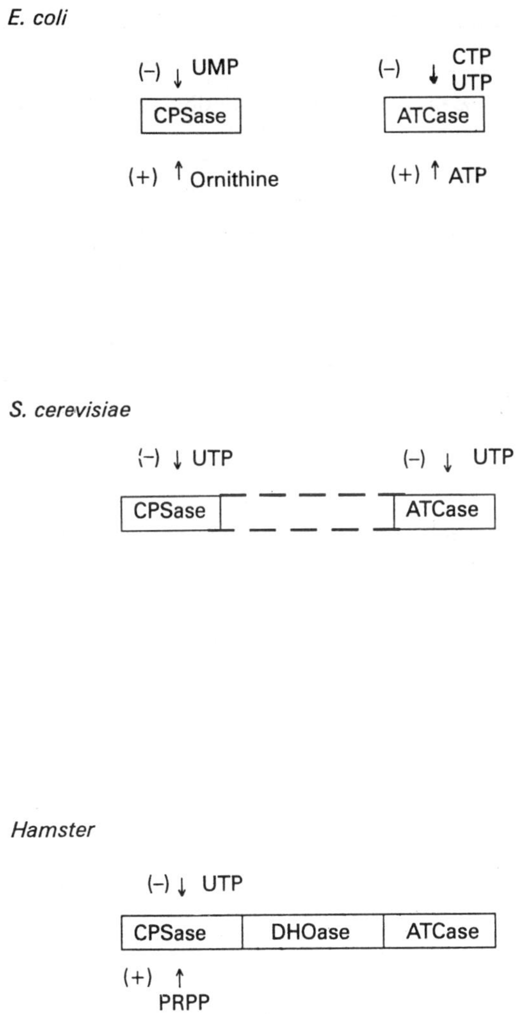

Fig. 3.6 *Evolution of the allosteric properties of the first enzymes of the pyrimidine pathway.*
Minus and plus correspond respectively to negative and positive allosteric regulation. The dotted line in the case of the multifunctional protein from *Saccharomyces cerevisiae* represents the pseudo-DHOase domain (see text).

The same phenomenon of gene-directed covalent association of enzymes is also observed for the 5th and 6th enzymes of this pathway, orotate phosphoribosyl transferase and orotidylate decarboxylase, which are independent proteins in *E. coli*, but are covalently linked in higher organisms. Thus, in higher organisms, the metabolic pathway is organised into two clusters of enzymes. In mammals, the enzyme dihydroorotate dihydrogenase is localised in the mitochondria, the optimal location for redox enzymes. This organisation illustrates the fact that, in the cell, intermediate metabolites need to migrate from one cellular compartment to another in order for biosynthesis to proceed.

3.5 Subcellular localisation of enzymes and metabolite compartmentation

The evolution of the cell is also accompanied by variations in the subcellular localisation of enzymes. This aspect of the organisation of the metabolic pathways is particularly remarkable in higher organisms, in which particular differentiated cells contain particular compartments and organelles.

In scientific literature cellular compartmentation is often referred to simply as compartmentation.

> ***Assignment 3.3***
>
> What advantages and disadvantages would arise from the localisation of specific physiological functions in specific organelles within a cell?

Over the past few years, numerous investigations have been aimed at the determination of localisation of enzymes in the cell, and it is a general observation that the enzymes which catalyse the different steps of a metabolic pathway are not all located in the same cellular compartment. This phenomenon is linked to the fact that, in the cells, many metabolic pathways are interconnected and share some intermediate metabolites. This fact explains the observation that the enzymes which catalyse oxido-reduction reactions are found either in the mitochondria or in membranes. A consequence of this organisation is what is called "metabolic compartmentation", whereby a metabolite is sequestered into a subcellular compartment and does not readily diffuse into other compartments. Furthermore, for a given metabolic pathway, this subcellular organisation of enzymes varies from one organism to another. Such is the case for the pyrimidine nucleotide pathway; prokaryotic organisms possess only one carbamylphosphate synthetase (CPSase) which provides carbamylphosphate for both the arginine and pyrimidine biosynthetic pathways; higher organisms possess two CPSases, one specific for the arginine pathway (CPSase-A) and another one specific for the pyrimidine pathway (CPSase-P). In higher organisms the carbamylphosphate pools synthesised by these two pathways are either fully or partially compartmentalised, depending on the subcellular localisation of these enzymes. In both *Neurospora crassa* (a filamentous fungus) and *Saccharomyces cerevisiae* (a yeast), the CPSase-P:ATCase complex is localised in the nucleus. In contrast CPSase-A and ornithine transcarbamylase, the first two enzymes of the arginine pathway, are located in the mitochondria in the case of *Neurospora crassa*, but are located in the cytoplasm in *Saccharomyces cerevisiae*. A consequence of this difference is that in *Neurospora crassa* the two carbamyl phosphate pools are strictly compartmentalised and this metabolite cannot be transferred from one metabolic pathway to the other, while in *Saccharomyces cerevisiae* some partial cross-feeding between the two pathways is possible. Even between closely related species, such differences in subcellular localisation of an enzyme can occur. This is the case for dihydroorotate dehydrogenase (DHOdase), the 5th enzyme of the pyrimidine pathway, in different yeast species; in *Shizosaccharomyces pombe* this enzyme is located in the mitochondria, while it is located in the cytoplasm in *Saccharomyces cerevisiae*. The difference in localisation of the enzyme in these two species is linked to different enzyme mechanisms (at least as far as the nature of the cofactor is concerned) and to different metabolic characteristics of these two yeast species.

It is an interesting phenomenon that intermediate metabolites must migrate from one compartment to another, and sometimes return for conversion into the final product. This observation illustrates the existence of a complex "intracellular metabolic physiology" which is still very poorly understood, but must be of general significance and can have very important consequences, for instance, in the practice of chemotherapy.

Metabolic compartmentation and the existence in the cells of non-communicating pools of the same metabolite is well documented in several instances for amino acids and nucleotides. As far as amino acids are concerned, in *Saccharomyces cerevisiae* aspartate, glutamate, leucine and isoleucine are present in high concentrations in the cytoplasm, while the other amino acids are essentially in vacuoles. The nature of the nitrogen source, the effectiveness of nitrogen assimilation, the rate of protein synthesis and the presence of high internal basic amino acid pools are important factors in the dynamic equilibrium of these cytoplasmic and vacuolar pools. In some mammals and in the protozoon *Tetrahymena pyriformis*, it has been shown that an exogenously added amino acid is incorporated into proteins in preference to the same amino acid generated endogenously (within the cell). In rat liver, amino acids generated by lysosomal degradation of proteins are used almost exclusively for the biosynthesis of intracellular proteins, whereas exogenously added amino acids are used for the synthesis of both intra-and extra-cellular proteins. In cultured muscle cells, leucine formed by the intracellular breakdown of proteins is preferentially used for protein synthesis while leucine coming from outside the cell is preferentially used for oxidation. Even *Escherichia coli* appears to be able to maintain two different amino acid pools, one derived from outside the cell and the other coming from intracellular metabolism.

These phenomena are even better documented in the case of nucleotides, which are compartmentalised. Several independent or slowly communicating intracellular pools are specifically used for DNA synthesis and RNA synthesis. In some cases it has been shown that the *de novo* and the salvage pathways feed different cellular compartments, and cytoplasmic and nuclear pools of ribonucleotides and deoxyribonucleotides exist. Compartmentation between the uracil coming from the medium and that provided by internal degradation of nucleic acids has been demonstrated. Elegant pulse-chase experiments using Novikoff hepatoma cells infected with Mengo virus showed that externally provided ^{3}H-uridine was readily incorporated into the virus RNA which is synthesised in the cytoplasm, whereas no labelling of the different kinds of RNA synthesised in the nucleus could be detected. In cultured hamster fibroblasts, external uridine bypasses the cytoplasmic pool to directly enter into RNA. The same phenomenon has been found in the alga *Euglena gracilis*. In this organism, it appears that external ^{14}C-labeled uracil goes directly into a pool which is immediately and preferentially used for RNA synthesis. More precisely, the large cytoplasmic pool is preferentially fed by the *de novo* pathway and used for mRNA synthesis, and there is a small cytoplasmic-nuclear pool fed by uridine (external and salvage pathway) which provides nucleotides for the biosynthesis of ribosomal RNA.

Compartmentation seems even more important for the precursors of DNA. The distribution of deoxyribonucleotides in the cytoplasm and the nucleus is unequal and varies during the cell cycle; the nuclear content of dCTP and dTTP increases markedly during DNA replication. The smaller thymidylate pool seems to exclusively serve DNA synthesis, while the larger pool seems to be involved in the repair of nuclear DNA and synthesis of the mitochondrial DNA. In addition, mitochondria have their own nucleotide pools which appear to equilibrate slowly with the cytoplasmic pool. It is obvious that this complex heterogeneity in the nucleotide content of the cells must be kept in mind for the success of chemotherapy using nucleotide analogues, since the analogue may not reach the target pool.

Problems in chemotherapy can also arise from subcellular location of either enzymes or isoenzymes involved in the metabolism of chemotherapeutic agents. Such a phenomenon might be responsible for some cases of resistance of acute leukaemia to arabinosylcytosine (AraC). The main mechanism of action of this nucleotide analogue requires its metabolisation into AraCTP. It has been shown that human cells possess two deoxycytidine kinases; one is located in the mitochondria and recognises only deoxycytidine. Differences between the intracellular levels of these two enzymes and in the pools of AraC might have important consequences on the production of AraCTP, and thus on the efficacy of this drug.

3.6 Compartmentation and channelling

The compartmentation of metabolites in the cell can result from different kinds of mechanisms.

> ***Assignment 3.4***
>
> What mechanism leading to compartmentation of metabolites has already been discussed in this chapter?

It can result from the lack of permeability of some cellular organelles to the given metabolites. In numerous cases, it probably results from enzyme channelling. The overall channelling process is complex and its definition controversial. This controversy arises from the fact that the same word "channelling" is used to designate several different phenomena. Therefore, we will restrict the term "channelling" to the process by which, in a multifunctional protein, an intermediate metabolite proceeds from the catalytic site where it is produced to the catalytic site where it is used as substrate, without diffusion. Tryptophan synthetase from the bacterium *Salmonella typhimurium* is an example of such a bifunctional enzyme, in which the intermediate metabolite, indole, is transferred from the catalytic site where it is produced to the second catalytic site, where it is used as substrate. In this process, indole is passed through a hydrophobic tunnel (Fig. 3.7).

As far as "channelling" (as defined above) is concerned, several mechanisms exist. The intermediate metabolite can be actively transferred from one catalytic site to the next one through a chemical step involving the formation of a covalent bond. Such a process is observed in the multifunctional proteins responsible for the biosynthesis of some peptide antibiotics. This is an extreme case and if, for instance, two sequential catalytic reactions are involved, they can be considered as two steps of the same reaction, even in terms of kinetics. Alternatively, owing to some structural features of the enzyme molecule or of its cellular environment, the intermediate metabolite can be directed to the next catalytic site by privileged or oriented diffusion. Such a mechanism can be called "microcompartmentation". These different processes can be differentiated on the basis of kinetic experiments.

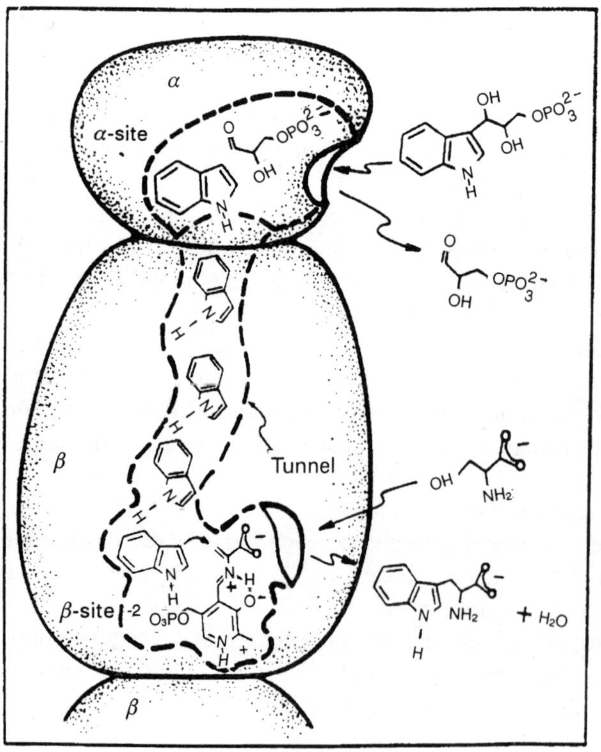

Fig. 3.7 *The channelling of indole in tryptophan synthetase.*
Tryptophan synthetase is made of two kinds of subunits. The α-subunit produces indole from indole glycerophosphate. The β-subunit condenses indole and serine to form tryptophan. The intermediary product indole is channelled from the α-site to the β-site through a hydrophobic tunnel.

Self assessment exercise 3.2

Answer true or false to each of the following statements:

(i) In yeast cells there is cross-feeding in pools of dihydroorotate as DHOase activity and the CPSase-P: ATCase complex are both located in the cytoplasm.

(ii) In the fungus *Neurospora crassa* there are separate cellular pools of carbamylphosphate, as CPSase-P activity is limited to the nucleus and CPSase-A activity is limited to mitochondria.

(iii) In mammalian cells there are two separate pools of deoxycytidine kinase due to compartmentation of AraCTP in these cells.

(iv) In the bacterium *Salmonella typhimurium* there is compartmentation of indole due to channelling within the tryptophan synthetase enzyme complex.

3.7 Catalytic and regulatory properties of enzymes inside the cell

For about a century, enzymologists devoted themselves to the difficult work of purification of enzymes to homogeneity. This approach was essential in order to determine the intrinsic properties of these biological catalysts. This tremendous effort resulted in the accumulation of knowledge that we have today about the structure of enzymes, catalytic mechanisms and the regulation of enzyme activity. However, this knowledge needs to be completed by investigations into the ways in which enzymes are functioning inside the cell. Many observations already indicate that intracellular environment has an important influence on enzyme function. This is particularly so in the case of regulatory enzymes, which appear to be very sensitive to many environmental factors and molecules. For instance, it has been shown that the activity of glutamine synthetase in *Escherichia coli* is sensitive to more than sixty metabolites. It is sometimes difficult for investigators to discern interactions which are of real physiological significance. This approach can be rationalised using a classification of enzyme properties.

(a) Specific catalytic and regulatory properties
These properties are clearly related to the role that a given enzyme plays in the cell: specificity towards substrates; susceptibility to feed-back inhibitors when they are end-products of the pathway to which the considered enzyme belongs; susceptibility to other effectors (inhibitors or activators) when they belong to a metabolic pathway whose activity interferes with that of the enzyme studied.

(b) Environmental properties
These properties result from the fact that during evolution, enzymes did not become progressively formed and selected in pure water, but in the complex, variable and concentrated intracellular medium. They were specifically devised to function in this complex environment, and it is not unexpected that these conditions in which enzymes were selected influence their properties.

(c) Irrelevant properties
These properties are intrinsic consequences of the structure that the enzymes possess. They are observed under the experimental conditions *in vitro* that enzymologists use, but do not play any significant role under the physiological conditions in the cell.

(d) Vestigial properties
These properties result from the existence, in some enzymes, of ligand binding sites which played a physiological role in a precursor homologous protein, from which the particular enzyme has been derived during evolution. Such "fossil" sites do exist, although in most cases, their activity has decreased over time by the lack of counter-selection of unfavorable mutations. For example, such a nucleotide binding site was detected in the isolated catalytic subunits of *Escherichia coli* ATCase.

This site is of no use for this organism, since it is not accessible in the native enzyme. It might be a vestige of a once effective allosteric site, since some organisms have an ATCase which is the equivalent to the isolated catalytic subunits of *E. coli* and yet are sensitive to effectors.

Self assessment exercise 3.3

Which of the following explain(s) (wholly or in part) why ATCase is a marker for cell proliferation?

(i) In *E. coli*, ATCase has a vestige of an allosteric site for nucleotides.
(ii) In tumour cells, ATCase activity is greater than in the cells from which the tumour formed.
(iii) In mammals, ATCase is associated with a CAD protein.
(iv) ATCase is involved in the *de novo* pathway of nucleotide synthesis.

Over the last few years, numerous methods have been developed in order to study the properties of enzymes *in situ* (inside the cell). Although such studies are difficult to perform within intact cells, some reactions can be followed without permeabilisation of the cells.

Assignment 3.5

What methods could be used to study reaction *in situ* without permeabilisation of cells?

Quantitative and time-resolved fluorescence scanning can be performed, and sophisticated equipment is now available which allows measurement of the scanning fluorescence spectrum and intensity in a single cell. NMR (nuclear magnetic resonance) analysis *in situ* has been increasingly used to follow some specific metabolic reactions or chains of reactions, especially where it has been possible to enrich the metabolite involved with isotopes providing specific signals (^{31}P, ^{13}C, and fluorinated metabolites). However, these sophisticated methods are of limited use, and most studies aimed at deciphering the behaviour of enzymes *in situ* have been made using permeabilised cells, into and from which substrates and products diffuse freely. An important consideration in experimental design is that the rate of this diffusion should not be limiting, as compared to the rate constants of the different steps of the enzymatic reaction under study. Numerous methods of cell permeabilisation have been devised to enable permeabilisation of different kinds of cells.

Assignment 3.6

What methods could be used for the permeabilisation of cells?

Although in many cases permeabilised cells cannot be distinguished from the normal cells upon examination by electron microscope, they cannot be considered as normal cells, and although the enzymes are studied in their physiological environment, the properties studied under these conditions are called *in situ* properties. However, in the best cases, the results obtained are of sufficient quality to allow the determination of enzyme mechanisms inside the cells. In many instances, significant differences from the properties of enzymes *in vitro* have been observed.

An important aspect of enzyme activity *in vivo* or *in situ* is that intracellular concentration of enzymes and substrates are far from the conditions in which the Michaelis–Menten theory is valid. In most cases, the intracellular concentrations of enzymes are much higher than the concentrations used in assays *in vitro* (which are of the order of 10^{-10}– 10^{-12} M). Under such cellular conditions, the assumption that the concentration of enzyme–substrate complex is negligible compared to that of the substrate does not hold true, and deviations from the hyperbolic saturation curves of enzymes are observed. These deviations can be of important physiological significance. It must not be forgotten that, considering that the volume of an *Escherichia coli* cell is approximately one cubic micrometer, the presence of a single molecule of an enzyme in this volume corresponds to a concentration of 10^{-8} M. Several rounds of gene expression may rapidly bring the enzyme concentration to values of the order of 10^{-5} to 10^{-3} M. This situation applies also to the cells of higher organisms, and in numerous cases the enzyme concentration is higher than that of its substrate. Under these conditions, enzyme kinetics are far from those predicted by the Michaelis–Menten theory.

Algorithms have been proposed to describe this situation and to predict the rates of enzyme reactions under these conditions. At such high concentrations, enzymes can display apparent co-operative behaviour which can play an important role in the physiological modulation of their activity. Again, permeabilised cells can be very useful for the study of enzyme kinetics at high concentration. The great difficulty which is inherent to this kind of study is that in the presence of a high concentration of enzyme the initial rate of reaction is difficult to measure, as the substrate becomes rapidly exhausted. Using permeabilised cells the rate of reaction can be slowed down simply by diluting the cell suspension, thus decreasing the concentration of enzyme in the reaction mixture without changing its concentration in the cells.

3.8 Adaptation of the cell to extreme conditions

A fascinating aspect of living organisms is their ability to adapt their structure and physiology to diverse environments and even to extreme conditions. In particular, microorganisms are able to colonise environments of extreme conditions of temperature, pressure, solute composition or radioactivity. The molecular constituents and especially the enzymes of these extremophile microorganisms have been extensively studied, with a view to using this knowledge for the development of certain aspects of biotechnology. These studies have been extended to the use of extreme conditions to study structure-function relationships in enzymes extracted from organisms which are living under normal conditions

(a) Temperature

> ***Assignment 3.7***
>
> From which natural environments could organisms adapted to high temperatures be isolated?

Some bacteria, including some archaebacteria, grow at very high temperatures in volcanic hot springs and in deep sea thermal vents. In the latter case, living bacteria have been isolated at temperatures as high as 105°C and some of these bacteria have an optimal temperature for growth which is very close to that value. It appears that the thermostability of enzymes from these organisms results from very discrete structural features; these enzymes have a structure which is more rigid than the structure of the proteins from mesophilic bacteria (organisms growing in 30°C to 40°C range), and a higher temperature is necessary to produce small conformational changes which are necessary for their catalytic activity. The comparison of amino acid sequences of proteins and enzymes from thermophiles with those of the homologous proteins and enzymes from mesophilic organisms indicates that thermophilicity is determined by some specific amino acid substitutions, the more frequent ones being glycine to alanine and serine to alanine. These substitutions are located in the hydrophobic core of the protein, where the more bulky alanine fills the space between residues and makes these hydrophobic regions more compact.

Assignment 3.8

What biotechnological processes would thermostable enzymes be applicable to?

At the other end of the spectrum, some organisms are able to grow at temperatures as low as 0°C (psychrophilic organisms) and these are also of interest to biotechnologists. Bacteria, lichens and algae living in polar regions are sometimes adapted to both extreme cold and desiccation, conditions found, for example, in the Ross desert in Antarctica. Some enzymes from these organisms are of great interest. Their use at low temperature for the enzymatic conversion of some crude extracts or natural substrates would avoid the interference with other enzymes such as proteases, which would not be active at such low temperatures. Little is currently known of the structural features responsible for the particular properties of these enzymes.

Assignment 3.9

What biotechnological processes would enzymes active at very low temperature be applicable to?

(b) Pressure

Some bacteria and higher organisms live in the deep sea, especially around volcanic thermal vents. These organisms are adapted to pressures as high as 700 atmospheres and they must have adapted to the influence that pressure has on all chemical and enzymatic reactions. These effects of pressure are governed by the law of Le Chatelier: any equilibrated reaction is accompanied by a change in volume. Pressure favours the side of the equilibrium for which the sum of the partial specific volume of the reactants is the smallest. The same applies to any conformational equilibrium of a protein, including allosteric transition. Numerous technological developments have been achieved during the past few years enabling cell growth, enzyme steady state and pre-steady state kinetics, quench-flow analysis, electrophoresis and NMR to be performed under high pressure. These developments have made the study of the effects of pressure on cells easier.

(c) Solute concentration

Cells living in an aqueous solution appear to adapt to extreme conditions of solute concentration. Some bacteria are able to grow in an extremely acidic medium. To do so, they have developed very efficient proton pumps that eliminate excess protons through their membranes against the steep concentration gradient of these protons. Some cells have adapted to extremely high ionic strength and studies of enzymes from these organisms have shown that the enzymes are enriched in charged amino acid side chains; this allows these molecules to bind more ions, thus ensuring that osmotic pressure within the protein will allow the capture of all the water molecules which are necessary to maintain the protein in its active conformation. The structure of a protein is the result of a set of interactions with the solute, and any change in the composition and properties of the solution (polarity, osmolarity, viscosity, dielectric constant) will have some influence on the conformation and activity of the protein. Advantage can be taken of this by investigators and biotechnologists. For instance, the use of cosolutes allows the stabilisation of the extreme T and R conformations involved in the cooperative properties of haemoglobin, glycogen phosphorylase and aspartate transcarbamylase.

(d) Radioactivity

Workers were surprised to observe that some algae were able to develop in pools where uranium bars were stored, where the algae were exposed to levels of radioactivity which would be lethal to other organisms. These algal cells acquired this resistance to radioactivity by developing extremely efficient enzymatic systems for DNA repair. This adaptation, which concerns both the amount and the efficiency of these enzymes, is reversible.

Self assessment exercise 3.4

A colleague says to you "An enzyme active at very low temperatures will always be more useful in a given enzymatic process than the equivalent enzyme active at very high temperatures, as the low-temperature enzyme would allow the process to be operated at temperatures at which contaminating microorganisms cannot grow and their degradative enzymes are not active."

Do you (i) agree or (ii) disagree? Explain your answer.

3.9 Conclusion

The cell is the basic unit of organisation of living organisms. All its remarkable properties of evolution, adaptation and differentiation leading to specialisation and compartmentation have potentials for the development of biotechnologies. Not only might entire permeabilised cells be used for biotechnological purposes, but also organelles such as mitochondria or chloroplasts isolated from cells. The rapidly developing knowledge of the integrated molecular events on which cell life relies will provide new ways for the development of biotechnologies.

3.10 Suggested further reading

1. Alberts, B., Bray, D., Lewis, J., Raff, M., Roberts, K. and Watson, J. (1983). *Molecular biology of the cell.* Garland Publish. Inc., New York.
2. BIOTOL Series (1992) — *The molecular fabric of cells.*
   — *Infrastructure and activities of cells.*
   Butterworth-Heinemann, Oxford.
3. Brock, T. (1985). Life at high temperature. *Science*, **230**, 132-138.
4. Corliss, J. and Ballard, R. (1977). Oases of life in the cold abyss. *Natl. Geograph. Magazine.*, **152**, 441-453.
5. Gounot, A.M. (1991). Bacterial life at low temperature: physiological aspects and biotechnological implications; a review. *J. Appl. Bacteriol*, **71**, 386-397.
6. Herdie, D.G. (1991). *Biochemical messengers: hormones, neurotransmitters and growth factors.* Chapman and Hall, London.
7. Herve, G. (1991). Semantics of channelling and compartmentation. *J. Theor. Biol.*, **152**, 131-132.
8. Ovadi, J. (1991). Physiological significance of metabolic channelling. *J. Theor. Biol.*, **152**, 1-22.
9. Reddy, G. and Pardee, A. (1980). Multienzyme complex for metabolic channelling in mammalian DNA replication. *Proc. Natl. Acad. Sci. U.S.A.*, **77**, 3312-3316.
10. Schwemmler, (1984). *Reconstruction of cell evolution: a periodic system.* C.R.C. Press, Boca Raton, Florida.
11. Sies, H. (Ed.) (1982). *Metabolic compartmentation.* Academic Press, London.
12. Srivastava, D. and Bernhardt, S. (1986). Metabolite transfer via enzyme–enzyme complexes. *Science*, **234**, 1081-1086.

3.11 Responses to assignments

3.1 Surrounding synthetic sites by a semi-permeable membranous vesicle means that: both exogenous substrates and synthetic agents (enzymes) can be concentrated within the vesicle leading to higher reaction rates in the reactions for which they serve as substrates; the internal environment can be regulated with respect to ionic strength, pH etc., thereby creating conditions optimal for synthetic reactions to take place; inhibitory factors in the environment can be excluded from reaction sites.

The main disadvantage is that if a particular metabolite or substrate does not readily pass through the semi-permeable membrane, a specific mechanism will be required to actively transport it across the membrane.

3.2 If enzymes involved in the catalysis of the sequential steps in a pathway are spatially separated within the cell, then the intermediate metabolites must pass from one enzyme to another (by diffusion or active transport mechanisms across membranes of organelles of higher organisms). If the enzymes involved are combined into one complex, then as the intermediate is formed by the reaction of one enzyme it is instantly on hand as the substrate for the next enzyme, without the need for intracellular transport.

3.3 If specific functions are contained within a specific membrane-bound organelle, then the advantages and disadvantages outlined in Response 3.1 above, apply. A further disadvantage is that if a product of reactions within one organelle are required as reactants for reactions in another part of the cell, then these reactants will need to pass freely out of the organelle, thus perhaps requiring specific transport mechanism across the membrane.

3.4 Compartmentation into organelles within a cell leads to metabolic compartmentation (see Section 3.5).

3.5 If the product of the reaction has specific and measurable properties (such as fluorescence) then its production may be measurable in a quantitative manner.

3.6 Methods which produce limited localised perturbations in structure of the cytoplasmic membrane, without complete disruption, would be suitable. Osmotic shock (rapid exposure to osmotic changes) and treatment with lipid solvents or detergents and are among treatments used for the purpose.

3.7 Volcanic springs and undersea volcanic vents are aquatic habitats where there would be heat without drying, and thus would be places to search for organisms adapted to growth at high temperatures.

3.8 High concentrations of sugars or starches produce viscous solutions which need to be heated (to $>60^{\circ}C$) in order to lower their viscosity for easy handling; enzymatic conversion of such solutions can thus be most easily performed at high temperatures. A recent example of application of a thermostable enzyme has been in PCR (polymerase chain reaction) where a thermostable DNA polymerase enzyme has been applied in reaction mixtures which need to be repeatedly heated to high temperatures.

3.9 Many commodities, such as pharmaceuticals or foods, are relatively unstable at room temperature or above, often due to degradative activity of enzymes (mostly produced by microbial contaminants). Microbial growth in pharmaceutical solutions can result in the release of endotoxins, which can induce febrile responses in recipients. For these reasons such commodities are held as far as possible at refrigeration temperatures (which minimise enzyme activity and microbial growth). Enzymes active at low temperatures could be used in enzymatic processing of such commodities at low temperatures, thus minimising the unwanted reactions described above.

3.12 Respones to self assessment exercises

3.1 The correct matching is as follows.
(i) .... (b)
(ii) .... (a)
(iii) .... (d)
(iv) .... (e)

3.2 The correct answers are as follows.
(i) False — the CPSase-P: ATCase complex is involved in synthesis of carbamyl aspartate (the substrate for DHOase), and thus would not lead to formation of separate pools of dihydroorotate or cross feeding between any pools of it.
(ii) True.
(iii) False — the statement would be true if 'deoxycytidine kinase' and 'AraCTP' were swapped around.
(iv) True.

3.3 The correct answer is as follows.
Statements (ii) and (iv) help to explain why ATCase is a marker for cell proliferation; remember that the *de novo* pathway is characteristic of dividing cells and that tumours are characterised by the uncontrolled (rapid) growth of cells within them.

Statements (i) and (iii) are correct in themselves but do help to explain why ATCase is a marker for cell proliferation.

3.4 The correct answer is (ii) — disagree. While it is true that low-temperature enzymes allow processes to be run at temperatures at which microorganisms cannot grow and enzymes cannot function, the same is true of high-temperature enzymes (in fact most microorganisms become rapidly killed and most enzymes become rapidly inactivated at temperatures above 60°C). In addition, high temperatures allow handling of concentrated (viscous) solutions.

4 Gene structure and expression

Chapter outline

4.1 Genes: the units of inheritance

"They are in you and in me; they created us, body and mind; and their preservation is the ultimate rationale for our existence ... they go by the name of genes, and we are their survival machines" — Richard Dawkins, British biologist, writing in his book *The Selfish Gene*, 1989.

The genotype of an organism consists of a set of units of inheritance called genes. The biological function of a gene is to preserve and express the genetic information encoded within it. The phenotype of an organism refers to the observable characteristics of that organism and results from the interaction of the organism's genes with the environment. Genes are normally very stable entities; if they were not, it would be impossible for genetic information to be passed accurately from generation to generation. Genetic stability is not absolute, however. Genes may occasionally become altered, and these changes are called mutations (see Section 4.11). Mutant genes are also stable entities and are inherited in the same way as normal, wild-type genes. The presence of a mutation may sometimes be indicated by an alteration of an organism's phenotype. Many genetic diseases of humans (sickle cell anaemia and phenylketonuria, for example) are caused by single mutations in a single gene. However, many mutations do not lead to an altered phenotype and can only be characterised after examination of a gene's structure at a molecular level. Many phenotypic traits are governed by more than one gene, or sets of genes.

Normal diploid cells such as the somatic cells of humans contain two sets of genes. One set is inherited from the male parent, the other from the female parent. The corresponding genes derived from each parent are called alleles. Together, the two alleles govern the phenotype of an organism. When two alleles are identical the organism is said to be homozygous; when they are different, the organism is heterozygous. In heterozygous organisms the phenotype depends upon the interaction of the two alleles. If one of the alleles is dominant, the other allele is recessive, and the phenotype of the heterozygous organism is dictated by the dominant allele.

Assignment 4.1

A phenotypic trait in a diploid organism is coded for by a single gene, designated as R. The dominant form of the allele is R, and the ressessive form of the allele r. The possible combinations of these allelles in a diploid organism are therefore:

RR
Rr
rr

For each combination, state whether the organism is heterozygous or homozygous, and whether the combination would result in the R or r phenotype.

For the majority of living organisms the genetic material is deoxyribonucleic acid (DNA; see Section 4.4). However, the genetic material of some viruses is ribonucleic acid (RNA). For many such viruses the RNA is converted to DNA copies within a virally-infected cell, so it could be said that most genes are defined in a message or code "written" in DNA.

Assignment 4.2

Bacteria are haploid organisms that multiply by asexual means. Bacteria therefore inherit their genes from only one parent, and have only one set of genes from that parent. They generally have only one copy of any given gene, not two allelic copies. Does this make genetic analysis of bacteria easier or more difficult than genetic analysis of diploid organisms? Explain your answer.

4.2 Linkage and genetic maps

The DNA of higher (complex) organisms is packaged into several chromosomes that are composed primarily of DNA and proteins. Different species of organisms have different numbers of chromosomes, and alterations in the number or morphology of chromosomes are frequently associated with genetic disorders. In Down's syndrome in humans, for example, each somatic cell has three copies of chromosome 21 instead of the normal two copies.

We can ask whether the same gene lies on the same chromosome and in the same position in all cells of the same species. Furthermore, are these positions inherited from one generation to the next?

Recombination is the process in which part of the DNA on one chromosome (representing a gene or genes) can be replaced by DNA from another chromosome or from extraneous DNA introduced into that cell. Analysis of phenotypic changes associated with such recombination events allows the frequency of recombination to be determined. If two genes are situated on different chromosomes, or are located far apart on the same chromosome it would be expected that each gene will recombine independently. If the genes are located close together on a chromosome, then there is a high probability that during a recombination event both genes will recombine at the same time. In this latter case the genes are said to be linked. For example, early studies with the fruit fly *Drosophila melanogaster* characterised genes that are closely and permanently linked on the sex chromosomes. We now know that the majority of genes do indeed occupy fixed positions on particular chromosomes in normal cells. Certain DNA sequences, known as mobile genetic elements (or jumping genes), are known to move from one chromosomal location to another, but a description of their properties and behaviour is outside the scope of this article.

We have seen above that genes located close together on a chromosome will have a high frequency of recombining together as a unit; the closer the genes are together, the higher this frequency will be. The frequency of independent or joint recombination allows a linkage map of the relative positions of genes to be constructed. Map units are measured in centimorgans (cM) after the pioneering geneticist Thomas Hunt Morgan, who first developed the idea of a genetic map in his work with *Drosophila.* One genetic map unit (m.u. or cM) is the distance between two genes that recombine with a frequency of 1%.

Individual map distances are additive for eukaryotic genes carried on the same chromosome. For example, if two genes X and Y map 5 cM apart, and gene Z is 3 cM beyond Y, then the map distance between X and Z will be 8 cM. The eukaryotic chromosomal genetic map is therefore linear. In contrast, the genetic map of prokaryotes is circular. It is comparatively easy to generate physical and genetic maps for prokaryotes, since their genomes are relatively small, and they grow and divide rapidly. However, it is much more difficult to construct genetic or physical maps for higher eukaryotic organisms. Their genomes are much larger and more complex, and only a small number of offspring can be obtained from each mating.

The first genetic maps to be constructed were those of the simplest of organisms: bacterial viruses (also known as bacteriophages), and one of the first of these maps to be derived was that of bacteriophage lambda. This is shown in Fig. 4.1. More recently, the use of gene probes and DNA sequencing (Chapter 5) has enabled the relative locations of genes and their nucleotide sequence to be readily determined. The mapping and sequencing of the entire genome of the bacterium *Escherichia coli* is nearing completion, and an ambitious international collaboration, The Human Genome Project, is underway to produce a genetic and sequence map of the human genome.

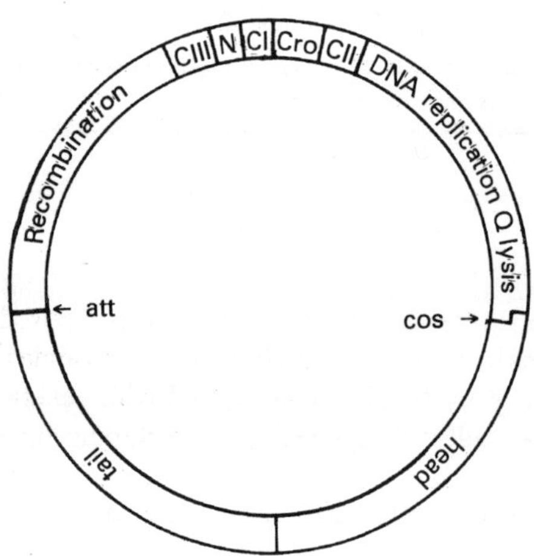

Fig. 4.1 *A simple genetic map of the bacteriophage lambda chromosome.*
Genes of related function are often found grouped together in clusters, and are in many cases coordinately regulated.

4.3 Overview of gene expression

4.3.1 *One gene, one product*

As long ago as 1908 the British physician Sir Archibald Garrod proposed that certain human diseases were due to "inborn errors of metabolism", and that these errors resulted from a lack of a specific enzyme to perform a particular biochemical reaction. He hypothesised that the lack of enzyme function was the result of inheriting a defective gene. A breakthrough was made in 1941 when Beadle and Tatum published an account of their work with the fungus *Neurospora*. They irradiated *Neurospora* with X-rays and then selected for X-ray-induced mutations that normally would have been lethal. Their selection system allowed them to identify *Neurospora* cells that would only grow when the growth medium was supplemented with vitamin B_6. X-irradiation had therefore altered (mutated) a gene in each of these cells that encoded an enzyme for vitamin B_6 synthesis. Beadle and Tatum concluded that a particular gene coded for the synthesis of one enzyme. Their "one gene, one enzyme" hypothesis became well established after numerous genetic and biochemical data were obtained in its support, and was subsequently modified to "one gene, one protein".

The one gene, one protein hypothesis was again modified when it was shown that many proteins consist of two or more separate polypeptide chains that are often encoded by different genes. A more accurate description of the relationship between a gene and its product then became: "one gene, one polypeptide". Even this description is not entirely accurate since not all genes encode protein: some encode polynucleotide RNA species such as transfer RNA (tRNA) and ribosomal RNA (rRNA). A strictly accurate expression is therefore: "one gene, one product". The discovery of the relationship between gene and gene product established the important link between classical genetics and biochemistry. This was an exciting time in biochemistry because metabolism could now be investigated using the powerful tools of the geneticist.

4.3.2 *The flow of genetic information*

How is the information held within DNA translated into protein products? In 1957 Francis Crick (who with James Watson had earlier determined the structure of DNA) laid the foundations for deciphering the genetic code. He proposed that DNA sequences and protein sequences are collinear, and that genetic information must be arranged in a linear manner along the length of a DNA molecule. The collinearity of gene and protein was subsequently established experimentally by the precise genetic mapping of mutations in both *Escherichia coli* and bacteriophage T4. Mutations (base changes) in a gene's linear sequence of DNA were shown to produce corresponding changes in the linear sequence of amino acid residues of the gene's protein product.

Crick's central dogma stated that the information held in DNA flows through RNA to protein, and that RNA acts as a "translator" of the genetic code. The central dogma can be represented thus: DNA $\leftrightarrow$ RNA $\rightarrow$ protein, where the arrows signify the direction of information transfer. Note the double-headed arrow between DNA and RNA. This is necessary because certain RNA viruses encode an enzyme (reverse transcriptase) that synthesises DNA from an RNA template, and hence transfers information from RNA to DNA. The human immunodeficiency virus (HIV) is a good example of such a virus. Following HIV infection of a human cell, the virus's RNA genome is reverse transcribed to yield a double-stranded DNA copy which is then inserted into (and therefore becomes part of) the cell's own DNA.

All genes undergo the first stage of gene expression, called transcription, in which the DNA template (Section 4.5) is copied into an RNA transcript. For some genes this is the end product of their expression. For protein-coding genes the RNA transcript, known as a messenger RNA (mRNA), undergoes the second stage of gene expression called translation. During translation an mRNA molecule directs the synthesis of a polypeptide with the help of ribosomal RNA (rRNA) and transfer RNA (tRNA) molecules. The amino acid sequence of a polypeptide is determined by the nucleotide sequence of the mRNA (which itself is derived from the DNA sequence of the gene from which it was transcribed). The information in the mRNA is translated in units of three nucleotides, called triplet codons, with each codon specifying a single amino acid. The genetic code specifies which amino acid is encoded by which triplet codon. Three of the 64 possible triplet codons result in the termination of polypeptide synthesis, and are therefore known as "stop codons". Transcription, translation and the genetic code will all be considered in more detail later in this chapter.

Self assessment exercise 4.1

Which of the following statements defines a gene most precisely ?

(i) A gene is a fragment of DNA that codes for the production of a phenotypic trait.
(ii) A gene is a fragment of DNA that codes for the production of a polypeptide or polynucleotide.
(iii) A gene is a fragment of DNA that codes for the production of a protein.
(iv) A gene is a fragment of DNA that comprises a unit of inheritance.

4.4 The structure of DNA

Both DNA and RNA share the same basic structure (Chapter 2). A phosphodiester backbone links together either deoxyribose sugar units (in DNA) or ribose sugar units (in RNA). Each sugar unit is covalently linked to a base which may be adenine (A), guanine (G), cytosine (C) or thymine (T) in DNA. RNA contains uracil (U) instead of thymine. The complex of sugar and base is termed a nucleoside, and when a nucleoside is linked to a phosphate group the molecule is termed a nucleotide. In the DNA strand on the left in Fig. 4.2 each nucleotide is connected to the phosphate group below it through the deoxyribose carbon designated 3' (pronounced "three-prime"), and to the phosphate group above it through a 5' ("five-prime") carbon. Notice that the opposite orientation is found in the strand on the right of the figure.

The polarity of any nucleic acid strand is defined by the direction in which the 5' to 3' links run in that strand. Double-stranded nucleic acids always pair in an anti-parallel fashion, in which the 5' to 3' sense of one strand is opposite to that of its partner. Conventionally, only one strand of a double-stranded DNA sequence is usually given, and nucleic acid base sequences are written in the 5' → 3'-direction, i.e., from the 5'-terminus at the left to the 3'-terminus at the right (e.g., 5' GAATTC 3').

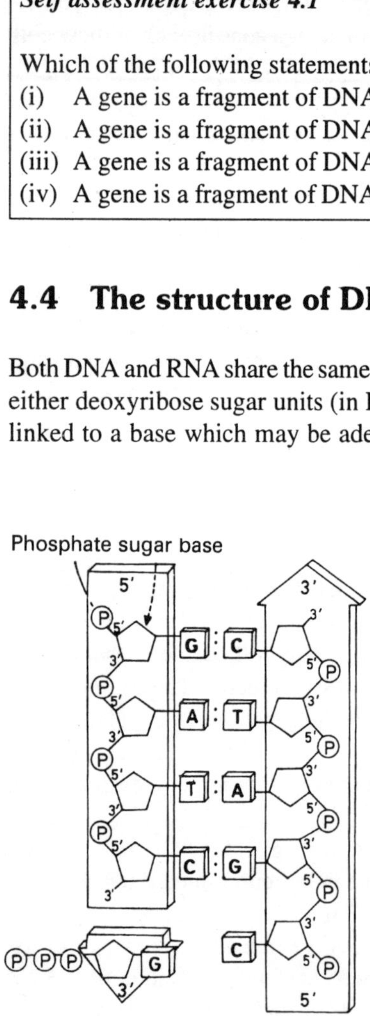

Fig. 4.2 *The anti-parallel, double-stranded structure of DNA.*
The backbone consists of alternating deoxyribose sugar units and phosphate groups. Directionality of the strands (represented by the arrows) is defined by the way each deoxyribose residue is attached to the phosphate groups by 5' and 3' linkages (see text).

Assignment 4.3

Write down the 5' → 3' base sequence of the right-hand strand of the segment of the DNA molecule depicted in Fig. 4.2.

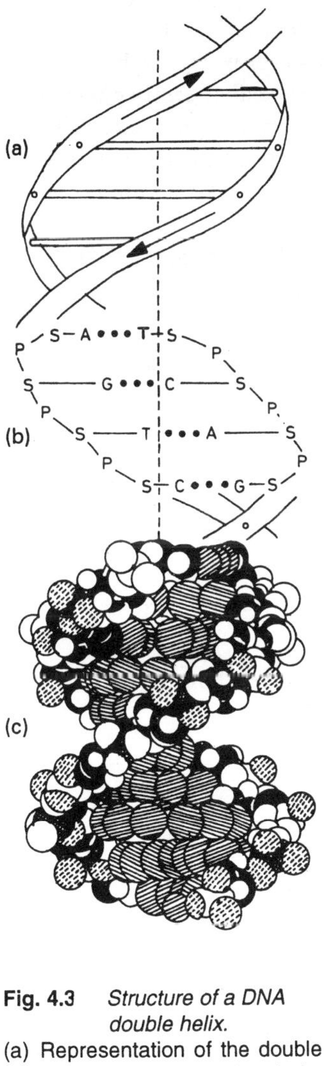

Fig. 4.3 *Structure of a DNA double helix.*
(a) Representation of the double helix showing stand polarity (arrows) and bonding across the helix between the two strands.
(b) Representation of the sugar (S)-phosphate (P) backbone of the strands with bases. (T, A, G, C) paired across the helix by hydrogen bonding (•••).
(c) Representation of the molecular structure.

DNA molecules are normally double-stranded (duplex) and consist of two anti-parallel chains wound around each other in a double helix (Fig. 4.3). The phosphodiester backbone is on the outside of the helix while the bases are on the inside. Guanine (G) is always found opposite cytosine (C) and adenine (A) is always found opposite thymine (T). GC and AT base-pairs are known as Watson–Crick base-pairs after the discoverers of the DNA double helix. Base-pairing specificity is dictated by the hydrogen bonds (H-bonds) which form between a purine (G or A) and an appropriate pyrimidine (C or T) base. A DNA fragment of length 10 base-pairs (bp) is therefore 10 nucleotides in length. A DNA molecule carries negatively charged phosphate groups at its surface. These negative charges are neutralised mostly by univalent (Na^+) and divalent (Mg^{2+}) cations as well as by positively charged proteins.

Double-stranded helical DNA molecules in solution may be unwound (denatured) by an increase in temperature. Specific, complementary base-pairing (G with C, A with T) contributes to the thermodynamic stability of DNA in two ways: (i) by hydrophobic base stacking, which results from the interactions between electron systems of the stacked bases; and (ii) by H-bonding between each base-pair. Since a GC base-pair is held together by three H-bonds and an AT base-pair by two H-bonds, the GC bond is stronger and hence the stability of the DNA double helix depends on the GC/AT base composition. The higher the GC content of a particular region of a DNA molecule, the more resistant it will be to loosening its double-stranded form (unwinding). As we will see later, unwinding and rewinding is crucial to DNA's biological activity.

One might think that because the nucleotide bases of DNA lie in the interior of the molecule they would be relatively inaccessible to proteins. Indeed, in many text books DNA is regarded as a store for genetic information that is *only* accessible after the helix has been unwound. However, it seems that recognition of DNA sequences by certain proteins can occur without local unwinding, and X-ray analyses of DNA-protein complexes have confirmed this.

4.5 DNA replication

If genetic information is to be faithfully inherited from generation to generation, the nucleotide sequence of DNA must be faithfully copied from the original parental DNA molecule to two daughter molecules. The clue as to how this duplication is achieved lies in the double-helical structure of DNA molecules. Since complementary DNA strands are held together by hydrogen bonding, their separation can be readily achieved and, when separated, each strand serves as a template for the synthesis of a complementary copy. The principle is illustrated in Fig. 4.4. After replication, the original (parental) DNA duplex will have given rise to two identical daughter duplex molecules.

Following DNA replication, one strand (i.e., half) of a daughter molecule has been conserved from the parental duplex, while the other strand has been newly synthesised. This mode of replication is therefore called semi-conservative. Although semi-conservative replication requires strand separation, the disruption of the double-helical structure occurs only transiently and only a small part of a chromosome is single-stranded at any given time. DNA replication is extremely complex and involves the coordinated participation of many proteins with many different enzymatic activities.

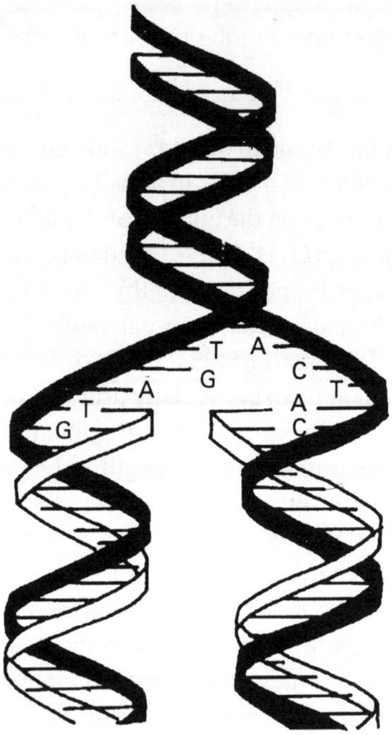

Fig. 4.4 *A mechanism for DNA replication hypothesised by Watson and Crick, and subsequently proven by the experiment of Meselson and Stahl.*
The parental DNA duplex (solid line) unwinds to expose the single bases of each strand. Each of the two parental single strands then acts as a template for the synthesis of a new strand (open line). The two double helices which are formed are therefore identical to the parental helix. Nucleotides are polymerised into the new strands by the enzyme DNA polymerase.

> ***Assignment 4.4***
>
> Consider the mechanism of DNA replication described above. Explain precisely how the two complementary strands of the original DNA duplex give rise to two duplex molecules of identical base-pair composition to the original.

4.6 DNA conformations

DNA molecules can adopt more than one conformation in solution depending upon the pH and ionic environment. Under physiological conditions the dominant conformation of DNA is known as the B-form, although some parts of the duplex may undergo a transition to other conformations. The rare A-form exists only in dehydrated states and differs from the B-form in that the plane of the base pairs is tilted 20° from the perpendicular to the helix axis. Diagrams of the A- and B-forms of DNA are shown in Fig. 4.5.

There is no strong evidence that other conformations (known as C-D-E- and F-forms) exist *in vivo*. The Z-form of DNA, shown in Fig. 4.6 influencing supercoiling (see below), is the only left-handed form of the helix and takes its name from the zigzag path that the sugar-phosphate backbone follows. Segments of DNA in which purine and pyrimidine residues alternate are able to form Z-DNA under certain conditions, and Z-DNA may be stabilised *in vivo* by binding Z-specific protein(s). Evidence that Z-DNA exists *in vivo* comes from the identification of anti-Z-DNA antibodies in patients with certain auto-immune diseases. It seems unlikely that these antibodies would be formed if Z-DNA were not present in the human body.

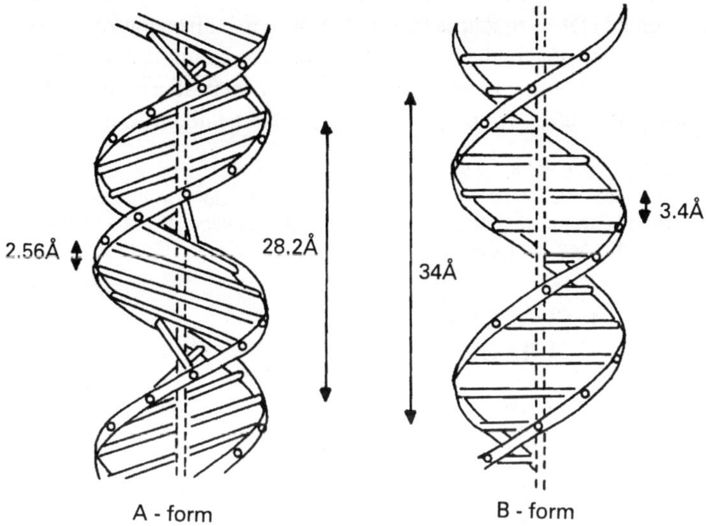

Fig. 4.5 *Diagrammatic representation of the A and B conformations of DNA.*
The figure shows their characteristic dimensions in angstrom units (Å), where 1 Å = 10^{-10} m.

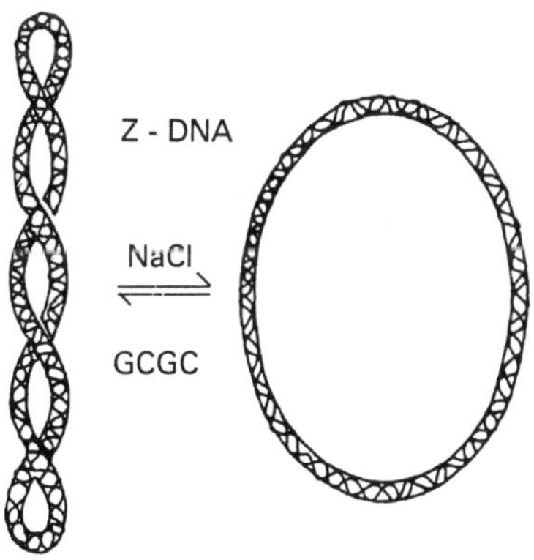

Besides the conformations described above, DNA may also form unusual structures that appear to be important in regulating gene expression. These include a triple helix conformation and a cross-shaped structure called cruciform DNA. Triple-helices (Fig. 4.7) are characterised experimentally by their susceptibility to digestion *in vitro* with S1 nuclease (an enzyme that specifically degrades regions of single-stranded DNA). Research is currently in progress to determine whether induction of a triple-helical conformation may be a mechanism to switch off gene expression.

Fig. 4.6 *Z-DNA.*
Regions of left-handed Z-DNA in circular DNA molecules (formed by alternating GC residues) influence the extent of DNA supercoiling.

Cruciform DNA takes its name from a locally formed cross-shaped segment (Fig.4.8). A DNA segment is only able to fold into a cruciform if the DNA sequence contains an inverted repeat, or palindromic sequence.

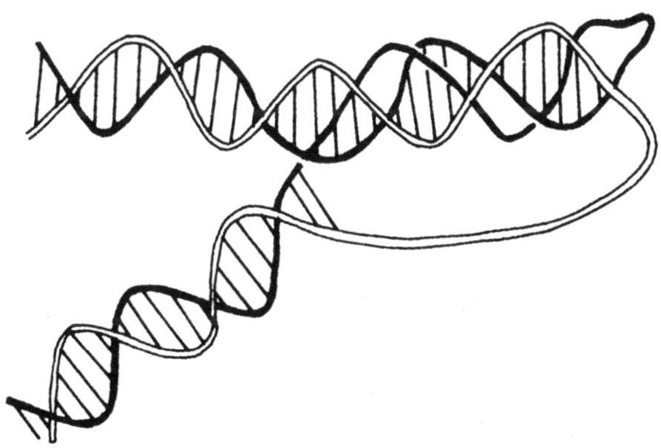

Fig. 4.7 *The triple helix conformation of DNA.*
Note the single-stranded region (which can be identified experimentally by its susceptibility to digestion with the enzyme S1 nuclease).

Inverted repeat sequences (palindromes) are symmetrically repeated DNA sequences that are the same in both chains of a double helix when read in the 5'- to 3'-direction. The example below shows a double-stranded 16 base-pair (bp) inverted repeat in bold typeface, with the centre of symmetry (dyad axis) indicated by a dash.

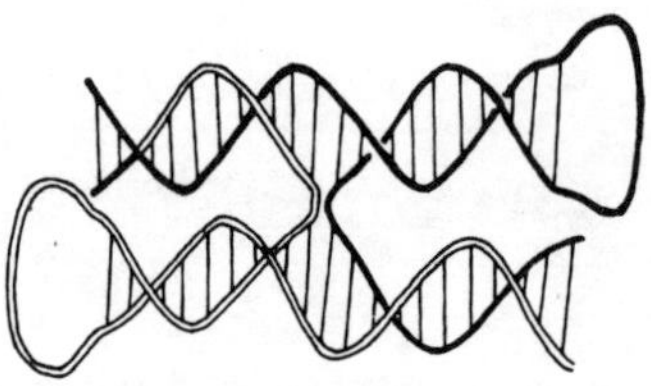

Fig.4.8 *Cruciform DNA.*
The structure consists of two "hairpin loops" which form in the region of an inverted repeat sequence. See text for explanation of the structure of inverted repeat sequences.

5' **TATACGAT-ATCGTATA** 3'
3' **ATATGCTA-TAGCATAT** 5'

In naturally-occurring DNA sequences the two halves of the inverted repeat are often separated by a non-inverted repeat sequence thus:

5' **TATACGAT**-GCCATCACCCG-**ATCGTATA** 3'
3' **ATATGCTA**-CGGTAGTGGGC-**TAGCATAT** 5'

Inverted repeat sequences are widespread in both prokaryotic and eukaryotic genomes, and cruciform structures have been postulated to be the sites recognised or induced by regulatory DNA binding proteins. There is good experimental evidence for their formation in *E. coli* cells.

4.7 DNA supercoiling

In vivo the majority of DNA molecules do not have free ends but are present as closed (circular) structures. When a circular DNA duplex is twisted around its axis, supercoils appear in the DNA, rather like the coiling seen when a rubber band is twisted. When B-DNA is exposed to forces that twist it in the opposite direction to the twist of the double helix, it forms negative supercoils. This allows the DNA molecules to relieve the twisting force by loosening the winding of the two strands around each other. Supercoiling can only occur in a closed circular structure and the greater the number of supercoils, the greater the torsion in the closed duplex. A DNA molecule cannot form a supercoil if one or both of the DNA strands are broken. If this occurs, the DNA supercoil unwinds to a relaxed form. Supercoils are classified in two ways. Negative supercoils twist the DNA around its axis opposite to the direction of the turns of right-handed helix. Positive supercoils twist the DNA in the same direction as the right-handed helix. It is thought that only negatively-supercoiled DNA exists *in vivo*, although positively-supercoiled DNA may be generated *in vitro*.

Negative supercoils favour local unwinding of the DNA, allowing processes such as transcription, DNA replication and recombination to take place. Negative supercoiling is also thought to favour the transition between B-DNA and Z-DNA, and to moderate the interactions of DNA-binding proteins involved in gene regulation.

4.8 Introduction to transcription

As we saw earlier, the synthesis of complementary RNA molecules on a DNA template is known as transcription and is the first stage in which the information contained within genes is expressed.

Transcription occurs in three steps: initiation, elongation and termination, and is catalysed by the enzyme RNA polymerase. Since the newly synthesised RNA is complementary to the template DNA strand, no loss or change of genetic information usually occurs. In a complex, differentiated organism only a small proportion of the total gene complement is expressed in any particular cell type. We need to understand how proteins are able specifically to recognise and interact with genes, and how this interaction influences RNA polymerase activity. Both nucleic acid/protein and protein/protein interactions are involved in controlling transcription.

4.9 Transcription and its control in prokaryotes

4.9.1 Initiation, elongation and termination

Initiation of transcription occurs after RNA polymerase has bound to a specific DNA sequence called a promoter (Fig. 4.9). The actual DNA sequence at a promoter site varies from gene to gene, but all promoters have a similar sequence known as the consensus sequence, which for *E. coli* has two distinct components: a "–35 box" and a "–10 box" as follows:

–35 box   5' TTGACA 3'

–10 box   5' TATAAT 3'

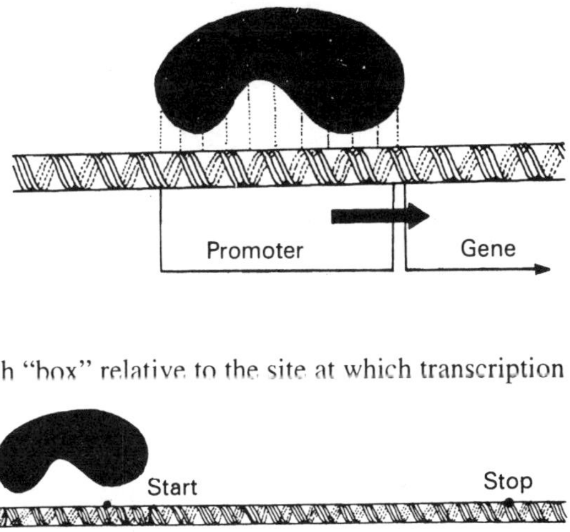

Fig. 4.9 *RNA polymerase recognition of a promoter sequence in DNA.*
The enzyme covers a length of DNA equivalent to about 6 turns of the helix. The direction of transcription is shown by an arrow.

The negative numbers refer to the positions of each "box" relative to the site at which transcription starts, which is designated as +1. The "+1" position therefore corresponds to the site at which the first nucleotide is polymerised into the encoded RNA molecule (Fig. 4.10).

These consensus sequences were determined by sequencing the DNA of a large number of *E. coli* promoters and identifying those bases which predominate at certain positions relative to known transcriptional start sites. As noted above, the sequence of any one promoter may differ from the consensus sequence and certain variations give rise to promoters with reduced efficiency of RNA polymerase binding. This in turn leads to lower levels of transcription and is an important factor in the control of gene expression. In order for transcription to take place the DNA double helix becomes unwound at the promoter site and the template

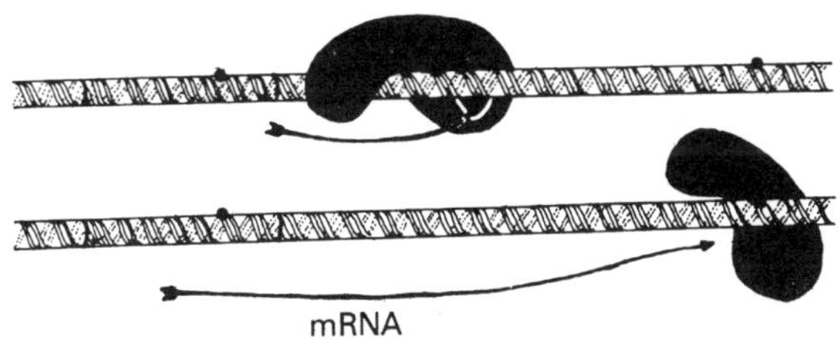

Fig. 4.10 *Transcription of a gene by RNA polymerase.*
The enzyme remains in close association with the DNA as it travels along the gene to the transcriptional terminator site. When the polymerase reaches a terminator sequence (not shown) it dissociates from the DNA, and the RNA strand is released.

strand is made available. Unwinding occurs only locally, and the rest of the DNA molecule remains in a double-helical state.

The second phase of transcription is called elongation. RNA polymerase moves along the DNA, reading the template information in the 3'- to 5'-direction and synthesising the complementary RNA chain in the 5'- to 3'-direction. An RNA–DNA hybrid is thus formed in the unwound regions. (Note that the requirement for all double-stranded nucleic acid structures to be anti-parallel is maintained here.) As the RNA polymerase moves further along the DNA template the newly synthesised RNA strand is displaced and the DNA duplex reforms behind the transcriptional complex (Fig. 4.11).

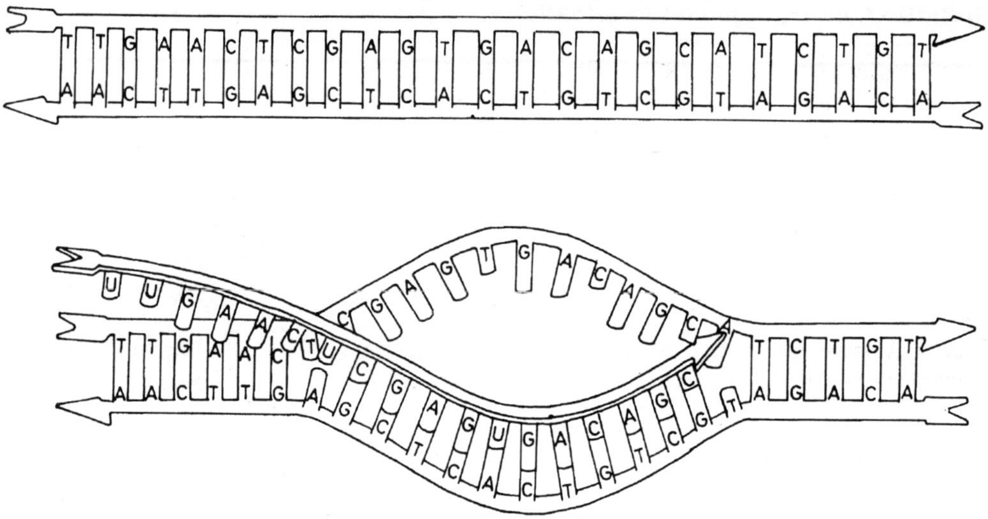

Fig. 4.11 *Local unwinding of the DNA duplex allows RNA to be synthesised on a DNA template.* Note (i) that RNA contains the pyrimidine base uracil instead of thymine, and hence forms UA base-pairs with the template DNA sequence, and (ii) that the RNA and DNA strands pair in an anti-parallel fashion.

> ***Self assessment exercise 4.2***
>
> What would be the RNA transcript of the DNA sequence 5' GCATTGGCAGCTA 3'?

The third and the last phase of transcription is called termination, and occurs when RNA polymerase recognises a specific DNA sequence known as a terminator. Certain *E. coli* terminators take the form of an inverted repeat sequence in the DNA template followed by a run of A nucleotides. After terminator recognition has occurred, RNA polymerase dissociates from the DNA template and transcription ceases.

Before RNA polymerase can start transcription it must be able to find promoter sequences in a genome of approximately 4×10^6 base-pairs in size. This searching process is highly efficient, and two main mechanisms have been proposed. In the sliding model the enzyme moves along the DNA and stops when it comes across a promoter. In the diffusion–displacement model, random diffusion of RNA polymerase occurs until the enzyme complex finds a promoter.

4.9.2 The composition of E. coli RNA polymerase

The mechanics of transcription have been studied intensively in *E. coli,* which contains approximately 7×10^3 molecules of RNA polymerase. This single enzyme species catalyses the synthesis of all types of *E. coli* RNA molecules (messenger, transfer and ribosomal). This is in contrast to the finding in eukaryotic cells where several different RNA polymerases have been identified (see later).

The complete *E. coli* RNA polymerase complex is composed of five protein subunits: two alpha (α)-subunits, one beta (β)-and one beta-prime (β') -subunit, and a sigma (σ)-subunit. Collectively, the complex is known as the holoenzyme (from the Greek word *holos,* meaning whole) and is represented as $\alpha_2\beta\beta'\sigma$. The $\alpha_2\beta\beta'$ core of the enzyme is required for the polymerising activity.

4.9.3 The roles of the RNA polymerase subunits

The σ-subunit is required solely for DNA sequence recognition at a promoter site. The α-subunits attach to the promoter sequence (following recognition by σ) while the β- and β'-subunits cooperate in catalysing RNA synthesis during transcription. Once transcription has been initiated, the σ-subunit dissociates from the rest of the enzyme after approximately 8 ribonucleotides have been polymerised into RNA, leaving the core enzyme to continue transcription. At initiation the RNA polymerase complex has an elongated shape and covers up to 80 bp of DNA; during movement along the DNA the core enzyme covers approximately 30 bp of DNA. The optimal rate of polymerisation is estimated to be approximately 40 nucleotides per second at 37°C.

The selective action of certain antibiotics on RNA polymerase subunits has allowed the initiation and elongation steps of transcription to be studied separately. Rifampicin inhibits initiation of RNA synthesis by binding to the β-subunit of RNA polymerase but it does not prevent subsequent RNA chain elongation. In contrast, streptolydigin inhibits elongation, but not initiation.

Since various *E. coli* genes are known to possess structurally different promoters, and since the presence of a σ-factor is a necessary requirement for the specific initiation of transcription, it would seem logical that different σ-factors could be responsible for controlling the specificity for, and hence the expression of, different groups of genes. A great deal of experimental evidence now shows that bacteria do indeed use different σ-factors to control the expression of different sets of genes. For example, when *E. coli* cells are grown at 42°C the pattern of transcription changes dramatically and a group of genes encoding "heat-shock proteins" is induced. The switch to the transcription of the heat-shock genes is achieved by substitution of the usual σ-factor, known as σ^{70} (MW 70,000; encoded by the *rpoD* gene) with another σ-factor: σ^{32} (MW 32,000; encoded by the *rpoH* gene). σ^{32} then directs RNA polymerase to the promoters of the heat-shock genes (which share a common DNA sequence recognition site, which is different to that of other genes). If the temperature is returned to 37°C, heat-shock gene expression is switched off and σ^{32} molecules are rapidly degraded. The use of different sigma factors is only one of many ways in which bacteria are known to regulate gene expression.

4.9.4 Induction and repression of E. coli genes: the lac operon

Unicellular organisms must be able to react rapidly to changes in their environment, and appropriate enzymes may need to be induced in response to changes in a cell's environment. Regulatory events that govern enzyme induction have been extensively studied in *E. coli.*

Induction and repression are the ways in which a bacterial cell adjusts its metabolism in the most economical way in response to rapid environmental changes. A substrate (or substrate analogue) which switches on transcription of the gene(s) required for substrate metabolism is called an inducer, while a molecule that switches off transcription of a gene or set of genes is called a repressor. Both inducers and repressors are highly specific in their mode of action. Molecules may act as efficient inducers even though they are not themselves a substrate for the induced enzyme activity. The enzyme β-galactosidase, for example, is very effectively induced by a non-metabolisable structural analogue of lactose, called IPTG (isopropyl-β-D-thiogalactose).

A common feature of bacterial genomes is gene clustering: genes whose products are related to one another are frequently coordinately expressed on one mRNA molecule from a cluster of genes called an operon. The lactose (*lac*) operon of *E. coli*, for example, contains the genes that the bacterium requires to metabolise lactose and is only switched on (transcribed) when the substrate for the gene products (lactose, or its analogue IPTG) is present in the cell. The behaviour of *E. coli* cells both in the presence and absence of lactose was extensively studied by Jacob and Monod, whose painstaking research culminated in the first demonstration of regulation in a bacterial operon.

The *lac* operon (Fig. 4.12) consists of a regulatory region and three structural genes: *lacZ, lacY* and *lacA*. The product of the *lacZ* gene (β-galactosidase) hydrolyses lactose to its glucose and galactose substituents, the product of the *lacY* gene (β-galactoside permease) is used to transfer lactose into the cell, while the role of the *lacA* product (β-galactoside transacetylase) remains obscure. The *lacZ, Y* and *A* structural genes are distinguished from a regulatory gene (known as *lacI*) which is also a constituent of the operon. The *lacI* gene's product is a tetrameric (4-subunit) repressor protein that prevents transcription of the entire operon by binding to a DNA sequence called the operator. The *lac* genes are controlled by negative regulation: they are normally switched off by the repressor protein, and this repressor protein must be removed following interaction with an inducer for transcription to be switched on.

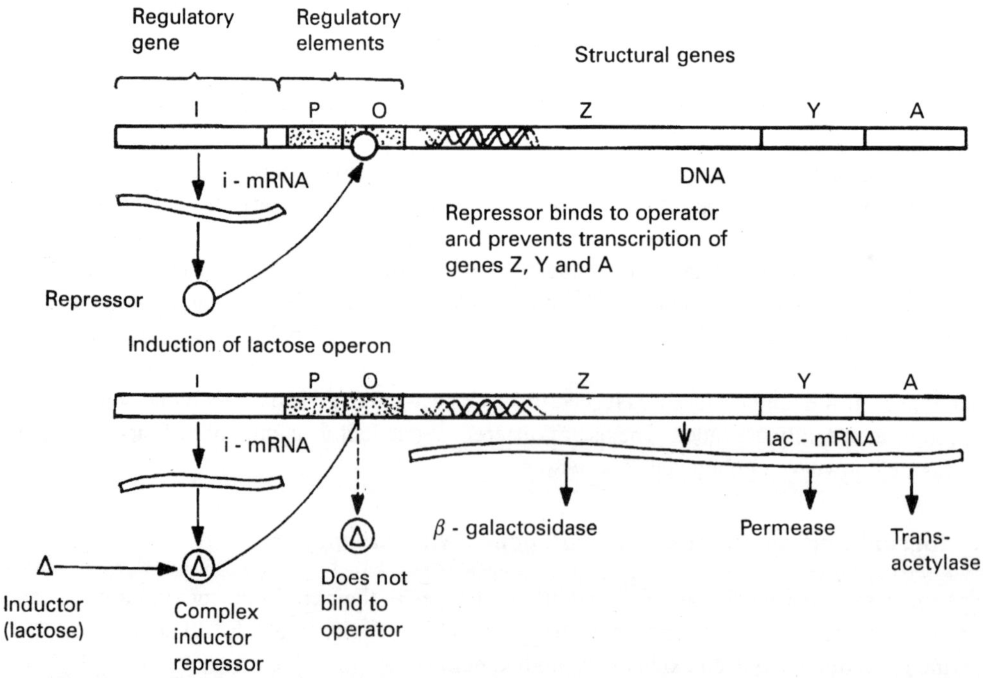

Fig. 4.12 *Repression and induction of the genes of the lac operon of Escherichia coli*. See text for details.

Figure 4.12 also illustrates how the *lac* operon is regulated. The operator is located between the promoter and the structural genes. It consists of a symmetrical inverted repeat DNA sequence (described earlier in Section 4.6) which is recognised by the repressor protein. The repressor protein has one binding site for the operator and one for the inducer. When the repressor binds to the operator sequence it blocks the initiation of transcription by physically preventing the RNA polymerase from moving along the DNA from the promoter site. When an inducer enters the bacterial cell it binds to the repressor and induces a conformational change in the protein in such a way as to decrease the affinity of the repressor protein to DNA. This renders the repressor–inducer complex unable to bind to the operator. Transcription now becomes possible since RNA polymerase is able to move freely along the DNA, across the operator site. This type of regulation, mediated by conformational alteration of a regulatory molecule, is called allosteric control.

Assignment 4.5

The normal condition for the *lac* operon is for it to be 'switched off' — that is in the absence of lactose the repressor molecule produced by the *lacI* gene binds to the operator and prevents transcription of the genes in the operon. What is the advantage to *E. coli* of this type of control over gene expression?

Self assessment exercise 4.3

Answer true or false to each of the following statements:

(i) A mutation (alteration in DNA sequence) in the *lacI* gene such that the repressor molecule produced can no longer bind to the operator region would lead to constitutive (continuous, even in the absence of an inducer) expression of the operon genes.

(ii) A mutation in the operator region such that the repressor molecule could no longer bind to it (the operator) would lead to constitutive expression of the operon genes.

(iii) A mutation in the *lacI* gene such that the repressor molecule produced can no longer bind to the inducer would lead to constitutive expression of the operon genes.

Self assessment exercise 4.4

Suppose, for a gene or an operon, a regulatory gene codes for production of a repressor protein molecule which has affinity for a particular molecule which is the end product of the reaction catalysed by the enzyme(s) coded for by the operon. In the absence of the end product molecule the repressor protein does not have affinity for the operator. In the presence of the end product molecule the repressor protein binds to it (the end product molecule) and the conformational changes brought about confer affinity for the operator on the repressor protein.

What will be the effect of (i) presence and (ii) absence of the end product molecule on expression of the gene(s)?

Induction of gene expression ensures that enzymes involved in substrate conversion are only synthesised when the substrate is present. Repression of gene expression by an end product of a metabolic pathway ensures that the end product is not over -produced. In these ways modulation of gene expression ensures that cell metabolism is not directed towards synthesis of enzymes or metabolites that the cell does not require.

4.10 Transcription and its control in eukaryotes

4.10.1 Eukaryotes encode several RNA polymerases

The control of transcription in eukaryotes is more complex than in prokaryotes. Multi-step mechanisms regulate the initiation of transcription, its specificity and its efficiency. Instead of a single RNA polymerase enzyme complex transcribing all genes (as in *E. coli*) eukaryotes possess several different RNA polymerase enzymes, each of which is responsible for transcribing different types of genes. RNA polymerase I (Pol I) transcribes rRNA genes, RNA polymerase II (Pol II) transcribes protein-coding genes, while RNA polymerase III (Pol III) is required for the transcription of tRNA genes. Eukaryotic RNA polymerases are composed of many proteins that interact both with each other and with multiple sites on the DNA.

4.10.2 The role of transcription factors

Along with RNA polymerase II, transcription of eukaryotic protein-coding genes requires the participation of proteins called transcription factors (TFs). Functionally, TFs may be subdivided into two groups: general transcription factors, GTFs, that are required to transcribe all Pol II-dependent genes, and sequence-specific transcription factors, SSTFs, which regulate transcription. GTFs and Pol II form stable multi-protein complexes near a transcription initiation site by interacting with basal (core) promoter elements (sequences). The eukaryotic core promoter element is called a "TATA box" because of the TATA base sequence it contains. It is located 25–30 base-pairs upstream of the transcription initiation site.

The presence of a core promoter element is sufficient to support a non-regulated, low (or basal) level of transcription following formation of a complex with GTFs. SSTFs, in contrast, bind to DNA sequences located upstream (to the 5' side) or downstream (to the 3' side) of core promoter elements and may either enhance or inhibit the level of transcription.

A basal level transcription *in vitro* from Pol II-dependent promoters requires the concerted action of at least six GTFs: TFIIA, TFIIB, TFIID, TFIIE, TFIIF and TFIIH. The first step in pre-initiation complex assembly involves binding of TFIID to the promoter at the TATA box. This is facilitated in some cases by TFIIA and creates a stable complex that subsequently promotes the stepwise assembly of the other general initiation factors and Pol II into the complex. TFIIA was demonstrated to function early in pre-initiation complex formation probably by altering the binding of TFIID. The requirement for TFIIA for basal level transcription varies, depending on the transcription system being studied.

In relation to proteinp-structure studies, the term domain refers to a distinct entity (3-dimensional shape) in the folding of a polypeptide chain. Discrete groups of mammalian transcription factors have now been categorised on the basis that they contain domains which are typical of those found in known DNA-binding proteins. Two of these domains, the helix–loop–helix (HLH), also known as the helix–turn–helix (HTH), and the leucine zipper (LZ) have been extensively studied. Both HLH and LZ domains are frequently found next to regions that contain an abundance of positively-charged amino acid residues. Mutational analysis has shown the DNA binding specificity to be located in this positively-charged region. For example, TFs lacking positively-charged domains are incapable of DNA binding and have no influence on transcription *in vitro*. The HLH and LZ domains have been shown to aid protein dimerisation between different members of the same HLH or LZ family. Besides the HLH (HTH) and LZ domains, other amino acid sequences have been postulated to be DNA-binding regions in many proteins involved in eukaryotic transcription and its regulation. Models of protein/DNA interactions have been proposed, and several are shown in Fig. 4.13. Many DNA-binding regulatory proteins from DNA-containing viruses have been characterised, but since most eukaryotic regulatory proteins are present in cells in very small amounts it seems likely that many more remain to be discovered.

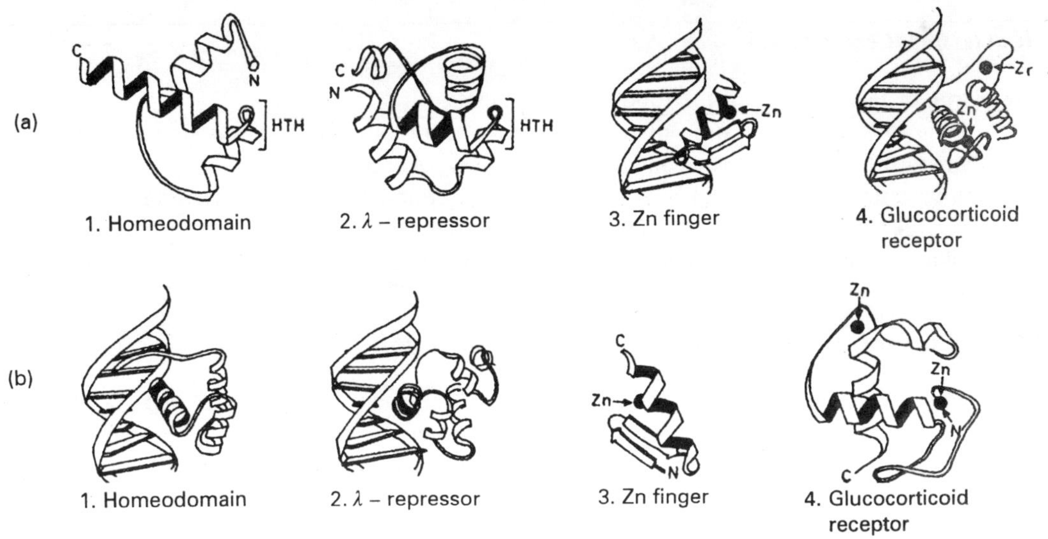

Fig. 4.13 *DNA binding proteins.*
(a) Diagrammatic representation of four DNA-binding protein domains.
(b) Association of the protein domains with the DNA helix.

Short stretches of DNA sequence in the promoter regions of eukaryotic genes specifically interact with transcription factors. The transcription factors then interact directly or indirectly with different components of the general transcription machinery. For a given gene, the combination of control elements and the availability of transcription factors play a key role in determining the level of transcriptional activity. Recently it was discovered that both activators and repressors may be encoded by the same gene, and two mechanisms have been proposed to account for this: (i) alternative splicing, and (ii) alternative initiation of translation. These are shown in Fig. 4.14. This finding has important implications for defining the final transcriptional response.

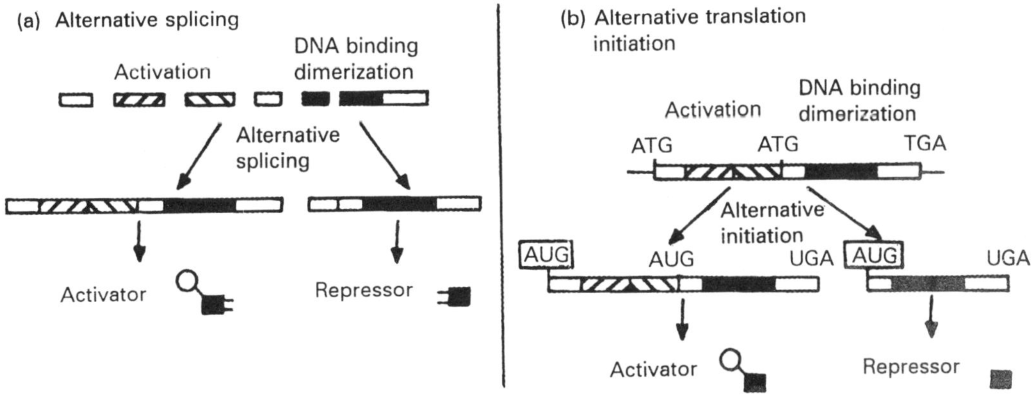

Fig. 4.14 *Control elements.*
The generation of either an activator or a repressor protein from the same gene by: (a) alternative splicing; (b) use of alternative translation initiation sites.

Self assessment exercise 4.5

Answer true or false to each of the following statements:

(i) In eukaryotic cells, initiation of transcription of protein-coding genes occurs after RNA polymerase I and transcription factors bind to a 'TATA' box.

(ii) In eukaryotic cells, the level of transcription of protein-coding genes is modulated (raised or lowered) by the binding of sequence-specific transcription factors to a 'TATA' box.

(iii) In eukaryotic cells, TFIIA functions by facilitating the binding of TFIID to a 'TATA' box.

4.10.3 Control of transcription by DNA methylation

Mammalian DNA is extensively methylated at C residues where they occur to the 5' side of G residues (represented as the sequence CpG, where p denotes the connecting phosphate group). Furthermore, the methylation of certain regions of DNA sequence in mammals appears to correlate with repression of transcription. Two models have been proposed to explain how methylation of DNA might result in transcriptional repression. The direct model postulates that a transcription factor is unable to bind in the region of methylated CpG sequences. The indirect model postulates that methylated DNA is bound by proteins which then prevent a transcription factor from interacting with its recognition site. The role of DNA methylation in transcriptional control remains unclear, however, and it is interesting to note in this context that transcription of the eukaryote *Drosophila* cannot be mediated by methylation since the DNA of *Drosophila* is not methylated.

4.10.4 Post-transcriptional editing of mRNA sequences

We are so used to the idea that nucleotide sequences are accurately copied from DNA to RNA that it is perhaps a little unsettling to learn that post-transcriptional editing of nucleotide sequences can occur, albeit only very rarely. Nevertheless, the effect of editing is to change the information held in the gene sequence. Editing has been demonstrated in mitochondrial mRNA in certain protozoa, and results in insertion and deletion of uridine residues. In other forms of mRNA editing, specific nucleotides are modified to alter the mRNA sequence, such as (i) the interconversion of C and U residues in plant mitochondrial and chloroplast transcripts, (ii) extensive deamination of adenosine in double-stranded RNA, and (iii) the alteration of a glutamine codon (CAG) to an arginine codon (CGG) in a mammalian mRNA. Codons are explained in Section 4.11. RNA editing in trypanosomes involves the addition and deletion of U residues within coding regions in several mitochondrial mRNA transcripts.

The information required for generation of edited mRNAs from pre-edited precursors is found in small RNA molecules called guide RNAs (gRNAs). Note that the RNA editing must be absolutely precise, otherwise a "scrambled" genetic message would result, but why does RNA editing occur at all? Wouldn't it be simpler for the *complete* genetic information to be held within the gene? A number of elaborate hypotheses have been proposed to account for this information "tinkering", but we still don't know the answers to these questions. We have a lot to learn.

4.10.5 Exons and introns in eukaryotic genes; RNA splicing

Earlier in this chapter (Section 4.3.2) we saw that DNA sequence and amino acid sequences were shown to be collinear. This relationship is not strictly true for the majority of eukaryotic genes. Eukaryotic and prokaryotic cells differ markedly in both the structure and the organisation of their genetic material. Whereas protein-encoding information of prokaryotes consists of linear runs of nucleotides, the coding information in the majority of eukaryotic structural genes is interrupted by stretches of non-coding DNA. The term "split genes" is often used to describe this organisation. The non-coding segments are called intervening sequences or introns. The expressed coding segments are called exons. Only exon sequences appear in the final mature mRNA following transcription and post-transcriptional processing. The removal of introns from a pre mRNA to yield mature mRNA for translation is called RNA splicing.

The selection and joining of exons is carried out following recognition of specific sequence elements by a spliceosome, a multi-component complex consisting of protein and RNA. The selection and joining of the exon sequences is, of necessity, a very precise process. Different splicing patterns in different tissues can result in the generation of several different proteins starting from the same pre-mRNA.

4.11 Translation and the genetic code

We have seen in Section 4.3.2 that genetic information flows from DNA through RNA to protein and that the sequence of bases in DNA is transcribed into RNA, which then specifies the linear order of amino acid residues in a polypeptide chain. Since proteins are constructed from a population of 20 different amino acids, the code which specifies them must have at least 20 different possible "words". Since there are only 4 different nucleotide bases in DNA (A, G, C, T) and correspondingly in mRNA (A, G, C, U), a one-base code could only specify 4 different amino acids. A two-base code could specify 16 amino acids (4×4), but this is still not sufficient. A three-base code is capable of specifying all 20 amino acids since there are 64 possible combinations ($4 \times 4 \times 4$) of bases in groups of 3. Information in mRNA is indeed translated in units of three nucleotides, and these units are called triplet codons. The assignment of particular codons to particular amino acids is called the genetic code, and is shown in Table 4.1. Examine the table carefully to see how it works.

Examine the first line across the table (under the heading). A codon UUU (first position U, second position U, third position U) codes for Phe. A codon UCU (first position U, second position C, third position U) codes for Ser. A codon UAU codes for Tyr and a codon UGU codes for Cys. Examine the fifth line across the table. A codon CUU codes for Leu, a codon CCU for Pro, a codon CAU for His and a codon CGU for Arg.

Of the 64 possible codons, 61 specify a particular amino acid, with the remaining 3 codons (UAA, UAG, UGA) being used as "stop" signals which indicate to the translation machinery that protein synthesis is complete. The codon AUG (which specifies methionine) is also known as a "start" codon since in almost all cases the start-site for translation is determined by an AUG in the mRNA. Note that only 2 amino acids (methionine and tryptophan) have a single codon; all the other amino acids have more than one. The code is said to be degenerate. Codon assignments are not random, however, and amino acids with similar properties tend to have closely-related codons. This is no accident: evolution has selected a code which minimises the effects that many single base change (point) mutations might have on protein function. Taking the amino acids leucine and isoleucine as examples: both have hydrophobic side-chains of similar size and hence substitution of one for the other (as a result of a mutation) may not impair the function of the protein encoded by the mutated gene. Inspection of the genetic code shows that changing CUA to CUU, for example, still results in a code for leucine; changing CUA to AUA would result in the substitution of isoleucine for leucine, and so on. The code on the mRNA is translated by complexes of rRNA and protein known as ribosomes, which move along the mRNA in the 5'- to 3'-direction synthesising a polypeptide chain from the amino-terminal end to the carboxyl-terminal end. This process was described in Chapter 2.

When the genetic code was finally deciphered in 1966 it was assumed that it was universal, i.e., that it was used by all organisms. It was therefore surprising to learn in 1979 that human mitochondrial genes use a slightly different code in which UGA (normally a stop codon) codes for tryptophan, AGG and AGA are stop codons, and AUA codes for methionine. Other rare differences have also come to light in last few years, but it is true to say that the code shown in Table 4.1 is followed by virtually all nuclear genes.

Table 4.1 The genetic code

1st position (5' end)	2nd position U	2nd position C	2nd position A	2nd position G	3rd position (3' end)
U	Phe	Ser	Tyr	Cys	U
	Phe	Ser	Tyr	Cys	C
	Phe	Ser	Stop	Stop	A
	Phe	Ser	Stop	Trp	G
C	Leu	Pro	His	Arg	U
	Leu	Pro	His	Arg	C
	Leu	Pro	Gln	Arg	A
	Leu	Pro	Gln	Arg	G
A	Ile	Thr	Asn	Ser	U
	Ile	Thr	Asn	Ser	C
	Ile	Thr	Lys	Arg	A
	Met	Thr	Lys	Arg	G
G	Val	Ala	Asp	Gly	U
	Val	Ala	Asp	Gly	C
	Val	Ala	Glu	Gly	A
	Val	Ala	Glu	Gly	G

Key to amino acid abbreviations:
Ala - alanine; Arg - arginine; Asn - asparagine; Asp - aspartate; Cys - cysteine; Gln - glutamine; Glu - glutamate; Gly - glycine; His - histidine; Ile - isoleucine; Leu - leucine; Lys - lysine; Met - methionine; Phe - phenylalanine; Pro - proline; Ser - serine; Thr - threonine; Trp - tryptophan; Tyr - tyrosine; Val - valine.

As early as 1955 Francis Crick had proposed that an "adaptor" molecule must be responsible for attaching to amino acids and aligning them on the mRNA template during translation on the ribosome. This indeed proved to be the case when tRNA molecules were identified for all 20 amino acids. For detailed descriptions of the machinery and mechanisms of translation the reader is recommended to consult the suggested texts in Section 4.13.

Self assessment exercise 4.6

(i) Referring to the genetic code given in Table 4.1, what would be the amino acid sequence of the polypeptide encoded by the following mRNA sequence?

5' AUGGUGGCCUAUCAUUAGGGGCUU 3'

(ii) What would be the effect on translation of the above sequence of a single base (point) mutation which gave rise to an A instead of a U at the twelfth base?

(iii) What would be the effect on translation of the sequence in (i) above, if an extra C were inserted between the third and fourth bases, i.e., between the two Gs at positions 3 and 4?

4.12 Concluding remarks

Deoxyribonucleic acid (DNA) is the molecule of heredity in all cellular organisms and must be faithfully replicated so that precise genetic information may be passed from mother to daughter cells and hence from generation to generation. It is useful to think of a gene as a length of DNA sequence that carries the code for a particular polypeptide chain, although this is not strictly correct; there are other types of gene that code for transfer RNA and ribosomal RNA molecules, for example. The relative positions of genes within any particular genome may be mapped by both physical and genetic methods, and mapping is becoming more accurate as the sequence of more DNA is determined. The phenotype of an organism is determined by its genotype: diploid organisms such as humans have two copies of every gene, whereas haploid organisms such as bacteria have only one. In the case of diploid organisms, the phenotype is determined by the dominant gene of the pair.

A DNA molecule consists of two polymeric strands wound around each other in the form of a double helix. A sugar-phosphate backbone is found on the outside of the helix, while the nucleotide bases project towards the centre. The four nucleotide bases of DNA are: adenine (A), guanine (G), thymine (T), and cytosine (C). Adenine forms hydrogen bonds with thymine (A–T); guanine forms hydrogen bonds with cytosine (G–C). These are known as complementary base pairs, and the sequence of these bases provides the genetic information. For DNA replication to occur the two strands must separate, each strand then acting as a template for the polymerisation of a complementary strand. This method of replication is called semi-conservative.

DNA is structurally a very dynamic molecule. It exists in various degrees of supercoiling and can adopt a variety of three-dimensional shapes that are determined by the nucleotide sequence at any particular region. The topology of a given DNA region is often important in the control of transcription, in which the genetic information stored in the DNA is expressed in the form of RNA. Regulation of transcription has been extensively studied in bacteria (and in *Escherichia coli* in particular), and the precise interplay of repressors, activators and inducers has been well documented. Transcriptional regulation in eukaryotes is less well understood. Progress is being made in determining the relative roles of a large number of transcription factors, their interplay with regulatory DNA sequences, and the possible role of DNA methylation.

The compartmentalisation of eukaryotic cells has allowed them to develop more sophisticated and complex methods of gene-expression control compared to those used by prokaryotes. For example, the splicing out of intervening (intron) sequences from between expressed (exon) sequences in the pre-messenger RNA allows for alternative splicing strategies to be carried out from the transcript of the same gene in different tissues. In very rare cases, editing of RNA transcripts is carried out by specialist gRNA molcculcs.

Proteins are synthesised on ribosomes which, with the aid of tRNA adaptor molecules, "read" the linear sequence of bases in mRNA in groups of three nucleotides. Each group of three nucleotides is called a codon, and the genetic code gives the meaning of each of the 64 possible triplet codon "words" The genetic code is degenerate, which means that most amino acids are encoded by two or more codons. Furthermore, the code has evolved so as to minimise the effects of single base (point) mutations on protein structure and function.

Over the years there have been many surprises for molecular biologists studying gene structure and expression. The discovery of split genes in eukaryotic genomes, the non-universality of the genetic code, and the post-transcriptional editing of mRNA transcripts are three good examples. No doubt many more surprises are in store as we attempt to explain the workings of a living cell at the molecular level.

4.13 Suggested further reading

1. Beebee, T. and Burke, J. (1992). *Gene structure and transcription* (2nd. edition). Oxford University Press, Oxford.
2. Berg, P. and Singer, M. (1992). *Dealing with genes: the language of heredity*. University Science Books, Mill Valley, California and Blackwell Scientific Publications, Oxford.
3. BIOTOL Series (1992) — *Genome management in prokaryotes.*
   — *Genome management in eukaryotes.*
   Butterworth-Heinemann, Oxford.
4. Dawkins, R. (1989). *The selfish gene*. Oxford University Press, Oxford.
5. Lewin, B. (1990). *Genes IV*. Oxford University Press, Oxford.
6. Micklos, D.A. and Freyer, G.A. (1990). *DNA science: a first course in recombinant DNA technology.* Cold Spring Harbor Laboratory Press, New York.
7. Ptashne, M. (1992). *A genetic switch*. Blackwell Scientific Publications, Oxford.
8. Singer, M. and Berg, P. (1991). *Genes and genomes*. University Science Books, Mill Valley, California and Blackwell Scientific Publications, Oxford.
9. Suzuki, D.T., Griffiths, A.J.F., Miller, J.H. and Lewontin, R.C. (1989). *An introduction to genetic analysis*, 4th edition. W.H. Freeman & Co., New York.
10. Watson, J.D. (1970). *The double helix: A personal account of the discovery of the structure of DNA*. Penguin Books Ltd., London.
11. Watson, J.D., Hopkins, N.H., Roberts, J.W., Steitz, J.A. and Weiner, A.M. (1987). *Molecular biology of the gene. Vols. I & II*. Benjamin/Cummings, Menlo Park, California.

4.14 Responses to assignments

4.1 The organisms of RR and rr genotype are both homozygous, but will exhibit different phenotypes: an organism of RR genotype will exhibit R phenotype, while the organism of rr genotype will exhibit r phenotype. An organism of Rr genotype is heterozygous, and will exhibit R phenotype since the R allele is dominant over the r allele. Therefore, organisms may have different genotypes (RR or Rr) but still exhibit the same phenotype.

4.2 The correct answer is that the genetic analysis of bacteria is easier. Using the examples of genotypes given in Response 4.1above, bacterial cells being haploid organisms, will be of either R or r genotype, and hence of R or r phenotype respectively. Hence, any organism of R phenotype must be of R genotype. This cannot be said of diploid organisms, in which organisms of R phenotype could be either of RR or Rr genotype. The study of the inheritance of genetic traits is therefore more straightforward in haploid organisms. This explains why bacteria have been used so widely in studies of genetics, gene structure and function, and in genetic engineering.

4.3 The 5' to 3' base sequence of the right-hand strand in Fig. 4.2 is CGATC.

4.4 The two anti-parallel strands of the original DNA duplex are complementary (G pairing with C, and A with T). Consider a GC base-pair. Where a GC base-pair occurred in the original (parental) duplex, on replication, the G in one original strand is replicated as its complementary base C, and the C in the other original strand is replicated as its complementary base G. The original GC pair therefore becomes replicated as a GC pair in both progeny (daughter) duplexes. The same process occurs for AT base-pairs. Refer to Fig. 4.4 for a diagrammatic representation of the mechanism of DNA replication.

4.5 The advantage to the organism is that in the absence of the inducer (normally the substrate for the enzyme(s) synthesised by the gene/operon) the enzymes are not manufactured by the cell; to synthesise enzymes for utilisation of a substrate when the substrate is absent is metabolically wasteful. This system ensures that only when the substrate is present in the cell environment are the enzymes required to use it synthesised.

4.15 Responses to self assessment exercises

4.1 The most precise definition is (ii). Neither (i) nor (iv) are precise definitions since many phenotypic traits are governed by the expression of more than one gene. (iii) is not a precise definition either since many proteins are composed of two or more polypeptides, which may be encoded by different genes. (A good example of such a "multi-subunit" protein is human adult haemoglobin, which consists of a complex of two α polypeptide chains and two β polypeptide chains.)

4.2 Remember that the DNA template is read from 3' to 5', so the transcript, written 5' to 3' according to convention, would be: 5' UAGCUGCCAAUGC 3'.

4.3 The correct answers are as follows.

(i) True — If the repressor protein cannot bind to operator, transcription will not be prevented.
(ii) True — If the repressor protein cannot bind to operator, transcription will not be prevented.
(iii) False — The opposite is in fact the case. If the repressor cannot bind the inducer, the repressor protein will bind to the operator, and the operon genes will not be expressed at all, even in the presence of the inducer.

4.4 The correct answers are as follows.

(i) In the absence of the end product (the repressor), the repressor protein does not have affinity for the operator and will not bind to it; transcription can occur and the enzymes produced generate the end product (repressor).
(ii) In the presence of the end product (repressor), the repressor protein does have affinity for the operator and will bind to it; transcription cannot occur.

4.5 (i) False — The statement would be true if 'RNA polymerase I' was replaced by 'RNA polymerase II', as RNA polymerase I is involved in transcription of rRNA genes, not protein-coding genes.
(ii) False — SSTFs bind to sequences upstream or downstream (on the DNA strand) of the 'TATA' box, not to the 'TATA' box itself.
(iii) True.

4.6 (i) The amino acid sequence of the polypeptide encoded by the mRNA sequence is: Met Val Ala Tyr His. Since the UAG codon which follows the codon for histidine is a stop codon, transcription would cease at this point.
(ii) Mutation of a U to an A at the twelfth position results in the sequence AUGGUGGCC UAACAUUAGGGGCUU. The effect of this is to change a tyrosine codon (UAU) to a stop codon (UAA). Hence, the product of translation would be terminated prematurely after the alanine, and the amino acid sequence would be: Met Val Ala.
(iii) The insertion of the extra base results in the sequence AUGCGUGGCCUAUCAUUAG GGGCUU. The effect of this is to shift the sense of reading frame from AUG, GUG, CCU, etc., to AUG, CGU, GCC, etc.

The encoded amino acid sequence becomes: Met Arg Gly Leu Ser Phe Gly Ala ...
Such mutations are called "frame-shift" mutations: a completely different amino acid sequence to the original has resulted after the point corresponding to the insertion of the extra base.

5 Gene technology

Chapter outline

5.1 Introduction

The techniques of modern gene technology have enabled enormous advances to be made in understanding how the genetic material of organisms is organised, how the expression of genetic information is controlled and how, sometimes, alteration of genetic information can give rise to genetically inherited disorders and/or cancer. In addition, they have made possible the bulk production of pharmaceutically important proteins. The process of cloning involves the production of multiple copies of a DNA fragment of interest by amplification either *in vitro* or *in vivo*. These modern approaches have largely overcome the tedious and expensive process of purifying proteins from organisms in which they are found only in small amounts (Table 5.1).

Table 5.1 Comparison of protein production and purification by classical methods and genetic engineering, using microorganisms

Classical approach	Genetically engineered protein
Grow the microorganism in a suitable medium	Grow the clone of cells bearing the engineered gene in suitable medium
Harvest cells and prepare crude cell lysate	Harvest cells and prepare crude cell lysate as a fusion protein
$(NH_4)_2SO_4$ precipitation	Affinity column chromatography
Extensive dialysis to remove the $(NH_4)_2SO_4$	Desired protein eluted as a fusion protein is cleaved
Ion exchange chromatography or gel filtration	Affinity column chromatography
The eluted protein from above is subjected to affinity column purification /HPLC	Yield is n(2% – 10%) where n can be increased several fold (30% – 40% of the total protein) by suitable choice of vector, growth conditions, etc.
The protein yield is 2% – 10% of the total starting material	

Overall, gene cloning requires two major steps:

(i) the gene of interest is isolated and incorporated into a small, self-replicating chromosome (called a vector because it 'carries' the gene into another cell) which can be an *E. coli* plasmid or a virus. The vector with an incorporated gene is called a recombinant vector;

(ii) the recombinant vector is then introduced into a suitable host cell. As the vector replicates, many identical copies of the cloned gene are produced.

5.2 Restriction endonucleases as tools for DNA fragment generation

In order to clone a DNA sequence that codes for a required gene product, the gene has to be removed from the host organism and moved to another easily manageable organism such as a bacterium.

In order to transfer (mobilise) the gene into the host microorganism, the gene must first be inserted into a suitable vector (DNA molecule).

Assignment 5.1

With your knowledge of DNA structure gained from reading the previous chapters, suggest what enzymatic activities would be necessary for inserting a gene into a vector molecule.

Restriction endonucleases are enzymes that make site-specific cuts in the DNA. Many restriction enzymes have now been discovered, the first of which was isolated from *Haemophilus influenzae* in 1970. Different restriction endonucleases present in different bacteria recognize different nucleotide sequences (Table 5.2).

When a DNA molecule is cut with a restriction enzyme, the ends or termini of the cut DNA fragment can be cohesive or blunt-ended depending on the enzyme (Fig. 5.1).

Table 5.2 Examples of some restriction enzyme specificities

Microorganism producing endonuclease	Endonuclease	Target site on DNA
1. *Bacillus amyloliquefaciens* H	Bam H I	G ↓ GATCC
2. *Escherichia coli* R Y13	Eco R I	G ↓ AATTC
3. *Haemophilus aegyptius*	Hae III	GG ↓ CC
4. *Haemophilus influenzae*	Hind III	A ↓ AGCTT
5. *Providencia stuartii*	Pst I	CTGCA ↓ G
6. *Thermus aquaticus*	Taq I	T ↓ CGA
7. *Serratia marcescens*	Sma I	CCC ↓ GGG

Note : ↓ Indicates the site of the enzyme cut within the specified nucleotide sequence.

Cutting with Eco R I

5'... G ↓ AATTC... 3'
3'... CTTAA ↑ G ... 5'

5'... G   AATTC....3'
3'... CTTAA   G... 5'

Cutting with Pst I

5'... CTGCA ↓ G... 3'
...G ↑ ACGTC...'

5'... CTGCA   G... 3'
3'... G   ACGTC... 5'

(a)

Cutting with Sma I

↓
5'... CCC GGG... 3'
3'... GGG CCC... 5'

5'... CCC   GGG... 3'
3'... GGG   CCC... 5'

(b)

Fig. 5.1 *Generation of cohesive and blunt-ended fragments using restriction enzymes. (a) Cohesive/sticky termini. (b) Blunt-ended termini.*
Cohesive fragments have protruding nucleotide termini, either at the 5' or 3' end and are termed cohesive/ sticky termini. Blunt-ended fragments bear no protruding ends.

> ***Assignment 5.2***
>
> Examine Fig. 5.1. Why are the termini (ends) of each DNA fragment produced by cutting with Eco R I or Pst I called cohesive or sticky ends?

When the vector DNA and the foreign DNA are both cut with the same restriction enzyme (e.g., Eco R I) the overlapping ends of the vector and foreign DNA are compatible and complementary. When the resulting DNA fragments are mixed together they form complementary base-pairs between overlapping, terminal single-stranded DNA sequences and the phosphodiester bonds between them are re-formed with the enzyme DNA ligase (Fig. 5.2). DNA ligase is isolated either from *E. coli* or bacteriophage T4. *E. coli* DNA ligase is NAD^+ dependent, while the enzyme from phage T4 is ATP dependent. The former is ineffective in ligating blunt-ended DNA fragments together while the latter can accomplish this only at higher concentrations.

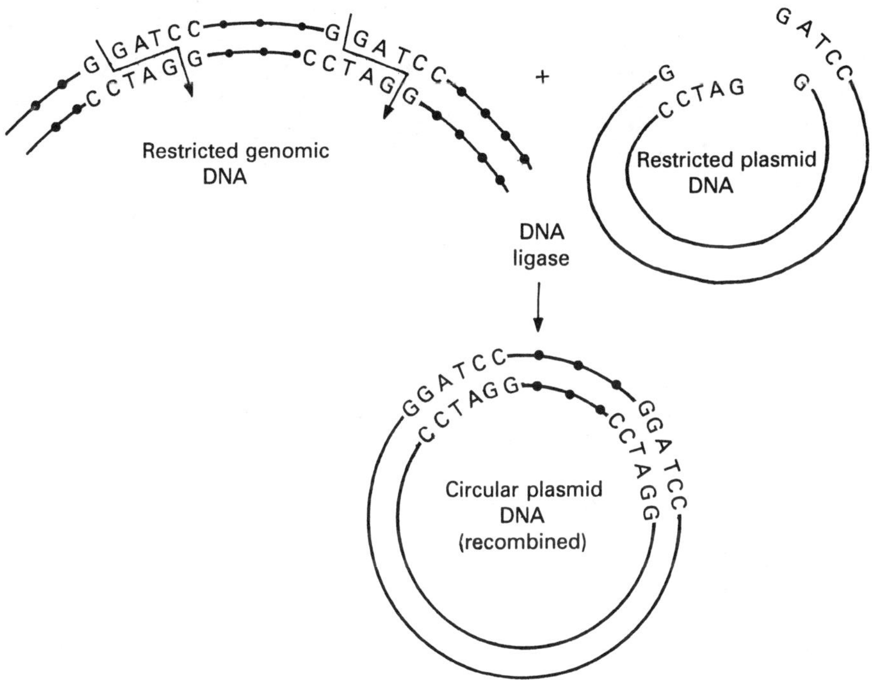

Fig. 5.2 *Restriction enzyme digestion of genomic DNA and plasmid vector DNA, followed by formation of a recombinant plasmid after annealing of complementary cohesive termini (sticky ends) and formation of covalent links by DNA ligase.*

5.3 Linkers and adaptors

Recombinant DNA and gene cloning experiments take advantage of the ability of restriction enzymes to cut DNA at particular positions. If suitable restriction enzyme sites are not available on the DNA to be manipulated, then these cleavage sites can be added as *linker* or *adaptor* molecules. Linkers are short double-stranded DNA molecules that contain restriction enzyme cleavage sites. These nucleotides can be ligated to the blunt ends of any DNA fragment and then cut with the specific restriction enzymes to generate DNA fragments with cohesive termini (Fig. 5.3).

Adaptor molecules are chemically-synthesised DNA molecules with pre-formed cohesive ends. They are required when the target site for the restriction enzyme used is present within the DNA sequence to be cloned.

In the homopolymer method oligo A(AA...AA) sequences and oligo T(TT...TT) sequences are added to the ends of foreign DNA and vector DNA molecules, respectively, by terminal transferase enzyme. When these two DNA molecules are mixed they anneal to create covalently closed circular recombinant molecules.

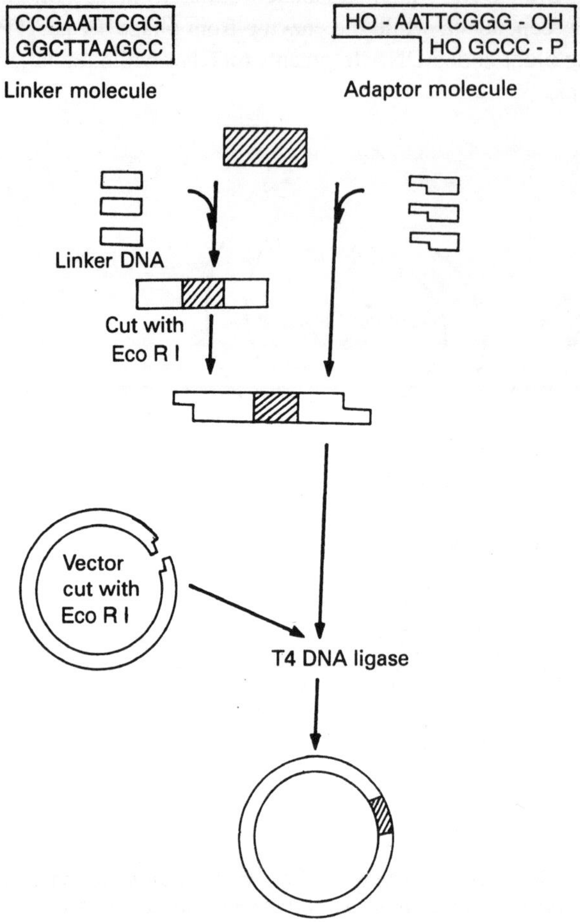

Fig. 5.3 *The use of linker and adaptor molecules in the formation of recombinant plasmids.*

5.4 The use of vectors as cloning vehicles

Vectors can be introduced into suitable host cells, where they replicate using the host's enzymes. A segment of foreign DNA which has been cloned (inserted) into the vector is therefore amplified along with vector DNA. Large amounts of the desired foreign DNA are therefore produced. The vast majority of vectors are plasmids, the remainder consisting of those based on bacteriophage lambda and M13. The major requirement of all vectors is an origin of replication for a given host cell in order that they may replicate autonomously (i.e., independently of the host's chromosome).

5.4.1 Plasmid vectors

Plasmids are examples of extra-chromosomal double-stranded circular DNA molecules that replicate autonomously. Plasmids are classified into different types. Conjugative or self-transmissible plasmids carry transfer (*tra*) genes that enable plasmids to transfer from one bacterium to another. Non-conjugative plasmids do not carry *tra* genes: they replicate autonomously but cannot transfer to another bacterium. Relaxed plasmids are maintained as multiple copies per cell. Stringent plasmids are present in limited number, typically 1–2 copies per cell.

A wide variety of plasmid vectors has been derived from the naturally-occurring plasmids of *E. coli*, such as ColE1 or from the related plasmid pMB1. Plasmid pBR322 is one of the most widely used multicopy cloning vectors of *E. coli* and is a hybrid between ColE1 and genes coding for resistance to the antibiotics tetracycline and ampicillin. ColE1 and pMB1 are maintained at 10–15 copies per cell, while some derivatives may have higher copy numbers (approximately 200 copies/cell). Amplification of the copy number up to 1000 copies/cell can be achieved by incubation of the host cells with the antibiotic chloramphenicol. This inhibits general protein synthesis (which is required for chromosomal replication) but does not inhibit plasmid replication and greatly increases the content of plasmid DNA. Low copy number plasmids are useful for cloning genes whose gene product may be lethal or toxic to the host cell when produced from a multicopy vector.

> ***Assignment 5.3***
>
> What would be the advantage of inserting genes coding for resistance to antibiotics into a vector molecule?

A plasmid vector should possess the ability to confer readily selectable phenotypic traits on host cells. In addition to antibiotic resistance genes, markers such as the *lac* or the *trp* genes are also used. In addition, the plasmid must possess single sites for a large number of restriction endonucleases, preferably within genes that code for a characteristic phenotype. Modern vectors carry a polylinker or "multiple cloning site". This is a short stretch of DNA sequence which includes nucleotide recognition sequences for a number of different restriction endonucleases.

There are two approaches for positioning the gene fragments in vectors: (i) non-selective insertion, and (ii) selective insertion into an indicator gene such that the gene function (e.g., antibiotic resistance) is lost.

5.4.2 Bacteriophage vectors

Double-stranded DNA bacteriophage vectors of *E. coli* are based on bacteriophage lambda. Bacteriophage lambda has a double-stranded DNA genome of approximately 49.5 kilobase-pairs (kb) which exists as a linear molecule with complementary single-stranded 5' ends which are 12 nucleotides long. Foreign DNA up to 25kb in length can be packaged into the lambda genome without diminishing the ability of the organism to infect *E. coli* and replicate. Although DNA packaging in the lambda system is only possible when the resulting genome size is between 78% and 105% of the wild-type genome size, large pieces of DNA can be introduced into lambda DNA because almost half of the lambda genes are dispensable and can thus be excised.

Single-stranded DNA bacteriophage vectors are based on the filamentous coliphages M13, fd and f1 and contain a 6.4kb single-stranded circular DNA molecule. These vectors are used for cloning of DNA fragments for the purpose of sequence analysis, for *in vitro* mutagenesis and for the preparation of single-stranded DNA probes (Section 5.9.2).

5.4.3 Cosmid vectors

Cosmid vectors possess an origin of replication, a selectable genetic marker (antibiotic resistance) and suitable cloning sites. They also possess a cohesive end sequence of lambda bacteriophage (called a *cos* site) that is essential for the efficient packaging of lambda DNA into the virus particle.

These regions are excised from the lambda phage DNA and ligated to small molecular weight plasmids that are of 5 kb size. This enables large DNA fragments to be cloned into cosmids (up to 45 kb). An *in vitro* packaging reaction packages recombinant cosmid molecules of the appropriate size into phage heads. Following adsorption of these phage particles onto suitable *E. coli* host cells, the cosmid vector circularises via the cohesive ends and replicates as a plasmid. Cosmids have been used for cloning long segments of eukaryotic DNA.

5.4.4 Shuttle vectors

Shuttle vectors are plasmids that are specially designed to replicate in different host systems. The origins of replication for different host systems such as *E. coli* or yeast are combined in one plasmid so that any gene introduced into the vector can be expressed in either of the host organisms.

> ***Assignment 5.4***
>
> Plasmids transform (enter) *E. coli* cells with much greater efficiency than is possible in organisms such as yeasts and the bacterium Bacillus. How would a shuttle vector with origins of replication for *E. coli* and Bacillus or yeast be of use in genetic engineering of Bacillus or yeast?

5.5 Expression of cloned genes in *E. coli*

Many factors are capable of influencing the expression of a cloned gene in *E. coli*: promoter efficiency, messenger RNA (mRNA) stability, translational efficiency, codon usage, plasmid copy number and plasmid stability. These are now considered in more detail.

5.5.1 Promoter efficiency

A promoter is a region of DNA involved in the binding of RNA polymerase to initiate transcription (see Chapter 4). A comparison of over 200 promoters has shown that the sequences at the –35 and –10 regions can be identified as consensus sequences. These are sequences in which each position represents the base most often found among the sequences compared. Expression of cloned genes may be maximised by cloning the required gene downstream of an efficient promoter such as *lac*, *trp* or *tac*, The *tac* I and *tac* II promoters are 11 and 8 times stronger than the *E. coli lac* promoter, respectively. Their –35 and –10 sequences are shown in Table 5.3.

Table 5.3 Sequences of some commonly used promoters

Sequence site	–35	–10
Consensus sequence	TTGACA	TATAAT
lac	TTTACA	TATATT
trp	TTGACA	TTAACT
tac I	TTGACA	TATAAT
tac II	TTGACA	TTTAAT

Some cloning vectors have been made that carry phage promoters such as T3, T7 and SP6 promoters, obtained from the coliphages T3 and T7 and *B. subtilis* phage SP6 respectively. The advantage of using these promoters is that only the inserted gene will be transcribed owing to the specificity of the viral

RNA polymerase for its promoter sequence; no other plasmid genes will be transcribed. Commonly, vectors have been constructed such that the DNA insert is flanked by promoters for two different viral RNA polymerases of phages T3, T7 or SP6, for example T3/T7; T7/SP6. The positioning of the two promoters at the opposite ends of the insert DNA allows either strand to be transcribed *in vitro* by the appropriate choice of RNA polymerase. These polymerases are highly specific in that they recognise only their individual promoter sequences as shown below:

SP6	GATTTAGGTGACACTATAG
T7	TAATACGACTCACTATAGGG
T3	ATTAACCCTCACTAAAGGGA

5.5.2 Stability of mRNA

The steady-state amount of mRNA transcribed from an inserted gene can be raised either by increasing the rate of mRNA synthesis or by decreasing the rate of mRNA degradation. Nucleotide sequences upstream of the initiation codon of gene 32 from T4 bacteriophage code for factors which have been shown to stabilise mRNA and this has been exploited by introducing these sequences into plasmids for producing foreign protein in large quantities.

5.5.3 Translational initiation

Sequences upstream of the AUG translational initiation codon are known to influence the efficiency of translation. The Shine–Dalgarno (SD) sequence, 5'-UAAGGAGGU-3', must also be present since it acts as a ribosome binding site. The four nucleotides following the SD region have also been shown to influence translational efficiency. This must be borne in mind when designing the construction of the recombinant plasmid.

5.5.4 Codon choice

Since there is more than one codon for most of the amino acids (Chapter 4), the availability of particular tRNAs in the cell can influence the expression of cloned genes in *E. coli*. A shortage of any tRNA can cause misincorporation of amino acids into the encoded protein.

5.5.5 Plasmid copy number

Increasing the amount of mRNA per cell can also be achieved by increasing the number of gene copies per cell. Therefore, the gene of interest should be ideally cloned into a high copy number plasmid such as pBR322. Replication of a ColE2 plasmid such as pBR322 is controlled by an RNA molecule called RNA I. This molecule inhibits chromosomal DNA replication by binding to a second RNA molecule called RNA II, which has a positive effect on ColE1 replication.

Plasmids can also be constructed in which the copy number is controlled by promoters whose activity is regulated by altering the temperature. Such vectors are referred to as runaway plasmid vectors. At 30°C the plasmid vector is present as a low copy vector. Increasing the temperature to 38°C results in loss of control of plasmid replication and a significant increase in plasmid copy number.

5.5.6 Plasmid stability

The growth rate of cells expressing high levels of recombinant proteins tends to be low.

Assignment 5.5
Explain why cells expressing high levels of recombinant proteins frequently have lowered growth rates.

If a mutant cell spontaneously arises in which expression of the recombinant protein is reduced, then the descendants of this mutant will have a faster growth rate and will be present in much larger numbers than the cells bearing the non-mutant recombinant plasmid. In addition, segregational instability due to defective partitioning can lead to plasmid loss. In other words when a cell divides, plasmids do not segregate to both daughter cells; one cell retains the plasmids and the other cell has none. The *par* region of the plasmid is responsible for plasmid segregation at cell division, and is found in naturally occurring plasmids. In plasmids vectors such as pBR322 this region is deleted. Hence, segregation of pBR322 occurs randomly and can result in plasmid-free cells under conditions of nutrient limitation. The problem of plasmid loss can be overcome by growing cells harbouring pBR322 in the presence of an antibiotic, resistance to which is encoded on the pBR322 molecule. This procedure is known as selection.

Plasmid instability can arise as a result of deletion, insertion or rearrangements of DNA fragments. Homologous (identical) repeated sequences can recombine (interchange). Insertion can be mediated by insertion (IS) elements or transposons (sequences which can jump from one location to another in DNA), although these are rare events.

5.6 Cloning hosts for the over production of proteins

Proteins may be synthesised in a cell as functional protein molecules or they may be synthesised as precursor molecules, which need to be converted to mature protein molecules by post-translational modifications. Post-translational modifications can include N- or O-linked glycosylation (addition of sugar residues), phosphorylation (addition of phosphate groups), amidation (addition of amino groups), formation of disulphide (sulphur–sulphur) bridges or proteolytic cleavage.

A number of host systems are now available for the production of recombinant proteins. The most popular host expression system is *E. coli* as it can be grown economically and its genetics are well understood. In addition, *Bacillus subtilis* is being increasingly used for recombinant protein expression.

Recombinant protein expression in prokaryotes has certain drawbacks, however. The proteins, when over produced, tend to aggregate to form inclusion bodies and have to be disaggregated and renatured to be made biologically active. Post-translational modifications are not carried out in prokaryotic systems since prokaryotes do not possess the enzymes necessary for phosphorylating or glycosylating proteins correctly. Nevertheless, proteins expressed in prokaryotic systems have been isolated and modified *in vitro* by adding the modification enzymes at a later stage.

Yeast (*Saccharomyces* spp.) is a favourite organism of brewers and bakers and its genetics have been extensively studied. It is easy to handle and its proteins can be secreted into the growth medium, which simplies purification of the sample. In addition, post-translational glycosylation of proteins is also possible. However, not all required modifications are possible and the composition of glycosyl residues in yeast differs from that of mammalian cells.

Assignment 5.6
In producing a human protein, such as insulin, in a recombinant microorganism such as yeast, why would it be important for post-translational modification (such as glycosylation) to be similar to that occurring in mammalian cells?

Expression of heterologous (foreign) proteins in insect cells using baculovirus vectors offers high-level expression and effective post-translational modifications similar to those of mammalian cells. A prospective vaccine for HIV has been prepared by producing one of the HIV glycoproteins in a baculovirus system. Mammalian cell cultures would provide the best choice of a cloning host since the protein synthesised would be identical to the native protein and thus will not be immunogenic to humans. Although mammalian cell culture systems are expensive, they have been used for the large-scale production of some therapeutically important products such as tissue plasminogen activator (Chapter 9). Mammalian cell lines may appear to be the choice for the expression of proteins requiring glycosylation. However, when recombinant erythropoietin was produced in two different cell lines, the carbohydrate composition varied with the cell line and was found to be different from that of native erythropoietin. A comparison of the different organisms used as cloning hosts is shown in Table 5.4.

Table 5.4 Comparison of cloning hosts

Cloning host	Usefulness as a cloning host	Applications
Escherichia coli	**Advantages**: can be manipulated easily; gene regulation is well studied; overproduction of proteins is possible owing to the availability of high expression vectors; scaling up of cultures is easy **Disadvantages**: the over-expressed proteins tend to form aggregates called 'inclusion bodies' of the denatured protein; foreign proteins may be easily degraded within cells; post-translational modifications do not occur	Manufacture of insulin, interferon, human somatotrophin
Saccharomyces cerevisiae	**Advantages**: easy to culture; large scale production by development of high-level expression systems; proteins can be glycosylated **Disadvantages**: glycosylation is not the same as in the mammalian system	Hepatitis B vaccine; human serum albumin
Mammalian cells	**Advantages**: post-translational modifications are appropriate; expression of the desired protein is good **Disadvantages**: culture of mammalian cells more costly and technically difficult than that of microbial cells	Factor VIII; tissue plaminogen activator

Self assessment exercise 5.1

Choose the correct completion to the following statement.

When vector DNA and foreign DNA to be inserted in the plasmid are both cut with an endonuclease that creates sticky ends, the foreign DNA is easily inserted into the plasmid because:

(i) the overlapping termini on the vector DNA are complementary and have a high affinity for each other;

(ii) the overlapping termini on the vector DNA and the foreign DNA are complementary and thus have high affinity for each other;

(iii) the overlapping termini on the foreign DNA have a higher affinity for the overlapping termini on the vector DNA than the termini on the vector have for each other;

(iv) the overlapping termini on the vector DNA have a higher affinity for the overlapping termini on the foreign DNA than the termini on the foreign DNA have for each other.

Self assessment exercise 5.2

In Column A are listed factors that are capable of influencing the expression of a cloned gene in *E. coli*. Match each factor with the appropriate action in Column B.

	Column A		Column B
(i)	Consensus sequence	(a)	The segregation of plasmids at cell division.
(ii)	SD sequence	(b)	The binding of DNA polymerase.
(iii)	*par* sequence	(c)	The binding of ribosomes to mRNA.
(iv)	RNA I	(d)	Misincorporation of amino acids.
		(e)	Increasing plasmid copy number.

5.7 DNA sequence analysis

In gene technology it is important to be able to sequence the nucleotides in a given polynucleotide chain; in this way the relationship between nucleotide sequence and function (as promoters, operators, genes and so on) can be determined and genes with particular functions can be characterised.

The first major hurdle encountered when contemplating DNA sequencing is that most naturally occurring DNA molecules are enormous (for example 3.8×10^6 base-pairs for the single chromosome of *E. coli*). It is therefore impossible to sequence the entire length of such a large molecule in one go.

Assignment 5.7

How could the impracticability of sequencing a large DNA molecule be overcome?

A major hurdle encountered in analysing the sequence of fragments of a DNA molecule generated as a result of action of restriction enzymes, is that the quantity of DNA to be sequenced is small (each cell will contain only one copy of a given segment of DNA per haploid genome).

Assignment 5.8

How could multiple copies of a given fragment of DNA be produced, in order to make sequencing of it possible?

5.7.1 *DNA sequencing by the dideoxy chain-terminator method of Sanger*

The dideoxy chain-terminator method requires single-stranded DNA templates. Bacteriophage M13 is the vector of choice. It infects *E. coli* and has been specifically engineered in the laboratory (i) to contain suitable restriction sites for cloning and (ii) to allow easy identification of phage that are carrying a cloned insert (these are known as recombinant phage). The M13 viral genome comprises a circular, single-stranded DNA molecule which is converted into a double-stranded circle inside the bacterial cell. The double-stranded replicative form acts as a template on which progeny single-stranded molecules are replicated inside the cell and this single-stranded DNA is packaged into viral protein coats and extruded from the cell to complete the life-cycle of the phage. Infection with M13 does not result in host cell lysis and so the host cell continues to grow and divide, secreting virus into the culture medium. M13 shows great flexibility in the size of the genome that it can accommodate within its protein coat. DNA can be inserted into the phage-derived double-stranded genome *in vitro* using recombinant DNA techniques; the insertion is then replicated along with the rest of the phage DNA sequences. Cultures of cells harbouring recombinant virus are grown and the viral particles isolated from the culture medium. The single-stranded template DNA is extracted for use in the dideoxy sequencing reactions.

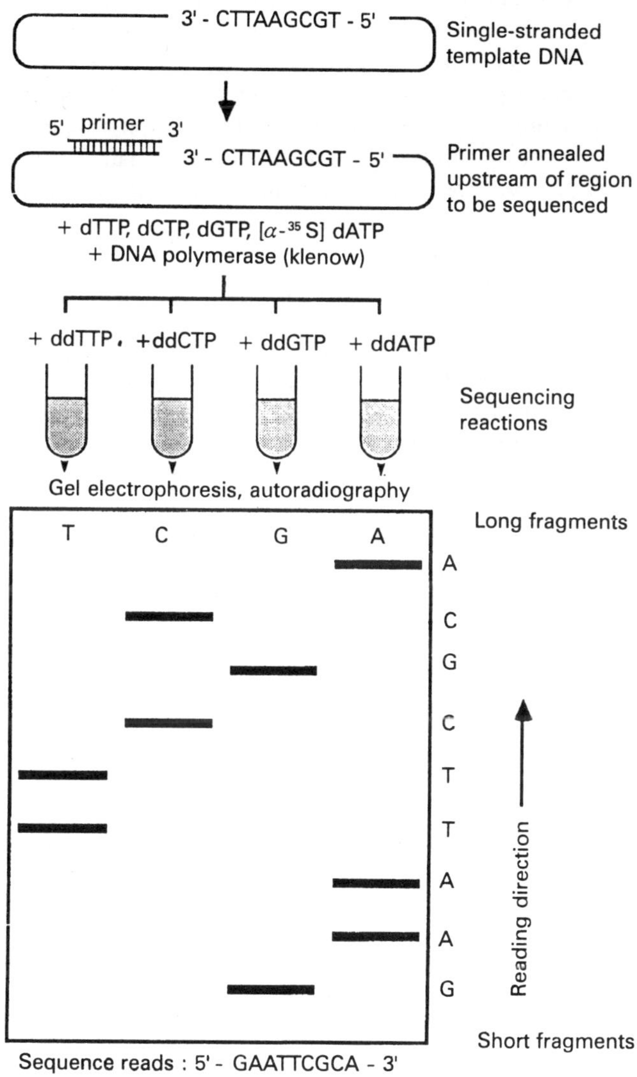

Fig. 5.4 *DNA sequencing by the dideoxy chain-terminator method of Sanger.*
Reproduced (with permission from Harwood Academic Publishers, GmbH) from: Robinson, A.C. (1991). *Clin, Chem, Enzym, Comms.*, **4**, 141–153.

A short, synthetic oligonucleotide, known as a primer, is annealed onto the single-stranded template M13 DNA just upstream of the restriction sites used for cloning (see Fig.5.4). When this complex is incubated *in vitro* with purified DNA polymerase and the four deoxynucleotide triphosphates (dNTPs), the primed site will act as a starting point for DNA synthesis in the 5' to 3' direction through the cloned sequence. If a dideoxynucleotide triphosphate (ddNTP) is present, synthesis stops when this molecule is incorporated into the growing chain. This happens because the ddNTP lacks a 3' OH group which is required for covalent linkage to the next incoming dNTP. Four separate reactions are therefore performed: in each reaction a different ddNTP is present in addition to all four dNTPs (dTTP, dCTP, dGTP, dATP). One of the four dNTPs (usually dATP) is radioactively labelled.

The ratio of dNTP to ddNTP in the reaction mix is carefully adjusted so that a dideoxy moiety is incorporated only occasionally in place of its deoxy homologue. Hence, each of the four reactions produces a population of partially-synthesised DNA fragments of different lengths, all sharing a common 5' end (defined by the primer), all of which are radioactively labelled and all of which are terminated by a dideoxynucleotide. A reaction is given the name of the base with which the 3' ends of the synthesised molecules are terminated. For example, all the newly-synthesised, radioactive molecules in the "T- reaction" containing ddTTP will be terminated at a T.

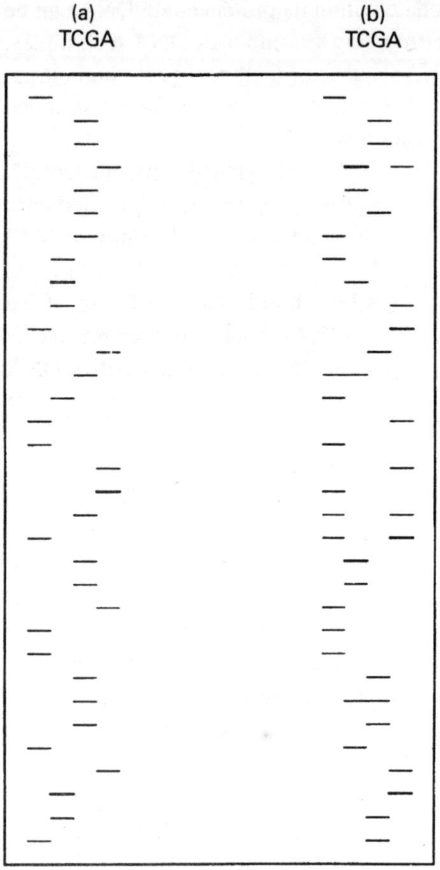

The DNA from these reactions is denatured and the products fractionated according to size in four different lanes (T, C, G, A) on a denaturing polyacrylamide gel which can resolve DNA chain length differences of one nucleotide. Autoradiography of the dried gel allows the sequence to be read directly. The order of bands on the autoradiograph is determined from the bottom of the gel, working towards the top (Fig. 5.4). This corresponds to the 5' to 3' sequence of the strand synthesised *in vitro* and is therefore complementary to the template cloned insert. All the bands in the T-track, for example, correspond to the position of an A in the template DNA sequence.

> ***Self assessment exercise 5.3***
>
> The accompanying Fig. 5.5 is a diagrammatic representation of a dideoxy sequencing autoradiograph for two clones (a) and (b). The direction of DNA fragment migration is from top to bottom and so the order of bands is read from bottom to top. Using your knowledge of the dideoxy method for DNA sequencing, write down the 5' → 3' DNA sequence of clone (a) and clone (b).

Fig. 5.5 *Diagrammatic represention of a dideoxy sequencing autoradiograph.*

5.8 The polymerase chain reaction

The polymerase chain reaction (PCR) has revolutionised molecular biology research. PCR amplifies a specified region of a DNA molecule without the requirement for time-consuming cloning procedures. In essence, amplification of DNA sequences by PCR is "cloning *in vitro*"

The PCR reactions are based on DNA replication *in vitro*. In PCR, the DNA sequence to be copied (plus the 15–30 bases flanking the sequence on either side) is mixed with a heat stable DNA polymerase (Taq DNA polymerase), the four deoxynucleotide triphosphates and suitable primers. Primers are short oligonucleotides (about 15 nucleotides in length) which are complementary to the flanking sequence at the 3' end of the DNA sequence to be copied. As the sequences flanking the DNA sequence to be copied are usually different, two different primers are usually needed, one for each 3' flanking sequence. Short oligonucleotides, such as primers, can be synthesised *in vitro* by automated procedures.

PCR involves the following steps:

(i) denaturation of the double-stranded DNA by heat, to generate two single strands, each of which serves as a template for DNA synthesis;

(ii) the temperature is lowered and the specific primers complementary to the 3' end of each of the single-stranded DNA strands are allowed to anneal to the template DNA at temperatures ranging between 37°C and 50°C;

(iii) DNA polymerase now acts on the single-stranded template starting from the primer annealed to the 3' end and extending the primer by complementary base-pairing and DNA synthesis. The complementary strand with its primer is similarly replicated;

(iv) in this first cycle of denaturation, annealing and extension, two copies of double-stranded DNA identical to the parent DNA molecule are produced;

(v) each of the two newly synthesised DNA molecules now re-enter step 1 of the procedure and the cycle of steps 1–3 continues until the required amplification of the DNA is achieved. The amplification is exponential. These steps are shown in Fig. 5.6.

As PCR involves raising the temperature to about 95°C to denature the DNA templates from their complementary strands, thermolabile DNA polymerase originally had to be added following each round of denaturation. This problem was overcome with the discovery of a thermostable DNA polymerase

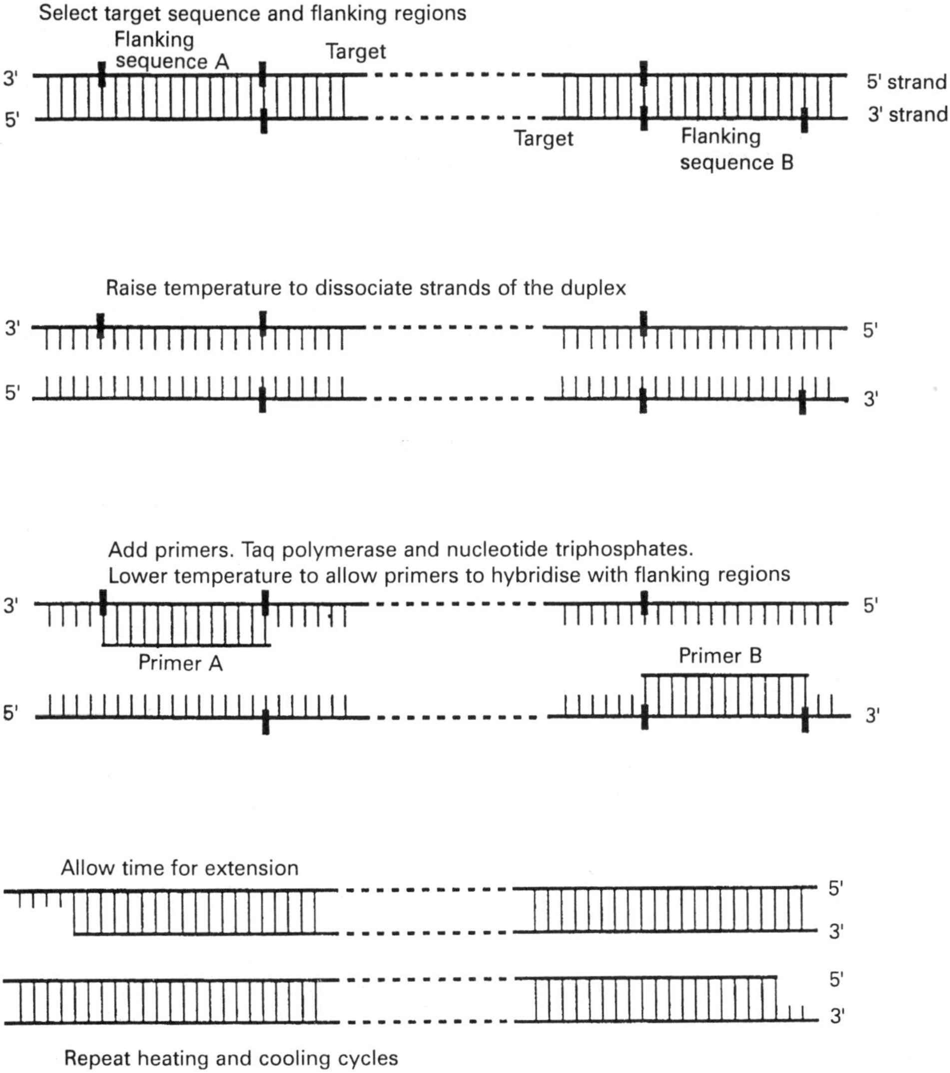

Fig. 5.6 *Principle of the PCR reaction.*

isolated from the bacterium *Thermus aquaticus*, known as Taq polymerase. Another thermostable DNA polymerase, Vent polymerase, is also available.

Since template DNAs will differ in their primer composition as well as in their nucleotide composition, careful adjustment of the conditions of individual PCR reactions is required. A US firm has patented the technology for a PCR machine that permits rapid changes of temperatures (+/– 0.1° C/sec) for a set time (ramping time). The use of Taq polymerase with PCR machines has made gene amplification by PCR a comparatively straightforward experimental procedure, although great care must be taken to prevent contamination of the sample to be amplified.

5.9 Applications of gene technology

We have seen in this chapter that gene technology enables genes to be copied and expressed in foreign cells and sequences. Some applications of this technology are presented here. Other applications are described in Chapters 8, 11, 12 and 13.

5.9.1 Production of pharmaceuticals from recombinant cells

(a) Recombinant insulin

Insulin is a hormone secreted by specialised cells in the pancreas called the Islets of Langerhans. It regulates sugar metabolism, and insufficient insulin or absence of production of insulin leads to low blood sugar levels and the disease diabetes. Diabetes affects a significant percentage of the world population and diabetics must receive daily injections of insulin in order to control the effects of this debilitating disease. Insulin for injection was previously isolated from the pancreas of slaughtered pigs and cows. Although these preparations were effective in patients suffering from diabetes, some patients developed antibodies against insulin owing to antigenic differences between the human and animal types.

Insulin consists of two short polypeptide chains, A and B, of 21 and 30 amino acids in length respectively, linked by two disulphide bridges. Insulin in mammals is synthesised as a pro-hormone, which contains an extra 35 amino acids (called the 'C peptide') which is not present in the mature insulin molecule. Post-translational proteolysis of the pro-hormone is necessary for the production of the mature insulin molecule. For the production of insulin in *E. coli,* the principal challenge was getting insulin assembled into this mature form.

The strategy to produce insulin as the first recombinant drug licensed for human use (Fig. 5.7) was as follows:

(i) genes for the mature A and B chains were first synthesised *in vitro*;

(ii) they were then cloned separately into the *E. coli* β-galactosidase gene and over expressed as β-galactosidase fusion proteins;

(iii) since a methionine codon had been inserted at the boundaries between the β-galactosidase gene and the insulin genes, the insulin chains are released by treatment with CNBr (cyanogen bromide) which specifically cleaves peptide bonds following methonine residues; the insulin chains were released intact as there are no internal methionine residues;

(iv) chemical cross-linking of the purified A and B chains yielded the active insulin molecule.

An alternative approach for insulin production is to over express the entire pro-insulin gene (containing the C peptide) in *E. coli* and, after purification of the protein, to cleave off the C peptide either by specific proteolysis or chemical hydrolysis. In a clever strategy, a hexa-peptide sequence (Arg–Arg–Gly–Ser–Lys–Arg) was substituted in place of the whole C peptide. This strategy made it possible to produce insulin following a two-step proteolytic reaction: an approach that preserves the N-terminal methionine of the insulin molecule.

Assignment 5.9

What steps would be involved in the production by synthesis *in vitro* of a gene for a mature insulin molecule?

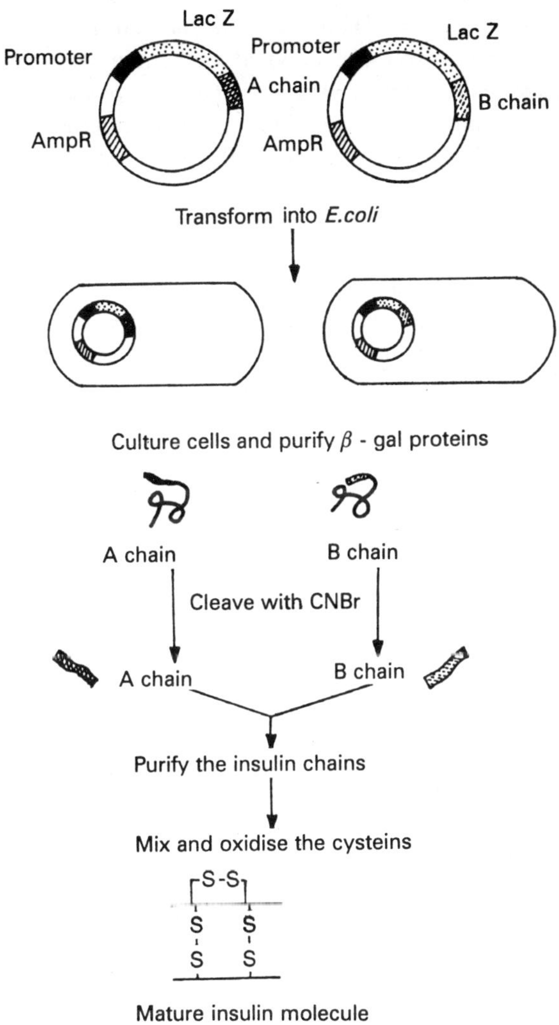

Fig. 5.7 *The generation of a recombinant insulin in E. coli.*

(b) Recombinant growth hormone

Growth hormone is produced by the pituitary gland and is involved in the regulation of growth and development. Deficiency of the hormone in children results in stunted growth, which explains why the condition is referred to as pituitary dwarfism. Treatment of such children involves regular injections of growth hormone extracted from the pituitary glands of deceased humans. Unlike diabetes, where injection of an equivalent animal hormone was found to be therapeutically effective, only human growth hormone (hGH) was found to be effective in children suffering from pituitary dwarfism.

hGH is synthesised normally as a large precursor molecule which contains a polypeptide sequence called a signal peptide involved in secretion across membranes. Human growth hormone is now available as a recombinant protein. It is produced in *E. coli* by linking the hGH coding sequence with a bacterial signal sequence. The bacterial signal peptide facilitates the secretion of the hormone into the bacterial periplasmic space (the space between the inner and outer bacterial cell membranes). Purification of the protein is therefore simplified since there are far fewer proteins in the periplasm than in the cell itself. The secreted hGH lacks the amino terminal methionine and is called "met-less hGH".

Recombinant growth hormones are now available for farm animals and are used fairly extensively in the USA. Their use leads to leaner meat and increased milk production. However, the ethics of the use of growth hormones is controversial as is the safety of food produced by such methods.

> ***Assignment 5.10***
>
> What advantages, from the point of view of transmission of diseases, would recombinant hormone have over human hormone extracted from pituitary glands?

(c) Recombinant vaccines

Recombinant DNA technology allows novel routes for the production of vaccines. For a vaccine to be effective, antigenic determinants must be exposed to the immune system. For recombinant vaccines, genes for the required antigenic determinants are cloned into suitable vectors and expressed in a host cell. This method offers several advantages over conventional methods of vaccine preparation. A number of recombinant proteins have been cloned with varying degrees of success, but a major problem has been the lowered level of immunogenicity of the recombinant proteins. Recombinant hepatitis B vaccine has been produced by cloning the gene for the surface antigen of the virus in yeast cells (Fig. 5.8).

Hepatitis B virus (HBV) is a 42 nm particle, called the Dane particle, consisting of a core containing the viral genome, surrounded by a phospholipid envelope carrying the major surface antigenic determinants (HBsAg). Infection with hepatitis B virus leads not only to the production of Dane particles but also to a dramatic overproduction of 22 nm large particles and filaments that contain the elements of the surface envelope. These filaments are about 1000 fold more immunogenic than the non-filamentous HBsAg protein. Hepatitis B virus is extremely difficult to grow in mammalian cell culture (Chapter 9), which is the conventional method of growing viruses for preparation of vaccines.

Recombinant hepatitis B vaccine was produced by cloning the gene for the surface antigen of the virus in yeast cells. The expression of the surface antigen in yeast results in the formation of virus-like particles. The yeast system with its complex membrane and ability to secrete and glycosylate proteins has made it possible to construct an autonomously replicating plasmid containing HBsAg coding sequences linked to the yeast alcohol dehydrogenase (ADH) I promoter (Fig. 5.8). The fragment to be cloned contains 6 base-pairs (bp) preceding the AUG which encodes the N-terminal methionine of mature HBsAg. This is then joined to the yeast ADH promoter cloned in the yeast vector pMA-56. The expression of hepatitis B surface antigen-coding sequences in yeast leads to the production of particles similar to the 22 nm particles seen in patients with hepatitis B infection. These particles were shown to be immunoreactive with anti-HBsAg antibodies. The formation of particles in the yeast vector suggests that the polypeptide sequence of the surface antigen molecule contains the necessary structure for assembly of the antigen as 22 nm particles within the cytoplasmic interior of the yeast cell. The similarity in structure of the yeast particle to the 22 nm particles isolated from HBV-infected cells, and its high immunogenicity, has made it possible for this recombinant product to be marketed as a vaccine for protection against HBV infection.

Foot and mouth disease (FMD) is a highly contagious disease of cloven-hoofed animals, especially cattle, sheep and pigs. The disease is caused by an RNA virus belonging to the picorna virus group which consists of a single-stranded RNA molecule of 8000 nucleotides surrounded by a capsid (coating) made of sixty copies of four proteins — VP1, VP2, VP3 and VP4. A number of vaccines have been developed, initially using the whole virus particle itself. The gene coding for VP1 was then isolated and a recombinant clone bearing this gene was expressed. Although the antigenic recombinant protein was stable and the immune response of vaccinated animals was good, the antigenicity of the recombinant protein was not as good as that of the virus itself. Unfortunately, the current vaccines are not totally effective on account of antigenic variation that is constantly

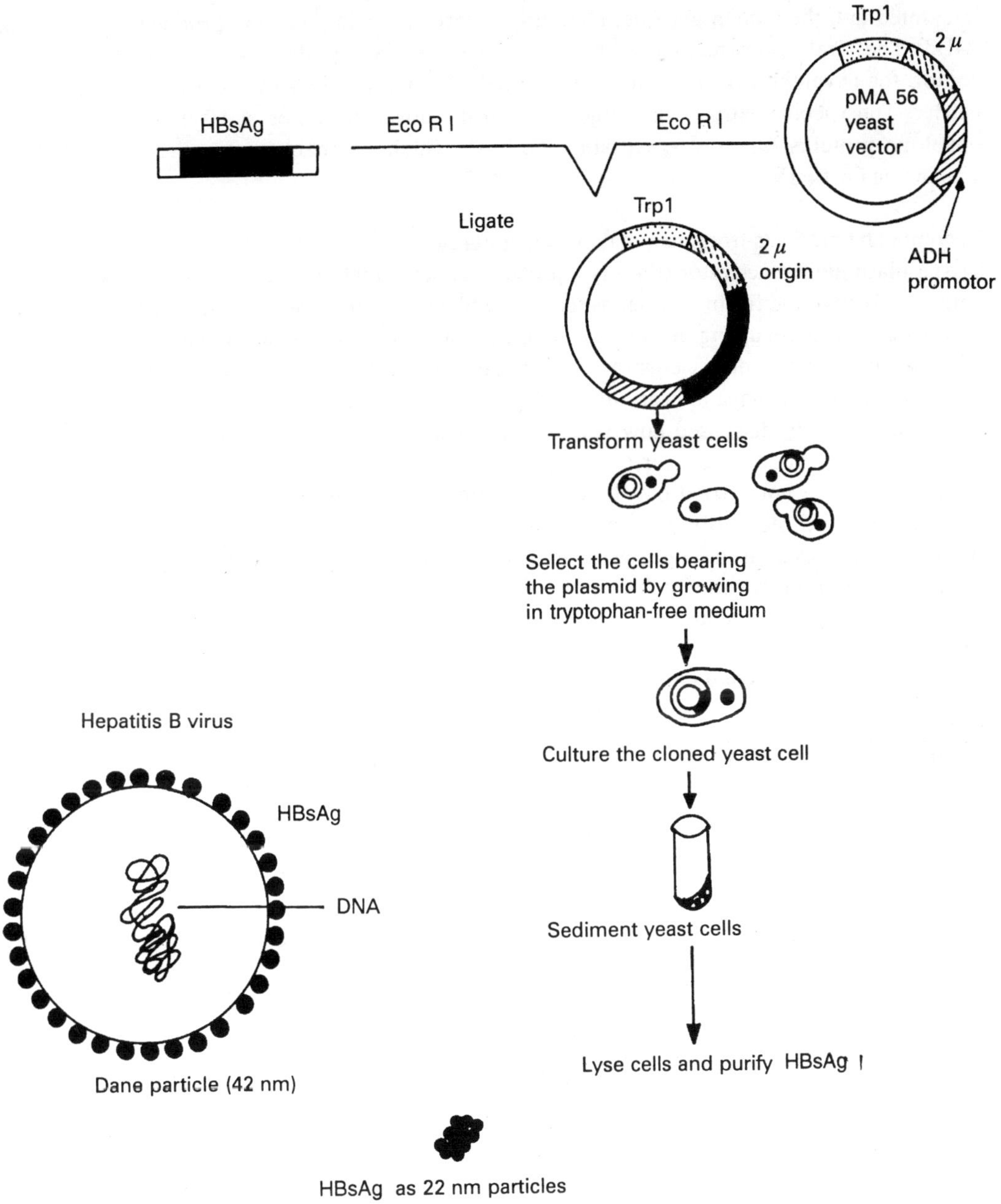

Fig. 5.8 *Expression of recombinant hepatitis B virus surface antigen (HBsAg) in a yeast expression system.*

occurring in the virus. Hence it is recommended that local strains are identified and vaccines developed from these.

AIDS (Acquired Immunodeficiency Syndrome) is caused by a single-stranded RNA virus, the Human Immunodeficiency Virus (HIV). HIV breaks down the immune defence systems of the body, exposing it to many forms of opportunistic infection. The progression of the disease is very rapid after an incubation period of many years, during which the virus is integrated into the human genome. Currently, efforts are being made to produce a vaccine for AIDS. The viral vector technique shows promise. Essentially, the genes of the HIV genome coding for envelope glycoproteins are inserted into a vaccinia virus vector which acts as a carrier. The vaccinia virus, when injected into the host, multiplies and synthesises the HIV envelope proteins, thereby sensitising the host. Such approaches may yield an effective vaccine against HIV.

(d) Recombinant products from mammalian cell cultures

Tissue plasminogen activator (tPA) is a protease enzyme that helps to dissolve blood clots. It helps to dissolve the fibrin clot deposited in the blood vessels supplying the heart muscle, that cause occlusion resulting in decreased blood supply to the heart muscles. Immediate administration of this drug helps to dissolve the life-threatening clots which would otherwise lead to permanent damage to heart muscle following a heart attack.

We have already discussed how several biologically active proteins of clinical interest are difficult to produce in bacteria and yeast because the appropriate post-translational modifications are not carried out. In such situations, mammalian cells have been utilised for expression. One recombinant protein successfully produced from human cell cultures (Chapter 9) is tissue plasminogen activator. Tissue plasminogen activator is now commercially produced from a mammalian cell line carrying a high expression vector, pSV2, as a fusion of mouse dihydroxyfolate reductase (DHFD) gene with the tPA gene. The recombinant protein is secreted into the culture medium since the tPA signal sequence (necessary for trans-membrane secretion) is cloned along with the tPA gene.

5.9.2 *Diagnosis of diseases*

(a) DNA probes in diagnosis

Recently, a great deal of interest has been generated in the use of DNA probes in diagnosis. DNA probes are single-stranded oligonucleotide sequences, complementary to a particular DNA sequence of interest, which are labelled with a radionuclide, enzyme or fluorescent molecule. This allows their detection. Radionuclides are detected by autoradiography (exposing the test sample to X-ray film-radiation is detected by the film), enzymes are detected by measuring the product formed by their action, and fluorescent molecules are detected by their emission of light when exposed to UV radiation or a laser. Many species-specific DNA probes consist of repetitive DNA sequences which are present in many prokaryotic and eukaryotic organisms, and which have been identified in malaria, filaria, tuberculosis and other parasites.

If a DNA probe hybridises with a nucleic acid, this demonstrates homology (the nucleic acid under test contains complementary sequences to the probe). Thus DNA probes are a convenient method of demonstrating the presence of particular nucleic acid sequences. In Southern blotting, DNA fragments are separated through agarose gels and the strands denatured (unwound) with alkali. The strands of DNA are then blotted by capillary transfer onto nitrocellulose membrane filters. The filters are vacuum dried and the radioactively-labelled, and denatured (single-stranded) DNA probes are added to the filter for hybridisation with homologous sequences. Autoradiographic development of the hybridised filter reveals the presence of any bound (hybridised) probe molecules and thus any complementary DNA sequences.

(b) PCR in diagnosis

PCR is capable of detecting even a single organism that has invaded the human host. The primers can be synthesised even with a limited amount of DNA sequencing of a previously uncharacterised gene, and can be used to amplify the region from DNA samples extracted from sera of patients with suspected infections.

Table 5.5 PCR as diagnostic tool

Disease		Causative agents
1. Bacterial Infections	a. Tuberculosis	*M. tuberculosis*
	b. Lyme disease	*Borrelia burgdorferi*
	c. Listeriosis	*Listeria monocytogenes*
2. Viral infections	a. Hepatitis	Hepatitis B Virus
	b. CMV infection	Cytomegalovirus (CMV)
	c. AIDS	Human Immunodeficiency Virus (HIV)
3. Genetic disorders	a. Cystic fibrosis	
	b. Haemophilia A	Factor VIII mutations
	c. Haemophilia B	Factor IX mutations
	d. Muscular dystrophy	
4. Parasite infections	a. Malaria	*Plasmodium falciparum/Plasmodium vivax*
	b. Filariasis/Onchocerciasis	*W. bancrofti, B. malayi/B. timori*
5. Cancer and Chemotherapy	a. Cancer markers	

Tuberculosis, a problem not only associated with developing countries, is also of serious concern in the developed world in patients with AIDS. Many AIDS patients harbour *M. tuberculosis* which is resistant to the common antitubercular drugs. PCR has been successfully employed as a diagnostic tool to detect very low numbers of this organism. Table 5.5 shows the diseases for which PCR may be used in a diagnostic role.

(c) Diagnosis of tropical diseases

Tropical diseases such as malaria, filariasis, schistosomiasis, leishmaniasis, tuberculosis, leprosy and trypanosomiasis affect millions of people worldwide. Tuberculosis and leprosy are caused by mycobacteria, while the others are parasitic in nature. Diseases such as filariasis and leprosy cause both physical and psychological damage. Symptoms range from mild fevers and physical disability to severe disfigurement, which render sufferers incapable of earning a living. The United Nations Development Programme/WHO-based Special Programme for Research and Training in Tropical Diseases (TDR) has implemented various programmes for field survey, diagnosis and treatment of these diseases. Product development, which includes development of diagnostic kits as well as new drugs and vaccine production, are the hallmarks of many of these programmes. As mentioned above, human malaria is widespread throughout the tropical world. It is caused by the *Plasmodium* species, mainly *P. falciparum* and *P. vivax*. The former strain has been cultivated in laboratory conditions in infected erythrocytes and many diagnostic procedures and potential vaccines have been investigated using this *in vitro* system to develop DNA libraries. Since *P. falciparum* causes cerebral malaria, especially in children, it is vital that a vaccine is developed for different forms of this infective agent.

Among the nematode infections, the major threat to humans comes from lymphatic filariasis. It is prevalent in Asia and Africa and is caused by the parasites *Wuchereria bancrofti* and *Brugia malayi* and is transmitted by mosquitoes. The related onchocerciasis is caused by the filarial *Onchocerca volvulus* and is transmitted by blackflies. In the chronic stages, the lymphatic filarial disease is characterised by gross swelling of the groin (hydrocele) or limbs (elephantiasis) while onchocerciasis results in blindness. In onchocerciasis, the nematode lives subcutaneouly and causes blindness by damaging retinal lens tissue. Construction of genomic libraries to search for antigens for diagnostic purposes has made significant progress in recent times. Advantage is also taken of the presence of repeat sequences present in the nematode DNA to develop PCR- based diagnostics.

Not only is it necessary to develop new and improved tools and solutions for disease control but it is equally important to develop diagnostic tests for the early detection of these infections. Diseases such as filariasis and leprosy can be effectively treated before they develop into the irreversible, chronic stages.

Self assessment exercise 5.4

Explain how PCR and gene probes could be used to diagnose the presence of hepatitis B virus in a sample of blood.

5.9.3 Insect control

So far we have been discussing the applications of genetic engineering to products that are of high commercial value. However, for all of us in developing countries, the development of biotechnology impinges on the needs of day-to-day living. Mosquitoes are a constant menace and are responsible for the transmission of diseases such as malaria, filaria and encephalitis. Environmental pollution has augmented mosquito breeding, and stagnant or polluted water outlets from houses and huts has exacerbated the problem. Similarly, in agriculture, while the "green revolution" has enhanced productivity of rice, wheat cereals, vegetables and fruits, the increased use of pesticides has become necessary for controlling insect pests.

The use of DDT in the 1970s eradicated mosquitoes from many countries. However, the eradication was often short-lived. DDT was shown to be carcinogenic and it accumulated in the food chain. In many developed countries this pesticide was banned. Later versions of chemical pesticides have also been shown to be harmful to mammals and have subsequently been banned. Currently there are about 25 banned chemical pesticides, but many developing countries continue to use them on account of their low cost.

Biopesticides are agents derived from microbial sources, principally from bacteria and viruses and the existence of microbes pathogenic to insect larvae has been known for several decades. However, their slow and specific mode of action and lack of suitable techniques for their production resulted in high production costs. However, the advent of genetic engineering and advances in molecular techniques have dramatically altered the scenario. Today biopesticide sales have attained 3% of the global chemical pesticides market and it is expected that they will reach 10% in ten years time. What has brought about this change? In particular the bacterial-based biopesticides have become extremely popular in view of their efficacy and from the knowledge of the genetic basis of their insecticidal activities. For mosquitoes and black flies, two major microbes have emerged: *Bacillus sphaericus* and *B. thuringiensis var. israelensis*. These bacilli form intracellular inclusion bodies at the onset of sporulation and when whole bacilli are ingested by the mosquito larvae, the bacilli dissolve in the alkalline pH of the gut and release the endotoxins contained in the inclusion bodies. These are present as multiple toxic peptides which bind specifically to the larval gut cells and dissolve the membrane by causing calcium ion leakage, eventually resulting in gut paralysis and death.

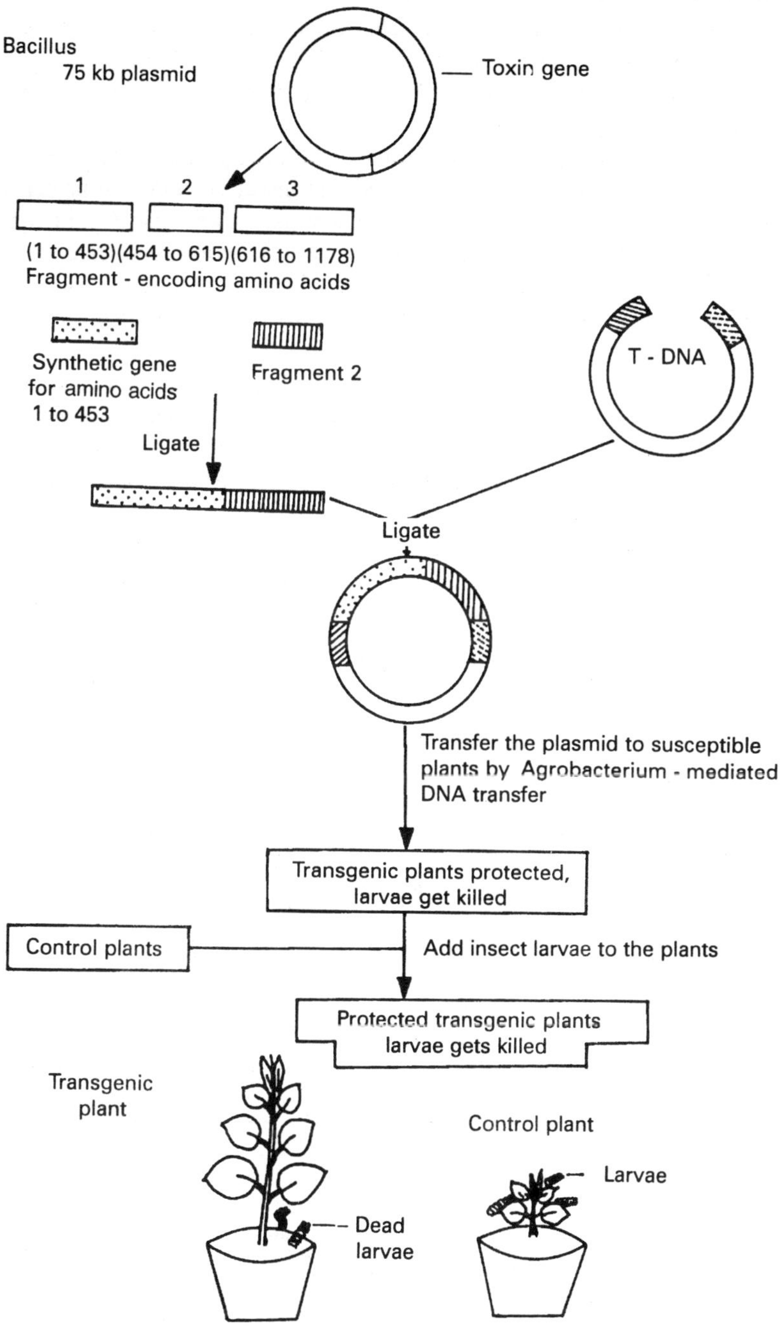

Fig.5.9 *Expression of Bacillus thuringiensis toxin gene in transgenic plants to protect them against insect damage.*

The control of breeding of mosquito larvae is achieved through the use of two bacilli strains (*Bacillus sphaericus* and *Bacillus thuringiensis var. israelensis*). Other insect-controlling biopesticides are derived from several strains of *B. thuringiensis* (such as *var. kurstaki*). These biopesticides are preferable biological alternatives to environmentally toxic chemicals.

There are many advantages in the use of biopesticides. This is because they are: (i) insect-specific and do not affect non-target organisms (unlike chemicals); (ii) bio-degradable; and (iii) development of resistance by the insects is expected to be slow in the case of bipyramidal crystal-shaped insecticidal proteins (ICPs).

ICPs are formed in different strains of *Bacillus* and are used as biocontrol agents for several species of insect pests. Many investigations have clearly indicated that insecticidal crystal proteins in different species of *B. thuringiensis* are encoded by large plasmids. It was also found that irreversible loss of these plasmids resulted in the appearance of acrystalliferous variants. In *B. sphaericus* the ICP genes are located on the bacterial chromosome. Fortunately, a variety of insect pests seem to be susceptible to different strains of *B. thuringiensis* (*B.t.*). Table 5.6 outlines the varietal spectrum of *B.t.* strains.

Since the mode of action of the ICPs is mediated by their specific binding to insect gut cells it is important that strategies are devised to prevent resistance development in insects. This is achieved by combining the different ICP genes in *B.t.* or other bacteria such as *Pseudomonas*, so that the possibility of resistance development is delayed. An important factor that may have contributed to the low frequency of insect resistance to *B.t.*strains is the low persistence of *B.t.* in the environment. Insect resistance may occur when the use of insect-resistant transgenic plants (Fig. 5.9) becomes widespread, since several generations of insects per year will be continuously exposed to crystal proteins, providing an ideal environment for resistance development. A strategy to overcome this problem would be to insert genes coding for several toxins into a plant.

Continued exposure of the toxin to the insect could be avoided by engineering the expression of a toxin gene only in certain plant tissues at certain growth stages, and only in response to insect feeding or only when induced by application of some elicitor.

Table 5.6 Genetic basis of ICP variations of *B. thuringiensis*

Action against insect species	Gene type
Lepidoptera	(cry I, A, B, C)
Lepidoptera and Diptera specific	(cry II)
Coleoptera specific	(cry III)
Diptera specific	(cry IV)
Cytolytic factor (haemolytic factor)	(cry A; cyt B)
Nematode specific	(cry V and VI)

Important target pests of *B. thuringiensis* ICPs

Order	Common name	Plants
Lepidoptera	Diamondback moth	Cabbage
	Armyworms	Cotton, vegetables
	Cotton bollworm	Cotton
	Cabbage looper	Cabbage
Coleoptera	Colorado potato beetle	Potato
	Alder leaf beetle	Alder
Diptera	Mosquitoes	Vectors of human pathogens
	Black flies	

5.9.4 Improved biological detergents

The power of genetic engineering to engineer natural products can be seen in many areas of research. One example is in the detergent industry, where new products with increased efficiency of cleansing action are being developed. It is now common practice to add proteolytic enzymes to remove protein stains from soiled clothing. Subtilisin, a serine protease produced by recombinant *E. coli,* is added to modern detergents but it cannot be used with bleach since the latter inactivates it. The methionine at position 222 in the enzyme has been found to be responsible for this inactivation. Genetic alteration has enabled this methionine residue to be substituted by other amino acids. Site-directed mutagenesis (a technique which allows mutations to be carried out inserting one or more nucleotides into a DNA sequence at a specific site, see below) produced mutant genes that expressed the enzyme protein

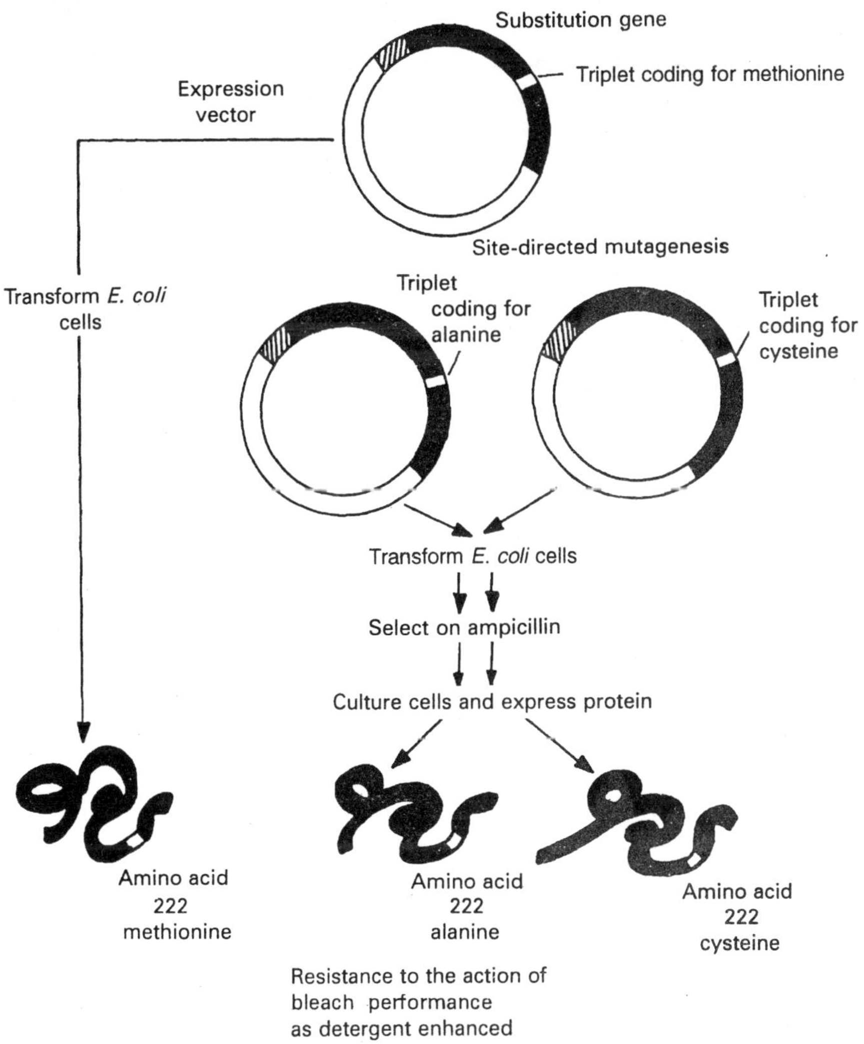

Fig. 5.10 *The use of site-directed mutagenesis (SDM) in the production of altered subtilisin enzyme for use in detergents.*

with substitution of the methionine residue with either cysteine or alanine. The alanine-substituted enzyme showed 53% activity as compared to wild-type subtilisin. This variant did not exhibit any sensitivity to bleach.

Site-specific mutagenesis or oligonucleotide-directed site-specific mutagenesis can be carried out (Fig. 5.10) on genes cloned in vectors that provide for the isolation of DNA single strands. An oligonucleotide of 12–15 nucleotides in length is synthesised that is complementary to the purified single-strand copy of the gene in the region of interest, but which contains one or more non-complementary or mismatched bases. The mismatched bases provide the desired mutant sequences. The synthetic oligonucleotide is annealed to the single-strand of the gene in the vector and a complementary strand is synthesised *in vitro* with DNA polymerase. DNA ligase is then used to seal the single-strand break left by the polymerase and the double-stranded circular DNA product is introduced into *E. coli* cells by transformation. The replication of the DNA molecules in *E. coli* produces a population of progeny DNA molecules, half of which carry the new or mutant DNA sequence. The oligonucleotide that was originally used to make the mutation can be used as a hybridisation probe to distinguish mutant from wild-type molecules, under conditions of high stringency. Modified enzyme molecules such as these are not only insensitive to the action of bleach, but show an increase in proteolytic activity compared to the wild-type protein.

5.9.5 Bacterial genomics

The techniques of gene technology are being applied to the characterisation and understanding of bacterial genomes.

Assignment 5.11
Why is the study of bacterial genomes important to humankind?

5.9.6 Gene therapy

Adenosine deaminase (ADA) is an enzyme of the purine metabolism which catalyses the deamination of adenosine and deoxyadenosine to inosine and deoxyinosine, respectively. In humans, deficiency of ADA activity is associated with a disorder called severe combined immunodeficiency (SCID). ADA deficiency is an extremely rare inherited immune disorder and is responsible for approximately one quarter of all cases of SCID. It is associated with defects in cellular and humoral immunity, affecting primarily the T-cells and to a lesser extent the B-cells (Chapter 10), with consequent severe immunological dysfunction. The molecular basis for ADA deficiency is the production of a catalytically defective enzyme molecule. It is a fatal disease as it leads to the intracellular accumulation of deoxyadenosine and its metabolites, primarily deoxyadenosine-5'-triphosphate, which are toxic to cells.

Children suffering from ADA deficiency have previously been treated either with conjugated bovine ADA or they have undergone bone marrow transplantation. With the advent of recombinant DNA technology, the ADA gene has been seen as a good candidate for gene therapy, as lymphohaematopoietic tissue can be readily obtained from bone marrow, manipulated outside the body and reimplanted. Retroviruses provide an efficient means by which this type of genetic manipulation can be done. Retroviruses are able to transfer their genetic information at high efficiency into eukaryotic cells and can be genetically manipulated to carry exogenous genes and hence act as vectors for gene insertion. Part of the retroviral life cycle involves the infection of susceptible cells, followed by the reverse transcription of its RNA into double-stranded DNA and insertion into the host cell genome.

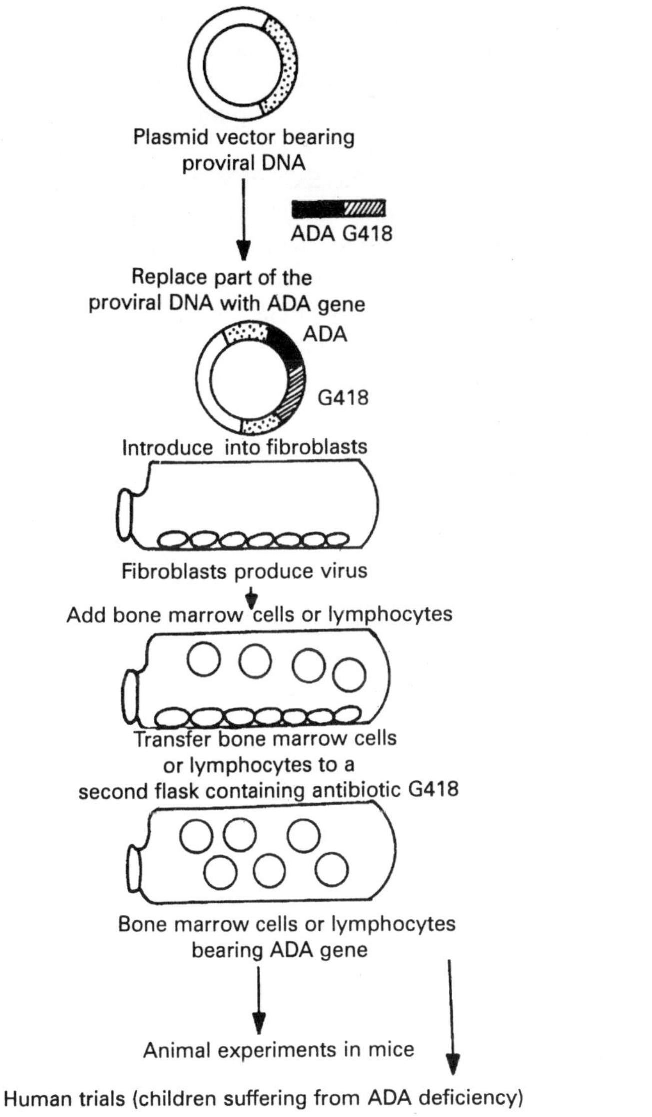

Fig. 5.11 *General scheme for introducing the ADA gene into bone marrow cells/lymphocytes.*

The general scheme for introducing the ADA gene into bone marrow cells/lymphocytes is outlined in Fig. 5.11. Proviral DNA from a retrovirus is cloned in a plasmid and some of the proviral DNA replaced with the ADA gene. For the purpose of selection, a gene coding for the G418 antibiotic resistance is cloned along with the ADA gene. This recombinant DNA is then introduced into fibroblasts in tissue culture. As the retroviral genome is defective, a fibroblast cell line harbouring the missing viral genes is used to produce the mature viral particles upon transfection with the recombinant DNA. Bone marrow cells are added to the growth medium when the mature viral particles have been released. These viral particles then infect the lymphocytes/bone marrow cells. The cells are subsequently removed and those carrying the antibiotic resistance gene selected by growth in the presence of G418. Bone marrow cells lymphocytes transfected with the ADA gene are then re-injected back into mice. Synthesis of human ADA can be detected.

The first human gene therapy experiments were carried out after presentation of these encouraging animal test results. Circulating lymphocytes were removed from two patients with ADA deficiency, transfected with the ADA gene and then re-injected. The activity of the ADA enzyme in the patients led to functional correction of the hypersensitivity to deoxyadenosine toxicity, although the effect was short-lived and no one has yet been cured of this deficiency by a gene therapy approach alone.

Since lymphocytes live for only a few months, this means that transfusions have to be performed frequently. In order to overcome this difficulty gene transfer into the stem cells of lymphocytes has been attempted in a 5-year-old child with the disease, involving transfection of stem cells *in vitro* and implanting transfected cells into the bone marrow.

5.10 Concluding remarks

Gene technology is a rapidly advancing technology that is improving our understanding of the structure and function of genetic systems, and providing the means for their manipulation and control. The potential benefits to humankind are enormous and will affect agriculture, human healthcare, animal husbandry, manufacture of biochemicals/biofuels, environmental protection/ remediation, food production and processing.

The development of this technology raises many issues. Is gene technology ethical and safe? Who controls it? Who should control it? Will the benefits of genetic engineering be overshadowed by any harmful effects that are not obvious at present? Will we end up raising only genetically engineered plants and animals? Will farm animals be dosed with huge amounts of recombinant hormones for better growth and milk production, which will be harmful to the human population consuming these products? Will gene therapy cure incurable inherited disorders? If so, will it introduce new problems?

All these questions must be carefully considered and will have to be put in a proper perspective. In most countries there are committees currently looking at the ethics of performing recombinant DNA experiments, and strict guidelines have been proposed for the handling of recombinant organisms.

5.11 Suggested further reading

1. Berg, P. and Singer, M. (1992). *Dealing with genes: the language of heredity*. Blackwell Scientific Publications, Oxford.

2. Micklos, D.A. and Freyer, G.A. (1990). *DNA science: a first course in recombinant DNA technology*. Cold Spring Harbor Press, New York.

3. Singer, M. and Berg, P. (1991). *Genes and genomes*. Blackwell Scientific Publications, Oxford.

5.12 Responses to assignments

5.1 The vector DNA molecule (which will, remember, be double-stranded) must be cut (both strands) at a given point, and the gene must be inserted at that point and the DNA strands rejoined (ligated). Endonuclease enzymes cause cutting of DNA molecules and ligase enzymes cause cut DNA molecules to rejoin. These enzymes can be produced in quantity from genetically engineering microorganisms, as outlined in Section 5.1.

5.2 The single-stranded ends of each fragment which overlap are complementary: for Eco R I the fragment overlap on one strand (the upper one on the diagram) is AATT and the overlap on the other strand (the lower one on the diagram) is TTAA. With Pst I, the fragment overlap is TGCA (upper strand) and ACGT (lower strand). Because of the natural affinity of complementary nucleotides (by hydrogen bonding, as outlined in Chapter 2), such overlapping sequences will tend to rejoin spontaneously (anneal) as if they were 'sticky'.

5.3 When recombinant vectors and host cells are mixed together, not all cells may become 'transformed' by a vector molecule. It would thus be an advantage to select for growth only those cells which have received a vector molecule. If the vector carries on it genes conferring resistance to particular antibiotics to which recipient host cells are sensitive, then any host cell receiving a vector molecule becomes antibiotic resistant. If that particular antibiotic is added to the medium in which recombinant organisms are to be grown, then only organisms which have received vector molecules will be able to grow.

5.4 If the objective in genetic engineering is to produce a population of yeast cells containing a plasmid carrying a foreign gene, it is easiest first to clone and propagate the recombinant plasmid in *E. coli*. Once a large quantity of cloned plasmids has been produced in this way, it can then be used to transform yeast cells.

5.5 Energy and cellular metabolites are directed towards synthesis of the recombinant protein at the expense of synthesis of other proteins (which are required for cell growth). In this way the growth rate of recombinant cells will be lowered.

5.6 Post-translational modification of a given protein may be important in determining the activity (enzymic, hormonal or other) of the protein produced. Thus, if the recombinant organism produces the protein with radically different modifications compared to human cells, then the activity of the recombinant protein may be radically altered.

5.7 We have seen that restriction endonucleases cut DNA at specific sites. Large DNA molecules can thus be cut up into smaller fragments using a restriction enzyme and the sequences of all the smaller fragments determined after the fragments have been separated. The sequence of the original DNA molecule can be determined by using different enzymes (each cutting the DNA at a different site) and examining the overlapping DNA sequences to produce a continuous sequence (known as a "contig").

5.8 The fragment in question could be inserted into a suitable vector and cloned (replicated as the vector replicates) in a suitable host organism.

5.9 The amino acid sequence of the insulin chain must be determined. From this, the DNA (codon) sequence that codes for this polypeptide can be determined. Using automated synthesisers, the desired sequence can be synthesised *in vitro* on an inert chemical matrix by a series of carefully controlled chemical reactions.

5.10 There is a potential danger that injecting hormone extracted from a deceased individual may transfer to the patient an infectious organism from which the deceased person was suffering. Examples of such organisms are hepatitis B and C viruses, HIV and the agent responsible for Creutzfeldt–Jacob disease (CJD).

5.11 Bacteria are of fundamental importance to humankind as they are involved in: diseases of humans, animals and plants; deterioration of food and other commodities; natural recycling of minerals in the environment; fertility of soils. Humans also use bacteria in: industrial production processes; waste effluent/sewage treatment plants; genetic engineering procedures; fermented foods. Thus an understanding of the genetic makeup of these organisms will help humankind to exploit and control them to even greater effect than is currently possible.

5.13 Responses to self assessment exercises

5.1 The correct response is (ii). Statement (i) is correct in itself, but does not explain how the foreign DNA would be inserted. (iii) and (iv) are incorrect statements; the affinity of complementary termini of DNA strands for each other is identical, irrespective of the origin of the DNA.

5.2 The correct matching is as follows.
(i) .... (c)
(ii) .... (b)
(iii) .... (a)
(iv) .... (e).

5.3 The DNA sequences are read from the bottom of the autoradiograph to the top.
Sequence of clone (a) 5' TCCATGGGTTAGGTGAATTCGATCCCGGGAGGT 3'
Sequence of clone (b)
Notice that at five positions in the autoradiograph there are bands in two tracks at the same position instead of just one. This difficulty is often encountered in reading "real" DNA sequencing autoradiographs and is caused by the DNA polymerase terminating DNA synthesis at certain positions on the DNA template even though a dideoxy-nucleotide triphosphate has not been inserted. This premature termination is frequently due to secondary structures (complex coiling) in the template DNA, which is sequence-specific. In these cases, the DNA sequence can only be read as "either/or". For example, when a band is present at the same position in both the C and the G tracks, the sequence is recorded as (C/G). The solution as to which base is actually at that position is usually found by sequencing the complementary DNA strand, which is obtained when the DNA fragment in question is cloned into the M13 vector in the opposite orientation. The sequence of (b) is therefore recorded as:
5' GGAACG(C/G)GTTTCC(T/A)(T/A)TATAT(T/C)GAACTTGC(C/A)GGT 3'

5.4 To the recovered blood sample would be added Taq enzyme, deoxyribonucleotide triphosphates, and priming sequences for a known sequence specific for heptatitis B virus. PCR is then induced by repeated cycles of raising and lowering the temperature.

Once copies of the heptatitis B-specific sequence have been amplified by PCR, the DNA is denatured (unwound) by heat or alkali and fixed (blotted) onto a suitable membrane. The membrane is exposed to single-stranded labelled probe, and the probe allowed to hybridise (anneal) with any sequences on blotted DNA which are homologous. Excess (unbound) probe is washed off the membrane, and the membrane treated in a way which demonstrates presence of the probe: for a radiolabelled probe, the membrane is placed on X-ray film; for an enzyme-labelled probe, the membrane is treated with an enzyme substrate that produces a coloured product when acted on by that enzyme; for a fluorescent-labelled probe, the membrane is illuminated by UV light and examined for light spots.

6 Protein engineering and design

Chapter outline

6.1 Introduction

A variety of molecules has evolved over a period of three billion years to yield the present-day proteins which carry out complex functions in the cells of all organisms. It is believed that random mutation of genes, and natural selection of those gene products which gave rise to a functional advantage in the struggle for survival of individual organisms, are responsible for today's large repertoire of complex proteins interacting in a complex way with other proteins as well as with other molecules such as DNA.

With the advent of molecular genetics and, in particular, techniques such as gene cloning (Chapter 5), we can now design new gene products and are not restricted to selecting useful genes that arise only from a rare event of natural mutation. It is possible that human beings can actively change the course of natural evolution.

Protein engineering is the foremost among the many recent advancements in biotechnology which reflect some of the exciting possibilities arising out of a deeper understanding of living systems. Protein engineering and design refer to the possibility of actually altering at will the structure of a protein in order to make it function 'better' in one way or another.

Protein engineering can be distinguished from protein design in the following way. In protein engineering an existing protein is changed, chemically or genetically, in order to alter its function in a predictable way. In protein design, the *de novo* design of a protein, to fulfill a desired function, is the aim of the process.

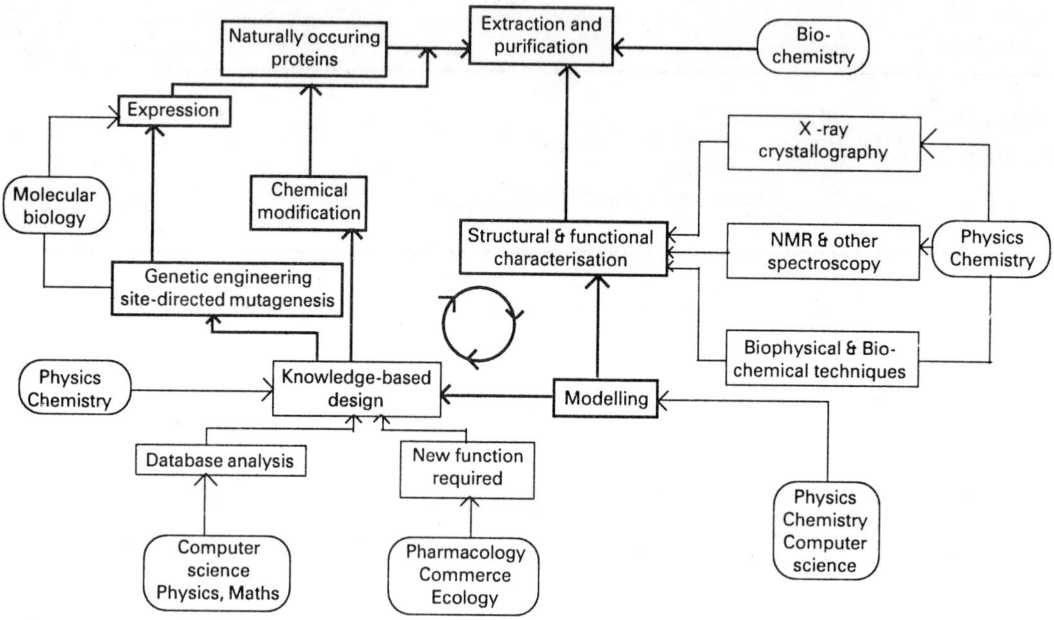

Fig. 6.1 *The protein engineering and design cycle, indicating the multidisciplinary approach required.*

Protein engineering is a multidisciplinary science. As illustrated in Fig. 6.1, the engineering and design flowchart requires the interaction of physicists, biophysicists, molecular biologists, organic chemists, biochemists and computer scientists. The starting point for a protein engineering project would probably be the identification of a protein with a specific function that needs to be changed in some way, either to make it faster, less energy consuming, more stable under harsh environmental conditions, or even to deactivate it, if it is harmful. Such a protein, once identified, has to be isolated and characterised biochemically. It is very important at this stage to have a good idea of the three-dimensional structure of the molecule, and to understand the relationship between structure and function.

Based on the structure and by an application of the known principles of protein conformation, modifications can be suggested which may confer the desired modulation of the properties of the protein. The changes are then incorporated into the protein by site-directed mutagenesis (see Section 6.3.2) or other biochemical or molecular biology techniques. The new protein has then to be tested for its activity and if, as almost invariably happens in the first trial, the results are unsatisfactory, the cycle will have to be repeated.

Protein design starts from a knowledge of the principles of protein structure. These are applied to design a sequence of amino acids which is presumed to take up a particular three-dimensional structure which will endow the protein with the necessary function. The amino acid sequence is synthesised as the appropriate DNA sequence which is then cloned into an expression system by recombinant DNA techniques. The expressed protein will then have to undergo tests which will be the further input for the next stage of the design process.

Assignment 6.1
Explain how a synthetic gene can be produced which codes for a particular peptide sequence.

Perhaps it is only fair to say, however, that both these techniques are at the stage of infancy. There have been, in the last decade or so, quite a few successful attempts at engineering proteins, but there is as yet no completely successful newly designed protein.

In this chapter, we first provide a brief overview of the principles of protein structure. Some of the most commonly used techniques in protein engineering are then described in an elementary way.

The chapter concludes with descriptions of some of the successful attempts at protein engineering and design.

6.2 Protein structure and function

Proteins are polymeric molecules built from about twenty different amino acids linked together in different sequences of length between a few tens and a few hundreds of amino acids. Though the twenty amino acids (see Fig. 6.2) vary in size, shape, charge, hydrogen bonding capacity and chemical reactivity, they have some crucial similarities, i.e., they are all α-amino acids. Moreover, the amino acids found in the proteins in natural systems are all L-amino acids. In human beings only ten of the twenty amino acids are synthesised within the body. These are Ala, Asn, Asp, Cys, Glu, Gln, Gly, Pro, Ser and Tyr. The other ten must be obtained from the diet and are therefore called the essential amino acids. The three- dimensional structure of a protein is well defined and is almost entirely specified by its amino acid sequence. In other words a protein with a given sequence will have exactly the same three-dimensional structure over a range of conditions of temperature, pH, ionic strength of the medium, etc. It is the three- dimensional structure of the protein which confers the biological activity on it. The same protein in a different or random conformation will not display the same activity. Function in the protein may thus be said to arise from the conformation which is the three-dimensional arrangement of atoms in the structure.

6.2.1 *Conformation*

It may appear that to attempt to find systematics in the structure of proteins is an almost hopeless task since the number of possible sequences is so enormous. However, theoretical and experimental studies over the past few decades have shown that protein structures can be reduced to a few elements or motifs which are arranged in different ways in different proteins. Each of the motifs themselves can further be understood in terms of the conformational possibilities of the amino acids. Ultimately it is the chemical nature of the atoms forming the protein which dictates the three-dimensional structure.

There are many chemical characteristics which are responsible for the motifs. One of them is the planarity of the peptide bond. One amino acid joins to another through the formation of the peptide bond. Studies have shown that the six atoms of the peptide group (Fig. 6.3) all lie in the same plane forming a rigid unit. There is, however, some amount of freedom of rotation about the bonds joining the peptide groups to the α-carbon atom. These rotations are designated ϕ and ψ. Steric hindrance from the side chains restricts rotation about ϕ and ψ quite considerably and the allowed range can be plotted in a so-called Ramachandran Plot (see Fig. 6.4).

Some of the values on the map can lead to regular, identifiable patterns of atoms if they are repeated over a number of residues in the protein chain. Such patterns are called secondary structures (in contrast to the primary structure of the protein, i.e., the amino acid sequence).

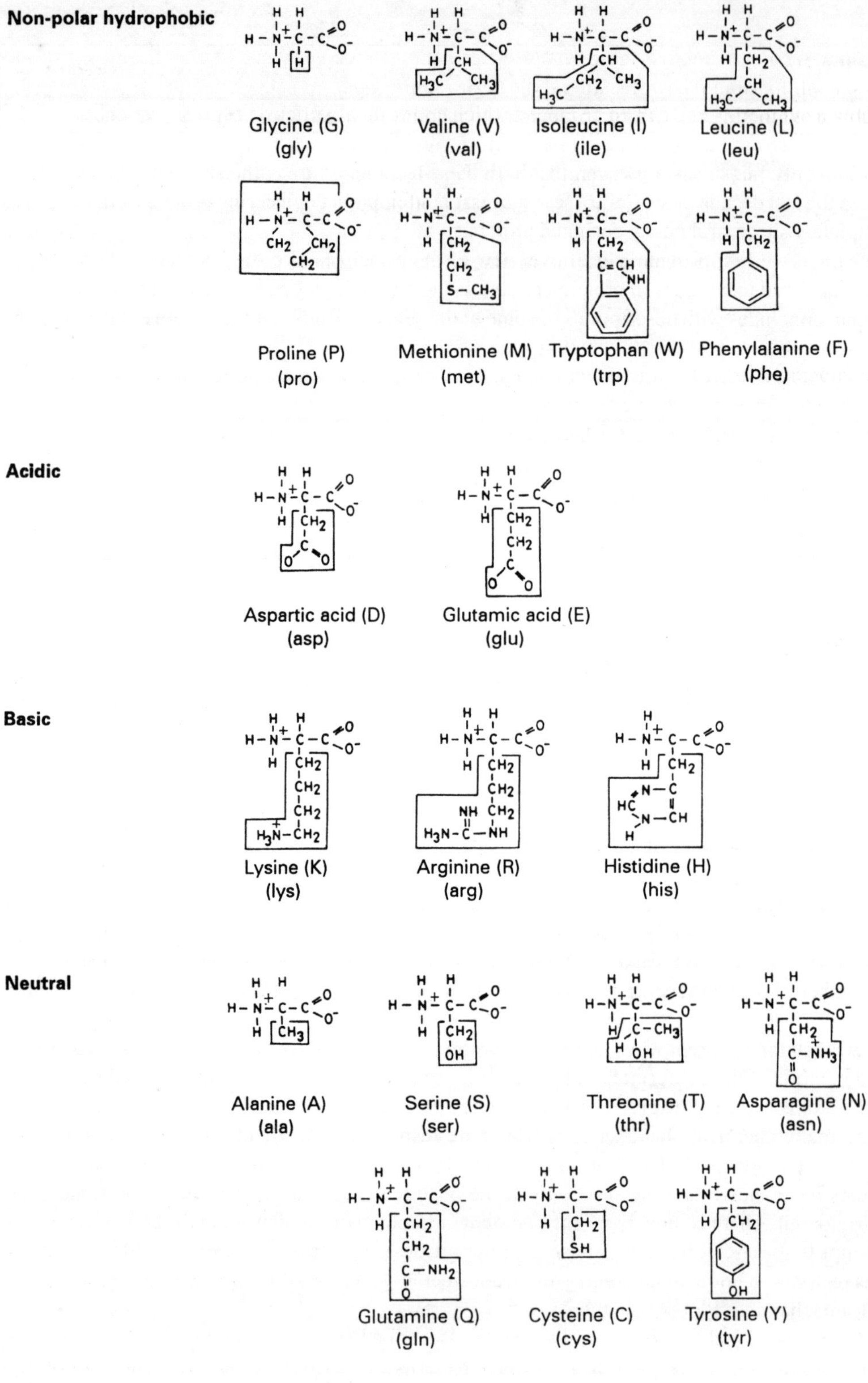

Fig. 6.2 *α-amino acids found in protein structures.*

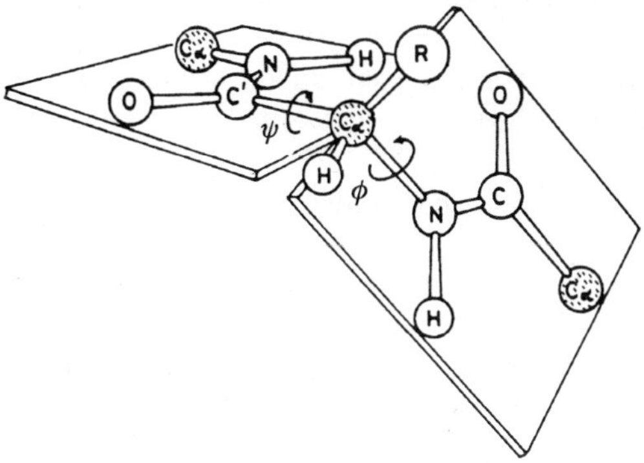

Fig. 6.3 *Chemical structure of the peptide bond.* The partial double bond nature of the peptide bond leads to the planar peptide unit. The conformation of the protein backbone can be specified in terms of the torsion (rotation) angles ϕ and ψ.

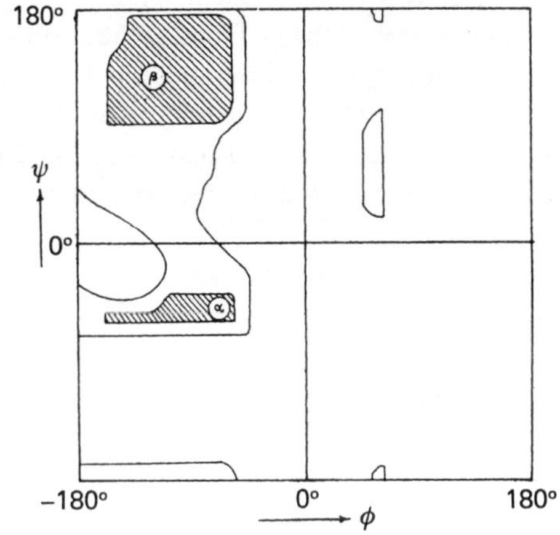

Fig. 6.4 *The Ramachandran plot.* An explanation is given in the box below.

Professor G.N. Ramachandran and the $\phi - \psi$ conformational map

The Madras school of structural biology, centred at the Department of Crystallography and Biophysics, University of Madras, India, was founded in 1950 by Prof. G.N. Ramachandran who was identified by his teacher, the Nobel Laureate Sir C.V. Raman, in response to an enquiry from the then Vice-Chancellor of the University, as the ideal person to head the physics research programme. By 1957 Prof. Ramachandran and his co-workers had established a worldwide reputation for the department in the area of structural biophysics. The jewel in the crown of their achievements at that time was the correct proposal of the triple helical structure of the fibrous protein, collagen.

Intense and detailed discussions between the Madras group and other groups abroad which had also been working on this structure led Prof. Ramachandran to a deeper understanding of the stereochemistry of polypeptide chains based on simple chemical and physical concepts. His group explored the conformational possibilities inherent in a polypeptide chain given the following model : (a) The peptide bond is rigid and will not allow rotation. (b) The molecule may be rotated about the other (single) bonds in the backbone, i.e., the $N - C_\alpha$ bond and the $C_\alpha - C'$ bonds. (c) All atoms are represented as hard spheres with specific radii. No inter-penetration is allowed. (d) There are no other attractive or repulsive forces.

In order to better display the results, the Madras group pioneered a representation of the conformational possibilities of a dipeptide which has since been called the 'Ramachandran Map'. Figure 6.4 shows a modern version of this map. The hatched areas represent the combinations of ϕ and ψ torsion angles allowed for any dipeptide (other than glycine, which has a much larger allowed region since it is the smallest of the amino acids). Also marked on the map are the combinations of the two torsion angles which lead to regular structures when repeated for consecutive residues along the polypeptide chain. A particularly noteworthy point is that the years have been very kind to the Ramachandran Map. It changed very little when the simple model of the peptide described above was replaced by a more complex one. When some degree of inter-penetration of the atomic spheres was accepted, then the allowed region expanded a bit, as shown in Fig. 6.4. Other complications introduced in the model did not change the map very much. When the map was originally proposed only a few protein structures were known and all the residues of these fitted very well into the allowed regions of the map. Today, even with the hundred fold increase in the number of known structures, over 98% of the residues in the proteins have their conformational parameters well within the limits suggested by Prof. Ramachandran over 30 years ago. The Ramachandran map is now used as a touchstone on which freshly-solved protein structures are tested before being published.

Some of the most common secondary structural patterns are described below.

(a) The α helix

When (ϕ, ψ) values of (–57°, –47°) are repeated, the polypeptide backbone chain of the protein is wrapped in a right-handed fashion around an imaginary cylinder in a helical arrangement with the side chains extending out. The α helix (Fig. 6.5) is stabilised by hydrogen bonds which are formed between the C–O group of each amino acid and N–H group of the amino acid, situated four residues ahead in the linear sequence. Each residue is related to the next one by a translation of 1.5 Å along the helix axis and a rotation of 100° about it, which gives 3.6 amino acids per turn and the pitch of 5.4 Å. α helices are essentially rod like structures, with a radius of 2.3 Å and are represented as such in some diagrams of proteins (Fig. 6.6). For example hair or wool is known, in some forms, as α keratin and exists in the α-helix conformation.

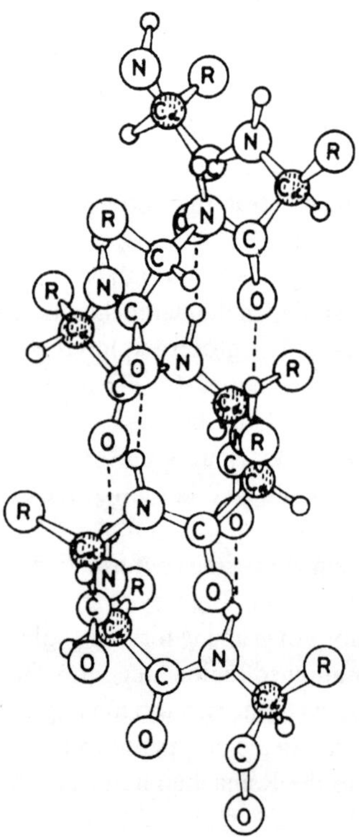

Fig. 6.5 *The α helix.*

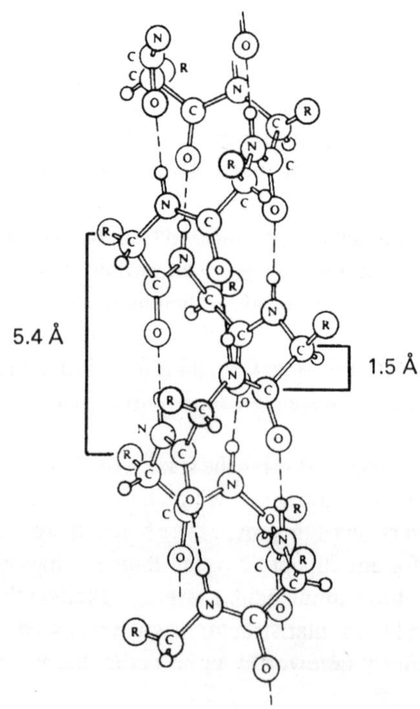

Fig. 6.6 *Crystallographic repeats in the α helix.*

(b) The β-pleated sheet

Extended polypeptide chains can associate by hydrogen bonding to form sheet like structures. These are planar and the polypetide chains may either be parallel or anti-parallel. The distance between adjacent residues, along the direction of the sheet, is 3.5 Å. The structure is stabilised by hydrogen bonds between N–H and C–O groups in different (or different parts of) polypeptide chains (Fig. 6.7). For example silk fibres exist in the β-strand or β-sheet conformation which gives them the observed tensile strength.

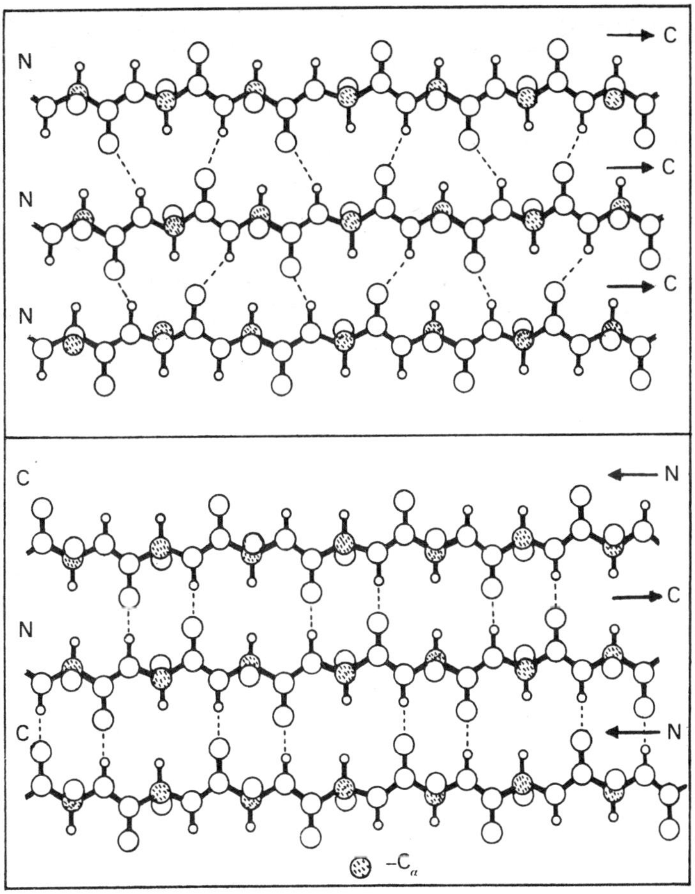

Fig. 6.7 *The β-strand/sheet conformation of the polypeptide chain.* Shown are both the parallel and the anti-parallel sheets.

(c) β turns or reverse turns

When the polypeptide chain has to reverse its direction, as happens many times in a globular protein, it does so through a structural motif called the β turn (Fig. 6.8). Reverse turns are abundantly formed in proteins. They largely occur at the surface of the globular protein.

(d) Other features

Other secondary structural elements are also seen in proteins. Some of these are variations of the α helix such as the 3_{10} helix and the ϕ helix, being narrower and broader than the α helix respectively.

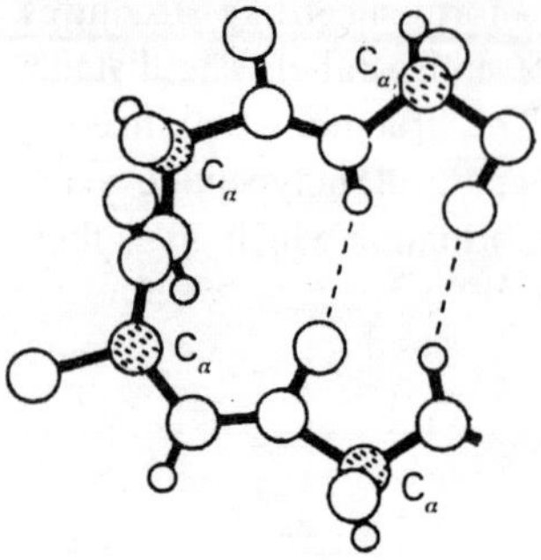

Fig. 6.8 *The β-turn conformation.*
Shown here is the type I turn. There are several 'types' of β turns.

The collagen helix is an extended left-handed structure that, in threes, goes to form the super helical fibres of collagen (Fig. 6.9).

Likewise, α helices are also sometimes found as coiled coils in a super secondary structural arrangement. Other super secondary structural motifs consist of specific arrangements of β sheets and/or α helices, such as the Rossmann fold (Fig. 6.10).

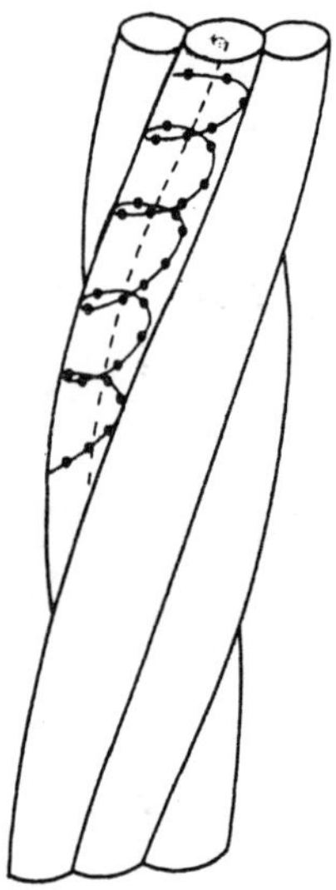

Fig. 6.9 *Schematic diagram of the collagen triple helix.*

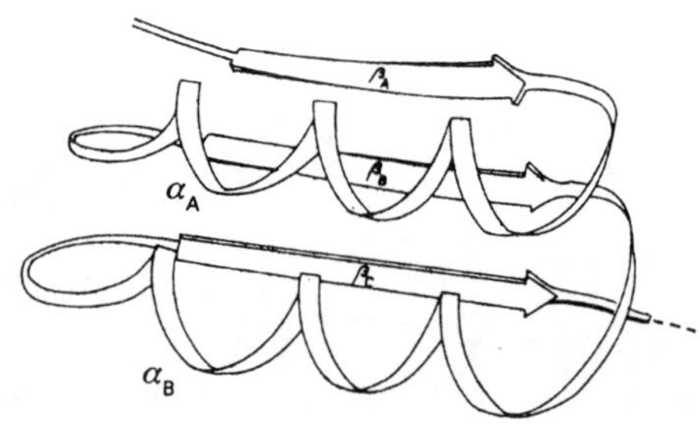

Fig. 6.10 *The Rossmann fold, a commonly seen super-secondary structural unit. It alternates between a β strand and an α helix along the chain. In three dimensions, the strands are arranged to make a sheet and helices are arranged together.*

6.2.2 Tertiary structure and structural domains

We have discussed the primary and secondary structure of proteins. The secondary structural elements are put together in specific ways to form the tertiary structure. Tertiary structure may also be said to refer to the three-dimensional relationship between amino acids that are far apart in linear sequence. The secondary structural elements are usually arranged in ways that may be recognised as structural domains

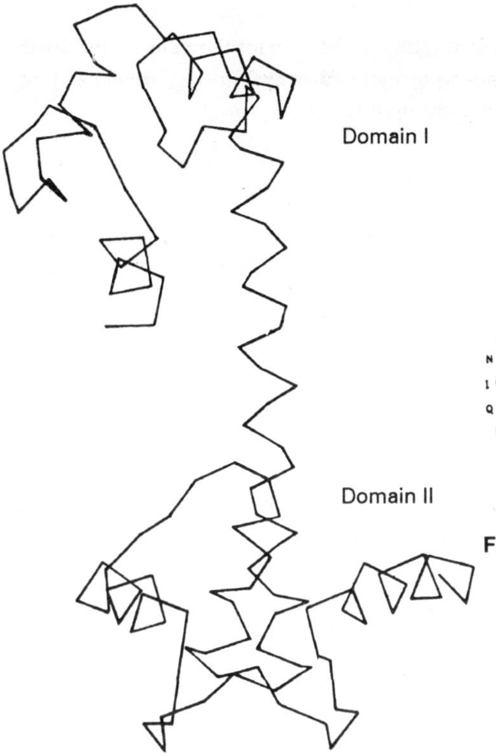

Fig. 6.11 *The structure of Troponin C, a calcium binding muscle protein. The two domains are clearly visible.*

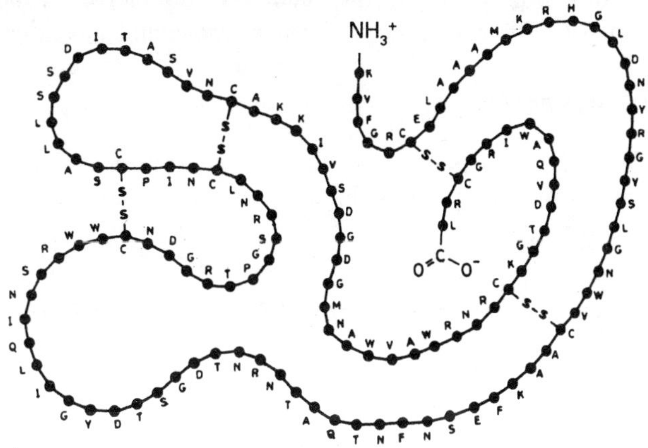

Fig. 6.12 *The sequence of lysozyme showing the two-dimensional arrangement of the disulphide bonds.*

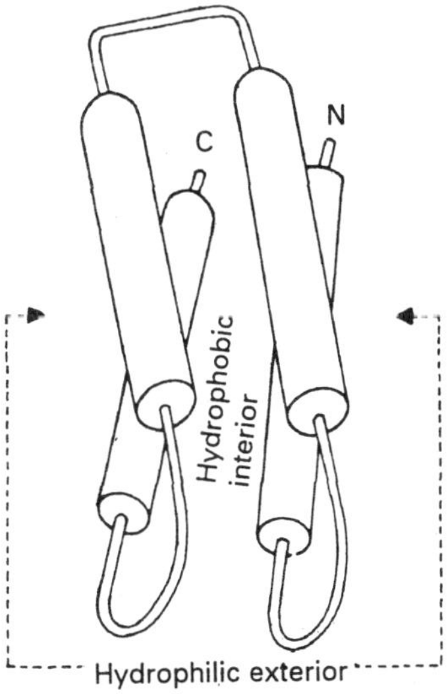

Fig. 6.13 *Four α helices arranged in a bundle.* This motif is common in proteins. Hydrophobic amino acids are to the interior and hydrophilic amino acids to the exterior.

within the protein (Fig. 6.11 and Fig. 6.16). Structural domains are thought to be the folding units of a protein. It is presumed, from experimental obser-vations, that each domain folds independently of the others, and then associates with other domains to form the globular protein. One may have domains consisting exclusively of α helices or β sheets or some combination of these.

The quaternary structure is the next higher level in the organisational hierarchy of proteins and refers to the way in which folded polypeptide chains are packed together in a protein that consists of more than one polypeptide.

The stability of the structure of a domain or of the quaternary structure may be traced to the chemical interactions both within and between domains, such as H-bonding, salt bridges, and, most importantly disulphide bonds.These disulphide bonds occur between cysteine residues which come close together in space. They are covalent bonds and bind the chains together strongly (Fig. 6.12).

Apart from disulphide bonds, among the principal interactions involved in stabilising three-dimensional protein structure are hydrophilic / hydrophobic interactions. Amino acid side chains can either be polar and hydrophilic or non-polar and hydrophobic.

Proteins fold in such a way that hydrophobic groups come together in the interior whereas hydrophilic amino acids occur on the surface in contact with the aqueous environment, as illustrated in Fig. 6.13. (In membrane proteins, the arrangement is exactly the reverse).

> ***Assignment 6.2***
>
> A long polypeptide molecule contains several domains which are hydrophobic, and several domains which are hydrophilic. How would this affect the folding of the molecule in an aqueous environment?

6.2.3 Relationship between structure and function

The varied activities of proteins are dependent on different aspects of the three-dimensional structure.

Enzymes are highly specific catalytic proteins. The specificity as well as the enzymatic activity of these proteins depends to a very large extent on the three-dimensional configuration of the various reactive groups in the protein, especially at the active site. The groups that form the active site come from different parts of the linear amino acid sequence and any change in the sequence and the consequent change in the folding pattern of the protein can lead to large changes in the activity. Both the so called 'lock and key' and 'induced fit' models of enzyme — substrate interactions require precise three-dimensional arrangements of the active site groups.

Transport proteins also require a specific three-dimensional structure in order to be functional. For example, haemoglobin (Fig. 6.14 and Fig. 6.16), one of the chief components of blood, is an oxygen carrier protein which has four subunits. Interactions between the subunits change substantially when oxygen binds to the 'haem' group in one of the subunits. These changes lead to precise alterations in the structures of the other subunits and this enables more efficient binding of further oxygen atoms.

Haemoglobin (α_2 β_2)

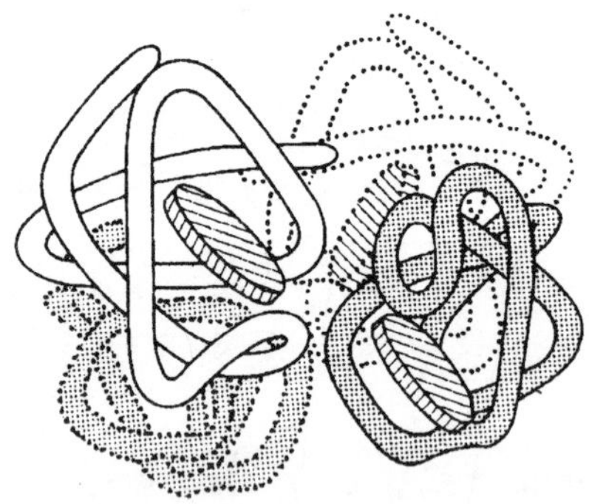

Fig. 6.14 *The haemoglobin molecule is built up of four subunits (polypeptide chains) placed together in a precise arrangement.*

Structural proteins depend largely on the specific sequence of amino acids and the consequent specific structure to give them the required strength and stability. For example, in collagen, every third residue is a glycine. This pattern is required for the stability of the collagen triple helix, since every third α-carbon atom is close to the superhelix axis. A compact hydrogen-bonded structure can be obtained only in the absence of a sidechain at the axis, i.e., if there is a Gly at this position.

It has been demonstrated over and over again that changes of certain crucial amino acids in a protein sequence can either enhance or diminish many of the properties of the protein such as stability, activity, etc. For example, a single change of the amino acid valine to glutamic acid in sequences of human haemoglobin leads to the disease called sickle cell anaemia, in which the molecules of the protein tends to stick together, thereby impairing its oxygen carrying capacity. This disease is genetically inherited and prevalent in African countries.

Self assessment exercise 6.1

Answer true or false to each of the following statements:

(i) Protein engineering involves taking an existing protein and modifying a part of this using chemical reactions.
(ii) Protein engineering involves modifying the gene(s) which code(s) for the protein in such a way that the structure of the protein produced from the gene is altered.
(iii) Protein engineering involves designing a polypeptide, synthesising a gene which codes for this protein, cloning the gene and producing the protein from the recombinant organism.

Self assessment exercise 6.2

Match each structure in Column A with the correct definition in Column B.

	Column A		Column B
(i)	Primary structure	(a)	Structural arrangement along a polypeptide chain caused by repeated patterns of atoms.
(ii)	Secondary structure	(b)	Structural rotation around a peptide bond.
		(c)	The sequence of amino acids in the polypeptide chain.
(iii)	Tertiary structure	(d)	The three-dimensional structure formed when a polypeptide folds upon itself.
(iv)	Quaternary structure	(e)	Structural arrangement between polypeptides which are associated into a macromolecule.

Self assessment exercise 6.3

Which of the following structures is/are stabilised by the effects of hydrogen bonding?

(i) Secondary structure (ii) Tertiary structure (iii) Quaternary structure

6.3 Methods in protein engineering and design

The design of new proteins and their construction is a multidisciplinary task and requires the combined efforts of scientists and technologists from many different disciplines. In the following few pages we will briefly describe some of the techniques used in the design cycle.

6.3.1 *Physical and computational techniques*

(a) X-ray crystallography

Much of the information described in the earlier section of this chapter was obtained through the technique of X-ray crystallography. In spite of rapid advances in NMR (see later), electron microscopy and computer modelling, X-ray crystallography remains the only method of obtaining a detailed, accurate, unbiased picture of a protein molecule in the case where no prior information is available about its three-dimensional structure. Quite obviously, such a detailed description of the molecule is the starting point for changes and modifications which will make the protein 'better'.

The technique of X-ray crystallography consists of crystallising the molecule, subjecting it to irradiation by a beam of X-rays and making a record of the three-dimensional diffraction pattern (Fig. 6.15). Analysis of these data through computer-intensive and extensive calculations results in a model of the protein which can then be subjected to refinement against the observed X-ray data. Finally the three-dimensional coordinates of each atom in the protein, as well as a number which quantifies its thermal vibration is obtained for further geometrical or other analysis. Such a refined model can be displayed and manipulated either as plastic or wooden models or on a computer graphics terminal.

In spite of many advances in the technology, there remain two rate limiting factors in the process of determining a protein structure through X-ray crystallography. The first, and the least understood of these is the growth of the crystals. In recent years, recombinant DNA technology and advanced purification methods have made available sufficient quantities of pure protein, which more or less ensures that crystals will be eventually obtained. Nevertheless it is not as yet possible to automatically grow crystals and a great deal of 'art' and 'craft' on the part of the crystallographer is required. Crystals, even when they are obtained, may not diffract well and protein structures obtained at low resolution (say 2.5Å) are common. At this resolution much of the detail is lost and one may have to make chemically educated guesses at the precise positions of many of the atoms, especially those with low atomic numbers such as hydrogen which scatter X-rays weakly.

The other main obstacle in the X-ray crystallographic process is called the 'phase problem'. It refers to the fact that the distribution of electrons in the crystal (or in other words the structure of the molecule) is obtained by a Fourier transform of the observed diffraction data. In order to carry out this mathematical procedure, however, it is necessary to know both the intensities and phases of the diffracted X-ray beams. The intensities can be measured, but the phases are impossible to measure directly and roundabout techniques are used to determine them. These phasing methods do not always work and a great deal of effort and intution is often required.

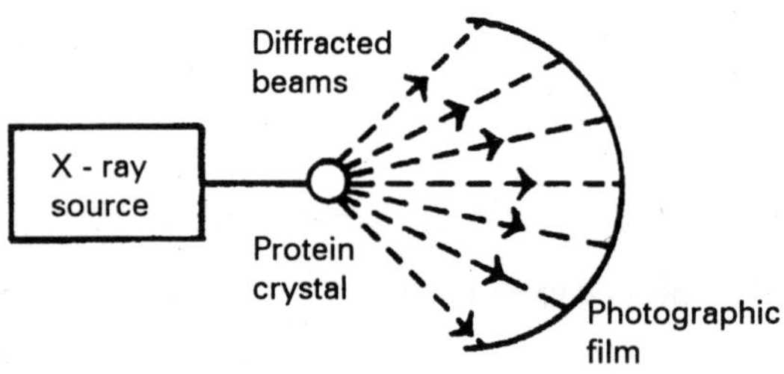

Fig. 6.15 *Schematic illustration of the technique of X-ray crystallography.*

The first protein structures were solved in the early 1950s by the heroic efforts of Max Perutz and John Kendrew in Cambridge. Since then protein structures obtained through X-ray crystallography have been reported at an ever increasing pace. There are now over one thousand five hundred individual structures known and collected in a computer database called the Brookhaven Protein Database.

There is a note of caution which has to be added in order to aid correct interpretation of the crystallographic results. Since the structures are determined in the solid state it is possible that the situation may differ in details in the living cell. NMR studies, which are performed, usually, in the solution state, are probably closer to the 'real thing'.

Structure and sequence databases

Extensive libraries of three-dimensional structures are now available. As already mentioned, the Brookhaven Protein Database at the Brookhaven National Laboratory in USA contains the coordinates and other information on all known protein structures in a relational database which helps the user to easily access the information. Entries are constantly updated and the Brookhaven Protein Database can be accessed through computer networks or can be obtained on magnetic tapes or CD-ROMs. A similar high-quality database is the Cambridge Crystallographic Structural Database (CSD) at the University of Cambridge, UK. This contains about 100,000 entries of the coordinates of the crystal structures of small organic molecules, i.e., those with a molecular weight less than about 1000. The Nucleic Acid Structure Database (NSD) contains the structures of fragments of DNA and RNA and is maintained at Rutgers University in USA. These databases are frequently used in protein engineering studies.

Apart from these structural databases there are sequence databases available either online on the computer networks or on other media. All the thousands of protein sequences known and being constantly discovered are gathered in the protein sequence databases known as Swissprot, PIR, etc. These databases exchange information amongst themselves so that it is usually sufficient to have access to one of them. Nucleic acid sequences are available in exponentially expanding numbers in the database maintained at the European Molecular Biology Laboratory, Heidelberg, Germany and also in the GenBank maintained by the National Library of Medicine, USA.

(b) Nuclear magnetic resonance

Nuclear magnetic resonance (NMR) is a technique of studying molecular structure in which the sample is placed in a strong magnetic field. It is then irradiated by a radio-frequency electromagnetic wave. Energy from this wave is absorbed by each of the nuclei of the atoms in the molecule according to its individual resonance frequency. The amount of energy which would be absorbed can be precisely predicted for every type of nucleus and the deviation from this value is a measure of the effect of other nuclei present in the molecule. Thus, for example, a proton or hydrogen nucleus attached to a carbon atom would absorb energy at a slightly different frequency from one attached to a nitrogen atom. Not all nuclei lend themselves to NMR. The most commonly used species are hydrogen, carbon, phosphorous and nitrogen.

The resonance frequency is not only modulated by atoms bonded to the nucleus (i.e., the 'J' coupling) but also by other atoms not bonded but within about 4.0 Å. The latter effect can be enhanced and observed more exactly through the nuclear overhauser (NOE) effect in which a particular nucleus is held in a strong electromagnetic field tuned to its individual absorption frequency. The effect of such irradiation on the other nuclei would then make it clear which of the others are close in space. The NOE is thus an excellent tool to study the three-dimensional conformation of the molecule.

Techniques such as NOE were extremely time consuming and could be applied only to small molecules until the advent of strong superconducting magnets, the Fourier transform spectroscopy technique and the multi-dimensional NMR technique. These advances in theory and technology allowed data collection and analyses for molecules such as proteins which contain, for example, several hundred hydrogen atoms and would therefore require the interpretation of spectra containing hundreds of peaks in a proton NMR spectrum. Two-dimensional NMR techniques such as COSY enable the assignment of hundreds of peaks at a time. NOESY is a technique which determines the through-space connectivities. Other techniques include those for analyses of heteronuclear interactions and three- and four-dimensional NMR spectra.

In recent years, especially, multidimensional nuclear magnetic resonance has gained enormously in power as a structure determination technique. Since NMR probes the nuclei in the molecule, and since it is more easily accomplished when the molecule is in solution, it is complementary to X-ray crystallography. One of the obstacles in the analyses of NMR spectra is that, as already mentioned, it determines the distance correlations between pairs of atoms in the molecule. This means that, while it is possible to produce a list of distances between atomic pairs, converting this list to a three-dimensional structure has still to be accomplished by computational methods. This is by no means an easy task and that is the reason why, as yet, *ab initio* determination of the larger protein structures purely through NMR is not possible. However many small- and medium-sized protein structures have been solved by NMR, including bovine pancreatic trypsin inhibitor, the structure determination of which was used as a vehicle to develop many of the techniques currently in use. NMR structures are also deposited in the Brookhaven Protein Database and there are now about fifty structures available.

NMR is also extremely useful in probing the dynamics of protein structure and in understanding the changes that may occur in the structure under changing conditions of pH, salt, ligand binding or temperature. Especially in protein engineering, NMR can be used to monitor the effect of sequence changes or other chemical modifications on the structure.

(c) Other spectroscopic techniques

These include UV and IR spectroscopy. Laser Raman spectroscopy is also extremely useful in probing some aspects of protein structure such as the hydrogen bonding patterns in the molecule. ORD-CD studies can be used to determine changes in protein conformation and also to determine relative content of particular secondary structural elements.

(d) Protein structure prediction

The prediction of protein structure from its amino acid sequence is a problem that has no complete or reliable solution as yet. A search for the optimal conformation of a polypeptide out of all those possible is a problem of extreme computational complexity. No algorithm exists which will allow such a search in reasonable computer time, no matter how powerful the computer. Prediction methods therefore do not depend on complete conformational searches but use known principles of protein structure. The chief among these principles is that proteins with homologous amino acid sequences are predicted to have similar three-dimensional structures. However in the case of a very short sequence, i.e., one which is only a few residues long, its ability to form a specific secondary structure is not only determined by the amino acid sequence but also by other residues close to it and by its environment.

A study of known protein structures has resulted in some rules which can be used to systematically predict the three-dimensional structure of an unknown sequence. Of these rules, the most frequently used are those of P. Y. Chou and G. D. Fasman, USA. This method is based on a statistical analysis of the protein structure database. The number of occurences of an amino acid in the α helix, β sheet and loop conformations was used to calculate the statistical parameters P(α), P(β) and P(turn) for each residue as a measure of its conformational preference. In the calculation of P(turn), preferences of the amino acids for specific positions in the turn were taken into account. The effect of neighbouring residues was also considered. A computer programe which incorporates these rules is available and for a given sequence will print a list of values of the average probability of finding a particular residue in a particular secondary structural motif. The list is scrutinised for stretches of the sequence with a high probability for, say, α helix. These are starting points for the secondary structure elements and are extended on both sides till a high probability for a different motif is encountered. Finally it may be possible to classify most of the unknown sequence in terms of the probable secondary structure.

Several other methods have been developed and others are constantly being developed and refined. The methods have been evaluated on known crystal structures of proteins and the success rate is about 60% at best. For example, when applied to adenylate kinase and T4 (phage) lysozyme, it was found that the conformational status of more than two-thirds of the residues had been correctly predicted. In another application the tertiary fold of the α-subunit of tryptophan synthetase was correctly predicted even before the X-ray crystal structure became available. The prediction was in close agreement with the X-ray structure excepting for one α-helical region consisting of seven residues. In spite of their limited accuracy these methods have proved useful in designing novel polypeptides for the recognition of antigenic epitopes, in the analysis of common motifs in leader sequences, etc. However, prediction of secondary and tertiary structure is still a hit and miss affair and does not produce a model close to the correct one every time.

(e) Computer modelling

Several software packages to implement the above structure predictions are available for a variety of computers, from PCs to mainframes. Most commonly used computers include the Evans and Sutherland range of graphics workstations and the Silicon Graphics range of machines. Popular software include DTMM (DeskTop Molecular Modeller) for PCs and FRODO and BioSym for the larger machines.

Apart from building, displaying and manipulating protein models, computer simulation is used to study the dynamic motion of proteins. Molecular dynamics algorithms are used to predict the conformational rearrangements that may occur in the course of the protein's activity. Groups important for activity and also the effect of chemical modifications can be predicted. Programme packages such as AMBER, CHARMm and X-PLOR are commonly used.

Self assessment exercise 6.4

Answer true and false to each of the following:

(i) To apply X-ray crystallography to determining protein structure, the protein is bombarded with X-rays, the absorption frequency of nuclei of atoms in the molecule is measured, and from this the structure of the protein molecule is determined.

(ii) To apply NMR to determining protein structure, the protein is bombarded with radio waves, the three-dimensional diffraction of waves is measured, and from this the structure of the protein molecule is determined.

(iii) To apply the method of Chou and Fasman to determining protein structure, the secondary structures of proteins of known amino acid sequence are studied, the statistical preference for given amino acids at given structural sites determined and a database produced; the most probable secondary structure of a given polypeptide sequence is determined by comparing it with the sequence/structure data in the database.

6.3.2 Biochemical techniques and molecular biology

(a) Site-directed mutagenesis

Mutation is a sudden change in the characteristics of an organism which may be transmitted to its offspring. This change can be either at the level of the structure of the gene (genetic mutation) or a change in the number of chromosomes (chromosomal mutations). Mutations can be caused by chemicals or by radiations and they can be lethal or beneficial. For example, as already mentioned, the mutation of a single amino acid in haemoglobin causes sickle cell anaemia. The same mutation is also beneficial in that it gives protection against a lethal form of malaria.

Site-directed mutagenesis involves specific alterations at the gene level so that an amino acid at a particular site in the protein is altered at will. The change of a single nucleotide in the gene sequence can change the amino acid sequence (Chapter 4) and hence the structure and consequently the function of the protein. Once the required change in the DNA sequence coding for the protein has been decided upon the new sequences can be synthesised chemically, amplified using PCR techniques (Chapter 5) and incorporated by recombinant methods into an expression system from which the newly altered protein can be harvested in the required quantities (Chapter 5). Since the alteration has to be highly specific, techniques have been devised both to make the changes in a controlled manner at the required sites and also to identify and isolate the correct mutants. The 1993 Nobel Prize was awarded to Michael Smith of Canada for pioneering the technique of site-directed mutagenesis. This technique is extensively used to study enzyme activity, protein folding and stability and it is obvious that it would play a major role in protein engineering.

Assignment 6.3

A DNA sequence -TATGCGCA**C**GTGCACGTA- in a cell is to be altered by site-directed mutagenesis, replacing a C (at position 9 on the sequence given) with a **T** (**C** is to be deleted and **T** inserted in its place). To achieve this, the sequence TGCGCA**T**GTGCAC is synthesised, amplified by PCR and introduced into the cell. Explain how recombination might lead to the required mutation (C→T) at the specified site.

(b) Recombinant DNA techniques
The process of genetic recombination involves the introduction of a piece of new DNA into an existing DNA molecule. Thus new combinations of unrelated genes can be constructed in the laboratory. Such a modified or recombinant genome can be introduced into suitable cells, amplified and expressed using the metabolic machinery of the host cell giving rise to new, sometimes 'chimeric' gene products.

6.4 Protein engineering and design — some achievements

6.4.1 Protein engineering in lysozyme

The pioneering studies on protein engineering in the protein lysozyme (Fig. 6.16) are described here. Lysozyme from bacteriophage T4 is 164 amino acids long and folds into two domains. There are two cysteine residues, Cys54 and Cys97, which are not cross-linked by a disulphide bridge. Different approaches were taken to engineer an enzyme thermally more stable than wild-type T4 lysozyme.

(a) Reduction in the entropy difference between folded and unfolded conformations
A reduction in the entropy difference between the folded and unfolded states would require a drastic reduction of the number of possible unfolded conformations. This can be achieved by introducing disulphide bridges in such a way that the native conformation is not altered very much. In the case of lysozyme T4 this was achieved by mutating the five amino acid residues Ile3, Ile9, Thr21, Thr142 and Leu164 each to cysteine in separate experiments. These residues were chosen since their mutations could lead to single, double or triple disulphide bonds in the protein. In addition to this, Cys54 was mutated to threonine to avoid formation of an incorrect disulphide bridge during folding. It was found that the introduction of the disulphide bridges decreased the number of unfolded conformations and hence the entropy difference between folded and unfolded conformations. This was reflected in the increased thermal stability of the engineered proteins. The effect was additive and mutants with a double disulphide bond had a higher melting temperature (T_m) and hence higher thermal stability than those with a single engineered bridge (T_m is a measure of the thermal stability of the protein). Similarly triple disulphide bonds engineered even higher stability into the protein.

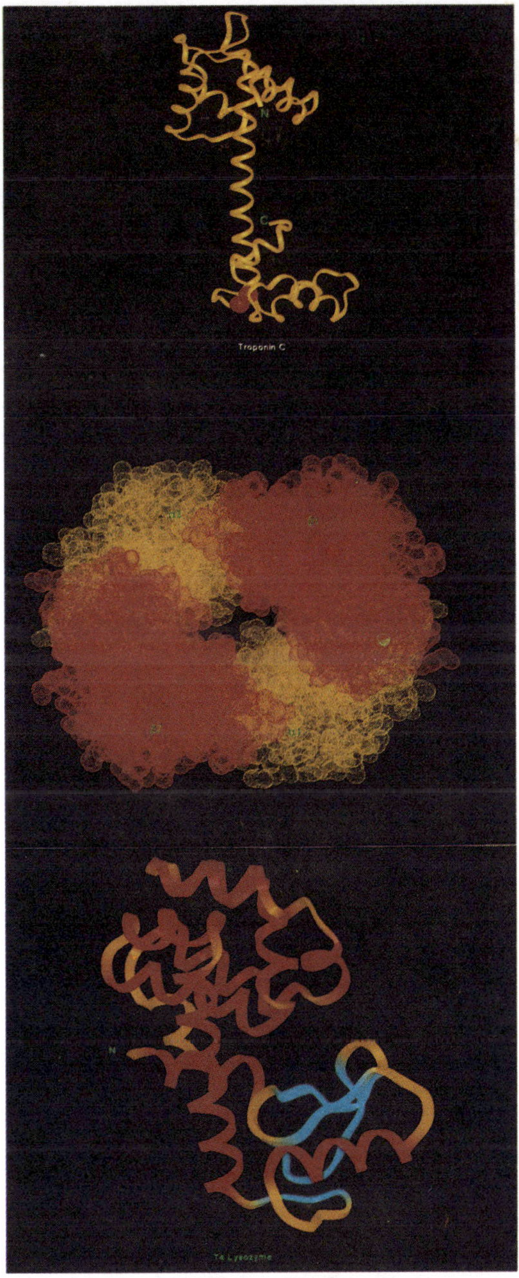

Fig. 6.16 *Three-dimensional structures of proteins.* Top figure is troponin C, the middle figure is haemoglobin and the bottom figure is T4 lysozyme.

(b) Imposing conformational constraints

Another method of engineering a protein to increase thermal stability is to replace glycine by any other residue or to increase the number of prolines in a given sequence. This is because glycine and proline have opposite effects on stability. The conformational freedom of glycine is much more than any other amino acid and hence a substitution of Gly by any other amino acid will decrease the number of unfolded structures. In T4 lysozyme, the substitution of Gly77 by alanine and Ala82 by proline increased the melting temperature T_m by 1° and 2° respectively.

(c) Stabilising the α helices

An α helix can be considered as a dipole with a positive charge at the amino terminus and a negative charge at the carboxy terminus (Fig. 6.6). Negative ions such as phosphate groups in coenzymes or substrates bind to the amino terminus of an α helix. If the helix is not part of a binding site then a negatively charged side chain is found at the N-terminus and a positively charged side chain at the other end of the dipole. Out of 11 α helices in T4 lysozyme seven are stabilised by the side chains. Of the remaining four, two were considered for stabilisation by using site directed mutagenesis to engineer additional side chain interactions. Ser38 and Asn144 were changed to aspartic acid. The effect was additive and the T_m increased by 2° for single mutants and 4° for double mutants in comparison to the wild type.

(d) Alterations in the hydrophobic core

Hydrophobic side chains tend to bury themselves in the interior of the protein in order to avoid contact with the aqueous exterior. This effect is a major determinant in the folding of proteins. Hence the filling up of existing cavities in the hydrophobic core of a protein should increase the hydrophobic interactions thus stabilising the enzyme. In T4 lysozyme there are two such cavities in the hydrophobic core of its α-helical domain. Site-directed mutagenesis was carried out in which Leu133 was substituted by phenylalanine and Ala129 by valine. Instead of showing the expected increase in stability, however, both the mutant proteins were less stable than the wild type. The crystal structures gave an explanation for this phenomenon. It was observed that in both the substitutions the new side chains adopted energetically unfavourable conformations introducing a strain in the structure which more than compensated for the energy gained by hydrophic interactions. Hence this method may be successful in certain cases but not in others.

6.4.2 *Charging tRNA with unnatural amino acids*

In this approach to protein engineering, the mediator tRNA molecules are charged (attached) with amino acids different from the ones which attach to them naturally (each particular tRNA has affinity for a particular amino acid — Chapters 2 and 4). A variant amino acid is thus incorporated in the protein in the normal way leading to a protein with properties different from the wild type. In a particular example, T4 phage lysozyme was modified by the incorporation of an aspartyl β-nitrobenzyl ester instead of Asp20 in the wild-type sequence. While the folding of the protein was not altered by this change, since the Asp20 residue was in the active site the mutant lysozyme (containing nitrobenzyl Asp) did not show the activity of the wild type. However, upon photo-irradiation at 315nm the mutant Asp was converted to the wild-type Asp and activity was recovered. This method therefore suggests a way to engineer proteins in a 'caged' form, to be released by irradiation.

6.4.3 *Protein engineering and molecular recognition*

(a) Molecular recognition and enzyme catalysis is generally controlled by the binding energy between the protein and its ligand (e.g., substrate). In order to modify such properties of the protein one needs to understand the interactions between the ligand and the protein. Such interactions have been studied in the enzyme, tyrosyl tRNA synthetase.

In these experiments the amino acid residues which form hydrogen bonds with the ligand were mutated one by one. These site-specific mutations would disrupt the bonds one at a time and hence their individual contributions to ligand binding could be estimated. It was found that not only was the conformation of the protein and the mode of binding of the ligand practically unaltered in spite of the mutations but also that the formation of the protein–ligand complex was not, in general, affected. This was explained as follows. Each hydrogen bond in the protein–ligand complex involves two groups — one in the protein and the other in the ligand. Before the formation of the complex both these groups bond with the surrounding water molecules. In the complex the hydrogen bonds with water are replaced by ones between the ligand and the protein. Thus each hydrogen bond in the complex replaces two such bonds in the uncomplexed state. If, in a mutant of the protein, a polar group involved in the hydrogen bond formation is replaced by a hydrophobic residue then this residue will not bond with water.

The binding energy of the protein–ligand complex involves the loss of the contribution from an equal number of hydrogen bonds as are actually present in the complex. In the complex of the mutant protein with the ligand, though there is a loss of one hydrogen bond due to the mutation, the overall difference in the binding energy between the complexed and uncomplexed states does not change significantly. The actual change in energy will vary in individual mutations and would depend on the precise nature of the interactions.

The studies also established that the contribution to the specificity from hydrogen bonds between uncharged residues was small as compared to that between charged residues. Thus one of the important driving forces for high specificity in complementary reactions between a protein and a ligand is the ionic interactions or formation of hydrogen bonds between the charged groups in the interacting area of the two components.

(b) Site-directed mutagenesis has been used to alter the specificity of substrate recognition of many proteins, such as aspartate aminotransferase, malate dehydrogenase, and the DNA-binding protein 434 repressor.

Semisynthetic specific nucleases have been constructed by making a chemical adduct of a nuclease with an oligonucleotide or a DNA binding protein. Chimeric, specific DNAases have been created by binding a 13-nucleotide and a 12-nucleotide pyrimidine oligonucleotide to K84C staphylococcal nuclease, the wild type of which is non-specific. The oligonucleotide sequence of the first chimera, called SN55, was 5'- CTTTTTCCTTCTC - 3' and was designed to bind at position 2488–2500 of the plasmid pUC19 via triplex formation and cleave between the ribosomal binding site and the –10 region of the promoter sequence upstream of the β-lactamase gene. The second chimera had the oligonucleotide sequence 5'-TTCCCTCTTTCC-3' and was designed to bind and cleave the sequence between 1028 and 1039 of the same plasmid. Experiments revealed a 60% efficiency of specific cleavage for these two engineered nucleases.

6.4.4 Engineering the pepsin class of enzymes

Acid proteases belong to one of the classes of enzymes that have been extensively studied through protein engineering due to their important clinical and commercial applications. The mammalian digestive enzyme pepsin belongs to this class and hence these enzymes are also referred to as pepsin-like enzymes.

Assignment 6.4
Name a commercial use for pepsin-like enzymes.

Chymosin is normally synthesised in the fourth stomach of the unweaned calf where it initiates milk clotting under acidic conditions. Protein engineering studies have been performed on this enzyme using site-directed mutagenesis. Three different mutations in the chymosin-B cDNA (cloned DNA) were made. The DNA was cloned and expressed in *E.coli*. The mutations were designed so as to affect both optimum pH and subsite specificity. The studies are still underway.

In other studies, mutant bovine chymosin was produced with a single mutation — the residue Val111 was changed to phenylalanine. The enzymatic properties of the mutated chymosin were studied and it was found that the reaction efficiency, as measured by the K_m values, of the mutant enzyme were approximately half those of the wild type. The mutant enzyme was crystallised and subjected to X-ray analyses. It was observed that the mutation leads to only local changes in enzyme conformation. The modelling of substrate binding showed that the mutated phenylalanine side chain occupies a part of the binding pocket, thus reducing the binding affinity for the substrate.

Such understanding of the mechanism of catalytic action and pinpointing the amino acids involved in substrate binding can lead to design of inhibitors for specific uses.

Renin is yet another important enzyme of the acid protease class. It catalyses the first and rate-limiting step in the conversion of angiotensinogen to the hormone angiotensin-II. The catalysis is highly specific and plays an important role in the regulation of blood pressure. Thus renin is a target for the design of inhibitor compounds to be used as drugs for controlling hypertension. It has been the subject of extensive protein engineering studies.

HIV-I proteinase, which is produced by HIV, has been found to be structurally homologous to the pepsin-like enzymes. This close similarity between the two structures is being exploited to synthesise antiviral drugs, by designing inhibitors to the pepsin-like fungal enzymes, which are expected to inhibit the HIV-I proteinase as well.

Subtilisin is an extremely important commercial proteinase and finds great use in the detergent industry as an agent for the removal of proteinaceaous stains (Chapter 5). It belongs not to the pepsin-like acid protease class of enzymes but to the serine protease class. It has been subjected to a series of improvements by protein engineering to enhance stability and activity at high temperatures, abnormal pH values, etc.

6.4.5 *Protein engineering through chemical modifications*

Chemical modifications of proteins have also been attempted to increase the stability of proteins to temperature and organic solvents. Cross-linking of lysine residues in the protein with chemical linkers is a commonly used method.

(a) Functional groups which are selectively modified in the proteins

Group	Residue
α — COOH	C-terminal
β — COOH	Asp
γ — COOH	Glu
—— SH	Cys
α — NH_2	N-terminal
ε — NH_2	Lys
—— OH	Ser, Thr
S — CH_3	Met
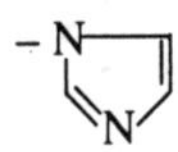	His

(b) Cross-linkers used

(i) Dialdehydes:

$$H-\overset{\overset{O}{\|}}{C}-(CH_2)_n-\overset{\overset{O}{\|}}{C}-H$$

e.g., n = 3 - Glutaraldehyde

(ii) Diimido esters:

$$CH_3-O-\overset{\overset{NH_2^+Cl}{\|}}{C}-(CH_2)_n-\overset{\overset{NH_2^+Cl}{\|}}{C}-O-CH_3$$

e.g., n = 4 - Dimethyl adipimidate (DMA)

(iii)

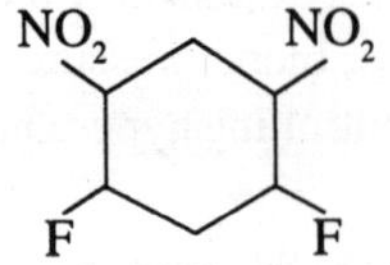

1,5 - difluoro - 2,4 - dinitro benzene (DNB)

Simple dialdehydes such as glyoxal, adipaldehyde and glutaraldehyde have been shown to act as protein cross-linkers. The most extensively used reagent is glutaraldehyde and it has been used to cross-link and stabilise protein crystals as well as proteins in solution. Haemoglobin, phospho-fructokinase, lactate dehydrogenase and insulin have been studied with glutaraldehyde. The chemical modification of lysine residues in proteins is based upon the ability of the amino group of this residue to react as a nucleophile. Lysine is the second strongest nucleophile in a protein (cysteine is the most reactive nucleophile). The protonated form of lysine is essentially unreactive and hence the modification reactions are performed at pH 8.0 to 9.0. Reagents which lead to non-specific modifications are of no use. Chemical modifications of proteins are used not only for studies on structure-stability relationships but may also be utilised to obtain artificially stabilised enzyme preparations for biotechnology.

Assignment 6.5

What would be the advantages and disadvantages of using an enzyme with increased thermostability?

An example of the engineering of a thermally stable enzyme by introducing disulphide bridges has been thymidylate synthetase (TS). A computer modelling procedure was used to search for suitable pairs of residues across which a disulphide bridge could be introduced. On the basis of the results obtained such bridges were created in TS by site-directed mutagenesis. These bridges were both intermolecular as well as intramolecular. A mutant TS which had symmetric bridges between the two subunits was stable and active even at 65°C, in contrast to the wild type which was inactive above 55°C.

Engineered proteins in industry

Some engineered proteins and the industry in which they find the greatest application:

Pharmaceutical industry

Pepsin	Elastase
Renin	Haemoglobin
Cathepsin-D	Dihydrofolate reductase
Thrombin	

Food industry

Chymosin	Subtilisin
Papain	Lipases

Detergent industry

Subtilisin	Chymotrypsin
	Trypsin

Self assessment exercise 6.5

Which of the following could be expected to lead to an increase in thermostability of an enzyme by reducing the possibility of conformational changes at high temperatures ?

(i) Insertion of cysteine residues at sites on the polypeptide so that disulphide bridges can be formed between them.
(ii) Decreasing the number of proline residues and increasing the number of glycine residues in the protein.
(iii) Alteration of amino acid residues such that cavities in the hydrophobic core of the protein are filled.
(iv) Using glutaraldehyde to cross-link amino acid residues on the protein molecule.

6.4.6 Protein design

The ultimate aim of protein engineering is to design an amino acid sequence which will fold into the desired structure and hence have the desired function. It appears that it is easier to design an amino acid sequence of a particular structure and function based on known protein sequences and structures than to predict the structure from a given sequence.

A globular, water soluble 'protein' without any specific function but with a very definite secondary and tertiary structure has been designed. This new protein consists of four α helices folded to form a compact bundle (Fig. 6.13). Four helix bundles have been noticed fairly commonly as a super-secondary structure motif in α proteins where the four helices pack in such a way that hydrophobic side chains form the central core and hydrophilic side chains are on the surface of the bundle. In the *de novo* design the amino acid sequence was chosen to be four 16-residue helix sequences connected by three short-loop regions. The sequence of all the four helix regions was

Gly-Glu-Leu-Glu-Glu-Leu-Leu-Lys-Lys-Leu-Lys-Glu-Leu-Leu-Lys-Gly

and the loop sequence was

Pro-Arg-Arg

The corresponding DNA sequence was synthesised, incorporated into *E. coli* by recombinant techniques and expressed in that system. CD spectroscopy of the designed protein confirmed the presence of a predominantely α-helical structure. However, the three-dimensional structure of this α-4 protein has not yet been determined to verify the prediction that the four α helices are arranged antiparallel.

A second interesting example of design of a protein is 'crystanova'. This protein was designed to resemble one of the family of eye lens proteins, namely the crystallins. An amino acid sequence, which is expected to fold in the β-barrel super secondary motif, was designed and the gene encoding this polypeptide was synthesised and expressed in *E. coli.* The purified synthetic gene product 'crystanova' is reported from preliminary investigations to fold incorrectly. However, the results obtained have proved useful for reanalyses of the relationship between protein sequence and structure.

6.4.7 Design of peptide and protein mimics

Several short oligopeptides such as enkephalin exhibit specific biological activity. These short peptides can assume multiple conformations due to low energy barriers separating them but only one conformation among these is compatible with the biological activity. It is therefore essential that synthetic peptides and protein mimics must have the correct conformation for their biological activity. This can be brought about by introducing stereochemical constraints through the incorporation of residues such as α-amino isobutyric acid, α- and β-dehydro residues, proline, d-amino acids, etc. These special residues would act to restrict the range of conformations that the peptide backbone can assume.

α- and β-dehydro amino acids have been found to occur naturally in peptides of microbial origin. Judicious placement of dehydro residues such as Phe, Leu, etc., can lead to the formation of specific structures such as β turns. Oligopeptides which contain α-amino isobutyric acid (Aib) adopt the α- or 3_{10}-helical conformation. In other words, Aib can nucleate a helix and can therefore be used in protein and peptide design.

6.4.8 Drug design

It has been established through protein crystallographic studies that the size and chemistry of an enzyme substrate is complementary to the active site of the enzyme. Computer modelling of the active site can suggest the nature of the molecules which are likely to interact with it, leading to the design of inhibitor molecules for the enzyme under consideration. This idea has been used in designing drugs such as methotrexate and trimethoprim which inhibit the action of the enzyme dihydrofolate reductase. Similarly, drugs have been designed based on the inhibitory mechanism of renin.

6.4.9 Development of peptide vaccines

Immune response is mediated through proteins known as immunoglobulins (Chapter 10) which recognise certain features of the surface proteins of the invading organisms. These antigenic sites are known as epitopes and consist of short peptide sequences. Recombinant DNA techniques have been used to synthesise oligopeptides which would mimic these epitopes, thus invoking immune response. These short peptides can be used in conjunction with carrier proteins as vaccines.

6.5 Concluding remarks

To summarise, protein engineering is a cooperative endeavour between organic chemists, biochemists, crystallographers, theoretical biophysicists, computer scientists and molecular biologists. The project could either be the engineering of desirable properties into extant proteins or the design and *de novo* synthesis of novel proteins. In either case the principles of protein structure have to be well understood, and correlations between the amino acid sequence and the three-dimensional structure of the protein have to be systematically studied and used in the design. Experimental and theoretical techniques such as X-ray crystallography, NMR, database analysis, site-directed mutagenesis and other molecular biology techniques have to be applied. This procedure has been followed in many cases and the resulting new proteins have been shown to possess the enhanced properties which were aimed at.

Exciting times are ahead for protein engineering. Advances in technology, and in understanding the principles of protein structure and design will allow us to design, engineer and produce novel proteins with novel functions. These may find applications in areas from ecology to health, from materials technology to bioelectronics. The impact of protein engineering on the future of mankind will be at least as large as that of the transistor, probably more !

6.6 Suggested further reading

1. Branden, C. and Tooze, J (1991). *Introduction to protein structure.* Garland Publishing Inc., New York.
2. Schultz, G.E. and Schirmer, R.H. (1979). *Principles of protein structure.* Springer-Verlag, New York.
3. Dickerson, R.E. and Geis, I. (1969). *The structure and action of proteins.* W. Benjamin Inc., California.
4. Richardson, J.S. (1981). *Advances in Protein Chemistry*, **34**, 167.
5. Blundell, T.L. and Johnson, L.N. (1976) *Protein cystallography.* Academic Press, New York.
6. Cantor, C.R. and Schimmel, P.R. (1980). *Biophysical chemistry. Vols I, II and III.* W.H. Freeman and Co., New York.

7. Oxender, D.L. and Fox, C.F. (Eds.) (1987). *Protein engineering*. Alan R. Liss Inc., New York.
8. Balaram, P. (1992). *Current Opinion in Structural Biology*, **2**, 845.
9. Schultz, P.G. *et al.* (1991). *Journal of the American Chemical Society*, **113**, 9398.
10. Alber, T. and Matthews, B.W. (1987). *Methods in Enzymology*, **154**, 511.
11. Subramanian, E. (1993) *Current Science*, **59**, 728.
12. Horn, A.S. and De Ranter, C.J. (1984). *X-ray crystallography and drug action*. Clarendon Press, Oxford.
13. Richardson, J.S. and Richardson, D.C. (1984). *Trends in Biochemical Sciences*, **14**, 304.
14. Fersht, A. and Winter, G. (1992). *Trends in Biochemical Sciences*, **17**, 292.

6.7 Responses to assignments

6.1 It was explained in Chapter 4 how the sequence in a polypeptide is determined by the sequence of codons (triplets of bases) in the DNA which codes for the polypeptide. It was explained in Chapter 5 how automated synthesisers can be used to manufacture DNA of given sequence. Thus, it is possible to sequence the protein, determine the particular codons and their sequence which would produce this protein (as a result of transcription and translation), and then synthesise the required DNA sequence using the automated synthesiser.

6.2 The hydrophobic domains will tend to come together in the centre of the molecule (away from the aqueous environment), while the hydrophilic domains will tend to arrange themselves on the exterior of the molecule (in contact with the aqueous environment).

6.3 Recombination can occur during DNA replication, when separate DNA molecules with the same (homologous) sequence can interchange sequences. In our example the sequences are

T G C G C A T G T G C A C
– T A T G C G C A C G T G C A C G T A–

During a recombination event, the synthesised DNA can replace its homologous portion in the cell DNA, and thus the mutation at the required site is brought about. By such means site-directed mutagenesis can replace a base (or bases) with another base (or bases), delete bases or add additional bases. In practice, synthesised oligonucleotides are incorporated into genes as outlined in Section 5.9.4.

6.4 Calf stomachs are steeped in brine to produce rennet, a commercial enzyme preparation used to promote the clotting of milk in the manufacture of cheese. The active ingredient in rennet is the enzyme chymosin, and this can now be produced from a recombinant (genetically engineered) microorganism. Several fungi produce pepsin-like enzymes, which have also been used in manufacture of cheeses suitable for consumption by vegetarians.

6.5 Thermostable enzymes have the following advantages: (i) acceleration of reaction rate — an increase of temperature by 10β results in 2 to 3 fold rise in the rate of chemical reactions catalysed by the enzymes; (ii) increased efficiency of enzymatic processes as a result of increased stability of reagents; (iii) the end products are more volatile and the viscosity of the solutions decreases at elevated temperatures; (iv) the end product is sterilised in the course of the reactions; (v) the reaction can be terminated by cooling; (vi) the enzyme is highly stable against denaturing agents such as concentrated solutions of urea or guanidine chloride, and against extreme pH values or against digestion by proteinases.

These disadvantages however are relatively minor and enzymes engineered to be more thermostable are useful in the food and detergent industry. In the food industry, for example, thermostable enzymes are used in the conversion of starch to sugar. In the detergent industry these enzymes are useful in removing stains made by organic substances like blood.

As against these advantages enzymes with altered thermostability may have the following disadvantages: (i) the elevated temperature leads to denaturation of the product or to acceleration of some side reactions such as high temperature oxidations and racemisation of amino acids and their derivatives; (ii) the increase in their stability may lead to decrease in their activity. These disadvantages, however, are relatively minor and enzymes engineered to be more thermostable are useful in the food and detergent industry.

6.8 Responses to self assessment exercises

6.1 The correct answers are as follows.

(i) True.
(ii) True.
(iii) False.
The statement is a definition of protein design, rather than protein engineering.

6.2 The correct matching is as follows.

(i) ...... (c)
(ii) ..... (a)
(iii) ..... (d)
(iv) ..... (e)

6.3 The answers are (i), (ii) and (iii). Secondary structures such as α helices and pleated sheets are stabilised by hydrogen bonding. Tertiary and quaternary structures are themselves determined by secondary structure, and thus stabilised by the stabling effect of hydrogen bonding on secondary structure.

6.4 The correct answers are as follows.

(i) False — The absorption frequency of nuclei in atoms is related to NMR, not X-ray crystallography.
(ii) False — The three-dimensional diffraction of waves is related to X-ray crystallography, not NMR.
(iii) True.

6.5 (i), (iii) and (iv) might be expected to lead to increased thermostability in a protein. (ii) might be expected to have the opposite effect — to increase thermostability glycine residues should be decreased and proline residues increased in number.

7 Enzyme technology

Chapter outline

7.1 Introduction

Enzyme technology is the technology associated with the use of enzymes as tools in industry, agriculture or medicine. Enzymes are biological catalysts derived from living systems and can be used *in vitro* to promote biochemical transformations.

Assignment 7.1
What is meant by the term catalyst?

Enzymes are proteins, composed of one polypeptide (amino acid chain) or two or more associated polypeptide chains. The sequence of amino acids in the polypeptide chain determines the primary structure of the protein. In living cells enzymes do not occur as a long polypeptide chain but as a structure having secondary, tertiary and even quaternary conformations (Chapter 6).

An enzyme is thus rather like a piece of string which has been randomly gathered up into a ball — the conformation of the ball in three dimensions represents the tertiary structure. The tertiary structure of an enzyme molecule is determined by the primary and secondary structure, and is influenced by environmental factors such as temperature, pH and ionic strength; changes in these environmental factors away from conditions found in living cells may lead to denaturation (distortion of the tertiary structural conformation).

The active site of the enzyme molecule is that region involved in catalytic activity. This site represents a small portion of the enzyme molecule (imagine only a small part of the ball of string) and may involve amino acids at various sites along the primary chain of the polypeptide(s). Normally the substrate on which the enzyme acts binds to the enzyme molecule at the active site. Molecules which bind to the enzyme molecule at sites other than the active site may (in binding to the molecule) cause conformational changes (distortion) in the tertiary structure of the enzyme. This may affect activity of the enzyme molecule.

The structure and conformation of the active site is not the only factor determining the activity of an enzyme molecule. The substrate for the enzyme (the molecule on which the enzyme acts) must have ready access to the active site; if the conformation of the molecule is such that access of the substrate to the active site is impeded, then enzyme activity will be prevented. Thus the tertiary structure in particular of the enzyme molecule must be such that it produces a suitable conformation to allow access of the substrate to the active site. Tertiary structure and enzyme activity can be influenced by factors such as temperature, pH and ionic strength of the environment of the enzyme. The situation is complicated further by the fact that some enzymes are not composed of a single polypeptide, but of two or more polypeptides, forming an enzyme complex. Such enzyme complexes may contain more than one active site, each active site having a different specificity (active on different substrates). The specificity of action of an enzyme on a specific substrate is determined primarily by the structure and conformation of the active site.

Of importance to the use of enzymes for various applications is the stability of the enzyme molecules. This is because the enzymes need to be stored between the period of manufacture and use, and in continuous processes using immobilised enzymes (see Section 7.5) enzyme activity is required over an extended period of time, perhaps several weeks. As proteins, enzymes will be subject to degradation by the activity of proteolytic enzymes (proteases) which may be produced by some microorganisms present in the enzyme mixture as contaminants. Alterations in pH, ionic strength or other interactions over a period of time tend to produce conformational changes (denaturation) in the enzyme molecules, leading to loss of activity. These factors need to be considered in the storage and use of enzymes. Conformation changes in enzyme structure caused by binding of molecules to the enzyme may also affect stability of the enzyme molecule.

7.2 Traditional uses of enzymes

The uses of enzymes in industrial processing goes back several decades. Some enzymes used have been of plant origin (for example, papain used as a meat tenderiser) or animal origin (for example renin used to clot milk during cheese manufacture). However, it has proved difficult to extract plant or animal enzymes on the scale necessary to meet demand in many industrial sectors, so for large scale applications microbial enzymes are favoured, as they can be produced in large-scale industrial fermentation units (see Chapter 8).

Microbial enzymes have been widely applied in the food and beverage industries: amylases to liquefy and hydrolyse starch in manufacture of alcoholic beverages, starch pastes and glucose syrups; proteases to hydrolyse proteins, tenderise meats, and clot milk to manufacture cheese; pectins to improve fruit juice extraction and prevent haze formation in fermented fruit juices; glucose isomerase in the production of fructose syrup as a 'high sweetener'. Microbial proteases and lipases (lipid [fat] digesting enzymes) have been produced on a large scale for inclusion in 'biological' detergents, with biological activity for removal of proteins and fats from clothing.

7.3 Immobilisation of enzymes

7.3.1 *Problems with the use of enzymes in free solution*

In traditional processes enzymes in free solution are mixed with a lot (or batch) of the substrate mixture, and the enzyme is then allowed to react with the batch of substrate and convert it to the desired product of the reaction.

> *Assignment 7.2*
>
> What disadvantages arise from the use of free enzymes in batch processes in industrial-scale conversion processes?

7.3.2 *Objectives of enzyme immobilisation*

The disadvantages of using enzymes in free solution in batch cultures have been outlined. Continuous processes are generally more desirable for industry, as they are more efficient and cost effective.

> *Assignment 7.3*
>
> Would the disadvantages outlined in Assignment 7.2 be overcome if enzymes in free solution were used in a continuous process (where a volume of substrate [and perhaps enzyme] were added continuously to the reactor in which conversion was to take place, and a volume of contents were continually removed for recovery of the product)?

Continuous systems can be run without the disadvantages outlined, if the enzyme can be retained or trapped within the reactor. In practice it is difficult to use molecular sieves (ultrafilters) to trap enzymes while allowing smaller molecules to pass through, so techniques have been developed to stick enzyme molecules onto larger insoluble supports; the large supports can be easily retained in a reactor by virtue of their large size and/or high density. The technique of anchoring an enzyme (or a cell) in or on such support material is called immobilisation, and enzymes (or cells) anchored in this way are referred to as immobilised enzymes (or cells). A reactor can be packed with the immobilised enzyme on/in its support (to form a packed bed reactor) and substrate percolated through; the immobilised enzyme converts the substrate to the product, which passes out of the reactor in the product stream, while the enzyme (attached to its support) is retained in the reactor. Thus efficient continuous systems for enzyme conversion can be operated.

Immobilisation of enzymes can have other advantageous effects. The immobilisation process can lead to increased activity and increased stability of the enzyme molecules (compared to activity and stability in free solution). Enzyme stability is vital for continuous processes, where the process needs to operate for as long as possible to keep non-productive downtime (the time between batches, when the reactor is being cleaned, etc.) to a minimum.

7.4 Methods of enzyme immobilisation

7.4.1 *Adsorption*

Adsorption is an easy way to perform enzyme immobilisation. Only contact between the enzyme solution and the support is needed (without the intervention of any chemical reagent which might denature the enzyme). Various kinds of supports are now available: mineral supports such as aluminium oxide, clay, glass; organic supports such as starch, modified sepharose and ion-exchange resins.

Adsorption of enzyme molecules onto the surface of the support involves bonds of low energy level such as ionic interactions, van der Waals forces and hydrogen bonds (Fig. 7.1). On account of this low level of energy of the bonds, the adsorption of the enzyme on the support can also be easily reversed and the enzyme can be removed (desorbed) by small changes in conditions, for example ionic strength, pH or temperature. Hence care must be taken to avoid the loss of enzyme from the support by desorption during the conversion process.

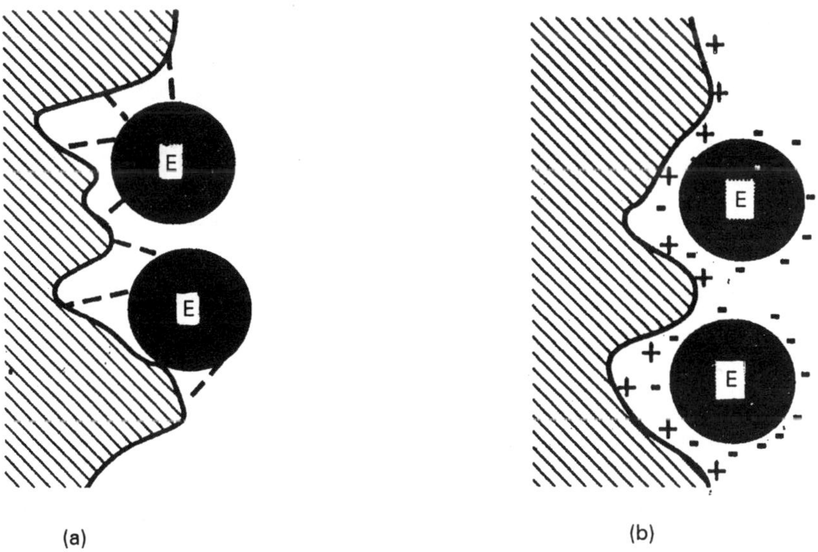

Fig. 7.1 *Adsorption by van der Waals (a) and (b) hydrogen bonding.*

> ***Assignment 7.4***
>
> In what circumstances, using adsorbed enzyme in a continuous reactor system, would the easy removal (desorption) of enzyme from the support prove to be an advantage?

7.4.2 Entrapment of enzymes

Immobilisation of enzyme molecules can be performed by the physical entrapping of the enzyme inside a polymer matrix, gel, or capsule. The support, or more precisely the matrix, varies in form and in nature but always the size of the pores in the matrix must be adjusted to prevent the loss of enzyme from the matrix. There are a number of ways in which enzymes can be entrapped.

(a) Inclusion in gels: The mechanism of polymerisation occurs in a variety of ways, for example by radical formation (polyacrylamide gel), by radiation (polyvinyl alcohol and polyvinyl pyrrolidone gel) or by light for a photosensible resin (polyethylene glycol dimethacrylate). Natural polymers (alginate, starch) are also used (Fig. 7.2a).

(b) Inclusion in fibres: A mixture containing the polymer and the enzyme is extruded to have the enzyme trapped in fibre format. The material used is usually polyacetate or cellulose (Fig. 7.2b).

(c) Inclusion in microcapsules: During formation of an emulsion containing a hydrophobic monomer in an organic solvent and a hydrophilic monomer and the enzyme in water, the hydrophobic monomer and the hydrophilic monomer polymerise at the air–water interface, forming a microcapsule with enzyme molecules trapped within it (Fig. 7.2c). Monomer mixtures used in microcapsule formation are: polyamine and polybasic acid chloride; polyphenol and polyisocyanate.

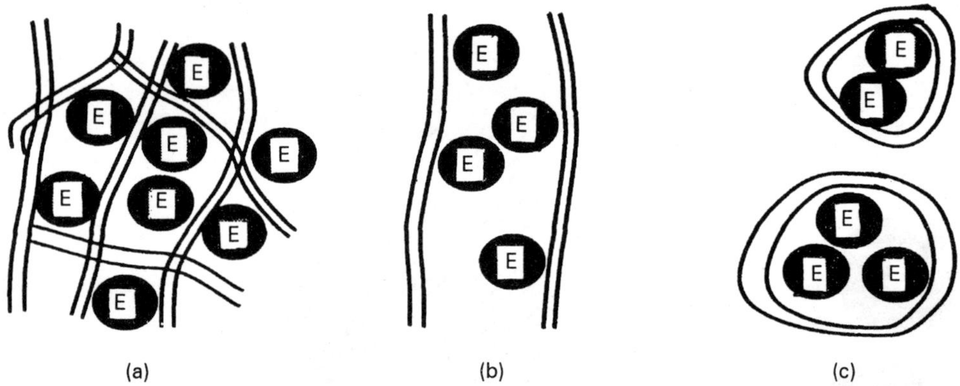

Fig. 7.2 *Immobilisation of enzyme molecules by entrapment.*

An understanding of the chemistry of polymerisation reactions enables the extension of these methods to any enzyme or mixture of enzymes, and the versatility of the form of matrix allows the process to be optimised.

> ***Assignment 7.5***
>
> What disadvantages might there be to immobilisation of enzymes using entrapment?

7.4.3 Direct covalent linkage

To immobilise an enzyme in a durable way, the approach frequently adopted is the creation of covalent bonds, with high-energy level, between chemical groups of the enzyme and chemical groups of the support (Fig. 7.3).

Inert supports may need pretreatment before covalent bonding with an enzyme is possible. Methods for direct covalent linkage are given below.

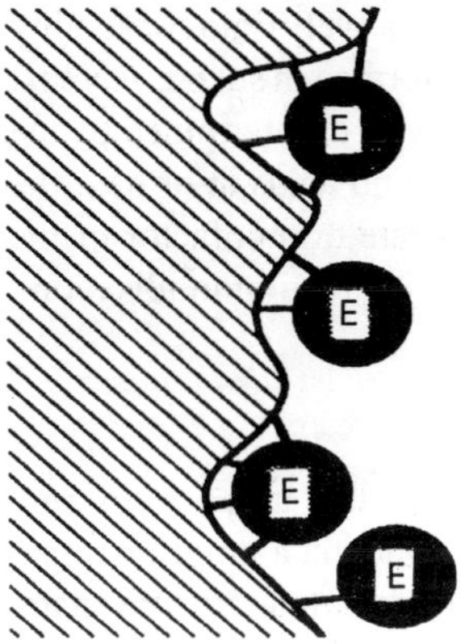

Fig. 7.3 *Covalent bonding.*

(a) Diazotation
This reaction involves bonding between the amino group of the support (for instance using aminobenzyl cellulose, amino derivatives of polystyrene, aminosilanised porous glass [glass treated with aminosilane] and a tyrosyl or histidyl group of the enzyme (Fig. 7.4a).

(b) Formation of peptide bond
The reaction occurs between the amino (or carboxyl) groups of the support and the carboxyl (or amino) groups of the enzymes (Fig. 7.4b).

(c) Group activation
Another popular method is the use of cyanogen bromide, which can be applied to a support containing glycol groups, i.e., cellulose, sephadex, sepharose (Fig. 7.4c).

(d) Polyfunctional reagents
In this method, a bifunctional or a multifunctional reagent such as glutaraldehyde is used to create bonding between the amino group of the support and the amino groups of the enzyme. Although the reaction mainly involves the amino group of the enzyme, reactions with groups such as guanidyl, histidyl, tyrosyl and sulfhydryl are not excluded. Some examples of supports used are, aminoethyl cellulose, aminoalkylated porous glass and albumin (Fig. 7.4d).

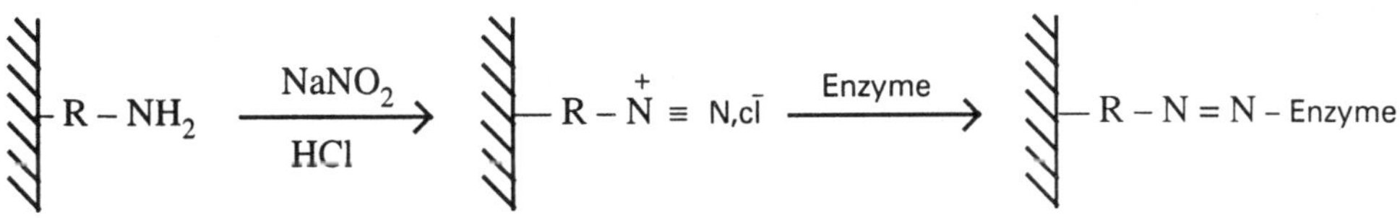

(a)

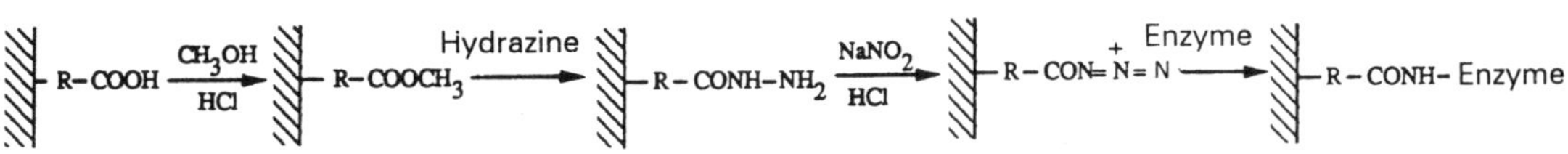

(b)

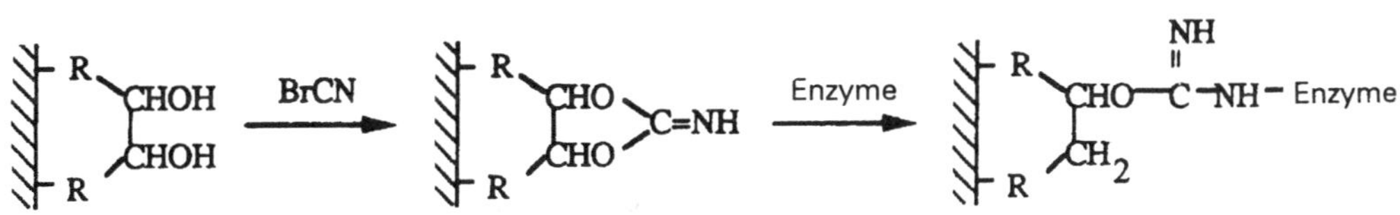

(c)

$$\text{Cellulose} - OCH_2CH_2NH_2 + OCN(CH_2)_6NCO + \text{Enzyme} \longrightarrow \text{Cellulose} - OCH_2CH_2\,NHCONH(CH_2)_6\,NHCOH - \text{Enzyme}$$

(d)

Fig. 7.4 *(a) Diazotation; (b) Peptide bond formation; (c) Group activation; (d) Polyfunctional agent.*

In contrast to the adsorption method, the high-energy level of covalent bonds is generally a guarantee of avoiding any loss of the enzyme from the support during the enzymatic process, even in the case of change of environmental conditions. Immobilisation by covalent linking requires the use of purified and thus expensive enzyme. The different steps make the immobilisation procedure relatively time consuming and the chemical reactions are often complicated. The covalent bonds must be formed in such a way that enzyme activity is conserved. It was explained in Section 7.1 that enzyme activity is determined by the configuration of the active site of the enzyme; covalent bonding involving amino acids at or near the active site is thus likely to reduce enzyme activity. The addition of a substrate or a cofactor of the enzymatic reaction can protect to some extent the active site of the enzyme during the immobilisation.

7.4.4 Cross-linking

In contrast to previously described methods, cross-linking is characterised by the absence of any support; covalent links are created between the various molecules of an enzyme via a polyfunctional reagent (Fig. 7.5). The different kinds of polyfunctional reagents are: hexamethylene diisocyanate, N–N′ polyme-thylene bis iodoacetinamide, N–N′ ethylene bis maleimide, diazonium salt, and glutaraldehyde. With glutaraldehyde, the form and the mechanical properties of the immobilised enzyme matrix can be easily modified just by a slight modification of the experimental conditions. The main advantage of the cross-linking method is its simplicity (it is sufficient to mix the enzyme and the reagent). The cost of immobilisation can be reduced by cross-linking a mixture of the enzyme with another protein such as serum albumin, gelatin, or haemoglobin or even by using unpurified enzyme. The principal disadvantage is the risk of denaturation of the enzyme by the polyfunctional reagent.

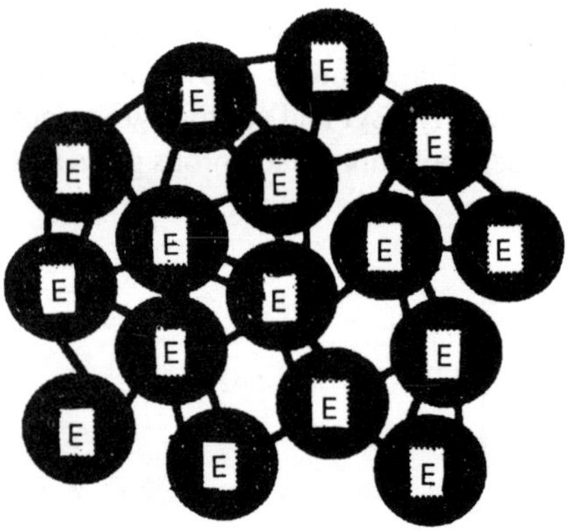

Fig. 7.5 *Cross-linking of enzyme molecules.*

Self assessment exercise 7.1

In Column A methods of enzyme immobilisation are given. Match each method from Column A with the appropriate characteristic in Column B.

	Column A		Column B
(i)	Adsorption	(a)	Chemical reaction is used to produce high energy bonding between enzyme and support.
(ii)	Entrapment	(b)	Chemical reaction is used to produce bonds between enzyme molecules.
(iii)	Direct covalent linking	(c)	Chemical reaction is not needed to produce bonding between enzyme and support.
(iv)	Cross-linking	(d)	Chemical reaction is used to polymerise support material.

> ***Self assessment exercise 7.2***
>
> In Column A methods of enzyme immobilisation are given. Match each method in Column A with the appropriate disadvantage in Column B.
>
> | Column A | Column B |
> |---|---|
> | (i) Adsorption | (a) Enzyme activity may be reduced by the action of chemical agents used in bond formation. |
> | (ii) Entrapment | (b) Enzyme easily becomes desorbed from support. |
> | (iii) Direct covalent linking | (c) Access of the substrate to the enzyme may be prevented by physical hindrance. |
> | (iv) Cross-linking | |

7.5 Bioreactors for immobilised enzymes in continuous conversions

7.5.1 Packed bed reactors

In a packed bed reactor (Fig. 7.6) the enzyme, immobilised on support material or crosslinked into particles, is packed into a reactor (usually columnar), and held in place by gravity or retaining mesh; the substrate is then pumped through. The substrate solution permeates through the bed of particulate matter, and the immobilised enzyme progressively converts substrate to product. The outflow of the reactor represents the product stream.

The aim of the process is usually to maximise output by passing as much substrate as possible through the reactor while minimising wastage (all of the substrate should be converted to product during passage through the reactor). In other words the aim is optimum conversion. To achieve this, enzyme activity must be maintained at maximal levels, and operating conditions (particularly flow rate and substrate concentration) must be optimal for the level of enzyme activity at any given time. In such a continuous system the level of enzyme activity over time will decline, particularly as a result of enzyme denaturation. It is thus also important to optimise conditions within the reactor for maintaining enzyme stability.

The fact that a plug-flow system such as a packed bed reactor does not result in a homogeneous internal environment means that it can be difficult for the operator of such a system to optimise conditions for enzyme activity and product formation. A further disadvantage is that in such reactors the bed may become distorted during operation, and open channels appear within the bed.

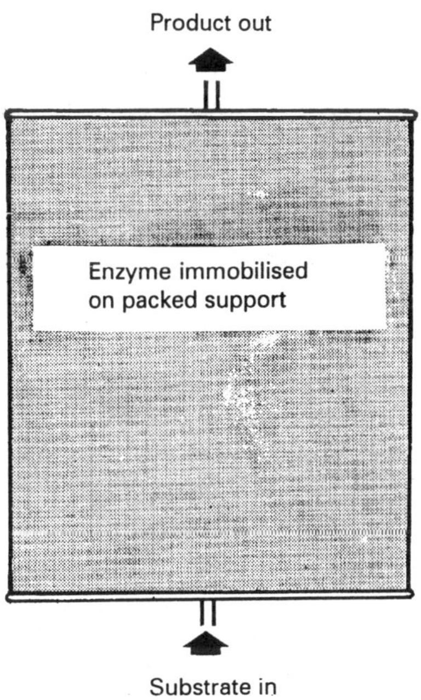

Fig. 7.6 *Packed bed reactor.*

> ***Assignment 7.6***
>
> Would conditions (for example, substrate concentration and product concentration) be homogeneous (similar in all places) within an operating packed bed reactor?

Assignment 7.7

What would be the effect of channel formation in a packed bed reactor?

7.5.2 Mixed reactors

In order to overcome the lack of homogeneity (and the problems associated with this) within a packed bed, reactors have been designed which use some form of agitation to mix up the contents of the reactor (in which the concentration of particles is reduced compared to a packed bed to allow free movement of particles). Agitation may be accomplished by stirring, or in a fluidised bed reactor the substrate is pumped rapidly into the reactor so that the upward flow of fluid maintains the particles in suspension.

Assignment 7.8

What would be the disadvantages of stirred reactors and fluidised bed reactors in continuous processes using immobilised enzymes?

With agitated systems operating in continuous mode, particle removal from the reactor can be prevented by: the use of filters (which may however tend to become clogged); passing the product stream out of the reactor via an expanded side arm in which the particles settle out (back into the reactor); immobilising enzymes onto supports (such as iron–nickel particles) which are magnetic and placing magnets around the outflow.

Mixed reactors of this type, if used in a continuous manner with high rates of throughput, often lead to unconverted substrate passing out in the product stream (in other words the degree of conversion of substrate may be low). This is particularly the case with mixed fluidised bed reactors where a minimum upflow rate is required in order to keep particles suspended. To overcome this, the substrate needs to be recirculated through the reactor; a batch of medium is recirculated again and again through the reactor until complete conversion is achieved.

Self assessment exercise 7.3

State whether each of the following characteristics relates best to either a packed bed reactor or a mixed reactor (assume both are working in continuous mode).

(i) Immobilised enzyme may easily be washed out of the reactor in the product stream.
(ii) Conditions can easily be optimised.
(iii) Unconverted substrate may be washed out of the reactor due to channel formation.

7.6 Applications of immobilised enzymes in industrial-scale conversions

Over the past few years, the use of enzymes has been the subject of increased interest, and more than one thousand original papers on the applications of immobilised enzymes have been published. Recently, numerous papers on the immobilisation of whole cells for industrial applications have also appeared. Judged by both the number of patents and the present industrial applications, Japan is ahead of all other countries in enzyme technology.

In 1969, Tanabe Seiyaku Co. were the first to develop a satisfactory industrial application for an immobilised enzyme: immobilised aminoacylase for the continuous production of L-amino acids from acyl-DL-amino acids. Since then, they have also developed industrial applications of immobilised microbial cells, and applied them in the continuous production of L-aspartic acid from ammonium fumarate, using immobilised cells of *Escherichia coli* with an increased aspartase activity.

Sugar and sugar syrup manufacture by enzyme technology has received much attention from food companies in the USA and over the years a large number of patents and papers on the industrial and technical aspects of this topic have been published in the USA, Japan and Europe. As a result of this work, fructose manufactured by using glucoamylase and glucose isomerase, is being produced on a large scale to replace sucrose as a sweetening agent.

The cost benefits of using immobilised enzymes compared to unimmobilised (native) enzymes is considerable. For instance, the production of high fructose syrup from corn starch is a rapidly expanding business and the production of high fructose syrup with the aid of immobilised enzymes is now approaching five million tons a year. The cost of the enzymatic transformation, using immobilised glucoamylase and glucose isomerase in continuous reactors, is 0.28% of the total cost of the final product. This cost is only one-tenth of that involved in enzymatic transformation in batch processes using native enzymes.

7.7 Enzymatic electrocatalysis

The term "enzymatic electrocatalysis" has been used to definite the synergy between the electrochemical and enzymatic oxido-reduction reactions in a configuration where the electrochemical and the enzymatic sites are close at the molecular level. The evolution of this concept has resulted in the deposition of chemically modified enzymes onto electrodes, which have become known as "biosensors" or "enzyme electrodes". The product of the enzymatic reaction is oxidised or reduced at this surface. Here, a new concept must be mentioned concerning the interaction between enzyme and electrode. This is the occurrence of an electron transfer directly between the active site of the enzyme and the electrode (direct transfer) or between a conducting polymer and the active site of the enzyme entrapped into it. In the last case, the mode of the electron transfer is:

Active site of the enzyme $\longrightarrow$ Conductive polymer $\longrightarrow$ Electrode

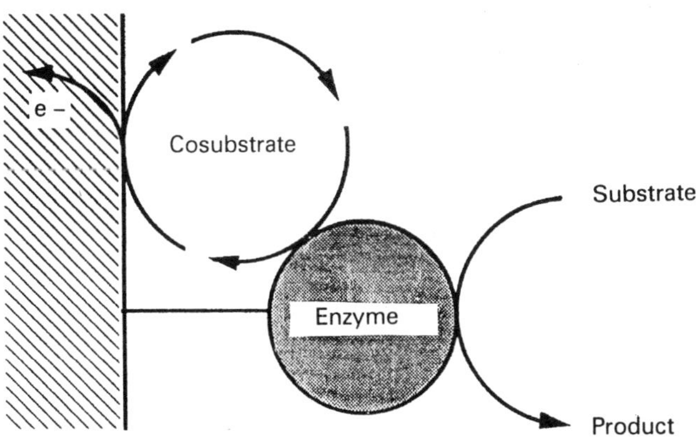

Fig. 7.7 *Enzymatic electrocatalysis.*

Normally, the coupling of the enzymatic and electrochemical reactions when the enzyme is covalently linked on the electrode surface, is occurring without intervention of direct electron transfer. Typically the enzyme is an oxidoreductase which catalyses the oxidation or reduction of the substrate into the product and at the same time the reduction or oxidation of the cosubstrate or the coenzyme, which reacts with the electrode (Fig. 7.7).

Examples of enzymatic electrocatalysis systems using a carbon electrode is the enzyme /substrate / cosubstrate combinations of: glucose oxidase / glucose / oxygen; lactic dehydrogenase / lactate / NAD, hydrogenase/ hydrogen.

7.7.1 General approach for the immobilisation of enzymes onto electrodes

To optimise the proximity between the enzyme and the electrode, an enzyme monolayer is covalently linked with the electrode surface by a two step procedure involving carbodiimide as condensation reagent. On carbon electrodes, an easy way to introduce functional groups, particularly carboxylic groups, is an electrochemical oxidation of the surface. The carboxylic groups of the electrode surface are then activated by carbodiimide. After rinsing to eliminate the excess of reagent, the electrodes are soaked in the solution of enzyme to perform the reaction between the activated groups of the surface of the electrode and the amine groups of the enzyme (Fig. 7.8).

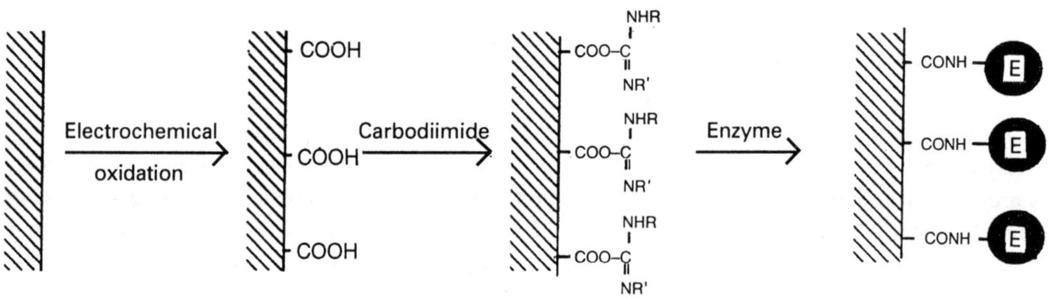

Fig.7.8 *Immobilisation of enzyme onto electrode.*

The geometry of the electrode is determined by the purpose of the work (Fig. 7.9).

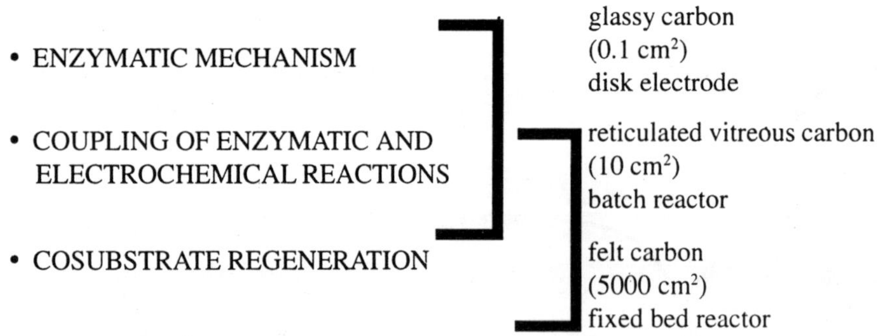

Fig. 7.9 *Electrode dimensions.*

7.7.2 Measurement of enzymatic activity

The activity of the enzyme determines the current across the carbon electrode which reflects the electron exchange. To have a good sensitivity, it is preferable to work at a fixed potential imposed between the carbon electrode and a reference electrode (such as a saturated calomel electrode), with a potentiostat.

A third electrode, an auxiliary platinum electrode is added to give a three electrode configuration (Fig. 7.10).

In the presence of the substrate of the enzymatic reaction, a flux balance is established at the interface of the carbon electrode between diffusion, enzymatic reaction and electrochemical reaction. The time needed to establish this flux balance is equivalent to the time of establishment of a diffusion-convection layer. This takes only a fraction of a second in the case of a rotating carbon disk electrode. An example of catalytic activity is the case of glucose oxidase immobilised on a rotating disk electrode of glassy carbon. The enzyme catalyses the transformation of D-glucose into gluconic acid with the reduction of the FAD prosthetic group of the enzyme. The $FADH_2$ can be re-oxidised by oxygen or other cosubstrate such as benzoquinone with the production of electrons.

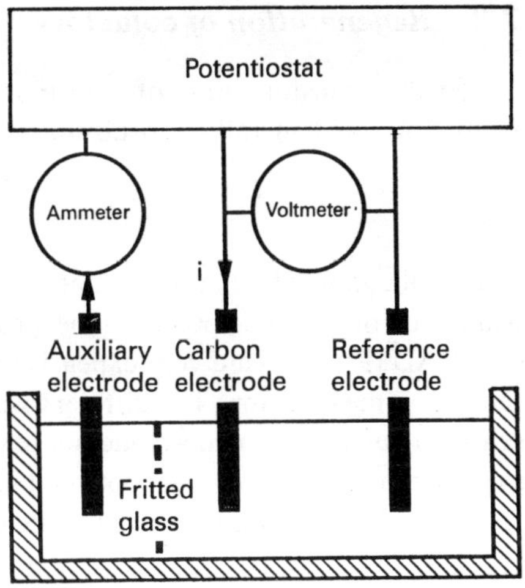

Fig. 7.10 *Three-electrode configuration.*

D-Glucose + Enzyme-FAD → Gluconic acid + Enzyme-$FADH_2$

E-$FADH_2$ + Benzoquinone → FAD + Hydroquinone

Hydroquinone → Benzoquinone $+2H^+ +2e^-$

Instantaneous catalytic current is then observed as a function of the concentration of the immobilised enzyme and glucose at the surface of the electrode (Fig. 7.11).

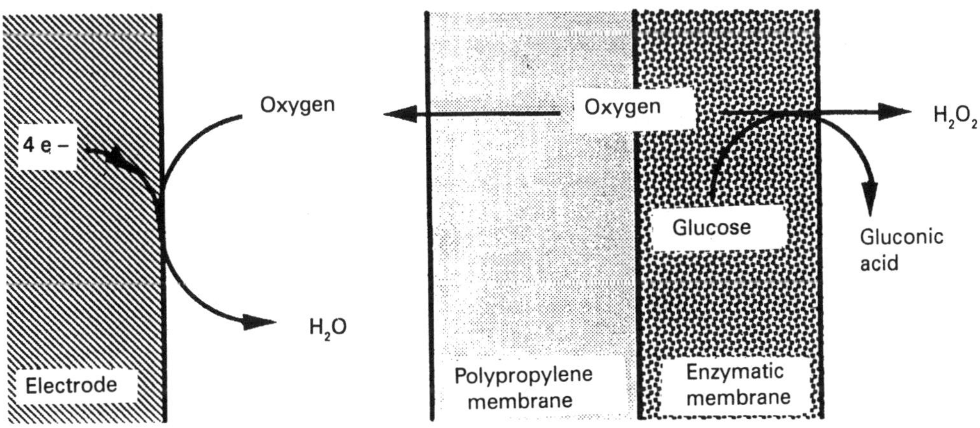

Fig. 7.11 *Instantaneous catalytic current.*

> ***Assignment 7.9***
>
> Summarise in your own words how electron generation and transfer permits the measurement of substrate concentration using an enzyme electrode.

7.7.3 Regeneration of cofactors

Consider the transformation of substrate, S, into product, P, by an enzymatic reaction involving a cofactor (or more generally a cosubstrate). The scheme of this reaction is:

$$\text{S+ Cofactor} \xrightarrow{\text{Enzyme}} \text{P+ Modified cofactor}$$

The application of such a reaction for practical purposes raises the problem of stoichiometric consumption of the cofactor during the production of P. Because many of the cofactors, (for example, flavin, NAD (P), ATP) are very expensive to produce, the viability of the process requires a means of regenerating the cofactor. The different ways of performing the regeneration can be illustrated in the case of the cofactor of oxido-reduction reactions, NAD. NAD regeneration has been the subject of a great deal of research due to the potential importance of reactions it is involved in. Important factors to be considered in cofactor regeneration are given below:

(a) The efficiency of the regeneration scheme
The efficiency of the regeneration reaction is the major point to examine in deciding whether or not the method of regeneration is practicable. Effectively any significant loss of active cofactor molecules during the regeneration step limits dramatically the number of times the cofactor molecules can be recycled. It is useful to take the example of two reactions of regeneration, one with an efficiency of 99% and the other with an efficiency of 99.99% (in other words in the first reaction 1% of the cofactor molecules are denatured during the regeneration). Thus, for an initial amount of cosubstrate of 1 mole, we can calculate the amount of product formed before the total inactivation of this cosubstrate pool; this gives 100 moles of product for the first regeneration reaction and 10,000 moles of product for the second regeneration reaction. This last value is usually considered to be the limit of economical viability for NAD.

(b) The rate of regeneration
To increase the productivity of a process (number of molecules of substrate transformed per unity of time), the overall rate of reactions must be optimised. It is thus essential that the regeneration reaction is not the rate limiting step (the rate of regeneration must exceed the rate of enzyme reaction).

(c) The simplicity of the scheme
The regeneration scheme must be compatible with the main enzymatic scheme of production. Moreover the quantity (and the price) of the added reagents must be as low as possible.

Several methods are available for cofactor regeneration and are considered below.

(a) Chemical regeneration
Alongside the enzymatic reaction which transforms a substrate into a product, a chemical reaction is responsible for the regeneration of the cosubstrate by a redox reagent. It is then useful in the case of NAD regeneration to have a redox reagent which is reoxidised by oxygen. Many kinds of reagents are cited in the literature, for example flavin mononucleotide (FMN), phenazine methosulfate or methylene blue. The advantages of chemical regeneration are the low cost and the stability of chemical reagents, and also the alternative with the choice of the reagent to regenerate both the oxidised or reduced form of the cofactors, NAD or NADH. The problems reside in the specificity of the chemical reaction and the intervention of undesirable secondary reactions. Also the introduction of the reagent complicates product purification even further.

(b) Electrochemical regeneration

The term electrochemical regeneration is reserved in this section for a scheme where the molecule of the cofactor reacts directly with the surface of the electrode for the redox electrochemical reaction (Fig. 7.12).

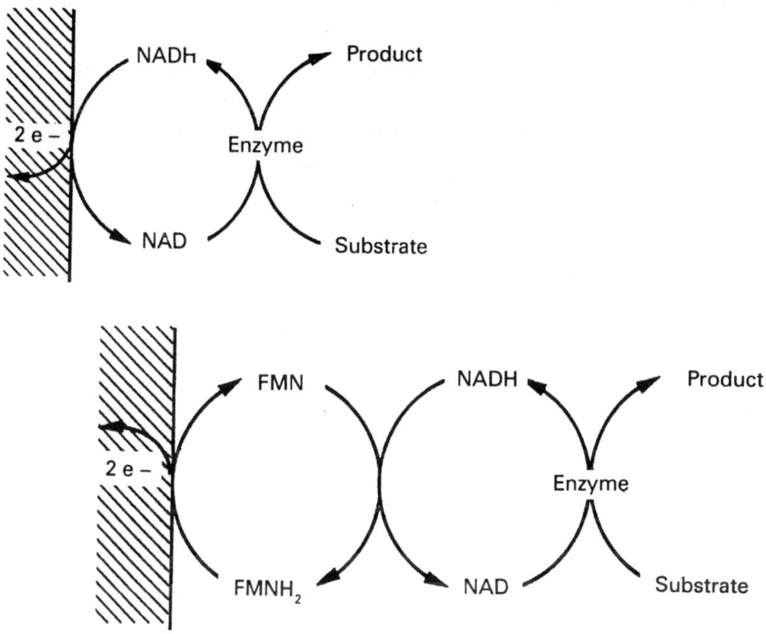

Fig. 7.12 *Cofactor regeneration.*

(c) Enzymatic regeneration

To benefit from both the specificity and the rapidity of enzymatic catalysis, a second enzymatic reaction can be used to perform the regeneration of the cofactor. Both the regeneration of NAD or NADH can be envisaged, depending on the choice of the enzymatic reaction of regeneration (Fig. 7.13).

(i) Method with coupled enzyme

In this method, a second and totally independent enzymatic system effects the regeneration of the cofactor, for example lactate dehydrogenase (LDH) or glutamate dehydrogenase. The cost of the technological process is increased by the cost of the second enzyme and the purification of the product is complicated by the three compounds of the enzymatic reaction of regeneration, enzyme–substrate product.

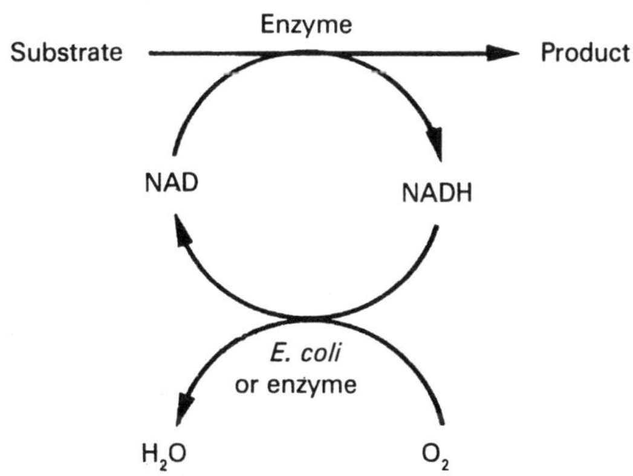

Fig. 7.13 *Enzymatic regeneration.*

(ii) Method with coupled substrate
In some particular examples, for example horse liver alcohol dehydrogenase (HLADH), the oxidation and reduction of the cofactor can be catalysed by the same enzyme utilising two different substrates. An interesting configuration is obtained by grafting the cofactor directly inside or very close to the active site of the enzyme. An increase in efficiency is then observed because both reactions occur in the same site without diffusion of the cofactor molecule.
(iii) Microbiological method
Instead of purifying the enzyme used for the regeneration, whole cells or a fraction of the microorganism having the enzymatic activity can be introduced in the process. One example is the use of membrane fraction of *Escherichia coli* bearing the NADH oxidase activity.

Assignment 7.10

What would the main advantage be of using whole cells or cell fractions of microorganisms in cofactor regeneration, compared to purified enzymes?

7.8 Basic principles of biosensors

In a biosensor the enzyme is simply entrapped with a dialysis membrane on the sensitive top of the detector or immobilised on it by covalent linking or cross-linking. At a given concentration of the substrate the enzymatic reaction gives a corresponding reaction rate which alters the local physical chemical environment of the detector. This alteration results in a response of the biosensor (Fig. 7.14). Depending on the construction of the biosensor, the response is of transient or stationary type.

Three main characteristics are important in determining the performance of the biosensor:

(a) The selectivity
This characteristic is brought about both by the specificity of the enzymatic reaction towards the substrate (for example amino acid oxidases recognise different amino acids) and by the selectivity of the detector towards the physico-chemical parameter. When the detector measures directly the concentration of the product of the enzymatic reaction, the improvement in selectivity is realised by the addition of a selective membrane between the enzymatic reaction and the detector.

(b) The sensitivity
Because the response of the biosensor is not linear in all the ranges of substrate concentration, it is necessary to define the range in which the concentration is

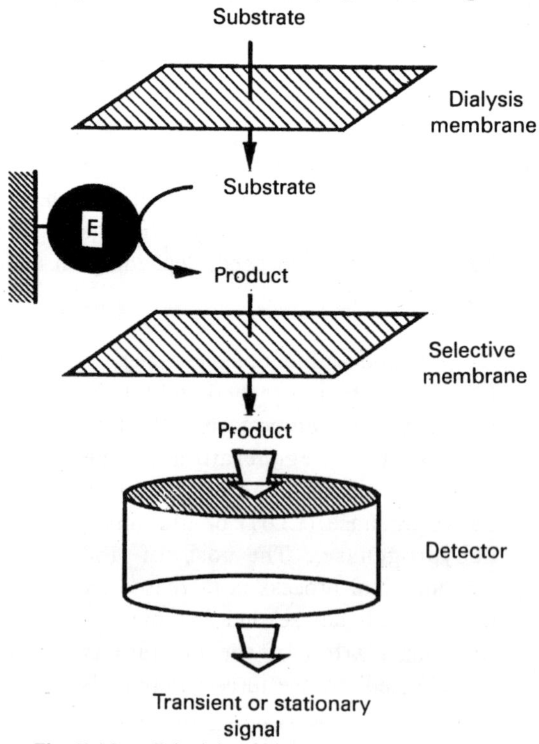

Fig. 7.14 *Principle of biosensor.*

determined with good precision. This range depends both on the detector type, particularly to its detection limit, and on the nature of the diffusion reaction of the enzymatic system.

(c) The stability
In addition to the need for general stability of the detector, the stability of the enzyme during measurement or even in storage is crucial. For this, immobilisation of the enzyme is not only a way to maintain the enzyme on the detector but also to stabilise its enzymatic activity.

7.8.1 Bioelectrodes

In the following discussion, the classification of biosensors is based on the type of detector. Despite the recent progress in this field, the majority of the biosensors have at this time a detector based on an electrode.

There are two principal types of electrodes employed in biosensors, namely the amperometric and the potentiometric electrode. In the ampero-metric electrode a potential is imposed between the sensitive part of the electrode and a reference electrode. At this imposed potential, a molecule intervening in the enzymatic reaction (substrate or product) can transfer an electron with the electrode surface to be oxidised or reduced. The current through the electrode reflects the concentration of this molecule, but because this molecule is consumed at the electrode surface, the current is also a function of the mass-transfer to the surface (Fig. 7.15).

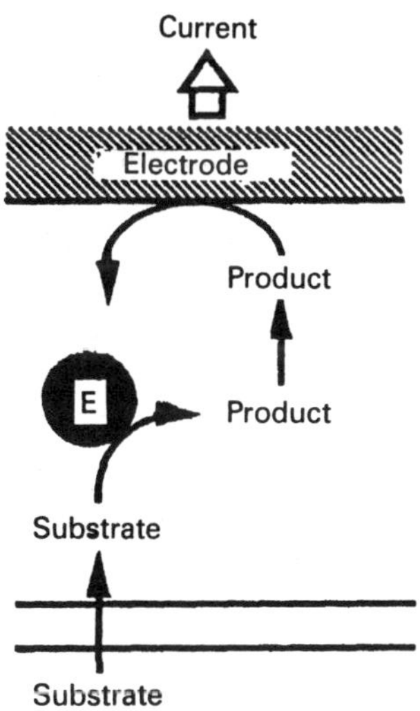

Fig. 7.15 *Amperometric biosensor.*

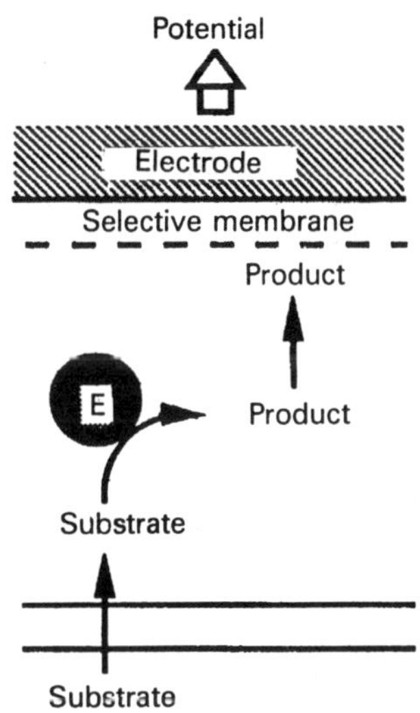

Fig. 7.16 *Potentiometric biosensor.*

In the potentiometric electrode, no current passes through the electrode, and the signal is then the potential obtained between the sensitive part of the electrode covered with a selective membrane and a reference electrode. No molecule is consumed at the electrode surface, so the signal is independent of the mass transfer to the active surface (Fig. 7.16).

7.8.2 Optrodes

A condensation of the words "optical" and "electrode", the optrode refers to the association of the new technology of fibre optics with the classic electronic method to follow an enzymatic catalysis by the change of the spectrophotometric characteristics of the solution. An example is the absorbance at the wavelength of 340nm with the transformation of NAD into NADH during dehydrogenase catalysis. In other words, the optrode is used to measure the absorbance of light of a set wavelength in a mixture of substrate, the absorbance being proportional to concentration of the product (adsorbing at the set wavelength). In essence it behaves as a miniature spectrophotometer. The range of the experimental wavelength depends on the material of the optic fibre, for example silica optic permits measurement in the ultraviolet range, whereas glass fibre is restricted to visible wavelengths. The size of the extremity of the optrode is down to submillimeter.

7.8.3 Immunochemical biosensors

In Chapter 10 it is explained that immunoglobulins (antibodies) are produced in the animal body in reaction to the presence of foreign substances (antigens). It is now possible to produce a batch of antibody (called a monoclonal antibody) which reacts speci-fically to one individual antigen. If the substance to be measured is itself antigenic, then a specific antibody against it can be produced and can be incorpo-rated into the biosensor system. The biosensor can in this way be rendered highly specific in its action.

The basic ideas of the immunochemical biosensor use immunological specificity with spectrophotometric detection. Two main procedures can be described, the sandwich method and the competitive method. In the first method, antibody (which reacts specifically with the substance to be measured [analyte]) immobilised at the surface of a microplate is incubated with the analyte. After rinsing, a solution containing the antibody linked to an enzyme is put in the microplate to react with the immobilised antibody–analyte complex from the first step. The spectrophotometric detection involves measuring a coloured product of the enzymatic reaction which reflects the number of analyte molecules bound to antibodies (Fig. 7.17).

In the second method, a mixture of the analyte and of an antigen coupled to the enzyme is added to the immobilised antibody. After rinsing, the enzymatic reaction is used to measure the number of antibodies which did not bind the analyte (Fig. 7.18).

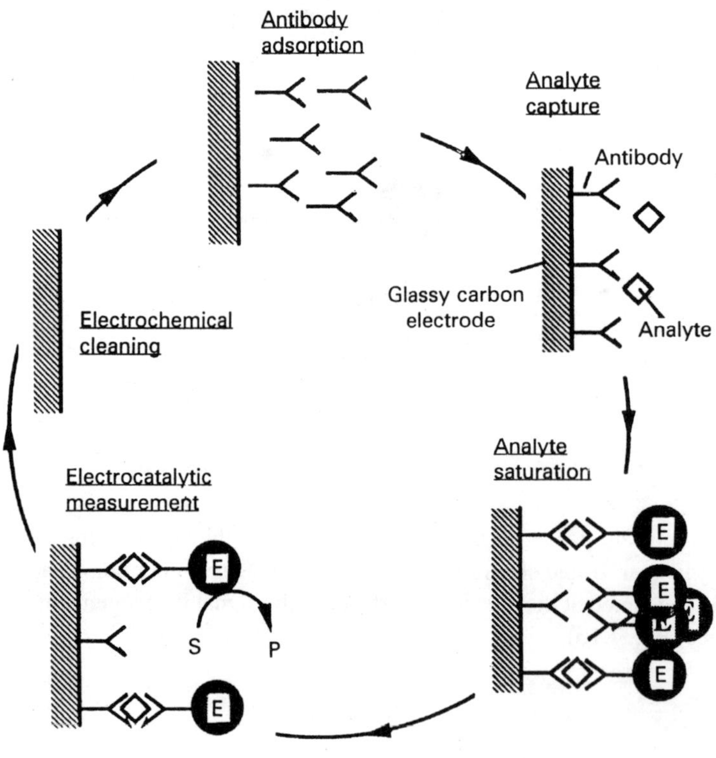

Fig. 7.17 *Immunochemical biosensors.*

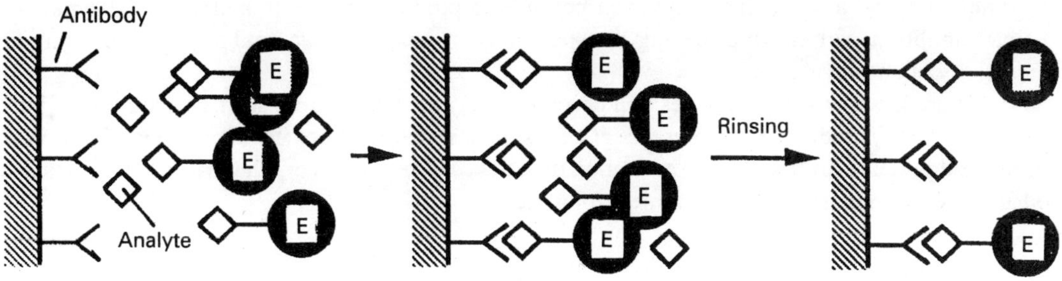

Fig. 7.18 *Competitive immunoassay.*

Self assessment exercise 7.4

A thermistor is a temperature-sensitive resistor fabricated with semiconductor materials; in other words, the electrical resistance of the thermistor is dependent on its temperature. Glucose oxidase oxidises its substrate (glucose) in an exothermic (heat generating) reaction. With these facts in mind, draw a diagram of a device using immobilised glucose oxidase and a thermistor to measure the concentration of glucose in a stream of liquid. Explain exactly how this system would function.

7.9 Concluding remarks

Enzymes offer the opportunity to carry out specific conversion of organic molecules, at low temperatures and relatively low cost. Such reactions are not only applicable to conversions on a large scale in industrial processes, but can be used on a small scale for other biotechnological processes (an example mentioned in Chapter 5 is the use of microbial restriction endonucleases in the process of genetic engineering). The development of immobilisation techniques has not only improved the efficiency of industrial-scale conversions, but also allowed the development of small-scale biosensors. Biosensors can be applied to the on-line (continuous, automated) measurement of compounds (the enzyme substrate) in situations such as fermentation processes (see Chapter 8) and the remote monitoring of locations for environmental pollutants.

Advances in enzyme technology have been encouraged by advances in other areas of biotechnology, for example:

(i) Techniques of genetic engineering have made it possible to improve the yield of an enzyme from a given type of cell (for instance by inserting into the cell multiple copies of the gene coding for that enzyme).

(ii) Using powerful computer programs and knowledge gained from fundamental research into the relationship between structure and function of enzymes, it is now possible to design novel structural features in an enzyme (Chapter 6). These features can then be introduced into the enzyme using specific techniques to alter the structure of the gene(s) coding for the enzyme.

(iii) The discovery of thermostable microorganisms living in deep sea volcanic vents has led to an understanding of how their enzymes are able to remain active at high temperatures and pressures. This should extend the range of use of enzymes in industrial processes (organic chemical synthesis is usually carried out in reactors at relatively high temperatures and pressures).

(iv) The specific binding between a monoclonal antibody and its antigen can be used to increase the range of enzyme applications when the enzyme is linked to the antibody, for instance immunological assays. It is now even possible to produce chimaeric molecules (abzymes — part antibody, part enzyme) having the specificity of binding of an antibody and the catalytic activity of an enzyme.

In the future it is likely that enzymes will become widely applied to areas such as: synthesis of bulk organic chemicals; production of fragrances (for cosmetics) and flavours (for foods); modification of oils, fats and proteins (for foods); detoxification of residues of pesticides or other harmful chemicals in food and water; monitoring the environment for toxic chemicals.

7.10 Suggested further reading

1. Iammoello, R.M. and Yacynych, A.M. (1981). Immobilized enzyme chemically modified electrode as amperometric sensor. *Anal. Chem.*, **54**, 1098–1101.
2. Leel, G. and Whitesidess, G.M. (1985). Enzyme catalyzed organic synthesis: a comparison of strategies for *in situ* regeneration of NAD from NADH. *J. Am. Chem. Soc.*, **107**, 6999–7008.
3. Legoy, M.D., Larreta–Garge, V., Le Moullec, J.M., Ergan, F. and Thomas, D. Cofactor regeneration in immobilised enzyme systems: chemical grafting of functional NAD in the active site of dehydrogenase. *Biochimie*, **62**, 341–345.
4. Lortie, R., Faassouane, A., Laval, J.M. and Bourdillon, C. Equilibrium displacement in acetaldehyde production using yeast alcohol dehydrogenase and electrochemical cofactor regeneration. *Biotechnol. Bioeng.*, **39** (1992),157–163.
5. Mosbach, K. (1976). Immobilized enzymes. *Methods in enzymology*. Academic Press Inc., **44**, New York.
6. Mosbach, K. (1987). Immobilized enzymes and cells (Part B). *Methods in enzymology*. Academic Press Inc., **135**, New York.
7. Mosbach, K. (1987). Immobilized enzymes and cells (Part C). *Methods in enzymology*. Academic Press Inc., **136**, New York.
8. Mosbach, K. (1988). Immobilized enzymes and cells (Part D). *Methods in enzymology*. Academic Press, Inc., **137**, New York.
9. Ngo, T.T. (1987). *Electrochemical sensors in immunological analysis*. Plenum Press, New York.
10. Romette, J.L., Yang, J.S., Kusakabe, H. and Thomas, D. (1983). Enzyme electrode for specific determination of L-lysine. *Biotechnol. Bioeng.*, **25**, 2557–2566.
11. Seitz, W.R. (1984). Chemical sensors based on fiber optics. *Anal. Chem.*, **56**, 16A-34A.
12. Thomas, D., Bourdillon, C., Broun, G. and Kernevez, J.P. (1974). Kinetic behaviour of enzymes in artificial membranes inhibition and reversibility effects. *Biochemistry*, **13**, 2995.
13. Thomas, D. (1978), Production of biological catalysts, stabilization and exploitation. Report edited by the Commission of the European Communities (*EUR 6079 EN*) Luxembourg.
14. Tran, N.D., Romette, J.L. and Thomas, D. (1983) An enzyme electrode for specific determination of L-lysine: a real-time control sensor. *Biotechnol. Bioeng.*, **25**, 329–340.
15. Turner, A.P.F., Karube, I. and Wilson, G.S. (1987). *Biosensors, fundamentals and applications.* Oxford University Press, **42**, Oxford.
16. Whitesides, G.M. and Wong, C.H. (1985). Enzymes as catalysts in synthetic organic chemistry. *Angew. Chem. Int. Ed. Engl.*, **24**, 617–618.

7.11 Responses to assignments

7.1 A catalyst is a substance which promotes (enhances/increases) a chemical reaction, without taking part directly in the reaction (as either a substrate or a product). In other words, the substrate is converted to the product(s) more quickly or easily when the catalyst is present, compared to when the catalyst is absent.

7.2 In batch systems, after the product has been formed from the substrate the batch of product is removed and processed. This means that the enzyme is lost and so a new lot of enzyme must be produced. Also the product is contaminated by the enzyme, and if the enzyme (or its activity) in the product is undesirable, expensive purification processes may be needed to remove it. Batch systems involve non-productive periods (called down time) between batches while the reactors are being cleaned, and so on (see Chapter 8) and so overall they are relatively inefficient; continuous processes are much more efficient (and thus more cost effective).

7.3 The continuous system would avoid non-productive periods between batches. However, the enzyme would be removed from the reactor in the product stream, meaning that contamination of product by enzyme and also enzyme loss from the reactor would still occur. These two problems could be overcome if the enzymes could be trapped and retained in the reactor by some mechanism.

7.4 In an immobilised enzyme reactor, used for a continuous process, the enzyme will gradually loose its stability and activity, and the process will need to be halted and a new batch of immobilised enzyme put into the reactor. If the enzyme can easily be desorbed from the support, the support can be recycled (the support from the reactor can be cleaned [enzyme desorbed], used to immobilise the next lot of enzyme, then replaced in the reactor).

7.5 The matrix, gel or sphere may be such that access of the substrate to the enzyme and egress of the substrate is impaired. The matrix, gel or capsule may not be stable in the environmental conditions used in the conversion process in the reactor. The conditions used to form the matrix, gel or capsules may lead to enzyme denaturation and loss of activity.

7.6 The answer is no. Packed bed reactors tend to act as plug-flow systems; essentially the fluid stream passes through the reactor from inlet to outlet with little back mixing. There is no stirring to promote even mixing. Thus, as the substrate solution passes through the reactor the concentration of substrate becomes progressively lower and the concentration of product progressively greater.

7.7 If a channel opens up in the bed, then the fluid stream passes more easily through the channel than through the bed, so fluid becomes directed along the channels (in which there will be no enzyme activity). If channels run in the orientation of inlet to outlet, then substrate will tend to pass through the reactor without conversion. Channel formation thus makes it more difficult to control the internal environment of the reactor and tends to reduce the level of product formation.

7.8 If you cannot think of any answer to this assignment, go over Assignment 7.3 again; with a mixed system it is likely that immobilised enzyme particles will be swept out of the reactor in the product stream, and thus enzyme activity lost from the reactor. To overcome this, a mechanism is required to withhold the particles within the reactor.

7.9 In simple terms, the immobilised enzyme catalyses a reaction which leads to the generation of electrons, and the electrons are transferred to and measured by the electrode. The number of electrons produced (and the reading obtained from the meter attached to the electrode) is proportional to the number of molecules of substrate being oxidised/reduced. Hence the reading obtained from the meter (attached to the electrode) is directly proportional to the concentration of the substrate in the solution in which the enzyme electrode is immersed.

7.10 The main disadvantages of using purified enzymes are cost (it is usually very costly to purify the enzymes [extracted from microbial cells]) and extraction and purification procedures may lead to loss of activity. Using whole cells overcomes both these problems. Using cell fractions is more complex than using whole cells (cells need to be disrupted and the required cell fractions collected by methods such as density gradient centrifugation) but this usually also overcomes both problems to some degree.

7.12 Responses to self assessment exercises

7.1 The correct matching is as follows.

(i) ... (c). (ii) ... (d).
(iii) ... (a). (iv) ... (b).

7.2 The correct matching is as follows.

(i) ... (b). (ii) ... (c).
(iii) ... (a). (iv) ... (a).

7.3 The correct answers are as follows.

(i) Mixed reactor.
(ii) Mixed reactor.
(iii) Packed bed reactor.

7.4 The device should look something like that presented below.

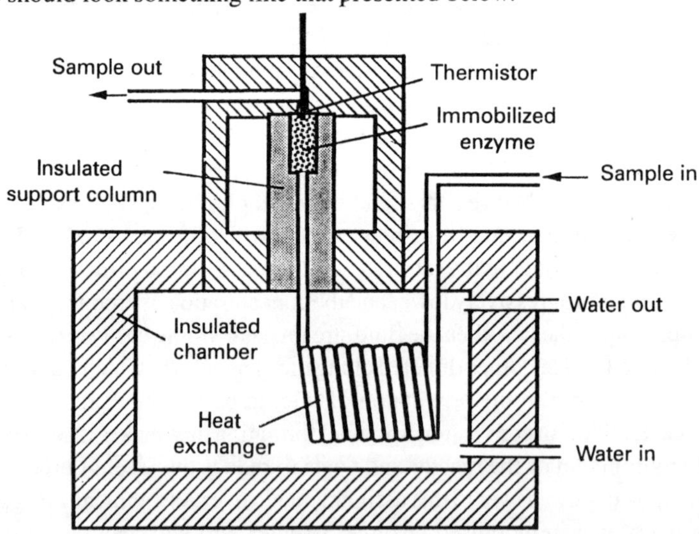

The stream of liquid (input) must at first be passed through a heat exchanger, where its temperature is adjusted to a given value. The liquid stream then passes through (or by) immobilised glucose oxidase (immobilised on a suitable support), and then bathes the thermistor. The resistance of the thermistor is measured by a suitable meter. At a set flow rate the increase in temperature of the liquid stream will be proportional to the number of glucose molecules oxidised (assuming enzyme is in excess). The increase in temperature can be related to glucose concentration by calibrating the device beforehand (noting the increase in temperature produced by glucose solutions of varying concentration).

8 Microbial biotechnology: exploitation of microorganisms for the production of chemicals

Chapter outline

8.1 Introduction

Humans have been exploiting microorganisms for many centuries for the production of fermented foods and beverages and for the making of bread. The production of alcoholic beverages became in many countries the first truly industrialised process. The technology developed for large-scale brewing was adapted for other anaerobic processes such as acetone and butanol in the early 1900s. With the discovery of penicillins and the urgent requirement for them in World War II, rapid developments were made in the 1940s and 50s in the technology of submerged fermentation of aerobic microorganisms in liquid medium under aseptic and environmentally controlled conditions. The large-scale culture of microorganisms in submerged culture in industry is loosely defined as fermentation. Fermentation processes have been/are used widely for the production of many microbial products (Table 8.1).

Table 8.1 Products from microbial fermentations

Sector	Product	Application
Organic chemicals	Ethanol	Fuel, solvent
	Butanol	Solvent
	Acetone	Solvent
Food/beverages	Amino acids	Dietary supplement, sweeteners
	Acetic acid (vinegar)	Acidulant
	Lactic acid	"
	Citric acid	"
	Mono-sodium glutamate	Flavour enhancer
	Nucleosides	"
	Yeast extract	Flavouring
	Lactic acid bacteria	Inoculant
	Yeast	"
	SCP	Protein food
	Mushrooms	Food
	Enzymes	Food processing
	Polysaccharides	Thickeners
	Alcoholic beverages	
	Fermented foods	
Health care	Vaccines	Prophylaxis
	Antibiotics	Therapy
	Vitamins	Dietary supplement
Agriculture	Insecticides	Biological pest control
	Root inoculants	Symbiotic nitrogen fixation
	Antibiotics	Animal husbandry
	Veterinary vaccines	"

In the strict sense, fermentation is a term used to denote the anaerobic utilisation of carbohydrates, whereas many industrial "fermentation" processes are in fact aerobic processes (carried out by aerobic microorganisms in an aerated environment). The vessels used in fermentation processes in liquid medium are called fermenters (or fermentors), and the technology associated with their design and operation is called fermentation technology. Waste-treatment processes, including anaerobic digestion (biogas production), are sometimes considered to be forms of fermentation as they involve the large-scale cultivation of microorganisms, but they involve different technology which is not discussed here.

In this chapter we are going to examine the design and operation of fermenters, the media which are used in industrial fermentations, the basics of product recovery, the kinetics of growth in culture systems, how culture systems can be operated to maximise output and how production organisms can be manipulated to maximise output.

8.2 The design and operation of a conventional fermenter

8.2.1 Design

Traditional fermenters used in brewing were open vats made of wood or slate or in more recent times of stainless steel. Modern brewing vessels are closed cylindrical stainless steel vessels with a conical base and dished (domed) top (Fig. 8.1).

Enclosing such vessels allows CO_2 to be collected and reused and prevents contamination of the contents from the air. External jackets allow temperature control (see Section 8.2.5).

For anaerobic processes such a simple system may be sufficient, and in the case of beer production the boiling of the medium during preparation, the low pH of the medium and the anaerobic conditions minimise contamination levels and allow non-aseptic operation. For most other processes, however, aseptic operation, aeration and environmental control are carried out.

Fermenters are tanks made from an appropriate grade of stainless steel. For industrial processes they range in volume from about 1000 litres (1 m^3) to about 1 million litres (1000 m^3). Conventional fermenters (Fig. 8.2) are cylindrical vessels with dished (domed) top and bottom. Often a circular toughened glass window (sight glass) is placed at the top of the fermenter (sealed into the fermenter wall with an air-tight [hermetic] seal), so that the contents of the fermenter can be viewed during operation.

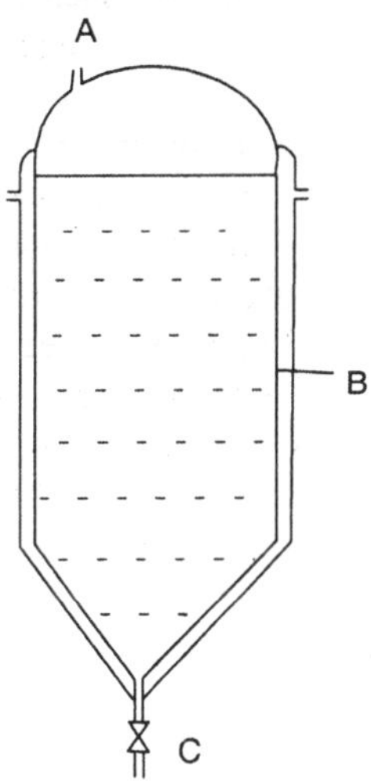

Fig. 8.1 *A modern brewing vessel.*
A – air outlet B – jacket
C – Valve on outlet

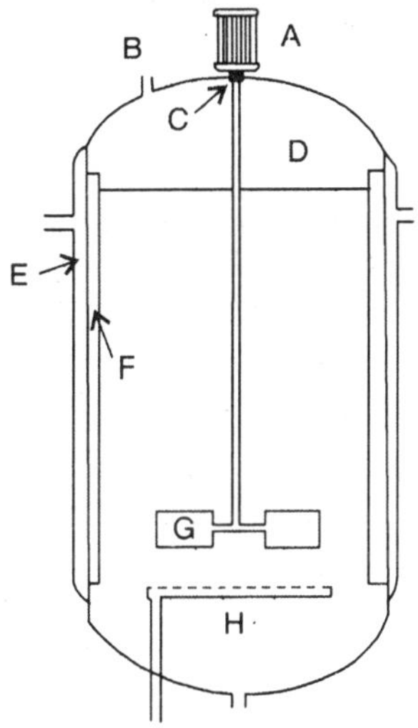

Fig. 8.2 *A conventional fermenter.*
A – motor B – air outlet C – seal/gland
D – headspace E – jacket F – baffle
G – impeller blade H – plate sparger

8.2.2 Sterilisation

Most industrial fermentation processes operate using pure cultures, and so the fermenter and its contents must be sterilised prior to inoculation with a pure culture of the production organism, and asepsis maintained thereafter.

> ***Assignment 8.1***
>
> What problems could be caused by the growth of contaminants during fermentation?

Sterilisation *in situ* involves heating the fermenter after it has been filled with the required volume of medium by injecting pressurised steam into a jacket or coil around the fermenter wall (Fig. 8.3). To ensure sterilisation, the whole system is heated to about 120°C and held at this temperature for about 20 minutes.

One problem with this method is that it does not use energy efficiently: it takes a relatively long time for the fermenter to heat up (so by the time the temperature reaches 120°C the contents may

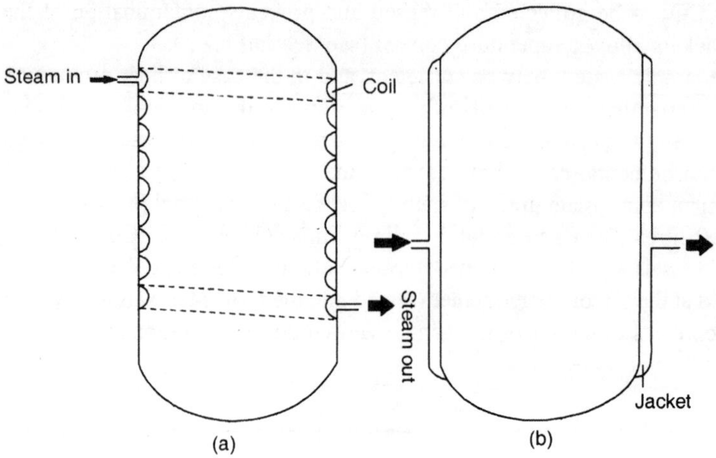

Fig. 8.3 *Methods for steam heating a fermenter. Pressurised steam is passed around (a) a coil built into the fermenter wall; and (b) a jacket (or sleeve) surrounding the fermenter.*

already be sterile) and all the while heat is being lost from the fermenter (which is not insulated). This means that heat (energy) is wasted. Another problem is that the long heating and cooling periods result in high temperatures for a relatively long time, and this can lead to destruction of vitamins and the precipitation of other medium components (when precipitated such components are insoluble and thus unavailable [unusable]).

These problems can be overcome by continuous heat sterilisation. In this method the empty fermenter is first sterilised by injecting it with pressurised steam. Medium is then run into the fermenter through a sterilising system. In this, the medium is heated rapidly to about 140°C, either by injection of pressurised steam directly into it or by passage through a heat exchanger heated by pressurised steam. Once heated, the medium passes through an insulated pipe (holding tube) where the temperature is maintained for long enough for sterilisation to occur (just one minute or so, depending on the temperature). The medium then flows through another heat exchanger system (cooled by cold water or refrigerated brine) in which it is rapidly cooled. The heat removed from the medium (heated water) can be reused in the fermentation plant.

Heating the medium to high temperatures for short periods can effectively sterilise the medium without causing undue denaturation of medium components, and can be more energy-efficient compared to sterilisation *in situ*.

Self assessment exercise 8.1

Answer true or false to each of the following:

(i) Contaminants in a fermentation medium can reduce output by using up some of the nutrients.
(ii) Contaminants in a fermentation medium can reduce output by inhibiting product formation.
(iii) Contaminants in a fermentation medium can reduce output by secreting penicillin.
(iv) Contaminants in a fermentation medium can reduce output by secreting enzymes which break down penicillin.

Self assessment exercise 8.2

Choose the correct completion to the following statement.

Sterilisation *in situ* is less efficient than continuous heat sterilisation because

(i) it uses a higher temperature for a longer time.
(ii) it uses a longer heating time during which heat is lost.
(iii) it uses a larger volume so it takes longer to cool the medium down.
(iv) it uses steam as a heating source.

Self assessment exercise 8.3

Answer true or false to each of the following:

(i) Continuous heat sterilisation causes more destruction of vitamins than sterilisation *in situ*, as the medium is heated to a higher temperature.

(ii) Continuous heat sterilisation is more efficient than sterilisation *in situ*, as less heat is used in sterilising the medium.

(iii) Continuous heat sterilisation is more efficient than sterilisation *in situ*, as the fermenter does not have to be heated by steam.

(iv) Continuous heat sterilisation causes less precipitation of nutrients than sterilisation *in situ*, as the medium is heated to a lower temperature.

8.2.3 Inoculation and sampling

Once the fermenter is filled to the required level and sterilised, it is inoculated with the production organism. The inoculum is prepared from a stock culture of a carefully selected high yielding production strain. Stock cultures are normally preserved by lyophilisation (freeze drying) or storage at very low temperatures in a deep freeze or liquid nitrogen. From a stock culture, an increasingly large volume of inoculum is prepared until the inoculum has a volume of 5–10% of the working volume of the production fermenter into which it is to be put.

After inoculation the fermenter is usually sampled regularly, so that the culture can be examined for signs of contamination and the concentration of product measured. Aseptic inoculation and sampling is performed using a steam-locked port, which is a T-shaped arrangement of piping with a valve in each arm (Fig. 8.4). When not in use the port is maintained in a sterile condition by passage of a gentle stream of pressurised steam (steam bleed) along the outer arm of the T (with the valve next to the fermenter closed). For inoculation or sampling the valve to the steam supply is shut off and the other valves opened, allowing liquid to pass in or out. After inoculation or sampling the valves are reset to their original condition and thus the system is resterilised.

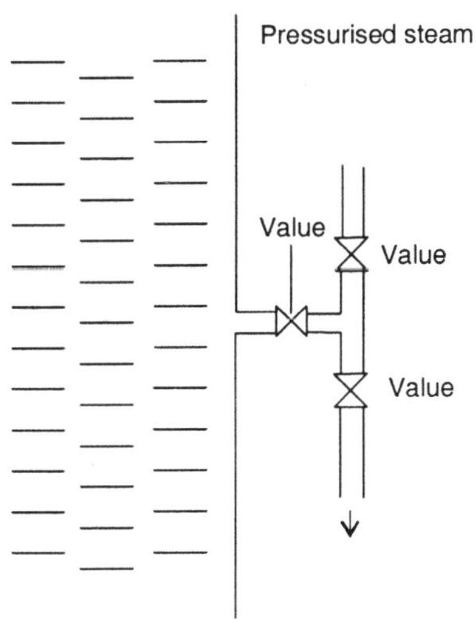

Fig. 8.4 *A steam-locked port.*

For inoculum preparation a smaller fermenter with a steam-locked port is used. The two vessels are linked by tubing or piping, which can then be sterilised by the steam bleed. After sterilisation the outer valves are shut and the inoculum pumped aseptically into the production vessel.

8.2.4 Aeration

Aeration of a fermentation medium has two functions: the supply of oxygen to the production organism and the removal of CO_2 formed as a result of aerobic respiration of the production organism. The aeration system is thus designed to promote the mass transfer of gases (gas exchange).

Oxygen is supplied from air compressed by a compressor and stored in tanks. It is passed into the fermenter through a flow-meter/regulating valve system (to regulate the flow rate), then a sterilising filter. It is introduced at the bottom of the fermenter through a sparger, which is an arrangement of pipework or a hollow plate perforated with small holes so that the air stream is introduced into the medium as small bubbles.

> ***Assignment 8.2***
>
> Explain why air is introduced into the fermenter as small bubbles.

The bubbles of air rise to the surface of the medium and take up space as they do so. At the surface the bubbles break up (disengage) releasing the gases they contain. To allow for these processes fermenters are not filled completely, but a space (the headspace) of about 20% of the capacity of the vessel is left free of medium. The actual volume of medium in the fermenter is called the working volume. Gases relea-sed into the headspace pass out of the fermenter through an air outlet at the top of the headspace. The air outlet is usually fitted with a sterilising filter to prevent aerosols of the production organism contaminating the workplace as well as to maintain asepsis within the fermenter.

The most simple aeration system involves sparging air in at the base of the fermenter (Fig. 8.5). The bubbles rise up producing upward flow of medium so this sytem is called the air-lift system. The bubbles displace medium in the liquid column up which they rise, thus reducing the density of that column of liquid. Liquid circulation is brought about by this difference in density of the liquid columns. A tube (draught tube) is usually placed in the centre of an air-lift fermenter to channel liquid flow around the system.

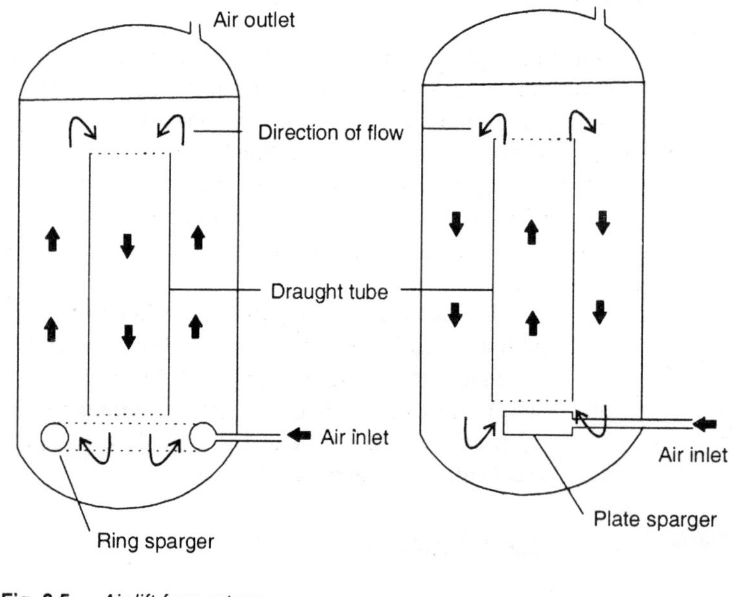

Fig. 8.5 *Air-lift fermenters.*

The advantage of the air-lift system is that it has no moving parts, making operation and maintenance easier. Its disadvantage is that the aeration capacity is relatively low. Aeration capacity of a fermenter under given operational conditions is a measure of the maximum rate at which oxygen can be dissolved into the medium. The rate at which the growing culture requires (uses) oxygen is called the oxygen demand. For aerobic organisms the aeration capacity should exceed the oxygen demand, otherwise oxygen depletion (starvation) will occur, inhibiting growth and product formation.

> ***Assignment 8.3***
>
> Will the oxygen demand be constant throughout the growth cycle in a fermentation process? Explain your answer.

The aeration capacity in an air-lift system is directly proportional to the air flow rate and the internal pressure.

In order to increase the aeration capacity of the system, the medium can be agitated (stirred) by paddles (impellers) driven by a motor (normally electric). The motor is placed directly above or below the fermenter and it drives a shaft which passes into the centre of the fermenter where impellers are attached to it (Fig. 8.2). There are various designs of impeller, a common one being the disc–turbine impeller, which consists of vertical vanes attached to a horizontal plate. There may be just one impeller (near the base) in a small fermenter, or several in larger fermenters. The aeration capacity in a stirred fermenter is proportional to the stirrer speed as well as the air flow rate and internal pressure.

Assignment 8.4

Explain how stirring brings about an increase in aeration capacity.

One major problem with this system is that it is necessary for the shaft to pass through the fermenter wall in such a way as to allow it to rotate freely while maintaining an air-tight and water-tight seal. The device used for this is called a gland or seal. Some seals use lubricated packing to produce a physical seal, whereas others rely on a steam bleed to prevent ingress of contaminants through the rotating mechanism.

Compared to an air-lift fermenter, the advantage of a stirred fermenter is that it can produce higher aeration capacities. The disadvantages are that it requires/uses energy for stirring and requires more maintenance.

Baffles in the form of narrow rectangular plates can be fixed to the internal walls of stirred fermenters (Fig. 8.2). They increase turbulence and thereby increase aeration efficiency, although this does increase the energy required for stirring.

Self assessment exercise 8.4

Answer true or false to each of the following:

(i) An air sparger's function is to increase the oxygen demand by producing smaller bubbles.
(ii) Impellers increase the oxygen demand by producing high shear forces.
(iii) Baffles increase the oxygen demand by increasing turbulence.
(iv) Air-lift fermenters produce lower oxygen demands than stirred-baffled fermenters.

Self assessment exercise 8.5

Answer true or false to each of the following:

(i) Fermenters are not filled completely with medium as room must be left for the air which is to be blown in.
(ii) The air blown into the fermenter is called the headspace.
(iii) Sight-glasses in fermenters allow you to look into the headspace.
(iv) Sight-glasses in fermenters allow the amount of medium in the headspace to be measured.

8.2.5 Control systems

During cultivation of microorganisms in a fermenter, parameters such as temperature, pH, dissolved oxygen concentration (oxygen tension) and nutrient concentration will tend to vary. In order to maintain optimal conditions for growth and product formation such variables can be monitored and controlled. Microprocessors and improved sensors enable the on-line (continuous, automated) measurement and control of important parameters in a more accurate and sophisticated way than was previously possible.

(a) Temperature
Temperature control is usually essential, and fermenters are normally equipped for both heating and cooling. If the ambient temperature is less than the required fermentation temperature then the contents of the fermenter will require heating and *vice versa*. As organisms grow, exothermic metabolic reactions may cause the temperature to rise, and cooling may be necessary.

Temperature control involves measuring the medium temperature using a thermistor (temperature-sensitive probe). This can be placed in a pocket (tube) welded into the vessel wall. The probe is linked to an actuator (control unit) which activates pumps to circulate hot or cold water around the fermenter jacket or coil as required. External temperature control involves circulating the medium from the fermenter around an external loop (pipework) in which is placed a heat exchanger.

(b) pH
As microorganisms grow they tend to change the pH of the medium, and so pH control is usually necessary.

Assignment 8.5

By what mechanisms might microbial growth alter the pH of the medium?

The pH of the medium is measured using a pH electrode. This must be in direct contact with the medium, and so is sealed through the fermenter wall using an air-tight/water-tight seal. The electrode is linked to a control unit which activates pumps to pump in acid or alkali as required. Sometimes an acid or alkaline medium component can be used, serving to correct pH and provide a source of nutrient.

(c) Oxygen tension
Oxygen tension (concentration of dissolved oxygen) in the medium is measured using an oxygen electrode, which must be in direct contact with the medium and so is sealed through the fermenter wall. The electrode is linked to a control unit.

Assignment 8.6

Which parameters could the control unit alter in order to adjust the oxygen tension?

(d) Nutrient concentration
It is often advantageous to limit the concentration of nutrients in a fermentation medium, either to prevent wastage (if the nutrient is in excess) or to prevent inhibitory effects of high nutrient concentration. Until fairly recently the on-line monitoring of nutrient levels was difficult, but recent developments in electrode technology (development of ion-selective and enzyme electrodes) mean the control of levels of nutrients such as ammonia (ammonium ions) and glucose is possible.

(e) Foam
The media used in industrial fermentations (see Section 8.5) tend to be produced from plant material, and are often rich in proteins which form stable foam when agitated (as during aeration). This means that froth or foam can build up in the headspace.

Assignment 8.7
What problems would be caused by foam in the headspace?

Chemical foam control involves addition of chemicals (antifoams) which lower the surface tension of the medium and so cause bubbles (foam) to collapse. Generally mineral oils based on silicone or vegetable oils are used as antifoams. The antifoam can be dosed (metered) in at set intervals. For a feedback control system a foam probe (metal rod) is fitted in the headspace at a certain level. When foam touches the probe, the control unit linked to it activates a pump to pump in antifoam until the foam subsides and no longer makes contact with the probe. Vegetable oils must be used where the product is destined for food use, and they have the added advantage that many organisms can also use them as sources of nutrient.

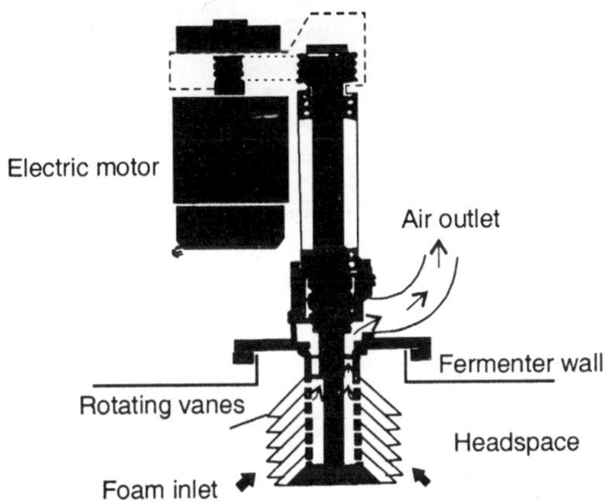

Fig. 8.6 *Mechanical foam breaker.*

Mechanical foam control uses a device called a mechanical foam breaker (Fig. 8.6). The device is positioned at the top of the fermenter (in which case the stirrer is positioned at the bottom). The electric motor drives a hollow shaft (sealed through the fermenter wall) attached to which, on the inside of the fermenter, is a stack of inverted funnel-shaped discs. To reach the air outlet the foam must pass between the rotating discs where bubbles are mechanically disrupted and the liquid thrown back into the fermenter.

Self assessment exercise 8.6

In Column A are various parameters which are controlled in fermentations. Match each with the appropriate sensor (probe) in Column B and the appropriate response mechanism in Column C.

Column A	Column B	Column C
(a) temperature	(1) metal rod	(i) activation of acid/alkali pump
(b) pH	(2) thermistor	(ii) increase/decrease in stirrer motor speed
(c) oxygen tension	(3) oxygen electrode	(iii) activation of vegetable oil pump
(d) foam	(4) pH electrode	(iv) activation of hot/cold water pump

8.2.6 Cleaning

Once fermentation is complete the fermenter is harvested (the contents are removed for processing). The fermenter is then turned round (prepared for the next fermentation). The time taken for turn round is called down time and it is essential to keep down time as short as possible as it is non-productive. Because of this, and the fact that the interior of large fermenters would be difficult to clean manually, automatic cleaning using high-pressure water jets from nozzles fixed into the fermenter is usually performed.

8.3 Variations in fermenter design

Not all fermenters are of the design described above. Some are spherical (Fig. 8.7a) with specialised stirrers and baffles to produce vigorous agitation in the spherical vessel. Loop fermenters (Fig. 8.7b) are of the air-lift type, but with one chamber (the riser) up which air and medium flow, and a separate chamber (the downcomer) down which medium returns. The expanded top of the fermenter slows the upward flow rate of medium, allowing bubbles to disengage more easily. Tower fermenters (Fig. 8.7c) are tall cylindrical towers; where air is simply bubbled up the column they are referred to as bubble–column fermenters. Usually baffles (perforated plates) are inserted at intervals along the tower.

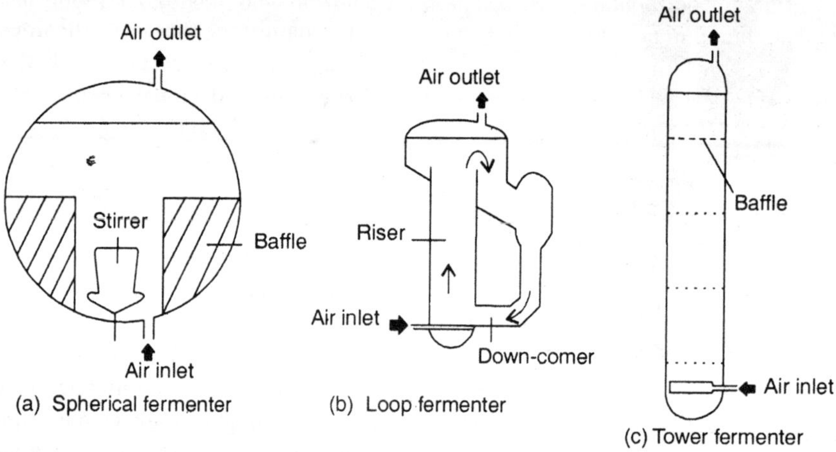

Fig. 8.7 *Variations in fermenter design.*

> ***Assignment 8.8***
>
> What function would baffles serve in a tower fermenter?

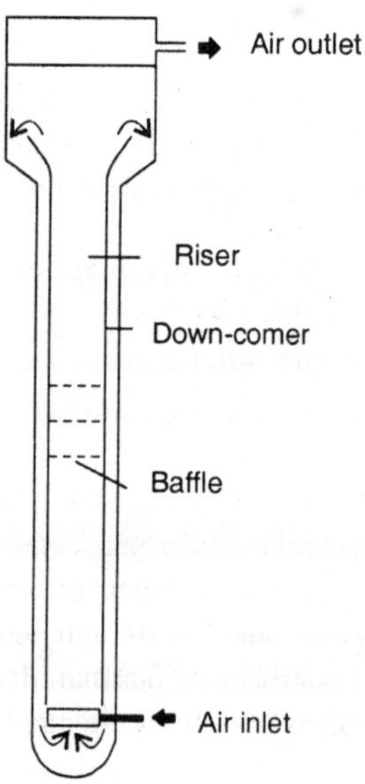

A pressure-cycle fermenter (Fig. 8.8) is a tower/loop fermenter of large dimensions: the column of medium is high (> 30 metres) so that high hydrostatic pressure is generated at the bottom of the column where air is sparged in. The high hydrostatic pressure increases the solubility of oxygen in the medium, increasing the aeration capacity. At the top of the riser the slow upward flow rate in the expanded top and lowered hydrostatic pressure promotes bubble disengagement, and CO_2 is expelled. The medium flows back down the downcomer and completes the cycle. The advantage of such a fermenter is that it produces high aeration capacities without having moving parts, but it is not possible to have small ones because a tall medium column is essential.

Fig. 8.8 *Pressure-cycle fermenter.*

Self assessment exercise 8.7

Answer true or false to each of the following:

(i) A pressure-cycle fermenter is a type of air-lift fermenter.

(ii) A loop fermenter has high aeration capacity by having high pressure on one side of the loop and low pressure on the other.

(iii) The pressure-cycle fermenter is called that because the medium cycles between high pressure at the top and low pressure at the bottom.

(iv) A loop fermenter is called that because the medium flows up one side of a loop and down the other.

8.4 Solid-substrate fermentation

Not all microbial production processes involve culture of the production organism in liquid medium; instead the organism can be grown on the surface of pieces of a solid substrate. This solid-substrate (or solid-state) fermentation (SSF) is an established traditional technology in many countries, producing commodities such as edible mushrooms, fungal-fermented foods, fungal-ripened cheeses and soy sauce. Before the development of processes in liquid culture, citric acid and some microbial enzymes were produced this way. Composting is regarded as a form of SSF.

Assignment 8.9

What advantages does SSF has over fermentation in liquid medium?

In practice SSF is generally operated as a non-aseptic process (which saves sterilisation costs), using pasteurisation of the solid substrate (usually by steaming) to reduce the number of competitors, coupled with heavy inoculation to ensure rapid establishment of the production organism, and low moisture content to depress the growth of organisms surviving pasteurisation.

It is important that the substrate is in a state such that, when it is heaped, the interstices (spaces between the pieces of substrate) remain large enough to allow air circulation (which promotes heat removal as well as gas exchange). Forced-air circulation or occasional

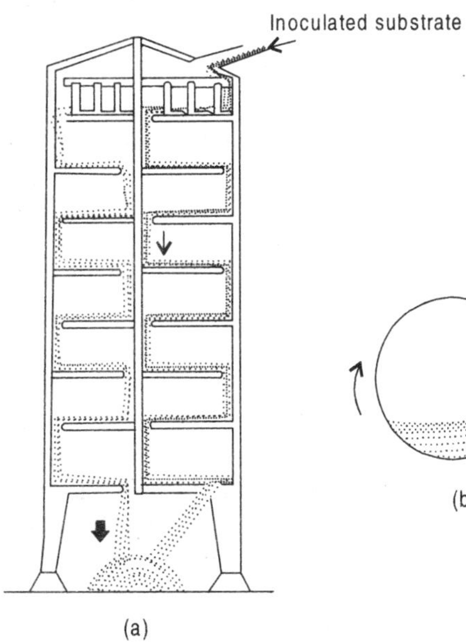

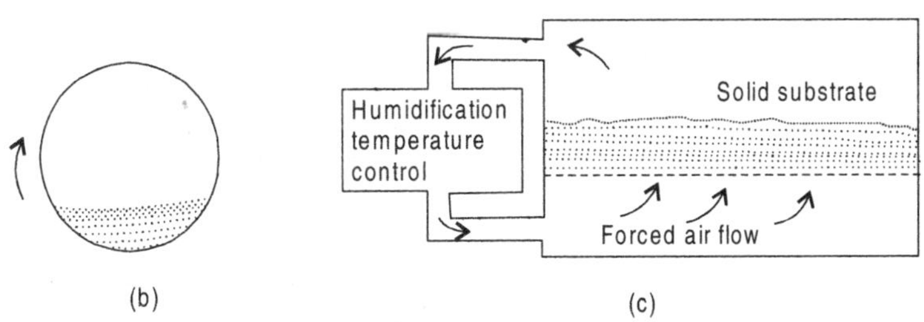

Fig. 8.9 *Reactors for solid-substrate fermentation.*
(a) Tower reactor. (b) Drum reactor.
(c) Forced-aeration reactor.

agitation of substrate may be required to maintain optimal conditions. In traditional processes shallow layers of inoculated substrate are placed in trays or baskets and incubated under conditions of controlled temperature and humidity. Agitation of substrate may be mechanical or manual. Modern processes may employ towers, rotating drums partly filled with substrate, or forced-air chambers in which humidified air is blown through the layer of substrate (Fig. 8.9). Alternatively inoculated substrate may simply be piled up in rows, and mechanically turned over at intervals.

One of the greatest challenges in biotechnology is the development of suitable technology to enable cellulosic wastes (domestic, industrial and agricultural) to be used as substrates for fermentation processes. It may be practicable and economic in the future to apply SSF to this, and large-scale mechanised processes for SSF have been developed, some of which are designed for continuous operation (Fig. 8.9a).

The major advantages of SSF compared to conventional fermentation in liquid medium are that SSF uses solids, is low-tech and low-energy technology, requires less capital investment, produces less waste effluent, and can make product recovery easier. Major disadvantages are that SSF is essentially limited in applicability to fungi (moulds) and the growth environment is difficult to regulate.

8.5 Media for industrial fermentations

In designing a medium for industrial fermentations several factors must be considered. For growth-linked products (primary metabolites) such as ethanol or citric acid, product formation is a direct result of the growth process and so a medium designed for good growth should generally be suitable for product formation. On the other hand, with secondary metabolites such as penicillin, product formation is independent of growth, so in this situation consideration must be given not only to requirements for growth but also to requirement for product formation (which may occur in the stationary phase). In addition, consideration must be given as to whether or not inducers are needed to induce product formation, and whether certain ingredients contain compounds which repress or feedback inhibit product formation. Not least, there are economic considerations: substrate costs contribute significantly to the overall production cost of a fermentation product, and so it is essential that ingredients are relatively cheap. Because of this, wastes from agriculture or other industries are especially suitable.

For most fermentation processes the production organism is heterotrophic, and so requires an organic source of carbon. For the majority of organisms the same material can also be used as a source of energy. Which substrate will be suitable will depend on the nature of the nutrient in the substrate and the physiological characteristics of the production organism: substrates such as molasses contain simple sugars (mainly sucrose) which are readily available to (in a form which is utilisable by) most microorganisms, whereas whey contains the milk sugar lactose which is only available to a limited number of organisms; locally produced starchy substrates may constitute available substrates for production organisms which are amylolytic, but for organisms which are not amylolytic such substrates will require saccharification (hydrolysis to simple sugars) in order to generate available nutrients.

Most production organisms are selected for their ability to assimilate ammonia (or ammonium salts) as a nitrogen source, as ammonia is a relatively cheap commodity in industrialised countries. Where ammonia is not readily available commercially, or for organisms not capable of assimilating it, organic forms of nitrogen are needed as substrates. For proteolytic organisms proteinaceous animal or vegetable matter may constitute an available substrate, whereas for organisms which are not proteolytic such material may require hydrolysis to convert proteins to available substrates (amino acids). Some substrates, such as distillers solubles (waste stillage from alcohol distillation) or corn-steep liquor (waste liquid from maize starch manufacture) are rich in available organic nitrogen compounds.

Many production organisms have a requirement for one or more growth factors (organic compounds such as vitamins which the organisms are unable to synthesise for themselves). These factors are usually prohibitively expensive to use in purified form, even in developed countries, and so crude sources must be sought. Yeast extract (formed by autolysis or hydrolysis of waste yeast from brewing) is, for example, a good source of certain vitamins which are commonly required as growth factors.

Self assessment exercise 8.8

Listed below all the compositions of three media A, B and C:

Medium A — A mash made from ground potatoes and fish meal (contains mainly starches and proteins but few vitamins).

Medium B — The effluent from a pineapple canning factory (containing sucrose and fructose, some vitamins but little nitrogen) supplemented with ammonium phosphate.

Medium C — A mixture of whey (containing lactose and protein) and yeast extract .

Fungus X produces a wide array of hydrolytic enzymes (amylases, proteases, pectinases and cellulases), uses most simple sugars (including lactose), requires few vitamins, and will utilise inorganic nitrogen.

Bacterium Y does not secrete hydrolytic enzymes, uses most simple sugars (except lactose) and requires organic nitrogen (it will not use inorganic nitrogen). It requires several vitamins.

On which medium/media would each organism grow well? (You can assume pH and osmolarity are suitable in all cases.)

Where the substrates are derived from plant or animal material they tend to have a suitable mineral content. Occasionally medium may need a supplement of those minerals required in greatest proportions (phosphate and sulphate).

The physiological diversity of microorganisms means that it may be possible for locally available substrates derived from renewable biomass resources to be used in any particular fermentation process. Achieving this should form an important part of any technology transfer, so that importation of commodities for fermentation substrates is avoided as far as possible.

8.6 Culture systems

In fermentation processes the microorganisms used have been carefully selected (and manipulated genetically if need be) to maximise yield of product (see Section 8.7). Fermenters used have been designed and engineered to provide optimal conditions for growth and product formation. These factors in themselves do not mean, however, that the output of product will be at optimal level for a given fermenter; it is important that the mode of operation of the fermenter (culture system used) is also such that output of product is optimised. There are four fundamental modes of culture that are applied in industrial fermentation processes: batch culture; fed-batch culture; semi-continuous culture; continuous culture. It is of value if the kinetics of growth of and formation of a particular product by a given organism in a given culture system can be understood, as this allows these processes to be modelled and thus optimised for a given set of working conditions.

8.6.1 Batch culture

Batch culture involves adding all nutrients to the fermenter prior to inoculation: no nutrients are added to, or medium removed from the fermenter during the culture phase. This system is thus a closed system in terms of nutrition.

> ***Assignment 8.10***
>
> What problems might be caused by a nutritionally-closed system?

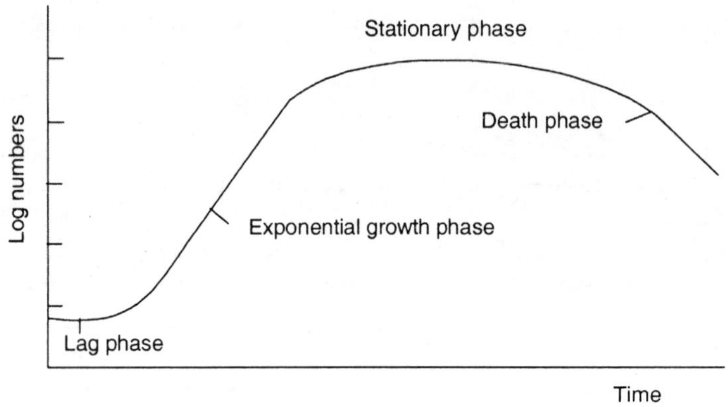

Fig. 8.10 *Growth curve for a unicellular organism grown in batch culture.*

If the biomass concentration (X), cell numbers (N), or another parameter directly proportional to biomass, are monitored throughout the growth cycle of a unicellular organism in batch culture, a plot of biomass against time can be made (Fig. 8.10). Initially there may be a lag phase, during which no growth of the organism will be evident. Lag phases occur when orga-nisms adapt to new environ-mental conditions (changes in temperature, pH, concentration of nutrients or growth inhibitors). In industrial processes it is important to minimise or avoid the lag phase as it is non-productive.

> ***Assignment 8.11***
>
> Suggest how the lag phase could be minimised or avoided.

Once any lag phase has been completed, organisms enter the phase of rapid growth. For unicellular organisms that divide by binary fission (division of the cell into two equally sized daughter cells), growth in this phase will be exponential, as outlined below.

If the initial number of cells is N_0, then after 1 generation the number will be $N_0 \times 2^1$, after 2 generations $N_0 \times 2^2$, after 3 generations $N_0 \times 2^3$ ... and so on.

Thus, in an exponentially growing culture, the number of organisms after a given time (N_t) will be given by

$$N_t = N_0 \times 2^n \text{ (where } n \text{ is the number of generations)} \qquad (8.1)$$

Microbiologists usually measure the growth rate of an organism in such circumstances in terms of the doubling time (t_d), or mean generation time (MGT), which is the time taken for a doubling in numbers or biomass to occur. However, t_d is inversely related to growth rate, and hence is not a suitable constant by which to model growth processes. The specific growth rate constant μ (pronounced mew, as in new) is a direct measure of the rate of growth of an organism, as outlined below.

If N represents the number of organisms at a given time, then the increase in numbers with time (i.e., the growth rate), (dN/dt), can be expressed as

$$\frac{dN}{dt} = \mu N \qquad (8.2)$$

where μ is the specific growth rate constant (the rate of increase in biomass per unit of biomass).

From equation 8.2, it can be shown that

$$\frac{\ln N_t - \ln N_0}{t_t - t_0} = \mu \qquad (8.3)$$

where N_0 is the biomass concentration at an initial time (t_0), N_t is the biomass concentration at a subsequent time (t_t), and ln denotes the natural logarithm.

In terms of $\log_{10}$, equation 8.3 becomes

$$\frac{\log_{10} N_t - \log_{10} N_0}{t_t - t_0} = \frac{\mu}{2.303} \quad (8.4)$$

This equation holds true for biomass concentration (X), as well as cell numbers (N).

Assignment 8.12

Using equation 8.4, suggest how μ for an organism growing in a batch culture can be calculated.

Assignment 8.13

Below are given data relating to the increase in biomass with time in a batch culture. Calculate μ from these data, using equation 8.4. Note that the units of μ are h^{-1}, so time must be expressed in hours.

Time (h)	Biomass concentration (kg dry wt m^{-3})	Time (h)	Biomass concentration (kg dry wt m^{-3})
0	0.33	24	14.66
3	0.50	28	22.60
7	0.92	30	25.95
12	2.15	32	27.08
18	5.50		

If μ for an organism is known, then useful predictions on its growth under the same conditions in batch culture can be made, using equation 8.4.

Assignment 8.14

Under given conditions, an organism grows with $\mu = 0.45\ h^{-1}$. Assuming exponential growth throughout and the same conditions, how long will it take for the population to increase from 1.30×10^4 cells m^{-3} to 2.56×10^5 cells m^{-3} under the same conditions?

The growth rate (μ) of an organism is not fixed, but can vary with environmental conditions such as temperature and concentration of particular substrates (nutrients). At low concentrations a given substrate can limit growth, according to Fig. 8.11. If the specific growth rate is calculated for each of the growth curves in Fig. 8.11, and a graph of μ against substrate concentration [S] plotted, the relationship is as outlined in Fig. 8.12.

Assignment 8.15

Summarise the effect of substrate concentration on growth rate.

The relationship between μ and [S] can be defined as

$$\mu = \mu_{max} \frac{[S]}{k_s + [S]} \quad (8.5)$$

where μ is the specific growth rate at a given [S], μ_{max} is the maximum possible growth rate, and k_s (the saturation constant) is the value of [S] which produced $\mu = 0.5\ \mu_{max}$.

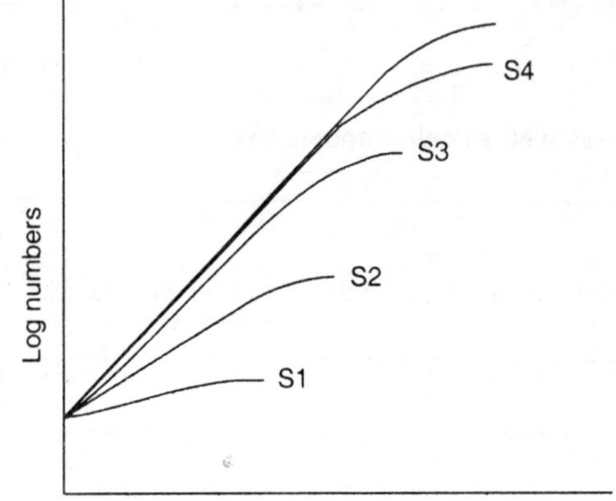

Fig. 8.11 *Growth curves for a unicellular organism grown in batch culture in media with low concentrations of a given substrate $S_1 < S_2 < S_3 < S_4 < S_5$.*

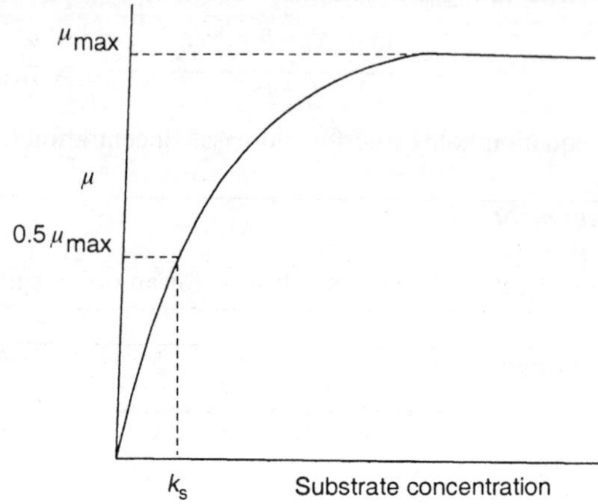

Fig. 8.12 *The effect of substrate concentration on the growth rate of a unicellular organism in batch culture.*

Assignment 8.16

An organism grows under given conditions on a given substrate with $\mu_{max} = 0.75h^{-1}$, and k_s with respect to that substrate is 0.01 kg m^{-3}. What will the growth rate of the organism be under the given conditions when the substrate concentration is 0.25 kg m^{-3}?

In a batch culture, nutrients are initially present at high concentrations ([S] >> k_s), so in equation 8.5

$$\frac{[S]}{k_s + [S]} \text{ is approximately 1 and thus } \mu \cong \mu_{max}.$$

If k_s is small, then it is only when [S] becomes small (at the end of the growth phase) that $[S]/(k_s + [S])$ becomes <1, and hence μ becomes $< \mu_{max}$.

From Fig. 8.12, it can also be observed that the substrate concentration also influences the total amount of biomass formed. The growth yield (Y_g) is the factor which relates the quantity of biomass formed from utilisation of a given quantity of substrate,

$$Y_g = \frac{X_t - X_0}{S_0 - S_t} \tag{8.6}$$

where X_0 and S_0 are biomass and substrate concentrations, respectively at an initial time and X_t and S_t are biomass and substrate concentrations, respectively, at a subsequent time.

Assignment 8.17

Below are given data relating to increase in biomass concentration of an organism with time and the concomitant decrease in a particular substrate with time. Calculate Y_g for growth of the organism on the substrate from these data.

Time (h)	Biomass concentration (kg dry weight m^{-3})	Substrate concentration (kg m^{-3})
0	0.33	55.0
3	0.50	
7	0.92	
12	2.15	
18	5.50	45.06
24	14.66	
28	22.60	
30	25.95	
32	27.08	3.56

Y_g for an organism growing on a given substrate under given conditions is fixed and is independent of time. If Y_g for an organism is known, then useful predictions on growth under the same conditions can be made, using equation 8.6.

Assignment 8.18

An organism grows on a given substrate in given conditions with Y_g = 0.45 (kg dry biomass per kg substrate). How much biomass will be produced by the utilisation of 10 kg substrate under the given conditions?

If X represents the biomass concentration at a given time, then the rate at which substrate is utilised (the change in substrate concentration [S] with time) is given by

$$\frac{d[S]}{dt} = q^X \qquad (8.7)$$

where q is the metabolic quotient for the given substrate (the rate of substrate utilisation per unit of biomass). It can be shown that

$$q = \frac{\mu}{Y_g} \qquad (8.8)$$

Thus, if Y_g for an organism is known, the value of q at a given growth rate can be calculated.

Assignment 8.19

Under given conditions 1 kg dry biomass of an organism is produced from utilisation of 2.2 kg of sucrose. What is the metabolic quotient (q sucrose) when the organism grows at a rate of μ = 0.55 h^{-1}?

For aerobic microorganisms, oxygen can be considered as an essential nutrient. Because of the relationship between the metabolic quotient, Y_g and μ (equation 8.8), if the values of q and Y_g are known for an organism growing on a given substrate under given conditions then the growth rate of the organism in culture under those conditions can be estimated by measuring the rate of utilisation of the substrate.

If the metabolic quotient for oxygen utilisation or carbon dioxide production is known for an organism under given conditions then growth rate in culture in those conditions can be estimated by measuring the rate of O_2 consumption or CO_2 production by the culture. Automated gas analysis of O_2 or CO_2

concentrations in the inlet and outlet gas streams allows O_2 consumption and CO_2 production rates, respectively, to be estimated.

From the point of view of industrial processes, important characteristics of a fermentation unit are productivity and output. The output of the system is the quantity of product produced per unit time, and the productivity is the output per unit culture volume (the quantity of product produced per unit time per unit volume). The productivity is a useful parameter by which to compare production efficiencies of systems having different volumes.

> ***Assignment 8.20***
>
> A fermenter, of working volume 40 m^3, produces Product Y in batch culture from a given substrate under given conditions. The final concentration of Product Y at the end of each batch culture (run) is 1.5 kg m^{-3}. The system is operated such that 50 runs are completed each year. Calculate: (i) the annual output of the system (Product Y produced per year in kg); (ii) the overall productivity of the system (Product Y produced in kg per year per m^3).

The final phase in industrial batch culture (Fig. 8.10) is the stationary phase, in which the growth rate (μ) is zero. Growth ceases because nutrients have become exhausted and/or growth inhibitors have accumulated. It can be seen that between this phase (where $\mu = 0$) and the phase of exponential growth (where $\mu = \mu_{max}$) there is a progressive slowing of growth rate (μ decreases with time). If such a decrease in growth rate occurs due to exhaustion of one particular nutrient, this is a manifestation of behaviour governed by equation 8.8.

For growth-linked products (including biomass) the stationary phase is non-productive (and thus a waste of time) and so the product is recovered as soon as possible after onset of this phase or even just prior to it. For secondary metabolites, product formation may continue into the stationary phase in which case fermentation periods need to be extended into it.

As outlined previously, batch culture involves not only periods after inoculation where lag phases or low-cell concentration result in little or no product formation but also non-productive down times between fermentations. Thus, the productive periods of a fermenter used batchwise may be relatively short compared to non-productive periods, and so overall output of product is low.

If the end product of the fermentation process is the biomass itself, the above considerations apply directly. If the end product is a primary metabolite (directly linked to growth) then product formation will be linked to biomass formation and substrate utilisation by a given factor, the product yield, which can be applied to the equations described above. If the end product is a secondary metabolite (unlinked to growth), then product formation may not be linked to biomass formation and substrate utilisation in the manner described above.

The advantages of batch culture are: simplicity of operation; opportunity of using down times for maintenance, lessening probability of mechanical breakdown occurring during fermentation; short fermentation period, reducing probability of mutation of the production organism or the formation of wall growth (large clumps of biomass formed by filamentous organisms on solid surfaces within the fermenter).

> ***Assignment 8.21***
>
> What problems could be caused by wall growth?

An important advantage of batch culture is that it fits in well with batch processing, which is the conventional way in which industries operate.

8.6.2 Fed-batch culture

A fed-batch culture is a batch culture in which one or more nutrient is fed into the fermenter during the fermentation period. In this way nutrients can be added at the same rate as they are used up, so excess of nutrient (and any inhibition resulting from this) can be avoided. An example of this is in the production of alkaline protease (an enzyme used in 'biological' detergents) by species of the bacterium *Bacillus*: batch feeding of nitrogen sources (ammonia, ammonium ions or amino acids) keeps these substrates at low concentrations and thus avoids the repression that these nitrogen sources exert on protease synthesis.

For fermentations with distinct phases, such as penicillin fermentation, the differing nutritional requirements in the distinct phases can be catered for by feeding in particular nutrients at the time they are required.

8.6.3 Semi-continuous culture

For products which are growth linked, such as ethanol or biomass, most of the product will be formed towards the end of the growth phase, when most growth occurs. If at the end of this phase a portion of the culture medium is removed from the fermenter and replaced by fresh medium (of identical composition and at the same temperature and pH), then growth of the organism will continue (without non-productive lag phases and growth phases). This process can then be repeated at appropriate intervals, and is known as semi-continuous culture. By using this system, down time is eliminated, lag phases and growth phases can be largely avoided, and the culture can be maintained in a condition of high biomass concentration and high rate of product formation. These factors result in higher overall outputs from semi-continuous systems compared to batch systems.

The disadvantages of semi-continuous culture compared to batch culture are that the system is technically more complex to operate and long culture periods mean that contamination, mutation, wall growth and mechanical breakdown are all more likely to occur.

Semi-continuous culture has been most widely applied in the production of ethanol.

8.6.4 Continuous culture

Continuous culture involves removing medium (culture) from the fermenter continuously and replacing this with fresh medium, usually at the same rate (so that the working volume remains constant). This is an open system (Fig. 8.13).

If the working volume of the fermenter is V m^3, and the flow rate in and out F m^{3}h^{-1}, then the dilution rate (D) is given by

$$D = \frac{F}{V} \tag{8.9}$$

The unit of D is thus per hour (h^{-1}).

In the type of continuous culture called chemostat culture, one

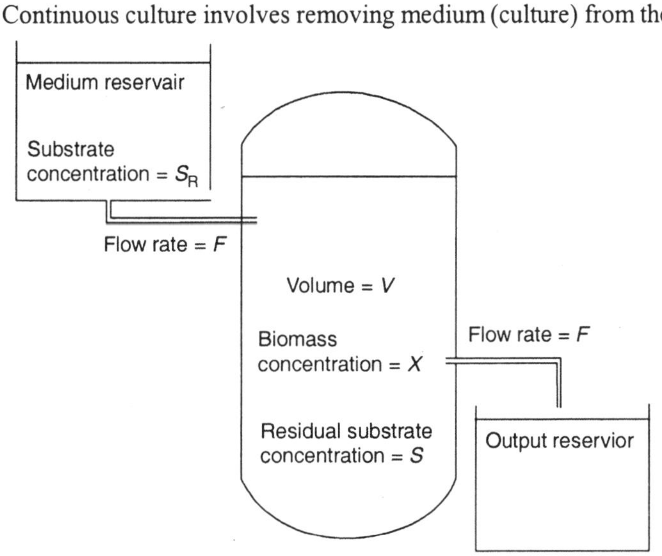

Fig. 8.13 *Chemostat mode of continuous culture.*

nutrient (usually carbon or nitrogen source) is present in the inflowing medium at a concentration low enough to limit the growth rate of the organism (Fig. 8.13 and equation 8.5). In such a system the rate of change in biomass concentration in the culture is given by the increase in biomass due to growth minus the loss of biomass carried out in the outflow. The increase in biomass per unit volume due to growth is given by μX (equation 8.2) and decrease due to loss in the outflow is DX (where D is the dilution rate of the system). Thus, in such a continuous system: rate of increase in biomass concentration

$$\frac{dX}{dt} = \mu X - DX \qquad (8.9)$$

Once the continuous system has been operating for a while a steady state develops, in which the biomass concentration remains constant ($dx/dt = 0$) (the biomass concentration is constant, so the rate of increase is zero); the amount of biomass lost from the fermenter by dilution (diluted out in outflowing medium) is replaced exactly by growth, so that

$$\mu X - DX = 0 \qquad \text{(from equation 8.10)}$$

and $$\mu = D \qquad (8.11)$$

Thus, in the steady state the growth rate is equal to the dilution rate. In this way the growth rate of the organism can be controlled: the faster the flow rate (dilution rate), the faster the growth rate. In the steady state the nutritional environment (nutrient concentration) also remains constant, making it easier to optimise conditions for product formation.

It has been demonstrated above that in the steady state in the chemostat the growth rate is equal to the dilution rate (equation 8.11). It has also been demonstrated that for a given organism in given conditions the growth rate is not infinite, but is limited to a certain maximum (μ_{max}). This leads to consideration of the effect on the culture of setting the dilution rate to a value larger than μ_{max}.

Assignment 8.22

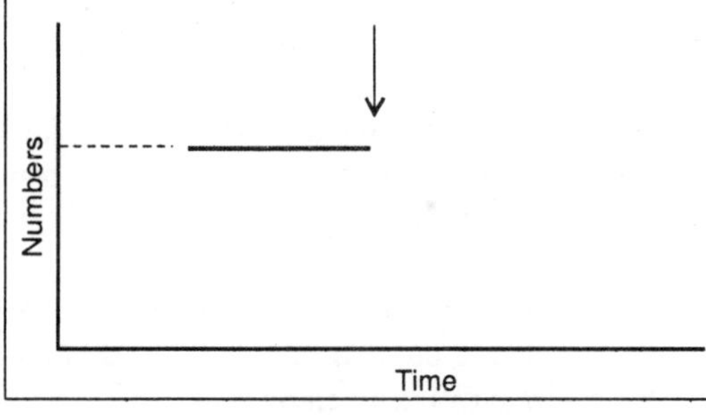

The graph shows the biomass concentration with time of an organism growing (in a steady state) in a chemostat culture at $D = 0.40\ h^{-1}$. μ_{max} for this organism under these culture conditions is $0.65\ h^{-1}$. At the time marked by the arrow the dilution rate is increased to $0.90\ h^{-1}$. Complete the graph. Remember, in simple terms the culture is being diluted at a rate faster than the maximum possible growth rate of the organism.

As continuous cultures are open systems, not all the substrate will be utilised; a proportion of the inflowing substrate will remain unused by the growing organisms and will pass out of the fermenter in the outflow (and is thus wasted).

It has been demonstrated that in the steady state in a chemostat $\mu = D$ (equation 8.11) and that

$$\mu = \frac{\mu_{max}[S]}{k_s + [S]}$$

substituting D for μ in equation 8.5 and rearranging gives

$$S = k_s\left(\frac{D}{\mu_{max} - D}\right) \tag{8.12}$$

where S is the concentration of residual substrate in the fermenter (and the outflowing medium). Thus, if the values of μ_{max} and k_s are known, then the value of S at a given dilution rate can be calculated. It is worthwhile to note that equation 8.12 does not contain the concentration of the substrate in the inflowing medium (normally denoted as S_R). This means that altering the concentration of substrate in the inflowing medium has no effect on the concentration of residual substrate in the fermenter or outflow (providing S_R is still growth-limiting).

> ***Assignment 8.23***
>
> Examine equation 8.12. Predict what the effect of an increase in the value of (i) k_s and (ii) μ_{max} would be on the value of S at a given dilution rate.

The rate of change in the residual substrate concentration in chemostat culture is given by

rate of change in residual substrate = rate of substrate input by inflow − rate of substrate output in outflow − rate of substrate consumption by organisms

or

$$\frac{dS}{dt} = DS_R - DS - \frac{\mu X}{Y_g} \tag{8.13}$$

where S_R is the concentration of substrate in inflow and S is the residual substrate concentration in the fermenter and the outflow. The equation for the rate of substrate consumption by organisms is derived from equation 8.8. Rearranging equation 8.13 gives

$$\frac{dS}{dt} = D(S_R - S) - \frac{\mu X}{Y_g}$$

In the steady state the concentration of residual substrate does not change with time, but is constant.

Thus $\frac{dS}{dt} = 0$, and so $D(S_R - S) = \frac{\mu X}{Y_g}$

and $\frac{D}{\mu}(S_R - S)\,Y_g = X$

Since in the steady state $\mu = D$, $\frac{D}{\mu} = 1$,

the biomass concentration in the fermenter and the outflow (X) is given by

$$X = Y_g\,(S_R - S) \tag{8.14}$$

Compare equation 8.14 to equation 8.6. They both signify that the biomass formed = the growth yield multiplied by the substrate used. In chemostat culture the substrate used is $S_R - S$ (the difference between the substrate concentrations in the inflow and outflow).

> ***Assignment 8.24***
>
> Consider equation 8.14. Suggest what the effect of an increase in the value of (i) Y_g, (ii) S_R and (iii) S would be on the value of X at a given value of dilution rate.

The output of biomass from a continuous-culture system is given by the rate at which medium passes out of the outflow (i.e., the flow rate, F) multiplied by the concentration of biomass in that outflow (i.e., X). Thus,

$$\text{output} = FX. \tag{8.15}$$

As $D = \frac{F}{V}$, $F = DV$, and so equation 8.15 becomes

$$\text{output} = DVX \tag{8.16}$$

The productivity of this system (output per unit volume) is thus

$$\text{productivity} \quad \frac{DVX}{V} = DX. \tag{8.17}$$

Assignment 8.25

An organism grows on a given substrate under given conditions with $\mu_{max} = 1.0\ h^{-1}$, $Y_g = 0.5$ kg dry biomass per kg substrate utilised and $k_s = 0.1\ kg\ m^{-3}$.

In chemostat culture $S_R = 1.0\ kg\ m^{-3}$, and the vessel volume = $10\ m^3$. For each value of D in the table below, calculate values for S (using equation 8.12), X (using equation 8.14) and output (DVX). Check your calculations for $D = 0.1\ h^{-1}$ with those in the table to make sure you are calculating correctly.

D (h^{-1})	S ($kg\ m^{-3}$)	X ($kg\ m^{-3}$)	output (kg dry biomass per hour)
0.10	0.011	0.495	0.495
0.20			
0.30			
0.40			
0.50			
0.60			
0.70			
0.80			
0.90			
0.95			

Plot these data on a graph, and from the graph answer the following.

(i) How do S and X vary with D?

(ii) At what value of D, is output at maximum?

(iii) What is the percentage substrate utilisation when output is at maximum?

Use all of the data above to answer the following.

(iv) How much substrate would be needed to produce 1000 kg dry biomass in the system, when it is operating with output at maximum? Remember that not all nutrient is used in biomass production; a proportion is wasted (unused in the outflow), so must be accounted for in the inflow.

(v) What would the concentration of substrate in inflowing medium (S_R) need to be in order for the biomass concentration at $D = 0.7\ h^{-1}$ to be $1.20\ kg\ m^{-3}$.

(vi) What would the biomass concentration be at $D = 0.60\ h^{-1}$ if S_R was $2.0\ kg\ m^{-3}$?.

(vii) If it takes 0.7 kg oxygen to produce 1 kg dry biomass, what would the rate of oxygen dissolution need to be in order to maintain biomass production with output at maximum? Remember that the productivity is the amount of biomass produced per unit volume per unit time (kg dry biomass m^{-3} h^{-1}). The aeration efficiency is the maximum possible rate of oxygen transfer (solution) into the culture medium (kg oxygen dissolved m^{-3} h^{-1}).

The advantages of continuous culture systems are that the non-productive periods (lag phase, growth phase and down time) are avoided when the systems are up and running, giving high outputs. The chemostat system, while up and running, has constant nutritional environment, making it easier to optimise conditions and thus maximise output.

The disadvantages of continuous culture are: the long culture period and associated problems; the complexity of operation; the failure of the system if the dilution rate is faster than the maximum growth rate of the organism; the fact that at high dilution rates a significant proportion of the substrate remains unused (is wasted).

Assignment 8.26

Suggest how a continuous culture could be operated to avoid (i) wash out and (ii) wastage of unused nutrient.

Continuous culture has been applied most widely in the production of microorganisms as food/feed protein (single cell protein, or SCP). The activated sludge effluent treatment system is a variation of it; it is a continuous process with biomass (sludge) feedback, enabling a greater volume of effluent to be treated compared to a plant of equivalent size without sludge recycle. Some continuous ethanol production processes have employed yeast recycle in order to boost output.

Self assessment exercise 8.9

Answer true or false to each of the following:

(i) Continuous cultures give higher outputs than semi-continuous cultures as they avoid down time and growth periods.

(ii) Fed-batch cultures give better control of the nutritional environment than continuous (chemostat) cultures, as medium can be fed in at controlled rates.

(iii) Mutation is more likely to occur in a semi-continuous culture than in a continuous culture because the culture period is longer.

(iv) Mutation is more likely to occur in a fed-batch culture than in a batch culture, because the culture period is longer.

(v) Wall growth is less likely to occur in a fed-batch than in a semi-continuous culture, as the culture period is shorter.

(vi) Batch cultures can give complete utilisation of substrates,whereas continuous cultures at high dilution rates cannot.

(vii) Fed-batch cultures can be used to grow up an organism in an environment where a nutrient is present at sub-optimal levels, whereas continuous (chemostat) cultures cannot.

Self assessment exercise 8.10

An organism grows in a continuous chemostat culture of 50 m^3 working volume, with sucrose as the growth limiting nutrient, at $D = 0.45\ h^{-1}$. The steady state biomass concentration is 3.74 kg dry biomass m^{-3} and residual sucrose concentration is 1.10 kg m^{-3}. The sucrose concentration in the incoming medium is 10.0 kg m^{-3}.

(i) What is the metabolic quotient?

(ii) What would the sucrose concentration in the input need to be in order for the output to be 35 kg biomass h^{-1}?

8.7 Strategies for maximising productivity

In all industrial production processes, not just fermentations, it is important to maximise productivity and output, and hence maximise profitability. Thus, all possible strategies are applied to maximising productivity and output in fermentation processes.

Self assessment exercise 8.11

Some strategies for maximising output from fermentation systems have already been described in this chapter. Name them.

8.7.1 *Examples from alcohol fermentation*

The production of *Saccharomyces cerevisiae* as an inoculum for bakers, brewers and distillers, is a well-established technology in which the yeast biomass is the required end product and as such it is important that we suppress ethanol formation (to avoid a drop in yield) by maintaining high oxygen tension. In some instances, airlift fermenters with low aeration capacity had to be used for the production of yeast biomass. In this case, fed-batch cultures in which the substrate (e.g., molasses) is fed at sub-optimal rates were used to restrict growth rate ($\mu_{max} < \mu_{max}$) and, in turn, oxygen demand. In this way, the oxygen demand was kept within the capabilities of the airlift fermenters, thus maintaining the desired aerobic conditions throughout the fermentation process.

In contrast, the production of alcohol requires anaerobic conditions and can be affected through one of two main pathways, namely: the *decarboxyclastic* and the *phosphoroclastic* pathways (Fig. 8.14). The yeast *Saccharomyces cerevisiae* employs the latter (*decarboxyclastic*) pathway that involves, in addition

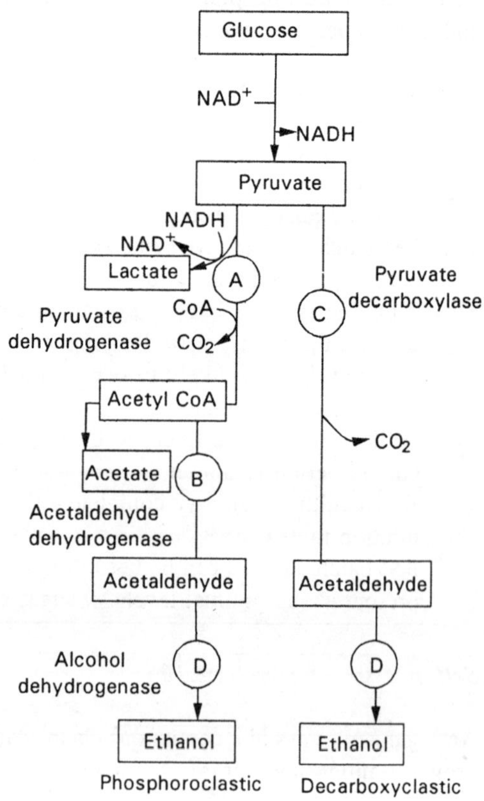

Fig. 8.14 *The metabolic routes employed by microorganisms for the production of alcohol. Note that the decarboxyclastic route employed by the yeast Saccharomyces cerevisiae involves two enzymes only, the phosphoroclastic route involves a whole host of different enzymes.*

to the glycolytic enzymes, pyruvate decarboxylase and ethanol dehydrogenase (Figure 8.14). However, any changes in the metabolic environment which renders the conditions aerobic will be accompanied by the diversion of carbon flux from pyruvate decarboxylase to pyruvate dehydrogenase to give acetyl CoA that is further oxidised through the Krebs (TCA) cycle, thus diminishing the yield of alcohol considerably. Consequently, a significant increase in growth yield is observed as more ATP, reducing power and biosynthetic precursors are made available to the organism as a result of pyruvate oxidation to CO_2 and H_2O through the enzymes of the TCA cycle and the electron transport chain. Thus, in order to maximise the conversion of carbohydrate to ethanol in the brewing processes, the medium (wort) should initially be aerated to facilitate biomass formation and once sufficient biomass is generated, aeration should be stopped. Conditions would rapidly become anaerobic and the biomass would convert the remaining carbohydrate substrates to ethanol. To this day, such practices are carried out in some brewing operations.

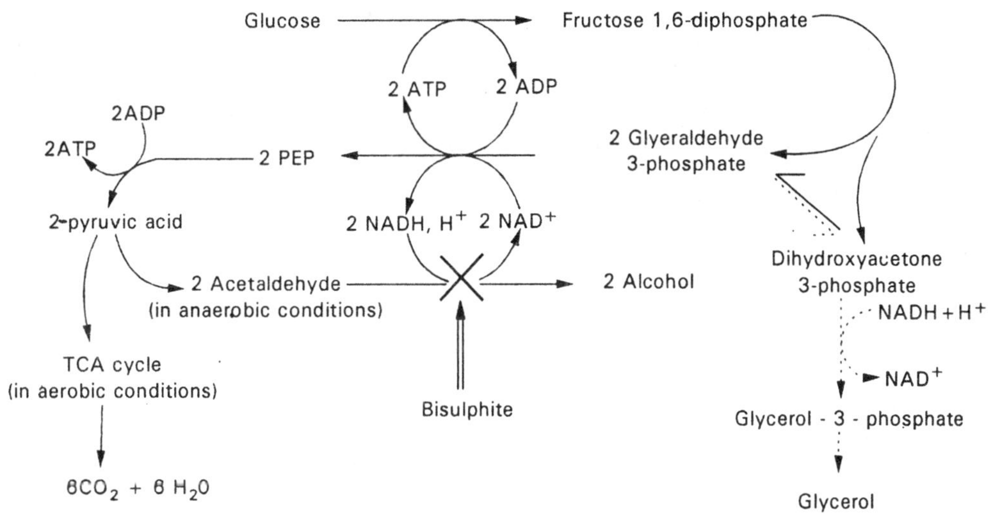

Fig.8.15 *Diversion of carbon flux from alcohol formation to glycerol production during the course of glucose fermentation by the yeast Saccharomyces cerevisiae. The addition of bisulphite arrests the available acetaldehyde thus preventing flux through alcohol dehydrogenase. The redox imbalance generated is redressed by the diversion of carbon flux to glycerol formation as depicted above.*

During World War I, a new strategy was developed in Germany for the production of glycerol to assist the war effort. When the British naval blockade prevented Germany from importing the vegetable oils necessary for the production of glycerol (which was required for the manufacture of ammunition), German biotechnologists were able to adapt yeast fermentation for the synthesis of glycerol rather than alcohol. In this case, bisulphite was added to specifically trap acetaldehyde in the form of bisulphite adduct which is not a substrate for alcohol dehydrogenase, thus blocking the reduction of acetaldehyde to ethanol (Fig. 8.15). Consequently, the intracellular level of NAD^+ drops significantly and such a dramatic change in the redox balance is redressed by the diversion of carbon flux to glycerol formation as illustrated in Fig. 8.15. Thus, flux is converted towards the formation of dihydroxyacetone phosphate and its reduction to glycerol 3-phosphate, which, in turn, is converted to glycerol through the activity of glycerol phosphatase. Although flux distribution under these circumstances favours glycerol formation, it does not support biomass formation and as such it is imperative that addition of bisulphite should be made only after sufficient biomass is generated. Through this process, *probably the first example of metabolic engineering*, Germany was able to produce well in excess of 100 tonnes of glycerol per calendar month.

When producing ethanol as a fermentation product (for use as a fuel, solvent or chemical feedstock) it is most convenient if a high concentration of carbohydrate substrate can be fermented to a high concentration of ethanol. However, both high carbohydrate concentrations and high ethanol concentrations inhibit yeast growth and ethanol production. This can be overcome by using fed-batch culture to feed in carbohydrate substrates at sub-optimal rates and using vacuum distillation to remove ethanol from the fermentation medium *in situ*.

8.7.2 Examples from amino acids production

Once an isolate that is capable of the production of the desired amino acid is isolated, the organism is then subjected to a very intensive program of random mutagenesis followed by selection of over-producers. Mutants that demonstrate higher level of amino acid production are further subjected to another round of mutagenesis until a producing strain is obtained. However, the degree of improvements in the organism's ability to produce a given amino acid may be restricted by tight regulatory control mechanisms such as feedback inhibition and repression of enzyme synthesis.

(a) Effect of feedback inhibition

Consider the following pathway that is employed by *Corynebacterium glutamicum* for the synthesis of L-lysine and L-threonine from a common intermediate, aspartic acid (Fig.8.16). During fed-batch fermentation, the production of these amino acids are very tightly regulated by feedback inhibition. To overcome such a limitation, the organism must be treated in a specific way so that such a feedback inhibition is elevated. The selection of mutants that are resistant to toxic analogues has proved very successful in the past, e.g., a mutant resistant to δ- (2amino-ethyl)-L-cysteine; a toxic analogue of lysine, is no longer subject to end product feedback inhibition

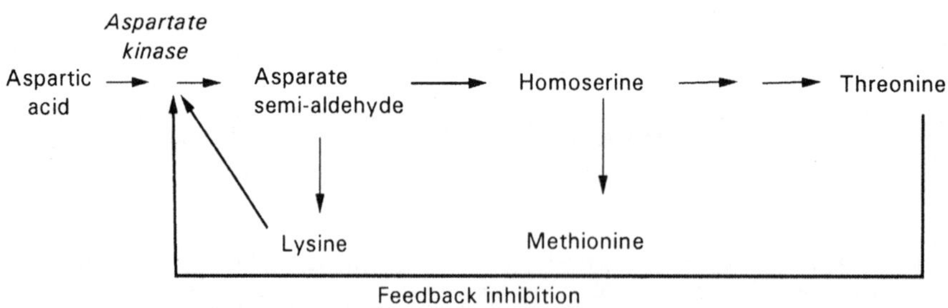

Fig.8.16 Pathway leading to biosynthesis of the amino acids lysine, methionine and threonine in *Corynebacterium glutamicum*

Each arrow in Fig. 8.16 represents a step in the pathway, and each step is catalysed by a particular enzyme. Each enzyme is, in turn, coded for by a particular gene. The end products of the pathway, L-threonine and L-lysine, exert feedback inhibition of their synthesis by allosteric interaction with the enzyme aspartate kinase, which catalyses convertion of aspartic acid into the first intermediate in the pathway. In other words, the end products bind to aspartate kinase enzyme molecules and thereby induce conformational changes in the active site of the enzyme molecules which lead to a decrease or abolition of enzyme activity.

> ***Assignment 8.27***
>
> Predict the effect on lysine and threonine synthesis of altering the gene coding for aspartate kinase (by mutation or site-directed mutagenesis) in such a way that the allosteric site of the enzyme is altered and the enzyme thereby rendered insensitive to allosteric inhibition by lysine and threonine.

> ***Assignment 8.28***
>
> Predict the effect on lysine and threonine synthesis of altering the gene coding for the enzyme homoserine dehydrogenase in such a way that the active site of the enzyme is altered and the enzyme thereby rendered inactive (incapable of converting aspartate semi-aldehyde to homoserine).

> ***Assignment 8.29***
>
> Predict the effect on lysine and threonine synthesis of altering the structure of the gene coding for the enzyme homoserine dehydrogenase in such a way that the enzyme structure is altered to render the enzyme sensitive to feedback inhibition by threonine?

(b) Effect of repression

Corynebacterium glutamicum also synthesises L-tryptophan as outlined in Fig. 8.17. Tryptophan not only exerts feedback inhibition on the enzyme DAHP synthase (which catalyses DAHP synthesis from PEP) but also exerts repression (at the genetic level, Chapter 4) of synthesis of the enzyme anthranilate synthase (which catalyses tryptophan synthesis from chorismic acid) and feedback inhibition on this same enzyme.

Attempts can be made to overcome feedback inhibition using the strategies described above. Repression, on the other hand, results from the binding of tryptophan to its specific repressor molecule (coded for by the specific regulator sequence of DNA) to form a complex; the complex binds to the specific operator sequence of DNA and thereby prevents transcription of the anthranilate synthase gene (enzyme synthesis is blocked).

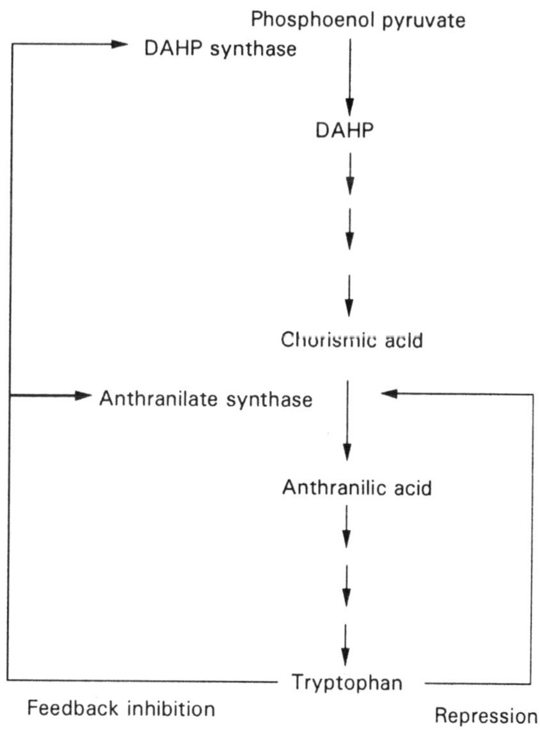

Fig. 8.17 *Pathway leading to biosynthesis of tryptophan in Corynebacterium glutamicum.*

The inhibitory effects of tryptophan on its own synthesis are exacerbated by the fact that tryptophan (in common with other anabolic metabolites) is not excreted by normal cells, thus leading to relatively high intracellular concentrations (of the inhibitory product). The strategy to overcome this in fermentations has been to induce organisms to excrete the product during fermentations, by inducing in them non-lethal alterations in membrane structure such that the cells become 'leaky'. Methods used have been mutation (of genes involved in membrane synthesis), starving the organism during fermentation of precursors essential for membrane synthesis (using fed-batch cultures to feed in precursors at sub-optimal rates) and addition to the fermentation medium of antibiotics or other agents that disrupt membrane structure and function.

> ***Assignment 8.30***
>
> What changes would need to be made in the DNA sequences of the organism in order to overcome repression by tryptophan on anthranilate synthase?

(c) Modulation of gene expression and gene dosage

The genes involved in the production of amino acids as well as antibiotics are generally clustered on the genome. The regulation of such a cluster of genes has been proved to be controlled positively in response to a specific activator molecule. Since the role of such an activator molecule is to activate the expression of the various genes in the cluster, cloning and over expression of such a *'master gene'* is likely to increase flux to product formation (Hunter, 1999). Cloning and subsequent expression of *'antisense genes'* is a very useful strategy for turning-off or diminishing the activity of harmful or unwanted enzymic activities. In this case, an antisense gene (a complementary sequence of DNA in which the sense strand rather than the antisense strand is transcribed, thus giving a mRNA sequence that is complementary to the mRNA sequence of the native gene, i.e., the gene that needs to be turned off) is cloned and subsequently expressed. The antisense mRNA will bind the native mRNA, thus forming a double-stranded molecule that can not be translated thus preventing the formation of the enzyme.

(d) Diversion of carbon flux from waste products to product formation

During industrial fermentation, a large fraction of the primary carbon source is wasted in the form of acetate and to a lesser extent lactate and α-ketoglutarate. For example, during glucose fermentation, some 20% of the total carbon is lost in the form of acetate. Inactivation of acetate kinase and/or phosphotransacetylase through mutation or the use of active site-directed inhibitors prevents flux from acetating excretion and this, in turn, is accompanied by a significant increase in the efficiency of carbon conversion to biomass formation (El-Mansi and Holms, 1989). Acetate, particularly its undissociated form, is a potent uncoupler of oxidative phosphorylation and as such conditions that promote flux to acetate excretion will, in turn, diminish flux for product formation as a large fraction of the carbon source would have to be diverted towards maintenance requirements. The diversion of carbon flux at a branch point using a stepwise increase in enzymic activity to divert flux through anaplerotic routes to diminish flux to wasteful flux is a valid approach for the direction of carbon flux away from co-product formation and into synthesis of the desired product (El-Mansi and Stephanopolous, 1999).

(e) Prevention of futile cycling

Flux through the central metabolic pathways and, in turn, product formation is influenced by a whole host of different factors including futile cycling. For example, any transient increase in the intracellular concentration of pyruvate in *Escherichia coli* may trigger a futile cycle in the central metabolic pathways at the junction of phosphoenolpyruvate (PEP). In addition to pyruvate dehydrogenase and malic enzyme, futile cycling at this junction involves the activities of PEP carboxykinase and pyruvate kinase (ATP generating reactions) on one hand and PEP-synthetase and PEP carboxylase (ATP utilising activities) on the other. The operation of this futile cycle may, in addition to ATP wastage, adversely affect the adenylate energy charge of the cell. Transient increąse in the intracellular level of pyruvate may also trigger futile cycling in other organisms. For example, *Klębsiella aerogenes* triggers a futile cycle involving two enzymes, the first is the NAD$^+$dependent pyruvate reductase which catalyses the transformation of pyruvate to D-lactate and the second is the FAD-dependent D-lactate dehydrogenase which completes the cycle as illustrated in Fig.8.18.

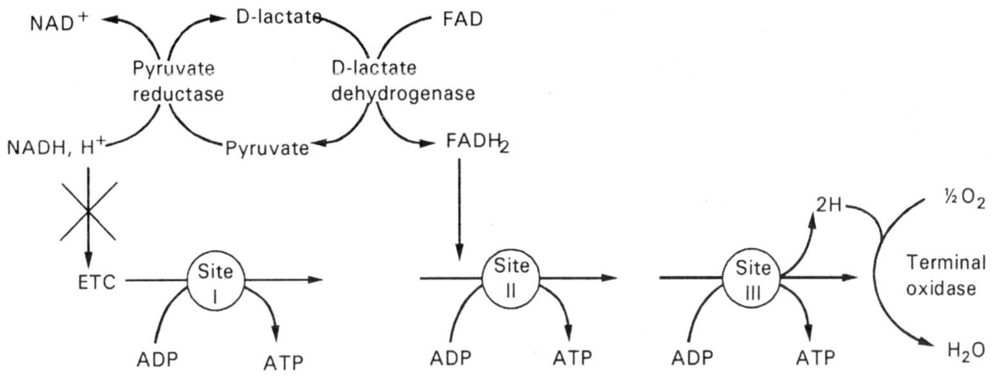

Fig.8.18 The role of pyruvate reductase and D-lactate dehydrogenase futile cycle in bypassing the first phosphorylation site in the electron transport chain.

Flux of carbon through these two enzymes provides a futile cycle in which NADH, H$^+$ is oxidised at the expense of reducing FAD, thus bypassing the first phosphorylation site in the electron transfort chain as illustrated above. It follows therefore, that any changes in the metabolic environment which lead to a transient increase in the intracellular level of pyruvate may trigger the aforementioned futile cycles and, in turn, energy dissipation.

The presence and function of glutamine synthetase and glutaminase in ammonia-limited cultures represents yet another possible futile cycle (Fig. 8.19) in

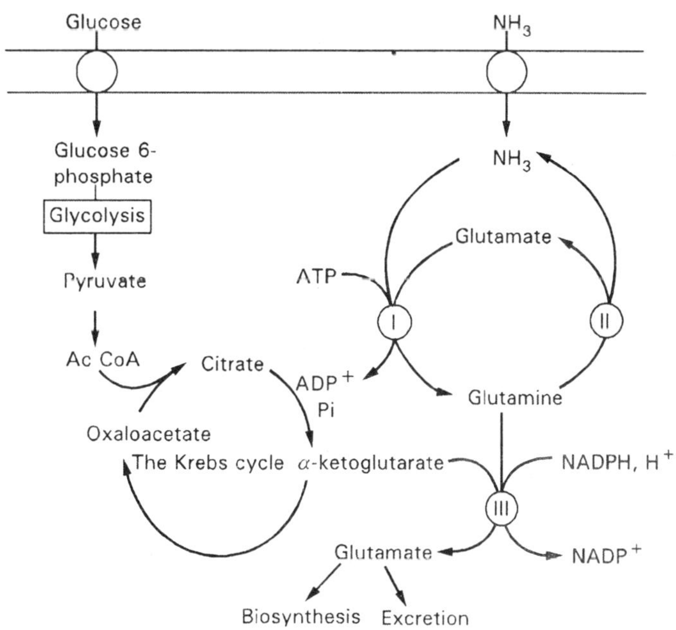

Fig. 8.19 *Metabolic inter-relationships among various enzymes of central and nitrogen metabolism, highlighting the role of futile cycling in energy dissipation as a result of flux through glutamine synthetase (I) and glutaminase (II). The biosynthetic role of glutamate dehydrogenase (III) is also highlighted.*

which the synthetase generates glutamate while the other affects its hydrolysis with a net loss of one molecule of ATP per turn. Such a futile cycle may play a significant role in the production of single cell protein (see the case study on SCP).

> ***Self assessment exercise 8.12***
>
> Consider the following pathway.
>
> ```
>     (Enz 1)          (Enz 2)        (Enz 3)
> A     →      B     →      C     →      D
>           ↓  (Enz 4)
>               (Enz 5)
>           E       →   F
> ```
>
> Control of the pathway is as follows:
> Product F exerts feedback inhibition on activity of Enzyme 1;
> Product D exerts repression on synthesis of Enzyme 2.
>
> Answer true or false to each of the following.
>
> (i) Synthesis of Products D and F could be stimulated by altering the gene that codes for Enzyme 1, so that allosteric site of the enzyme becomes insensitive to interaction with Product F.
> (ii) Synthesis of Product F could be stimulated by altering the gene that codes for Enzyme 2 so that the allosteric site of the enzyme becomes insensitive to interaction with Product D.
> (iii) Synthesis of Product D could be depressed by altering the gene that codes for Enzyme 4 so that the active site of the enzyme becomes altered and the enzyme rendered inactive.
> (iv) Synthesis of Product F could be stimulated by adding excess of D to the fermentation medium.

8.7.3 *Optimisation of phenylalanine production: A case study*

Assuming that no unknown by-products are being formed and that the metabolic routes and the precursors required are well defined, the theoretical yield coefficient for the production of any given product can be calculated stoichiometrically. For example, in *E. coli*, the biosynthesis of phenylalanine is affected through the common aromatic pathway leading to chorismate, which is then converted to phenylalanine through the activities of three different enzymes. The precursors required for phenylalanine biosynthesis are therefore phosphoenolpyruvate (PEP), erythrose 4-phosphate (E4-P), glutamate, NADPH and ATP. Stoichiometrically, the overall reaction for phenylalanine formation can be presented as follows:

$$\text{2PEP} + \text{E4-P} + \text{2NADPH} + \text{ATP} + \text{glutamate} \rightarrow \text{phenylalanine} + CO_2 + \text{2-oxoglutarate} + \text{2NADP}^+ + \text{ADP} + \text{4Pi} \quad (8.18)$$

Assuming that E4-P is generated through the activities of the transketolase and transaldolase reactions of the hexose monophosphate shunt (HMS) and that PEP is generated through glycolysis and is required for the uptake of glucose and its activation to glucose-6-phosphate, the overall reaction of phenylalanine production can be described as follows:

$$\text{X-glucose} \rightarrow \text{2 PEP} + \text{E4-P} + \text{X-pyruvate} \quad (8.19)$$

Taking the number of carbon atoms in each of the above molecules into consideration, the carbon balance of the above equation can be resolved as follows:

$$X(6) \rightarrow 2(3) + 1(4) + X(3) \quad (8.20)$$

This, in turn, can be solved as follows:

$$X(6) \rightarrow 6 + 4 + X(3)$$

$$X(6) - X(3) = 6 + 4$$

$$X(3) = 10$$

$$X = 10/3 = 3.33 \text{ moles of glucose.}$$

According to the above stoichiometry, 3.33 moles of glucose (MW 180) are required for the production of one mole of phenylalanine (MW 165.2) giving a maximum theoretical yield co-efficient of 0.275 g of phenylalanine per g of glucose.

The production of phenylalanine is a classic example in which metabolic and genetic engineering can be used to improve the efficiency of carbon conversion to desirable end product. Considering the reaction highlighted in equation 18.18, which clearly indicates that the biosynthesis of phenylalanine requires the participation of 2 moles of NADPH which are generated through the oxidation of pyruvate via the Krebs' cycle at the level of isocitrate dehydrogenase (ICDH), the overall reaction can be rewritten as follows:

$$\text{X-glucose} \rightarrow 2\ \text{PEP} + \text{E4-P} + 2\ \text{Pyruvate} \quad (18.21)$$

Solving the above equation in the same way as for equation 18.19, the carbon balance under these circumstances will become 2.66 moles of glucose per mole of phenylalanine and the overall stoichiometry will therefore be:

$$2.66 \text{ moles of glucose} \rightarrow \text{One mole of phenylalanine} + 7 \text{ moles of } CO_2$$

Further consideration of equations 18.20 and 18.21 reveals that 1.33 moles of pyruvate are wasted through oxidation to CO_2. Diversion of pyruvate to PEP formation rather than CO_2 production becomes therefore a suitable target for intervention by metabolic engineers and its effect on the efficiency of carbon conversion to phenylalanine can be assessed in the same manner as described above. In this case the diversion of 0.66 mole of pyruvate to PEP is required for every mole of phenylalanine produced and such a transformation can be achieved through the activity of the enzyme PEP synthase. Since this enzyme is not expressed in *E. coli* during growth on glucose, the use of recombinant DNA technology to construct a recombinant strain that is capable of expressing such an enzyme affords a good example in which metabolic and genetic engineering work in concert.

Metabolic engineering of a given pathway, such as that described above, requires the determination of metabolic flux distribution among various enzymes of the pathway. However, a complete description of metabolic flux requires knowledge of the kinetic properties of each enzyme together with the intracellular level of various intermediates. Such information is, of course, rarely available and recent efforts to overcome or bypass those limitations have been based on steady-state analysis and the metabolic requirements for cell growth. Once the information required has been gathered, an appropriate mathematical framework can be used to calculate the metabolic flux distribution among various enzymes of the network. This approach has proved valuable in improving the efficiency of carbon conversion in amino acid production.

8.7.4 Single cell protein (SCP) production: A case study

One of the first applications of genetic engineering to increasing output of a fermentation process was in the production of bacterial biomass as a source of feed (fodder) protein. The organism used, *Methylophilus methylotrophus*, was an obligate methylotroph (as its name suggests) — using only 1-C compounds such as methanol as carbon and energy source. In the production process the organism was grown in a methanol : salts : ammonia medium. The organism assimilates ammonia as outlined in Fig. 8.20.

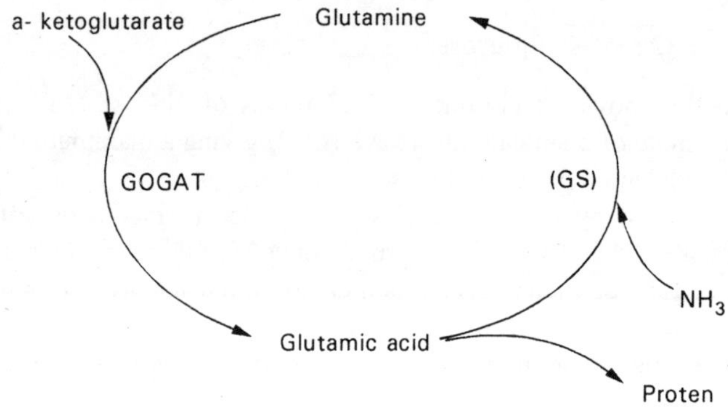

Fig. 8.20 *Assimilation of ammonia in Methylophilus methylotrophus.*

Ammonia is accepted by glutamic acid in an energy (ATP)-consuming step and is converted to glutamine. This reaction is catalysed by glutamate synthetase (GS) and can be written as:

$$\text{glutamic acid} + NH_3 + ATP \xrightarrow{GS} \text{glutamine} + ADP + Pi \qquad (18.22)$$

Glutamine then reacts with α-ketoglutarate to form 2 molecules of glutamic acid. This reaction is catalysed by glutamate oxoglutamate aminotransferase (GOGAT) and can be written as:

$$\text{glutamine} + \alpha\text{-ketoglutarate} + NAD(P)H \xrightarrow{GOGAT} 2 \text{ glutamic acid} + NAD(P)$$

One molecule of glutamic acid is recycled to accept ammonia and the other becomes available for protein synthesis.

Other bacteria such as *E. coli* assimilate ammonia by incorporating it directly into α-ketoglutarate in a reaction catalysed by glutamate dehydrogenase (GDH). This reaction can be written as:

$$\alpha\text{-ketoglutarate} + NAD(P)H + NH_3 \xrightarrow{GDH} \text{glutamic acid} + NAD(P)$$

This bypasses the energy-consuming GS step of the GS/GOGAT pathway (equation 18.22) used by *Methylophilus methylotrophus* and is thus more energy efficient.

The strategy was to use genetic engineering to introduce a more efficient GDH system from *E. coli* into *Methylophilus methylotrophus*. The strategy proved worthwhile as the yields of the engineered organisms were about 5% greater than the parent organism in culture.

Assignment 8.31

Select the correct completion to the following statement.

It is necessary to introduce the GDH-recombinant plasmid into a GOGAT-deficient mutant

strain of *M. methylotrophus* because:
(i) GDH would otherwise compete with GOGAT for ammonia;
(ii) GOGAT would otherwise compete with GDH for α-ketoglutarate;
(iii) GDH would otherwise compete with GOGAT for glutamine.

Summary

The examples discussed in this section illustrate how several strategies have been applied to maximising output from fermentation processes: from empirical optimisation of culture environment to manipulation of chemical and physical conditions based on sound understanding of the effect on cell biochemistry and physiology; from alteration of genotypes of production organisms by the non-specific methods of mutation to the highly specific methods of site-directed mutagenesis and genetic engineering. Underpinning these strategies is the advancement in fundamental knowledge of: bioreactor design; the kinetics of growth and product formation in various culture systems; the metabolic pathways of production organisms, the mechanisms by which they are controlled, and the methods by which the control mechanisms can be manipulated to advantage at the genetic level. It is also worthwhile to remember that the advances discussed above are themselves underpinned by advances in computer technology.

8.8 Downstream processing (product recovery and processing)

The efficient and economic fermentation of a product is not the end of a successful production process: the product must be recovered from the fermentation medium, purified if necessary and processed and packaged with equal efficiency and economy. Product recovery and purification is known as downstream processing. Although summarised very briefly here, the technology associated with downstream processing is just as important as technology associated with the fermentation process itself, and the operation of any fermentation production process integrates both technologies.

The one factor that distinguishes processing in fermentation industries from that of chemical industries, is that in fermentation processes the concentration of product at the final stage of synthesis (end of fermentation) is always relatively low.

Assignment 8.32

Why do fermentations yield relatively low concentration of product?

Downstream processing operations are summarised in Fig. 8.21. If the product is volatile it may be possible to distil it directly from the harvested culture without pretreatment. Distillation can be carried out at reduced pressure (to lower the temperature required for volatilisation) and continuous stills are available. It may be possible to distil the product directly from the fermenter at reduced pressure during fermentation, so preventing accumulation of product, and any inhibition high product levels may cause (as previously described for ethanol).

For other processes the first step in downstream processing is usually biomass separation from the culture medium (spent medium). If the product is the biomass (SCP or vaccines), the biomass is recovered for processing and the spent medium is discarded (and may be useful as substrate for other processes such as biogas generation). If the product is intracellular (such as vitamins and some enzymes) the biomass must be recovered so it can be disrupted (disintegrated) in order to release the product before further

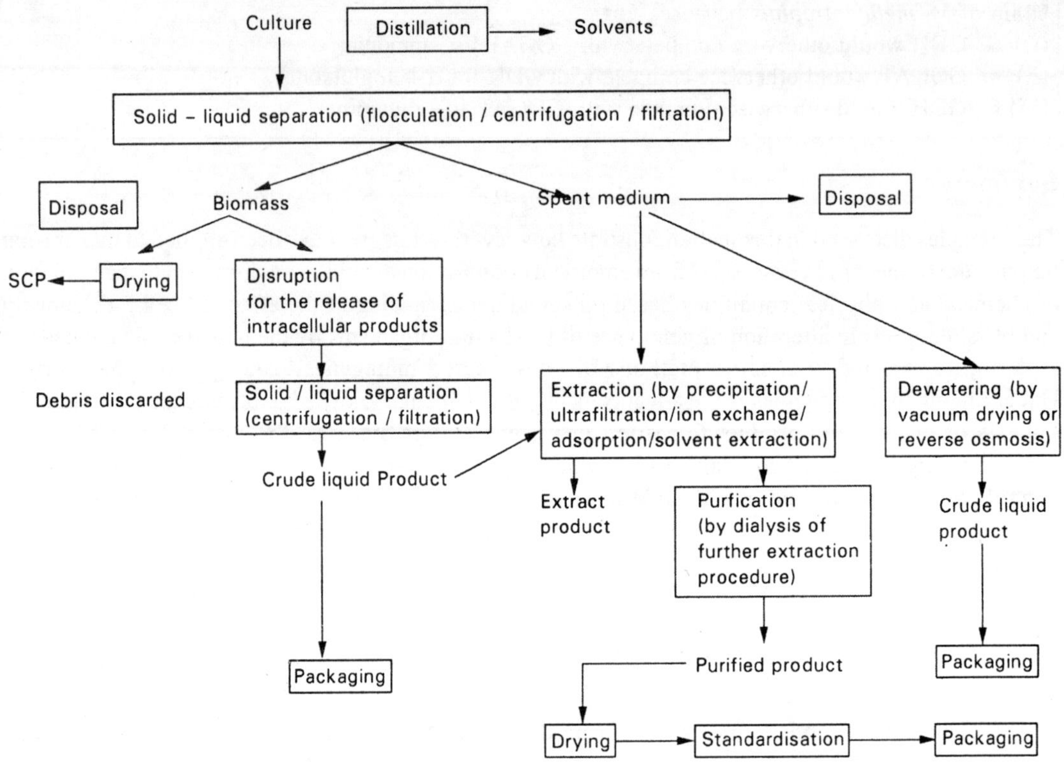

Fig.8.21 *The stages in downstream processing.*

purification. If the product is extracellular (such as antibiotics or some enzymes) the biomass is removed from the medium as it would clog up equipment used in subsequent processing: the clarified spent medium is then processed and the biomass discarded (and perhaps used as food/feed). Biomass separation usually involves the use of continuous centrifuges or continuous filters (rotary vacuum filters).

Rather than process a large volume of spent medium with very low concentration of product, it can be more efficient and economic to reduce the volume by removing water and so concentrate the product. This process is achieved by vacuum drying or reverse osmosis, and is termed dewatering.

Extraction procedures vary according to the physico-chemical nature of the molecules of the product, and to some extent on the preference of the manufacturer. For intracellular products the procedures are basically the same, after the cellular debris resulting from disintegration has been removed by centrifugation and/or filtration.

With SSF the product can be extracted with water or other solvent from residual solids, and the solution/suspension formed in the ways described above. One advantage in this instance is that it may be possible to efficiently extract the product in a relatively small volume of solvent, making further processing easier.

8.9 Concluding remarks

The ability to grow microbes on a large scale has led to the establishment of important industries, both old and new. Anaerobic processes, based largely on traditional methods, provide fermented foodstuffs, organic chemicals and fermented beverages. The economic value of these products is considerable (about 15,000 million $US in 1980 for alcoholic beverages worldwide), and in Scotland whisky production has been over a long period the most important sector of manufacturing industry in economic terms. Social impacts of alcoholic beverage consumption can be considerable, and not always beneficial. The products of the first large-scale aerobic processes, antibiotics, revolutionised treatment of bacterial diseases and altered attitudes to them, and contributed (along with vaccines) to increased life expectancy.

The efficiency of fermentation processes has benefitted greatly from advances in the design and operation of bioreactors and downstream processing equipment and genetic manipulation, much of which has been made possible by advances in computer technology. In addition, recent developments in understanding the biochemistry of industrially important organisms has dramatically extended our understanding of the biosynthetic pathways and the regulatory control mechanisms employed by microorganisms for the production of various metabolites. The new trends in molecular genetics and metabolic engineering which take into account the regulatory control mechanisms employed by microorganisms for the partition of carbon flux at various branch points in the metabolic network are expected to increase the efficiency of any given biotechnological process, thus advancing the frontiers of fermentation technology. Recent developments in genetic engineering and protein engineering herald new possibilities for the fermentation industry, with the production of non-microbial products using genetically engineered organisms and of "designer" proteins geared to a specific application. Major sectors of industry such as bulk chemicals, fine chemicals, food, health care and environmental management can expect to benefit from all these developments. With depletion of fossil fuel reserves at some future date, the production of liquid fuels and organic chemicals by fermentation of substrates from plant biomass may become the accepted way of producing liquid fuels and many organic chemicals, turning biomass producers of tomorrow into the equivalent of the oil producers of today.

The scenario in developing countries will depend on many factors. The availability of plant biomass as substrate is crucial to the development of a fermentation industry. Any technology transfer must consider the appropriateness of the technology, as well as environmental, economic and social issues. Patenting of processes and organisms has sometimes been seen as a barrier to developments in developing countries, and the currently confused state of patenting issues surrounding new technologies such as genetic engineering exacerbates this situation. However, moves such as the establishment of various research centres dedicated to biotechnological problems show the determination of developing countries to share in the fruits of new developments.

8.10 References and suggested further reading

1. *American Chemistry Society Symposium* (1990). Downstream processing and bioseparation: recovery and purification of biological products.
2. Atkinson, B. and Bernard, G. (1991). *Biochemical engineering and biotechnology handbook.* Macmillan.
3. Bastin, G. and Dochain, D. (1990). *On-line estimation and adaptive control of bioreactors.* Elsevier.
4. BIOTOL Series (1992) — *Bioprocess technology: modelling and transport phenomena.*
   — *Operational modes of bioreactors.*

— *Bioreactor design and product yield.*
— *Product resources in bioprocess technology.*
— *In vitro cultivation of microorganisms.*

Butterworth-Heineman, Oxford.

5. Bu'lock, J. and Kristiansen, B. (Eds.) (1987). *Basic biotechnology*. Academic Press, London.
6. Bushell., M.E. (Ed.) (1988). Computers in fermentation technology. *Progress in industrial microbiology*, **25**.
7. Crueger, W. and Crueger, A. (1990). *Biotechnology: a handbook of industrial microbiology.* W.H. Freeman and Co., New York.
8. Da Silva, E.J., Dommergues, Y.R., Nyns, E.J. and Ratledge, C. (1987). *Microbial technology in the developing world.* Academic Press, London.
9. El-Mansi, E.M.T. and Holms, W.H. (1989). Control of carbon flux to acetate excretion during growth of *Escherichia coli* in batch and continuous cultures. *J. Gen. microbiol.* **135**, 2875–2883.
10. El-Mansi, E.M.T. and Stephanopolous, (1999). Flux control analysis: basic principles and industrial applications. In *Fermentation microbiology and biotechnology: modern concepts.* (E.M.T. El-Mansi and C.F.A. Bryce, editors). Taylor and Francis. London, in the Press.
11. Fish, N.M., Fox, R.I. and Thornhill, N.F. (1989). *Computer applications in fermentation technology: modelling and control of biotechnological processes.* Elsevier.
12. Ghose, T.K. (1989). *Bioprocess engineering*. Ellis Horwood.
13. Hunter, I.S. (1999) Microbial synthesis of commercial products and strain improvement. In *Fermentation microbiology and biotechnology: modern concepts* (E.M.T. El-Mansi and C.F.A. Bryce, editors). Taylor and Francis, London, in the Press.
14. Lee, J.M. (1992). *Biochemical engineering*. Ellis Horwood.
15. McNiel, B. and Harvey, L.M. (1990). *Fermentation: a practical approach*. IRL Press.
16. Stanbury, P. (1995). *Principles of fermentation technology*. Elsevier Sciences, Amsterdam.
17. Sinclair, G.C. and Kristiansen, B. (1987). *Fermentation kinetics and modelling*. Taylor and Francis.
18. Smith, J. (1988). *Biotechnology*. Edward Arnold, London.
19. Trevan, M.D. *et al.* (Eds.) (1987). *Biotechnology: the biological principles*. Open University Press, Milton Keynes.
20. Ward, O.P. (1989). *Fermentation biotechnology*. Open University Press, Milton Keynes.

8.11 Responses to assignments

8.1 Contaminants may compete with the production organism for nutrient (substrate), leaving reduced quantity of substrate for growth of the production organism, and so resulting in reduction of product formation. Contaminants may produce a metabolic product which inhibits growth of the production organism and/or product formation. Contaminants may produce an enzyme which degrades the product.

8.2 Aeration involves the exchange of gases between the air and the medium. This exchange occurs at the interface between the two phases (the bubbles' surface). The smaller the bubbles are the greater their surface–volume ratio becomes, and the more efficient gas exchange will be.

8.3 The oxygen demand is directly proportional to the metabolic rate of the cells in a culture and the cell concentration. Thus, oxygen demand increases in line with biomass (microbial cell) concentration in the rapid phase of growth and slows down as the growth rate slows towards the stationary phase.

8.4 Gas exchange is enhanced by producing small bubbles of air. Stirring produces high shear forces in the medium which break up larger bubbles, maintaining small bubble size. Gas exchange is also enhanced by long residence time of bubbles (the time it takes for the bubbles to disengage after introduction into the medium). The swirling motion of the medium caused by stirring increases the residence time of bubbles. In these ways stirring increases aeration capacity.

8.5 Many microorganisms produce organic acids or CO_2 as end products of metabolism. Many also use alkaline components of the medium (such as ammonia) as nutrients. These activities tend to lead to decrease in pH. Many microorganisms use acidic components of the medium (such as citrate or acid phosphates) as nutrients. Ammonia may be liberated as a result of protein metabolism. These activities tend to increase the pH.

8.6 The aeration capacity is proportional to air flow rate and stirrer speed: increasing either of these will tend to increase oxygen tension and *vice versa*.

8.7 Air is being blown into the fermenter and it comes out through the air outlet. Foam can fill the headspace and be blown out of the outlet. This would contaminate the workplace and also result in loss of medium (loss of product). For these reasons foam must be controlled.

8.8 Baffles increase turbulence in each compartment of the fermenter and increase residence time of bubbles. Both these effects increase aeration capacity.

8.9 The major advantage is that SSF can make use of substrates that are not easily converted to liquid form, for example starchy or pectinaceous materials tend to form viscous solutions when heated and cellulosic material is insoluble. The conversion of such materials to liquid medium can involve time, effort and expense.

8.10 Obviously with such a system enough nutrients must be added prior to inoculation to sustain growth throughout the complete culture phase. This means that excess nutrient is initially present and this may inhibit or repress product formation. Towards the end of the growth phase, when most growth occurs and much of the product may be formed, the nutritional environment changes rapidly as nutrients become depleted.

8.11 Lag phase occurs because organisms have to adapt to new conditions after subculture, or rejuvenate themselves after a period of senescence. By growing the inoculum up under identical conditions (of medium composition, temperature and pH) as those in the production fermenter, and ensuring that at the time of inoculation the inoculum is in the rapid growth phase, the lag phase can be largely avoided.

8.12 If the biomass concentration, or cell concentration, or another parameter directly related to these is measured over time in the culture, and the values obtained are converted to $\log_{10}$ and plotted on a graph against time, then the slope of the graph in the straight line portion of the graph (when growth is exponential) will yield the value of m.

8.13 If the values for biomass concentration (X) are converted to $\log_{10}$ and plotted against time, a straight line portion of the graph will be apparent between 3–28 hours. Choose a point at the start of this straight line portion (X_0) and read off the time (t_0). For instance if t_0 = 7 hours, X_0 = 0.92 kg m^{-3}. If t_t = 24 hours, N_t = 14.66 kg m^{-3}.

Substituting in equation 8.4, $$\frac{\log_{10}(14.66)-\log_{10}(0.94)}{24-7}=\frac{\mu}{2.303}$$

or $$\frac{1.166 - (-0.027)}{17} \times 2.303 = \mu$$

or $$\frac{1.193}{17} \times 2.303 = \mu$$

Thus $\mu = 0.16\ h^{-1}$

8.14 Substituting in equation 8.4, $$\frac{\log_{10}(2.56 \times 10^5) - \log_{10}(1.30 \times 10^4)}{t_t - t_0} = \frac{0.45}{2.303}$$

Thus $$\frac{t_t - t_0}{1.294} = \frac{2.303}{0.45}$$

or $$t_t - t_0 = \frac{2.303}{0.45} \times 1.294$$
$$= 6.6 \text{ hours.}$$

Thus it will take 6.6 hours for the biomass to increase by the stated amount.

8.15 At low substrate concentrations the substrate concentration is proportional to growth rate (as the substrate concentration increases the growth rate increases). Above a certain substrate concentration, increase in substrate concentration does not affect growth rate, so that growth rate in these conditions is at a maximum.

8.16 Substituting in equation 8.5

$$\mu = 0.75 \times \frac{0.25}{0.01 + 0.25}$$
$$= 0.75 \times \frac{0.25}{0.26}$$
$$= 0.75 \times 0.96$$

Therefore, $\mu = 0.72\ h^{-1}$.

8.17 Using equation 8.6, $X_0 = 0.33$ and $X_t = 27.08$

$$Y_g = \frac{27.08 - 0.33}{55.0 - 3.56}$$
$$= \frac{26.75}{51.44}$$
$$= 0.52 \text{ kg dry biomass per kg substrate.}$$

8.18 Using equation 8.6, $$0.45 = \frac{X_t - X_0}{10}$$

$$0.45 \times 10 = X_t - X_0 \text{ or } X_t - X_0 = 4.5.$$

Thus, 4.5 kg dry biomass will be produced from 10 kg substrate.

8.19 Using equation 8.8, $q = \frac{0.55}{1/2.2}$

$$= \frac{0.55}{0.45}$$

Thus q = 1.22 kg substrate per hour per kg biomass.

8.20 (i) The output of Y from each run will be $VX = 40 \times 1.5 = 60$ kg. 50 runs are completed each year so the annual output of $Y = 60 \times 50 = 3{,}000$ kg or 3.0 tonnes.

(ii) The overall productivity $= \frac{\text{output}}{\text{working volume}} = \frac{3000}{40} = 75$ kg of Y m^{-3} $year^{-1}$

8.21 The large clumps of biomass can break away into the medium. This can change the rheology (flow characteristics) of the medium (affecting aeration capacity) or lead to clogging of pipes or valves (making sampling and harvesting difficult). The interior of large clumps may become starved of nutrient or oxygen leading to lowered product yield and output.

8.22 Organisms cannot grow as fast as they are removed by dilution, so the biomass concentration will decrease (exponentially) until it completely disappears from the fermenter. This phenomenon is referred to as washout. The dilution rate at which wash out just occurs is referred to as the critical dilution rate (D_{crit}).

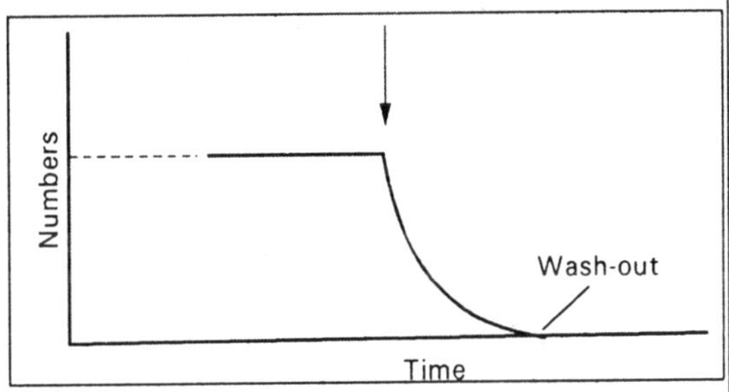

8.23 k_s is directly proportional to S, thus an increase in k_s will result in an increase in S at a given dilution rate. μ_{max} is inversely proportional to S, thus an increase in μ_{max} will result in a decrease in S at a given dilution rate.

8.24 (i) Y_g is directly proportional to X, so an increase in the value of Y_g leads to an increase in X. (ii) The same is true for S_R. (iii) An increase in S leads to a reduction in the value of S_R which in turn leads to a decrease in X (see also Response 8.23 above for the influence of k_s and μ_{max} on X).

8.25 (i) By plotting a graph it can be seen that the residual substrate concentration (S) increases from a low level as the dilution rate increases, until at high dilution rates the rate of increase becomes greater. Because S and X are inversely related, the change in X with D is a mirror image of that for S, with a decrease in X with increase in D, and rapid decrease at high dilution rates.

(ii) The output increases upto $D = 0.7\ h^{-1}$, after which it declines rapidly. Maximum output occurs at $D = 0.7\ h^{-1}$.

(iii) At $D = 0.7\ h^{-1}$, the residual substrate concentration is 0.23 kg m^{-3}.
Thus the substrate used is $S_R - S = 1 - 0.23 = 0.77$ kg m^{-3}.

Percentage utilisation is thus $\frac{S_R - S}{S_R} \times 100 = \frac{0.77}{1} \times 100 = 77\%$.

Thus, at optimum output substrate utilisation is 77%.

(iv) As $Y_g = \dfrac{\text{biomass formed}}{\text{substrate used}}$, substrate used $= \dfrac{\text{biomass formed}}{Y_g}$

Thus, substrate needed to produce 1000 kg biomass $= \dfrac{1000}{0.5} = 2000$ kg.

However, it has been calculated above that at $D = 0.7\ h^{-1}$ substrate utilisation is only 77% (23% is unused and passes out in the outflow). Thus at $D = 0.7$, the quantity of substrate needed to produce 1000 kg of biomass would be

$$2000 \times \frac{100}{77} = 2597 \text{ kg.}$$

(v) As $Y_g = \dfrac{\text{biomass formed}}{\text{substrate used}}$, substrate used $= \dfrac{\text{biomass formed}}{Y_g}$

Thus, substrate used to produce 1.20 kg biomass m^{-3} = (1.20/0.5) = 2.4 kg m^{-3}. The amount of substrate $(S_R - S)$ needed to produce a biomass concentration of 1.20 kg m^{-3} is thus 2.4 kg m^{-3}; thus, $(S_R - S)$ must equal 2.4 kg m^{-3}.

As at 0.7 h, S is 0.23 kg m^3,

$$S_R - 0.23 = 2.4$$

$$S_R = 2.4 + 0.23 = 2.63 \text{ kg } m^{-3}$$

Thus, for X to be 1.20 kg m^{-3} at $D = 0.7\ h^{-1}$, S_R would have to be 2.63 kg m^{-3}.

(vi) At $D = 0.6\ h^{-1}$, S is 0.15 kg m^{-3}.

As $X = Y_g (S_R - S)$

$$X = 0.5\ (2.0 - 0.15)$$

$$= 0.5\ (1.85)$$

$$= 0.925$$

Thus, at $D = 0.6\ h^{-1}$, the biomass concentration would be 0.925 kg m^{-3}

(vii) The productivity of the system at D = 0.7 h^{-1} is given by DX or $0.7 \times 0.385 = 0.269$ kg $m^{-3}\ h^{-1}$.
The quantity of oxygen needed to support this biomass production would thus need to be $0.269 \times 0.7 = 0.188$ kg $O_2\ m^{-3}\ h^{-1}$.

8.26 (i) If the biomass is recovered from the output stream (for example by centrifugation) and returned to the fermenter, washout will be prevented. Incidentally, with biomass feedback the fermenter can be operated at higher dilution rates and with higher outputs than would otherwise be possible.

(ii) Wastage of unused nutrient can be overcome by recycling the medium from the output stream after separation of the biomass. This also saves water. Such a procedure may not be possible if the medium contains antagonists to growth or product formation.

8.27 If the aspartate kinase molecule could be altered in such a way as to make it insensitive to feedback inhibition, the effect would be to stimualte synthesis of lysine and threonine.

8.28 If the homoserine dehydrogenase molecule is rendered inactive, the effect will be to block synthesis of homoserine from aspartate semi-aldehyde. The effect of this will be to stimulate synthesis of lysine, as carbon flux is diverted away from homoserine formation and into lysine synthesis. Synthesis of homoserine, and thereby synthesis of threonine and methionine, will be prevented; thus these mutant organisms will require homoserine (or threonine and methionine) in the growth medium if they are to grow well, as they are no longer able to synthesise these amino acids for themselves. Such mutants would be referred to as threonine/methionine auxotrophs.

8.29 If homoserine dehydrogenase was rendered sensitive to feedback inhibition by threonine the effect would be to depress synthesis of threonine and stimulate synthesis of lysine.

8.30 Tryptophan synthesis is prevented by the binding of tryptophan to its specific repressor molecule to form a complex, and the binding of the complex to the operator thereby preventing transcription of the anthranilate synthase gene. This could be prevented by either altering the DNA sequence of the regulator so that the repressor molecule formed from it was altered in structure so that it no longer bound to tryptophan to produce a complex, or by altering the DNA sequence of the operator so that the tryptophan: repressor complex could no longer bind to it.

8.31 The correct answer is (ii). Both GDH and GOGAT use α-ketoglutarate as a substrate. α-ketoglutarate is directed only through the GDH step.

8.32 Microorganisms have to be grown in relatively dilute solutions of nutrient, as high concentration of nutrient can inhibit growth due to generation of high osmotic pressure, or these nutrients or metabolites produced as a result of their utilisation (perhaps the product itself) can inhibit growth and/or product formation by repression or feed back inhibition of enzymes or by other inhibitory effects on the cell. While excess nutrient can be prevented to some extent by the use of fed-batch culture, the accumulation of inhibitory metabolites cannot be prevented. For these reasons dilute nutrient concentrations must be used, and as this nutrient is the source of both biomass and metabolites it stands to reason that product concentration will also be low.

8.12 Solutions to self assessment exercises

8.1 The correct answers are as follows.
- (i) True. (ii) True.
- (iii) True (if the production organism is sensitive to penicillin).
- (iv) True (if the product was penicillin).

8.2 The correct completion is (ii).

8.3 The correct answers are as follows.
- (i) False — continuous heat sterilisation produces less destruction of vitamins than sterilisation *in situ*.
- (ii) True.
- (iii) False — the fermenter has to be sterilised before the medium is run in.
- (iv) False — continuous heat sterilisation does cause less precipitation of nutrients than sterilisation *in situ*, because the overall heating is less. The heating temperature is, in fact, higher.

8.4 The correct answers are as follows.
- (i) False. (ii) False. (iii) False. (iv) False.

All these would be true if 'oxygen demand' was replaced with 'aeration capacity'. Oxygen demand is a measure of how much oxygen the organisms in a culture need.

8.5 The correct answers are as follows.
   (i) True.
   (ii) False — the headspace is the space above the medium.
   (iii) True.
   (iv) False — gas is in the headspace, not medium.

8.6 The correct matching is as follows.
   (a) (2) (iv) (b) (4) (i)
   (c) (3) (ii) (d) (1) (iii).

8.7 The correct answers are as follows.
   (i) True.
   (ii) False — the pressure in loop fermenters does not differ in this way.
   (ii) False — pressure is high at the bottom, not at the top.
   (iv) True.

8.8 *Fungus* X — should grow well on all media. The enzymes it produces should allow it to grow on medium A. It should grow on medium C, using protein and yeast extract as a source of nitrogen.
Bacterium Y — would not grow well on any of the media. It would not grow well on medium A as it does not secrete enzymes to break starch and proteins down. It would not grow well in B as it cannot use inorganic nitrogen. It cannot grow well in C as it cannot use lactose (in whey).

8.9 The correct answers are as follows.
   (i) False — semi-continuous cultures also avoid down time and growth periods.
   (ii) False — continuous cultures give best control of the nutritional environment.
   (iii) False — both culture systems have long culture periods.
   (iv) False — the number of generations would be the same (for a given amount of nutrient) — it just takes longer with a fed-batch culture.
   (v) True. (vi) True.
   (vii) False — continuous cultures in the chemostat use growth limiting (sub-optimal) nutrient concentrations.

8.10 (i) The metabolic quotient $(q) = \dfrac{\mu}{Y_g}$.

The growth rate is equal to D (equation 8.11), so $\mu = 0.45\ \text{h}^{-1}$.

Y_g can be calculated from equation 8.14, as

$$3.75 = Y_g\ (10.0 - 1.10).$$

Thus $$\frac{3.75}{(10.0-1.10)} = Y_g = 0.42 \text{ kg biomass per kg sucrose.}$$

Thus, the metabolic quotient $(q) = \dfrac{0.45}{0.42} = 1.07$ kg sucrose per hour per kg biomass.

(ii) The output of the system is given by DVX (equation 8.16).
Thus, for output of 50 kg biomass h^{-1}, $35 = 0.45 \times 50 \times X$.

or, $$\frac{35}{0.45 \times 50} = X = 1.56$$

For X to be 1.56 kg m^{-3}, according to equation 8.14, $1.56 = 0.42\,(S_R - 1.10)$

or, $$\frac{1.56}{0.42} + 1.10 = S_R = 4.81 \text{ kg m}^{-3}$$

Thus, for the output to be 35 kg h^{-1}, the incoming sucrose concentration would need to be 4.81 kg m^{-3}

8.11 Strategies to increase output that have been discussed already are:

(i) using pure cultures of the production organism;

(ii) formulating medium to stimulate product formation while avoiding inhibition of product formation;

(iii) minimising non-productive periods;

(iv) optimising the chemical and physical environment;

(v) using semi-continuous, continuous or continuous cultures with biomass recycle where possible.

8.12 The correct answers are as follows:

(i) True.

(ii) False. Product D does not inhibit activity of enzyme 2, thus this strategy does not apply.

(iii) False. Blocking activity of enzyme 4 would stimulate synthesis of D.

(iv) True.

9 Bioprocess technology: exploitation of animal cells

Chapter outline

9.1 Relevance of animal cell culture

Many highly valuable substances used in medicine and research are based on naturally occurring biomolecules that are produced by animal cells or cells genetically engineered using recombinant DNA techniques. The complex structure of these biomolecules, resulting from special post-translational modifications such as formation of glyco- and lipoproteins, carboxylation and phosphorylation, correct folding or disulphide bonding, determines their biological activity. Up to now such modifications can only be performed by the natural producers: the mammalian cells. Over the past several years the spectrum of products from cell culture has rapidly expanded. The success of genetic engineering of mammalian cells has led to increased productivity of cells as well as to new products. Table 9.1 gives examples of the use of mammalian cell products in the medical field.

The increasing demand for these substances necessitates the application of new and large-scale bioreactors for animal cell culture. The requirements imposed by such production scales often result in nutrient and cell concentrations far higher than those in tissue culture flasks.

This implies that cell culture technology, having been for a long time more art than science, must exploit the increased knowledge in biological, chemical and technical disciplines to design a scientific-based process for controlled large-scale cultivation. The aim is to achieve maximum productivity by ensuring predictable and optimal cellular function.

Table 9.1 Some examples of animal cell culture products of medical value

Products	Applications
Plasminogen Activators	
Urokinase-type plasminogen activator	Acute myocardial infarction, acute stroke,
Tissue-type plasminogen activator	pulmonary embolism, deep vein thrombosis
Recombinant plasminogen activator	
Erythropoietins	
Erythropoietin-a	Anaemia resulting from cancer and chemotherapy
Erythropoietin-b	Anaemia secondary to kidney disease
Vaccines	
HIV vaccines (gp120)	AIDS prophylaxis and treatment
Malaria vaccines	Malaria prophylaxis
Polio vaccines (Salk, Sabin)	Poliomyelitis prophylaxis
Human Growth Hormones	Human growth deficiency in children, trauma-
hGH	related infections, renal cell carcinoma
Somatotropin	Chronic renal insufficiency, Turner's syndrome
Monoclonal Antibodies, Therapeutic	
Anti-lipopolysaccharide	Treatment of sepsis
Murine anti-idiotype/human B-cell lymphomas	B-cell lymphoma
Monoclonal Antibodies, Diagnostic	
anti-fibrin 99	Blood clot (imaging)
Tcm-FAb (breast)	Breast cancer
PR356 CYT-356-in-111	Prostate adenocarcinoma

9.2 Historical development

The first recorded attempt to grow animal cells in culture is attributed to Ross Harrison in 1907, who was able to cultivate frog embryonic nerve cells using the hanging drop technique. This work was extended by Borrows and others, who established techniques for cultivation of a wide range of mammalian cells. Good proliferation of cells was obtained by using a mixture of chick embryo extract and plasma. The extract provided both growth factors and nutrients, whereas the fibrin clot in plasma served as an anchor for cell attachment.

The use of trypsin for cell disaggregation from tissue explants was a major innovation to obtain cell suspensions, which allowed single-cell cultures, thus distinguishing the technique of "cell cultures" from "tissue cultures". The discovery of antibiotics in the late 1940s led to the development of improved cell-culture techniques, reducing the risk of microbial contamination which was a major problem with the use of biological fluids and extracts. During this period, many human carcinoma cell lines such as the HeLa cell line were isolated and grown in culture. Ender's discovery in the late 1940s, that viruses can be propagated in cell cultures and used as vaccines, led to the development of large-scale animal cell culture. Polio vaccine production in the 1950s was the beginning of animal cell culture as a developing technology. Further milestones were the development of the first synthetically-based culture medium by Earle and co-workers and the development of microcarriers for large-scale anchorage-dependent cell culture.

9.3 Basics of mammalian cell culture

9.3.1 Establishment of cells in culture

Assignment 9.1
What is meant by the following terms: *in vitro/in vivo*; anchorage-dependence; cell suspension; confluency; contact inhibition; monolayer; passage; primary culture; cell line?

The first step in establishing cells in culture is to dissociate organs or tissues into a single-cell suspension. Such a suspension can be used as a primary culture or as a starter culture to attempt to initiate a continuous, i.e., immortal, cell line (Fig. 9.1).

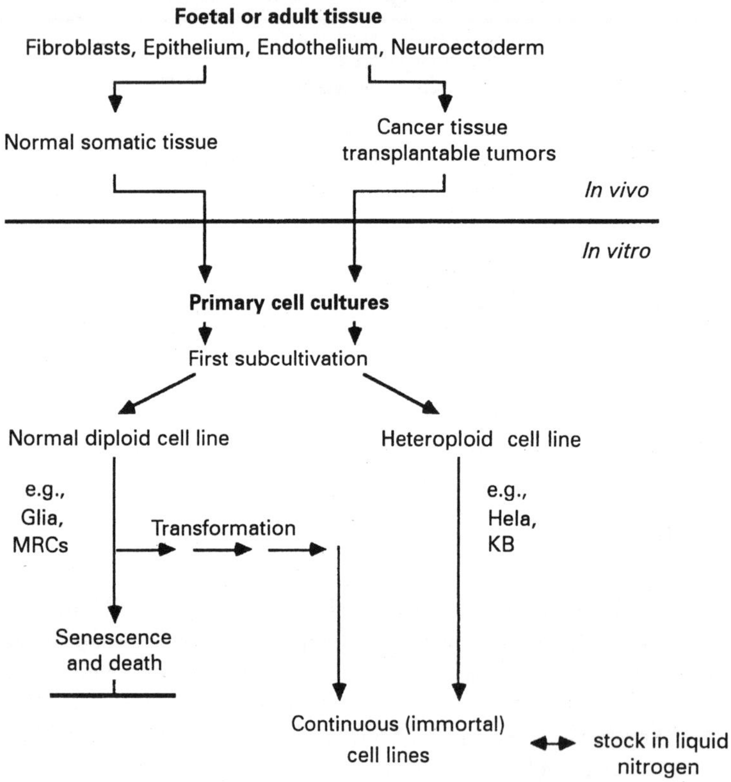

Fig. 9.1 *Derivation of cultured cells from somatic tissue (modified after Spier and Griffiths, 1985).*

Assignment 9.2
What treatments can be used to prepare single-cell suspension from organs or tissues?

The heterogeneous mixture of different cell types in resulting single-cell suspensions is transferred into special culture dishes or flasks with serum-containing medium providing near physiological conditions. Some cell types survive, attach to the surface, spread and proliferate until confluency. However, transformed (cancer-like) cells may be able to grow in multilayers or as a free suspension. For subcultivation, cells are diluted at a certain ratio with fresh medium and passaged into a new culture flask. For subculture, adherent cells are detached from the culture vessel by short exposure to trypsin. During subcultivations, the conditions *in vitro* produce a mostly undefined selection pressure; as a consequence, a certain cell type (e.g., fibroblast) is selected. Normal diploid cells will die after a certain number of generations, although during the cultivation time cells *in vitro* can also be transformed (spontaneously or chemically) to an undefined extent resulting in immortal, cancer-like cell lines. These cells proliferate infinitely with a doubling time of about 10 to 25 h. As an example for such cell lines, Fig. 9.2 shows adherent and suspension culture Chinese hamster ovary (CHO) cells.

The possibility of storage in liquid nitrogen makes established cell lines, like strains of microorganisms, generally available. Over the years they have been extensively characterised with respect to their kinetics of growth and metabolism, morphology, genetics, etc., and they represent well investigated host organisms for further genetic engineering leading to highly efficient producer cell lines.

An important cell type for the production of antibodies are hybridomas (Chapter 10), which result from fusion and hybridisation of B- or T-lymphocytes with cancer cells (lymphomas, myelomas). Initially after fusion the fused cells have two nuclei, but after mitosis cells become mononucleated. In hybridoma cells many genes of either progenitor cell may be expressed.

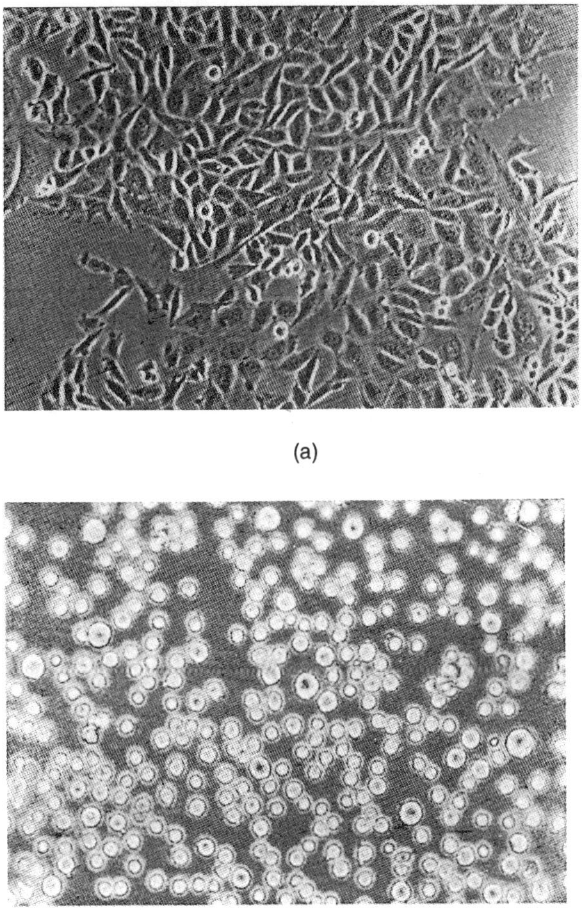

(a)

(b)

Fig. 9.2 *(a) Adherent and (b)suspension cultures of Chinese hamster ovary (CHO) cells.*
The adherent cells are the serum-dependent original cells.
The suspension culture is composed of cells of a transformed serum-independent cell line, derived from the serum-dependent cells.

Assignment 9.3

Explain why lymphocytes are hybridised with tumour cells for the production of antibody-producing cell lines.

In comparison to primary cells *in vivo*, transformed cells *in vitro* exhibit many differences such as:

(i) enhanced growth and proliferation;
(ii) altered chromosome number (aneuploidy);

(iii) different cell shape and organisation of microfilaments;
(iv) different growth factor requirements and response to regulator molecules;
(v) different energy metabolism;
(vi) ability to translocate.

However, not every cell line expresses all such altered characteristics. Human melanoma cells in culture, for example, show contact inhibition and cannot translocate.

Self assessment exercise 9.1

Which of the following has/have been produced commercially from mammalian cell cultures?

(i) Antibacterial antibiotics
(ii) Antibodies
(iii) Insulin
(iv) Antiviral vaccines
(v) Lactate dehydrogenase
(vi) Human growth hormone
(vii) Plasminogen activator.

Self assessment exercise 9.2

Match each of the cell/culture types in Column A with the appropriate definition in Column B.

Column A	Column B
(i) Cell line	(a) Cells on a solid surface as a confluent sheet of cells one cell deep.
(ii) Anchorage-dependent cells	(b) Cells growing as isolated unattached cells in agitated liquid medium.
(iii) Suspension culture	(c) Cells capable of growth in serum-free medium.
(iv) Monolayer	(d) Cells which have the capacity for indefinite growth in culture.
	(e) Cells only growing when attached to a solid surface.

9.3.2 Patterns of nutrient metabolism in vitro

In principle, nutrients can be used for energy production as well as for biosynthesis. The main energy and carbon sources in cell culture media are glucose and glutamine.

As indicated in Fig. 9.3, these two substances are used in a complementary way. This means that their metabolites can flow in basic pathways alternatively, depending mainly on their availabilities. Over a range of glucose to glutamine ratios from 1 to 11 in the medium, the uptake rate and utilisation by the cells also correspond to this ratio.

(a) Glycolysis and respiratory bottleneck

In contrast to normal cells *in vivo*, transformed cells exhibit high glycolytic rates due to fully activated or deinhibited key glycolytic enzymes, together with an increased number of glucose transporters in the cell membrane. Thus, large amounts of pyruvate are produced, connected with the generation of ATP. At the same time, transformed cells have fewer mitochondria with defects in the malate–aspartate shuttle. Furthermore, the high cytosolic ATP levels cause an additional downregulation in respirative activity. In order to regenerate reducing equivalents (oxidation of

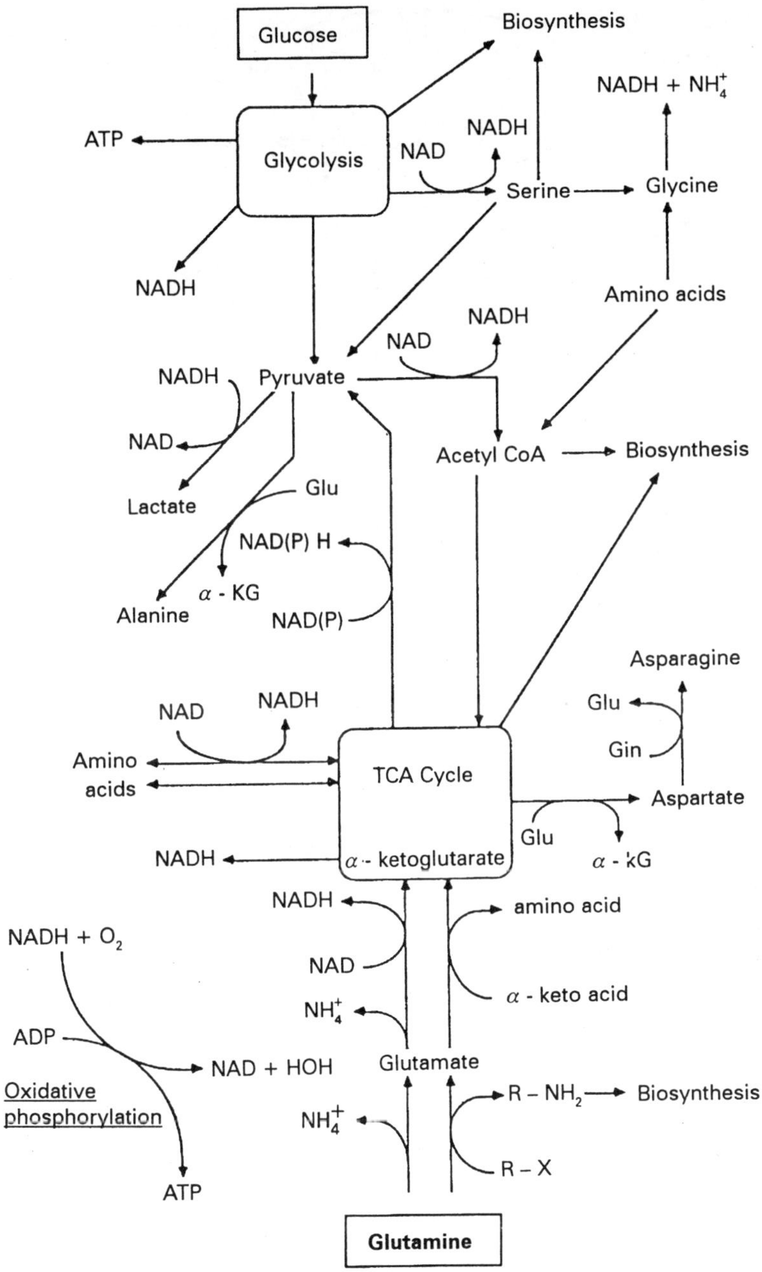

Fig. 9.3 *Schematic diagram of animal cell metabolic pathways.*

NADH to NAD^+) the cells use pyruvate as final electron acceptor by reducing it to lactate, which is secreted. This phenomenon is similar to ethanol production in yeast. By analogy, the model of a respiratory bottleneck can be formulated (Fig. 9.4).

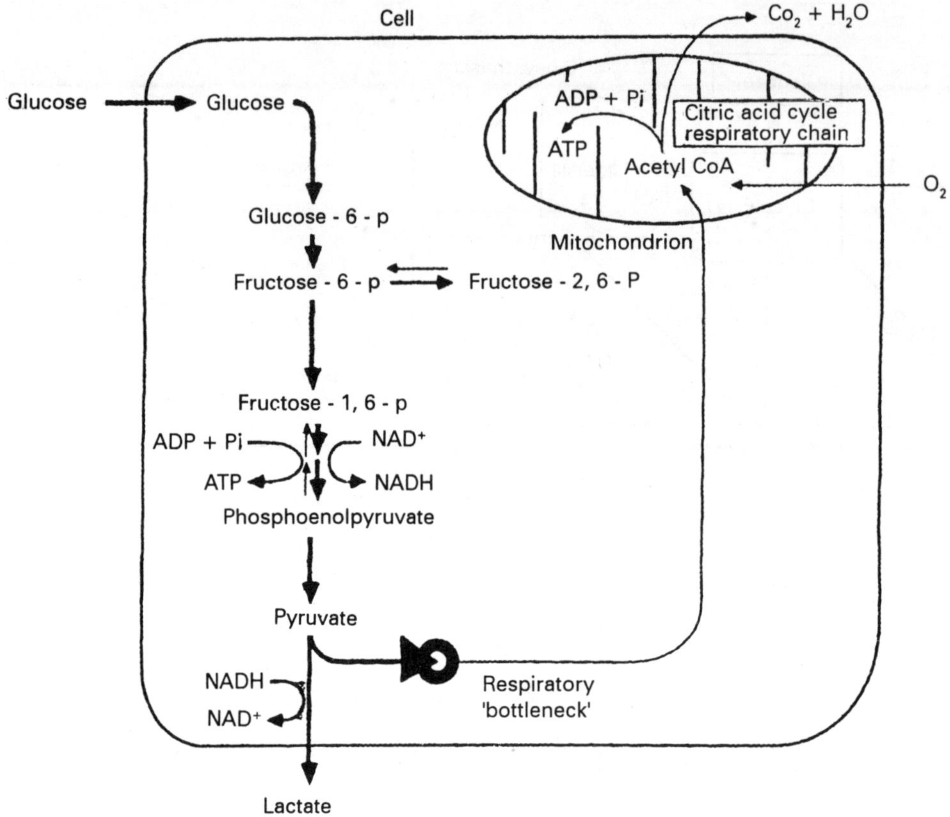

Fig. 9.4 *Respiratory bottleneck in transformed cells.*

(b) Glutamine metabolism

Normal cells *in vivo* synthesise glutamine. In contrast, tumour cells and cells *in vitro* exhibit large glutamine consumption rates. Most of the glutamine carbon skeleton enters the TCA cycle as α-ketoglutarate. Depending on the cell line, it is further oxidised to an extent of 35 to 55% for energy generation in the respiratory chain. The consumption of glutamine leads mainly to formation of ammonia but also to lactate and alanine as by-products. Ammonia in particular is toxic for cells; it is formed not only by cellular metabolism but also to a significant extent by spontaneous decomposition of glutamine in the medium. In serum-free medium, which lacks the detoxifying activity of serum components, ammonia is about ten times more toxic than in medium containing serum: for example concentrations of about 1 mM can reduce the specific growth rate by 50%. This puts a limit on the concentration of glutamine that can be used in culture media. Lactate is less toxic: here the tolerance of the cells is about ten times higher than towards ammonia (depending on the cell line; 10 to 100 mM lactate causes a reduction of 50% in specific growth rate).

Attempts to reduce ammonia production by partial substitution of glutamine with more stable asparagine have been performed with success. Other strategies use intermittent feeding schemes (small glutamine additions every 12 hours) to keep glutamine and thus ammonium concentrations low during the process. By this method ammonium production could be reduced by 40%. Also, technical solutions to the removal of ammonium from culture medium by passing it out of solution as gaseous ammonia using a microporous hydrophobic membrane (membrane distillation) are possible. On the acceptor side of this membrane an acidic stripping solution, vacuum or sweeping air removes the ammonia and assures a maximal ammonia flow due to a maximal concentration gradient over the membrane. As a new alternative, the replacement of glutamine by more stable dipeptides, formed with alanine or glycine, has been successfully performed. At 37°C these derivatives are sufficiently stable over a month and are even heat-sterilisable.

9.3.3 *Crucial cellular requirements of cells growing in vitro*

Basic problems in mammalian cell culture technology result from characteristics of the ceiis: in contrast to microorganisms, the cells are not suited to survive independently. To maintain basic cellular functions and thus productivity *in vitro*, technical equipment together with physical and chemical conditions have to meet the complex cellular requirements.

Often modern bioreactor systems for animal cell cultivation originate from reactors for the cultivation of microorganisms, with control loops for agitation, temperature, pH and oxygen partial pressure; however, they are modified to cope with the special requirements of mammalian cells.

(a) Mechanical and physico-chemical conditions

(i) Fluid mechanics and gas–liquid interphases

Compared to microorganisms, mammalian cells are relatively large (diameter 15 m; protein content 1000 times the amount of an *E. coli* cell) and complex in structure. They have many heavy organelles within a susceptible cytoskeleton and they lack a cell wall.

Assignment 9.4
Animal cells lack cell walls. Explain how this affects our ability to grow them in fermenters designed for the growth of microorganisms.

An example of an optimised system (with high mixing efficiency but with low shear forces) is a compact-loop bioreactor with a draught tube and marine impellers (see Fig. 9.8). Depending on the medium and cell type, stirrer tip speeds up to 1.5 ms^{-1} are tolerated.

For the cultivation of hybridoma cells, air-lift bioreactors are usually applied. For more fragile cells the liquid/gas interphases may cause problems; cells are ruptured on contact with gas bubbles or they burst as the bubbles burst on reaching the surface. In such cases gas exchange has to be performed by surface aeration, within cell free areas of the medium (application of external loop or spin filter technique), or indirect *via* bubble-free gas exchange through silicon membranes or tubings in the culture.

Assignment 9.5
Describe clearly, witn diagrams if need be, how gas exchange could be performed by (i) surface aeration, (ii) external loops and (iii) bubble-free gas exchange through tubing.

(ii) Temperature

Animal cells are sensitive to temperatures higher than 37°C. Aqueous solutions have low thermal conductivity, and with a low convection rate in animal cell cultures due to gentle mixing, high temperature gradients may occur with heating using electric heaters. Thus low gradient heating systems, mostly based on the water-jacket principle and with a high surface area, are favourable. In a compact-loop bioreactor this can be achieved by using a thermostated draught tube.

(iii) pH

Depending on the cell line and medium type, optimal cellular function requires an external pH of 7.0 to 7.6. For pH control, a dual system with acid and base is used.

Self assessment exercise 9.3

By what mechanisms would the growth of animal cells alter the pH of the culture medium?

The CO_2/bicarbonate buffering system is widely used in animal cell culture medium. Bicarbonate is added to the medium and CO_2 is added to the gas phase used to aerate the culture. At a given concentration of bicarbonate the pH can be controlled by adjusting the partial pressure (concentration) of CO_2 in the gas phase (CO_2 can be added to the gas stream used to aerate the medium). This is often preferable to pH control by addition of acids such as hydrochloric acid, as addition of such acids can increase the osmolarity of the medium to a point where growth of cells is inhibited and it may also cause corrosion of stainless steel components of the fermenter. At high lactate concentrations, however, the buffer system can be exhausted, in which case additional base must be supplied (isotonic NaOH, KOH).

(b) Chemical conditions

In contrast to many microorganisms, mammalian cells are auxotrophic for many substances and thus require them in the growth medium. The optimal function of such cells (and thus productivity, product safety and so on) is highly dependent on specific chemical and physico-chemical conditions in their microenvironment. Also, cells in culture do not have the complete detoxifying systems that they are provided with by special organs *in vivo*. As a result, cell cultures, especially under serum-free conditions, are very sensitive to traces of various toxic substances.

(i) Selection of materials being in contact with the culture

Stainless steel or glass are suitable basic reactor materials. Packings and tubings made from polymeric materials should not release toxic substances such as polymer additives. In most cases, ethanol-treated silicon or PTFE (polytetrafluorethylene) are suitable replacements

(ii) Selection of medium components, preparation of water

In addition to the above-mentioned criteria, all of the 40 to 70 defined chemical substances (inorganic salts, glucose, amino acids, vitamins and other low molecular weight compounds) in culture media should be of highest purity and cell culture tested. Essential trace elements, except iron, are present in sufficient amounts as impurities even in these fine chemicals, and so generally no separate addition is necessary. The selection of complex additives such as sera and growth factors is more problematic as these additives cannot be standardised and differ in composition from lot to lot.

Water constitutes the major component in cell culture medium. In order to avoid incalculable and also unfavourable chemical influences originating from impurities, it should be of highest quality and freshly prepared; in general, redistilled water is suitable for medium preparation. For large scale purposes, modern systems combining reverse osmosis, activated charcoal filters, ion-exchange and ultrafiltration units, provide good substitutes; large amounts of water which are practically free of minerals or contaminating organic compounds can be freshly prepared at relatively low cost.

(iii) Serum-containing and serum-free media

In contrast to chemically defined media in which components are known and present in known concentrations, serum represents the "black box" in culture media. Up to now, the components of serum have not been fully characterised and their action on different cell types is only partly understood. Serum contains about 1000 different proteins, with some of them having growth promoting or other beneficial characteristics.

In general, foetal or newborn animal sera are added to an amount of 5–20% to defined basic media. Table 9.2 lists a number of advantages as well as disadvantages offered by the use of serum.

Table 9.2 Advantages and disadvantages of serum as a supplement to cell culture media

Advantages
- contains a complete set of essential growth factors, hormones, attachment and spreading factors, binding and transport proteins
- binds and neutralises toxins (metals and pyrogens)
- contains protease inhibitors
- increases buffering capacity
- provides trace elements and other nutrients

Disadvantages
- not chemically defined
- variable composition from lot to lot
- can be a source of contamination: viruses, mycoplasmas, prions (responsible for diseases such as bovine spongiform encephalitis [BSE] and scrapie)
- components may bind, inactivate, antagonise or mimic the action of added medium ingredients
- presence of proteases can cause decreased cell and product yields
- increased difficulties and costs of downstream processing
- most expensive component in cell culture media

Due to the severe disadvantages of serum, mainly its high cost, many efforts have been undertaken to replace it, especially for cultures on larger scale. Proposed solutions have been the use of serum-fractions or addition of a number of different growth factors and hormones to the basic medium. However, as these supplements as crude extracts also originate from serum or other poorly defined material, the ultimate goal is cell culture in a fully-defined serum- and protein-free microenvironment. This implies, however, the further modification of cells to obtain autonomous cell lines able to synthesise and to secrete their own growth factors, or having receptors that are activated without the regulatory substances being present. Figure 9.5 presents a schematic diagram of the autocrine induction of cell growth in transformed cells.

New cell lines capable of growth in serum- and protein-free medium usually exhibit a reduced growth rate, lower maximal cell density and also lower genetic stability. However, an advantage of such cell lines is that their nutrient requirements can be characterised exactly with respect to all essential components by analysing the kinetics of utilisation of each component. This enables design of tailor-made minimal essential medium for each cell line and process mode. Cells and microenvironment are not considered to be different process components, but as unity; on the one hand cells shift their pheno- and genotype to adapt to the medium, on the other hand the operator designs a medium to fulfil cellular requirements with respect to the process goals. In an optimised system, growth characteristics equivalent to those of the serum-dependent progenitor cells may almost be achieved. Also, due to minimal selection pressure in such a medium, remarkable genetic stability occurs. In order

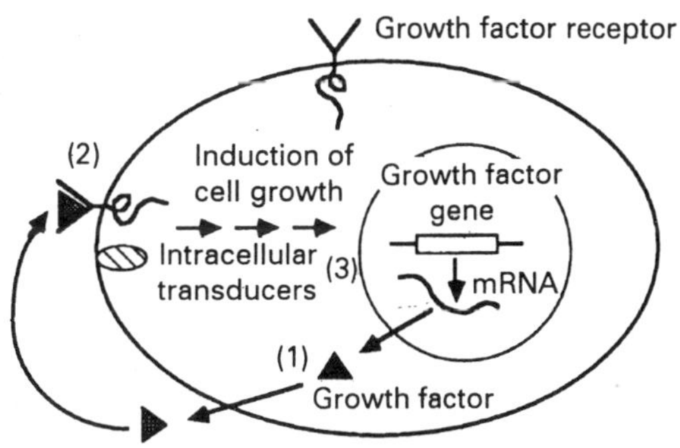

Fig. 9.5 *Autocrine induction of cell growth in transformed cells.*

to achieve optimal control on cellular functions and genetic stability over the whole cultivation period, a chemostat culture with control systems based on the on-line analysis of all essential microenvironmental and also suitable biochemical or biological parameters would be necessary; this represents a major goal of process development.

Assignment 9.6

Explain the ways in which genetic engineering could be used to overcome the problem of the need to add highly expensive serum to cell culture medium.

Assignment 9.7

Explain what is meant by chemostat culture.

Self assessment exercise 9.4

During culture of a particular cell line, alkali addition was necessary to control (raise) the pH of the medium when the cells were growing in serum-free medium, but was not necessary when cells were growing in serum-supplemented medium. Explain this. All operational parameters, nutrient concentrations and gas concentrations were the same in each case.

Self assessment exercise 9.5

Explain how the on-line measurement of ammonia concentration in medium, coupled with control of rate of addition (feed) of glutamine, can optimise conditions for the growth of cells in culture.

Self assessment exercise 9.6

Answer true or false to each of the following statements:

(i) CO_2 gas is added to the air mixture used to aerate mammalian cell cultures, in order to buffer the medium (against changes in pH), in conjunction with bicarbonate added to the culture medium.

(ii) Foetal calf serum is added to culture medium for the cultivation of many cell types in order to provide undefined growth promoting factors.

(iii) In serum-free medium, cells exhibit increased growth rates, higher cell density and increased genetic stability.

(iv) Purified water is required for mammalian cell culture medium, as untreated water may contain substances inhibitory to cell growth.

9.4 Mass cultivation of cells in bioreactor systems

In principle, cells can be cultivated under two basically different principles: (i) as freely suspended cells, or (ii) as an immobilised culture on solid phases or embedded in suitable materials.

The choice of cultivation technique depends mainly on whether the cells are capable of growth in an adherent or anchorage-independent condition. In suspension culture the cells are growing freely dispersed in a liquid medium which is free of any other solids. Sedimentation of cells is prevented by medium agitation. In contrast, adherent cells are attached to and spread on a solid surface and are thus cultivated as an immobilised culture. Cells normally growing in free suspension may be induced to attach to and spread on a surface after suitable pretreatment of the surface with attachment factors such as fibronectin or vitronectin.

Similar to the cultivation of microorganisms, the mode of operation can be batch or continuous and variations of them such as repeated batch, fed-batch, or continuous culture with cell retention. Figure 9.6 gives an overview of the large array of techniques for the cultivation of microorganisms and higher eukaryotic cells. It shows the impressive progress of the last few decades in reactor and process design.

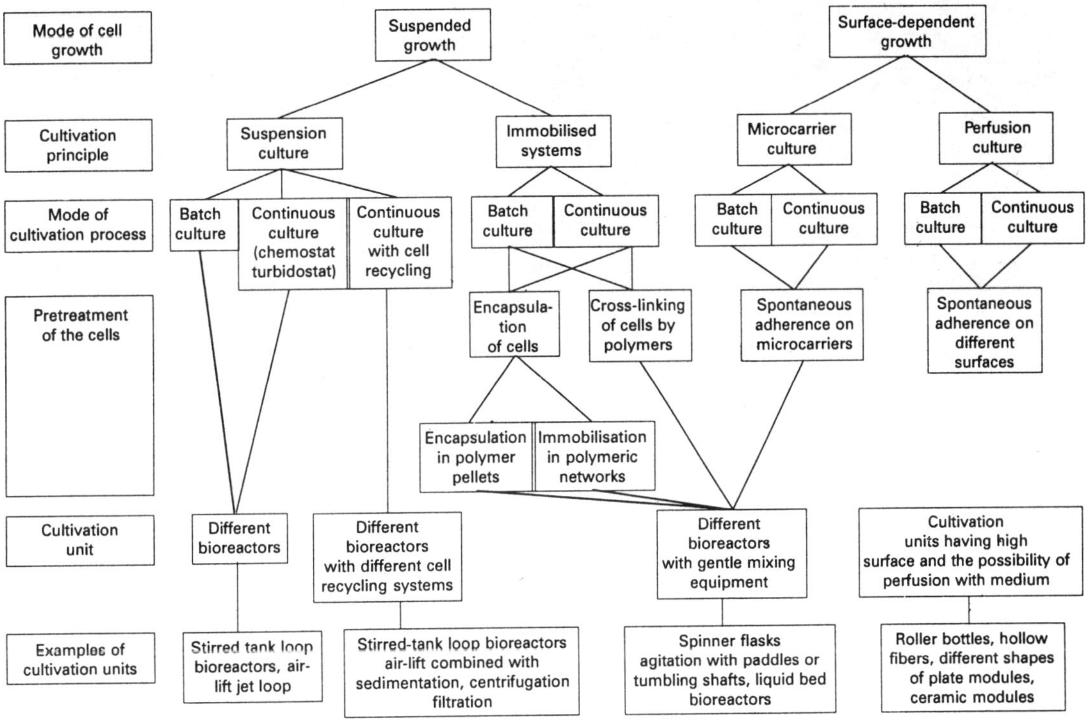

Fig. 9.6 *Techniques for the cultivation of microorganisms and higher eukaryotic cells.*

9.4.1 Integrated suspension culture

Reliability and predictability of the cell culture process are important preconditions to ensure quality and hence activity and safety of the desired product. This requires the precise control of the biological and biochemical funcions of all producer cells in the bioreactor. This can only be achieved by optimal control of a fully defined cellular microenvironment affecting all physical, chemical, biochemical and biological parameters. As each living cell permanently generates gradients due to its metabolism, it is evident that the conditions of the microenvironment will change compared to the bulk conditions, resulting in a system with practically incalculable transient or steady-state conditions. As the process is controlled on the basis of on-line analysis of the bulk conditions, an optimised process must keep the conditions in the reactor as homogeneous as possible. This means minimising the distances of slow transport phenomena (diffusion in the unstirred medium layer around the cells) by optimising convective transport. The only system to ensure these conditions is a bioreactor with freely suspended cells in an agitated liquid medium of low viscosity.

Assignment 9.8

Which type of culture system would be most likely to give homogeneous culture conditions—cell suspension culture or immobilised culture? Explain your answer.

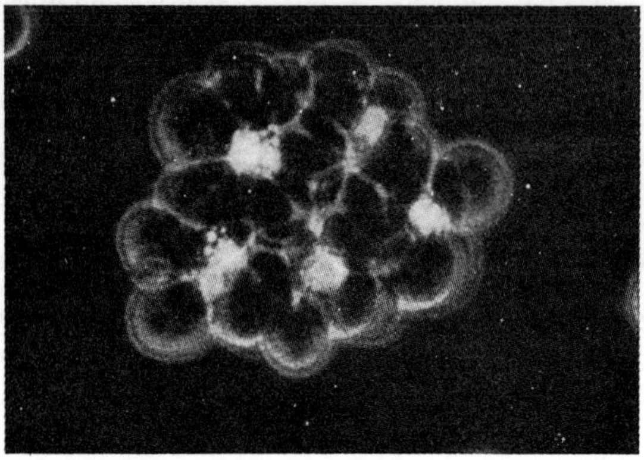

Fig. 9.7 *Cell aggregates in a suspension culture of CHO cells.*

However, single cells in suspension tend to form aggregates (Fig. 9.7). The DNA released from dead cells has been identified as a mediator of cell–cell adhesion. Furthermore, in stirred bioreactors the cell clusters can be disiupted by high shear forces. These reactions result in additional cell death and ultimately in an increasing amount of released DNA causing new aggregation. The process can be prevented by the addition of DNA-degrading enzyme (DNase). However, as cell death results mainly from unfavourable culture conditions, preven-tion is best achieved by strictly maintaining optimal growth conditions.

Assignment 9.9

What effect would aggregation have on the homogeneity of the cellular environment?

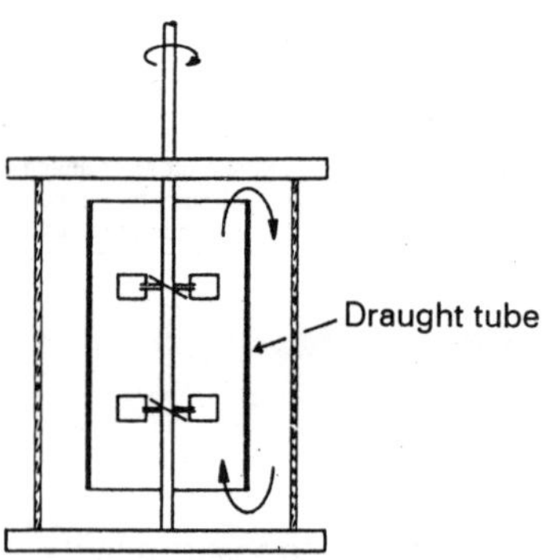

Fig. 9.8 *Compact-loop reactor.*

One of the most frequently used laboratory-scale vessels for suspension culture is a magnetically stirred spinner flask. This system has the advantage of simplicity and cheapness. However, it is not suitable for scale up as the large number of single units required are time consuming to operate, inhomogeneously mixed, and badly suited for process control. Hence, a single-unit process in a well adapted bioreactor is required. As mentioned previously, a compact-loop bioreactor with marine impellers would satisfy these conditions (Fig. 9.8).

As described previously, in addition to the routine control loops for physical conditions, an optimal bioreactor operation also relies on the integrated monitoring and control of chemicals, biochemical and biological parameters.

(a) On-line monitoring of chemical and biological parameters in suspension culture

In spite of slow growth rates exhibited by animal cells as compared to bacteria or yeasts, unfavourable metabolic states can occur rapidly following even small disturbances of the culture. However, the effect on cell density or productivity may be delayed by hours or even days. To overcome this problem efficient control loops based on quick and reliable registration of the changes in conditions are needed. Therefore, a future goal is the on-line monitoring and control of all essential parameters.

As glucose and glutamine are the main carbon and energy sources and their metabolic by-products (lactate and ammonia) exert a strong influence on cellular growth and productivity, control loops for these substances are of prime importance. The monitoring of glucose, glutamine and ammonia can be carried out by on-line flow injection analysis (FIA).

In future, with FIA or other automated analytical on-line systems (gas chromatography, HPLC) concentration curves of different chemical parameters will be generated not by interpolation of two values per day, but by the high frequency (on-line) taking of measurements themselves. Such new tools enable an efficient process control.

Similar automated systems (e.g., stop flow FIA) can be used to determine enzyme activities as biological parameters. Some of these enzyme activities have been demonstrated to correlate with cell density. For example, lactate dehydrogenase (LDH) activity is known to serve as a measure for viable and dead cells. The advantage of indirect determination of biomass emerges in particular when direct methods of biomass estimation (Coulter counter or optical procedures) cannot be applied (as in immobilised systems), or result in large errors due to dead cells.

(b) Continuous high density culture by application of cell-retention systems
In contrast to the ease with which biomass may be retained in immobilised systems, the retention or recycling of freely suspended cells in continuous systems is more problematic. In order to minimise the loss of valuable biomass and to increase productivity, cells are recovered from spent medium which is removed from the reactor by basically two different techniques: sedimentation and filtration (Table 9.3).

By application of these techniques viable cell densities are increased from about 10^5 ml^{-1} to more than 10^7 cells ml^{-1}.

Table 9.3 Evaluation of different basic techniques for cell recycling

Principle	Technical equipment	Evaluation/problems
sedimentation	cell settler	cells spend a long period in the settler without process control; negative influences can occur from pump and tubings
	continuous centrifuge	new technique with promising results
filtration	cross-flow filter	high-flow rates used to avoid blocking of filter can cause cell damage; negative influences can occur from pump and tubings
	spin filter	*in situ* method; degree of cell retention and filter blocking is dependent on mesh size/spinning ratio

Assignment 9.10

Explain how retention/recycling of cells in continuous culture of suspended cells leads to greater productivity.

9.4.2 Immobilised cell-cultivation systems

(a) Evaluation of materials for cell adhesion
The most common materials used in microcarrier beads have a charged surface: DEAE sephadex (diethyl–aminoethyl type of cross-linked dextran) is positively charged, whereas polystyrene is

negatively charged; gelatin beads can be slightly positive or negative, depending on the isoelectric point of the raw material and the method of preparation, and for sufficient cell adhesion they must be coated with fibronectin. In contrast, beads of native collagen (Verax process) do not require any treatment to enhance cell attachment. It seems that surface charge does not correlate with cell adhesion, rather the important parameter for cell adhesion is the surface energy of the material.

Most plastic materials in their native state show values of surface energy too low for cell adhesion (30–40 ergs/cm^2). After surface treatment (e.g., corona discharge process), their surface energy rises to 56 ergs/cm^2 or more, and cell adhesion becomes possible. Glass and ceramics, which have a naturally high-surface energy, are well suited as cell carriers.

(b) Fixed-immobilised beds

Laboratory-scale cultivation methods for anchorage-dependent cells include the use of tissue culture flasks and roller bottles. However, these systems have a limited application for large-scale production.

Assignment 9.11

Explain why the use of lab-scale tissue culture flasks is not suitable for large scale production of products from animal cells.

Assignment 9.12

What is immobilisation, and what are the advantages of using immobilised cell systems?

In order to increase surface area for mass cell cultivation, different materials for cell attachment are packed in different types of bioreactors. A bioreactor operated with support material packed within it, through which medium is perfused, is called a packed bed reactor. This is analogous to packed bed reactors used with immobilised enzymes (Chapter 7). Among the various materials that have been developed and evaluated, glass offers several advantages; this material can be used in its native form and it is inert, reuseable and cheap. Of various forms, such as spheres, rings, tubes, rods and helices, spheres (3–7 mm) have the advantage of stable packing with a regular channel size running through the bed.

The use of glass spheres for the manufacture of a cell product was initiated by Spier and co workers. For larger scale a 100 litre reactor with 50 kg of 3 mm diameter spheres (total surface area of about 20 m^2) was used for virus production. However, the system worked with a productivity of only about 50% compared with systems of 1 to 10 litres.

A method of increasing process intensity is to substitute solid beads with porous, open matrix beads. Among different possibilities, siran sintered glass carriers have a considerable potential. They have a large surface area per unit volume (74 m^2/l) and are at the same time mechanically strong enough to allow packing into a fixed bed. Another advantage originating from controlled porosity is that not only anchorage-dependent but also suspension cells can be entrapped in the system. For hybridoma cells this system allows a ten-fold increase in cell density over conventional stirred and air-lift reactors. However, at bed volumes over one litre such high cell densities cause severe problems due to nutrient depletion and waste product accumulation originating from unfavourable fluid dynamics in fixed bed reactor columns. Here, corrective medium transport depends on maintaining plug flow, which means that radial- or backmixing is avoided.

> ***Assignment 9.13***
>
> Would the concentration of substrates, oxygen, waste metabolites and product(s) be the same (homogeneous) throughout a fixed bed?

Other problems of fixed-bed systems are:
(i) inoculation of the system leads to a gradient of initial cell density;
(ii) cell layers detach during long-term cultivation due to lack of nutrients caused by partial blocking of the system;
(iii) difficulties in harvesting cells containing intracellular products (viruses);
(iv) lysis of cells *in situ* leads to blocking of the system due to the filter action of the bed.

(c) Hollow fibre systems
Conventional hollow fibre filtration devices have been applied to high density cell cultures: thin, hollow ultrafiltration fibres bundled together in a cylindrical cartridge provide a large surface area for perfusion of nutrients into the extracapillary space where the cells are maintained in densities up to 10^8 per ml. Low molecular weight nutrients and by-products diffuse back across the membrane, whereas valuable serum components can be placed in and retained in extracapillary space with the cells, thus reducing process costs. Whether or not the product passes through the filter tubes can be controlled by the choice of pore size of different membranes (10,000 to 100,000 molecular weight cut-off).

> ***Assignment 9.14***
>
> Explain, with the aid of diagrams if need be, how a continuous process operates with a hollow-fibre reactor.

In the case of retention, the accumulating product can be highly concentrated, although product degradation and feedback inhibition of cellular productivity may cause problems. Also, as in all high density systems, drastic gradients can occur due to insufficient medium circulation. As a consequence the behaviour of the cells in such a culture may be very unpredictable. Cells under optimal process conditions as well as starving and dead cells (or clusters) are all in the same reactor system. All analysis for bioreactor control and for estimation of the status of cells can only be performed as an integral part of the system as a whole, and may thus be of limited value. Process development and scaling up must be performed empirically in such circumstances.

(d) Ceramic matrices
In the Opticell system mammalian cells are immobilised by adsorption to a ceramic matrix. The constructive principle originates from a car exhaust catalyser; many small square channels, each with a cross sectional area of about 1.5 mm^2, run the length of a cylinder (Fig. 9.9).

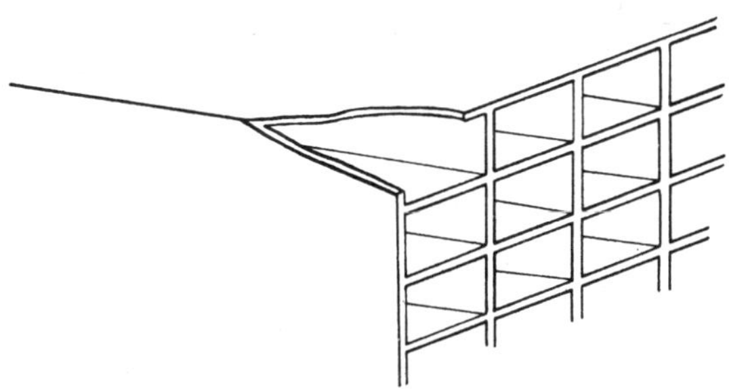

Fig. 9.9 *Ceramic matrix showing channels.*

They form a rigid cell support for adherent cells (smooth surface) as well as for attachment of anchorage-independent cells (porous

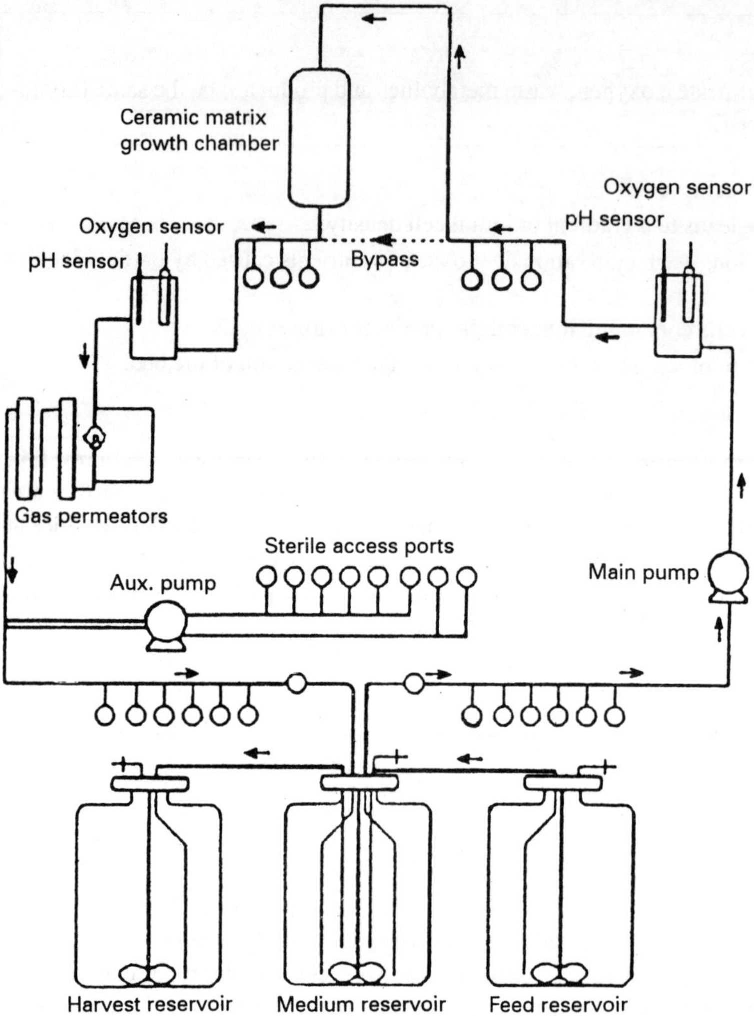

ceramic). The matrix is built into a perfusion loop provi-ding for the constant recirculation of the medium. Secondary loops provide for continuous replenishment with fresh medium and continuous removal of spent medium containing the product. Figure 9.10 provides a schematic diagram of the system.

Fig. 9.10 *The Opticell system.*
In the closed-loop system of medium recycle, pH and oxygen tension are measured before and after the medium has perfused the matrix culture of cells. Gas-permeator systems maintain oxygen and carbon dioxide levels at controlled levels.

The on-line evaluation of biomass within the system is performed indirectly by correlation with the oxygen consumption rate of the culture. Together with data on rates of nutrient utilisation and formation of metabolic waste products, the state of the culture can be estimated. Despite the empirically-based process control, this system has potential for scale up, up to a hundred-fold. However, disadvantages as described for the hollow-fibre reactors must be taken into account, particularly the blocking of channels during long-term cultivation.

Nevertheless, the overall productivity of such systems is high; for example, a comparison of IgG (immunoglobulin-G) production by hybridoma cells in spinner flasks (IgG titer 45μg ml^{-1}) and in the opticell system (IgG titer 185μg ml^{-1}) demonstrates the potential of this method of cultivation.

(e) Fluidised bed systems

In order to avoid concentration gradients within the reactor, porous microcarriers have been used as cell supports in suspension in fluidised bed systems. The density of the particles is adapted to that of the medium by incorporation of inert weights; as a consequence the particles can be fluidised with a minimum of energy input by medium flow.

> ***Assignment 9.15***
>
> What is a fluidised bed system?

The basic substances used are agarose, dextran, gelatin, collagen or open lattices of polyurethane foam. Beads of native collagen are widely used in the Verax process (Fig. 9.11). The native collagen has beneficial properties for cell-matrix interactions. Cell densities up to 5×10^8 ml^{-1} can be achieved, thus even in small perfusion reactors the process is efficient and leads to high productivities.

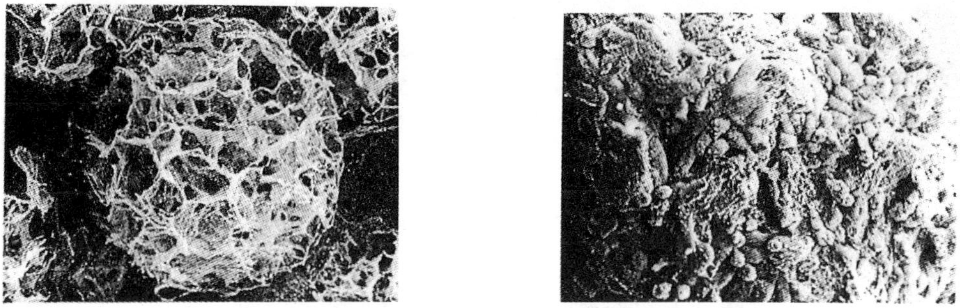

(a) (b)

Fig. 9.11 *Scanning electron micrographs of Verax collagen microspheres. (a) Empty microsphere showing 500 µm pores. (b) Part of a microsphere with growing CHO cells.*

A fluidised bed, as described, has the advantage over a fixed-bed system in that under optimal conditions gradients are minimised to the scale of single beads. As growing cells do not block the porous structure, transport of nutrients into and cell products out of spheres is enhanced by medium flow resulting from tumbling movements of spheres in the medium stream. A schematic representation of the system is shown in Fig. 9.12 and a diagram of the reactor is given in Fig. 9.13.

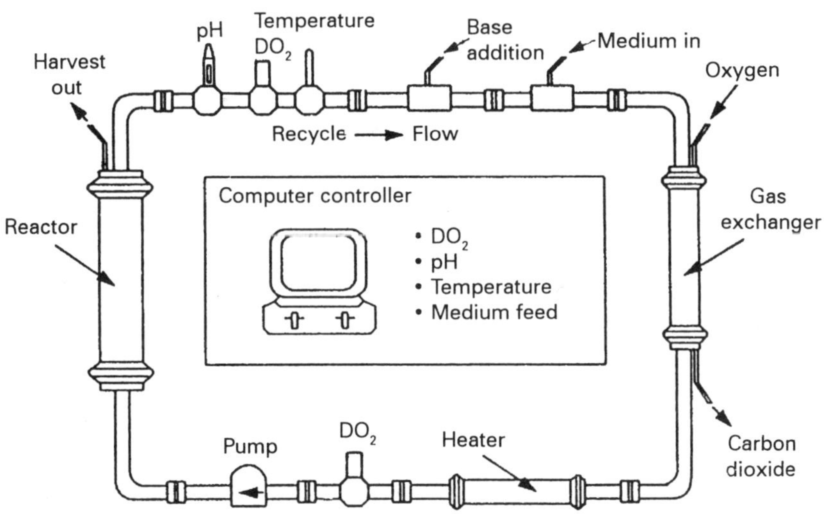

Fig. 9.12 *Schematic representation of a fluidised bed reactor system.*

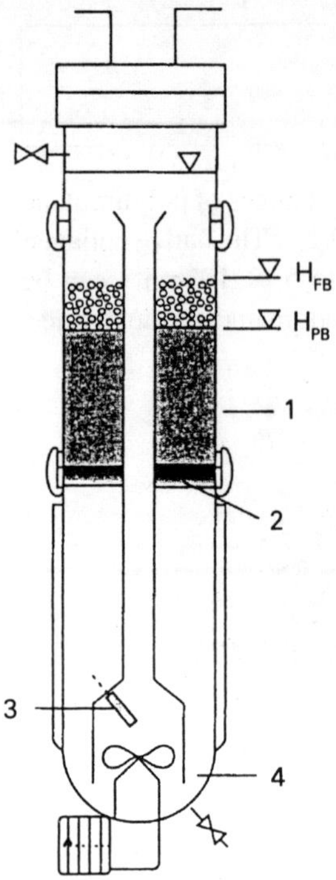

Fig. 9.13 *Fluidised bed reactor (Vogelbusch, Austria).*
1 – fluidised bed 2 – distributor
3 – oxygenator 4 – agitator
H_{FB} – liquid height of fluidised bed
H_{PB} – height of packed (non-fluidised) bed

Self assessment exercise 9.7

Match each of the items in column A with the appropriate description in column B.

	Column A		Column B
(i)	Packed bed reactor	(a)	Continuous culture using immobilised cells on freely moving dextran beads in an upflow reactor with medium recirculation.
(ii)	Matrix culture	(b)	Continuous culture using immobilised cells on freely moving dextran beads in an air-lift fermenter.
(iii)	Fluidised bed reactor	(c)	Continuous culture using cells immobilised on/trapped between stationary glass beads in an upflow reactor.
(iv)	Microcarrier culture	(d)	Continuous culture using cells immoblised on/ trapped within inert sponge-like material.
		(e)	Continuous culture using cells immobilised on the outside of porous ultrafilter fibres.

Self assessment exercise 9.8

Explain the problems caused by cell death in:

(i) a suspension culture;
(ii) a ceramic-matrix reactor;
(iii) a packed bed reactor (3 mm glass spheres).

> ***Self assessment exercise 9.9***
>
> Explain why a ceramic-matrix reactor employs an external gassing/degassing system.

> ***Self assessment exercise 9.10***
>
> Answer true or false to the following statements:
>
> (i) Anchorage-dependent cells attached as a monolayer to microcarriers in a compact-loop reactor would be in a more homogeneous environment than a suspension culture of anchorage independent cells in the same reactor.
>
> (ii) Cells attached to/trapped between the support in a packed bed reactor would be in a more homogeneous environment than cells attached as monolayers to the support in a fluidised bed reactor.
>
> (iii) Cells attached as a monolayer to the support in a fluidised bed reactor would be in a more homogeneous environment than cells attached as a monolayer in a ceramic-matrix reactor.
>
> (iv) Cells attached as a monolayer within a hollow-fibre reactor would be in a more homogeneous environment than cells attached as a monolayer to microcarriers in chemostat culture.

> ***Self assessment exercise 9.11***
>
> Which system offers the more simple technical solution to maintaining high cell concentrations in continuous cultures: immobilised cell systems such as ceramic matrix or suspension culture with biomass recycle? Explain your answer.

9.5 Product formation

In general, two types of kinetics of production are observed.

9.5.1 Growth-related production

An example of this is t-PA (tissue-type plasminogen activator) from CHO or melanoma cells, which express the desired protein mainly during the synthesis (S) phase of the cell cycle. In this case the process goal is to keep proliferative activity and cell density in a maximal state.

9.5.2 High production rates from cells having reached the stationary phase

An example of this is antibody production by hybridoma cells which often occurs in the state where cells are arrested in the late G1 phase or when they are even in the G0 phase due to unfavourable culture conditions. As many hybridoma cell lines are unstable with respect to antibody production, they must be periodically recloned to select for highly producing cells. This instability provides an incentive to design a process where cells, after having reached their maximal density, are maintained for a long time at high viability in a non-proliferative state. However, the often described high 'production' rates in dying cultures are more likely to be the result of product release from lysed cells than from biosynthesis.

9.6 Large-scale purification of pharmaceutical products

The steps of downstream processing include all purification and handling steps of the product from the crude cell-culture medium until storage as pure substance in a vial. Controlled pharmaceutical procedures have to assure that a native (unadulterated) substance reaches the final vial with a high lot-to-lot consistency. The standard quality demand for the final product is high, for example:

(i) there should be no structural changes brought about by unfavourable downstream conditions;
(ii) extraneous protein, DNA, and endotoxin levels must fulfill the purity specifications of the corresponding national administration guidelines;
(iii) the structure and activity in the final container for application should be stable;
(iv) the product should be free from high molecular weight aggregates that could reduce activity and induce harmful immunogenic side effects;
(v) the product should pass sterility tests (including tests for mycoplasmas and viruses).

> ***Assignment 9.16***
>
> What problems can be caused by (i) extraneous protein, (ii) DNA and (iii) endotoxin in a product from cell culture?

The first crucial procedure is the need for rapid removal of the product from the unfavourable degradative conditions existing in the harvested culture medium. From this point of view, continuous harvesting processes with short product residence times, dealing with high medium flow rates from high density reactors, are favoured. In order to minimise proteolytic degradation and aggregation processes, the medium is usually cooled down to 4°C as soon as possible after harvest until the following procedures are carried out.

Before purification, a 50 to 100 fold concentration of the product by ultrafiltration with hollow fibre, plate and frame, or spiral cartridge systems is performed. Hydrophilic membranes are preferred to minimise hydrophobic protein-surface interactions and to thus minimise conformational distortion of product molecules. Multiple ultrafiltered concentrates are pooled to obtain a larger lot size. After adjusting salt concentration and pH, purification is performed by means of chromatographic methods (e.g., ion exchange, or affinity chromatography). Where cell culture medium contains low levels of serum (1%), or is serum-free, purification steps are made easier.

All steps of processing must be performed under highest possible standards of hygiene. This necessitates sterile filtration of the harvested media and of all other solutions used. Adequate cleansing procedures should protect all equipment, especially resins in the columns, from microbial contamination.

> ***Assignment 9.17***
>
> What problems can be caused by microbial contamination of the product stream in downstream processing?

9.7 Concluding remarks

In general, two different process strategies for large-scale production of biochemicals by mammalian cells are applied.

Firstly, the cultivation of cells can be carried out as an immobilised culture resulting in economically worthwhile productivities. For fixed-bed systems, however, the price to pay with respect to product quality, and hence safety, is high: largely non-homogeneous conditions in culture permit only empirical evaluation of process strategies with poor possibilities for control. A more controlled process is the use of microcarriers in fluidised bed systems in which gradients of nutrients and waste products should only occur over small distances (along the sphere radius); as these systems are highly productive, they are the most favoured of the immobilised systems.

Secondly, the more sophisticated process strategy relies on tuning cellular function by controlled microenvironmental conditions. This method requires defined and reproducible conditions and hence results in a predictable and safe process. However, all cells within the bioreactor must be maintained under the same conditions, which implies the application of suspension culture technology. Here, future strategy centres around high performance bioreactors such as integrated chemostat suspension culture with monitoring and control of all essential culture parameters. Further process developments are directed towards strategies to obtain synchronised, controlled cycling cells. This would allow application of high-efficient feeding and harvesting strategies.

9.8 Suggested further reading

1. Anon. (1988). *Animal cell biotechnology*. Academic Press, London.
2. Biotol Series (1992) — *In vitro cultivation of animal cells*. Butterworth-Heinemann, Oxford.
3. Butler, M. (1987). *Animal cell technology, principles and products*. Open University Press, New York, 2–13.
4. Freshney, R.I. (1987). *Culture of animal cells: a manual of basic techniques*, (2nd ed.). Liss, New York.
5. Ho, C.S. and Wang, D.I.C. (Eds.) (1991). *Animal cell reactors*. Butterworth-Heinemann, Stoneham, MA, 445–478.
6. Leist, C.H., Meyer, H.P. and Fiechter, A. (1990). Minireview: potential and problems of animal cells in suspension culture. *J. Biotechnol.*, **15**, 1–46.
7. Spier, R.E. and Griffiths, J.B. (Eds.) (1988). *Animal cell biotechnology*. Academic Press, London.
8. Springer, T.A. (Ed.) (1985). *Hybridoma technology in the biosciences and medicine*. Plenum Press, New York.

9.9 Responses to assignments

9.1 *In vitro* means outside the living animal, in effect in culture. *In vivo* means within the living animal. Anchorage-dependence is the property of cells whereby those cells will only grow (multiply) in culture when they are attached to a solid surface. Anchorage-independent cells, in contrast, will grow in free suspension, in the same way as microorganisms; cells growing in this way form a cell suspension. Anchorage-dependent cells usually grow as a layer one cell in thickness (a monolayer) until all available surface is covered; the monolayer of cells covering the surface is said to form confluent growth, and the property of the cells which arrests their growth when they make contact with surrounding cells is termed contact inhibition. A primary culture is the culture of cells taken directly from a living animal. Passage means subculture. A cell line is a related group of cells which have the capacity to grow (multiply) in culture indefinitely.

9.2 Organs/tissues can be disrupted by mechanical actions, such as chopping or mincing. Alternatively, enzymes such as proteases and hyaluronidases which break down intercellular cementing substances can be used to disaggregate cells from organs/tissues.

9.3 Lymphocytes, which produce antibody, cannot be grown in culture using techniques currently available. Myeloma cells are lymphocytic cells which have become cancerous (they have developed the capacity for uncontrolled growth). By fusing lymphocytes with myeloma cells which have lost the ability to produce antibody, an immortal cell line can be produced, the cells of which express both the ability to produce a given antibody (inherited from the lymphocyte progenitor cell) and capacity for indefinite growth in culture (inherited from the myeloma progenitor cell).

9.4 Microorganisms have rigid cell walls and so can withstand, without damage, the shear forces developed by gassing and agitation in conventional fermenters. Animal cells lack cell walls, and as such can be easily damaged by such shear forces. Thus bioreactors used for large-scale cultivation of animal cells usually employ specialised aeration systems.

9.5 Surface aeration involves passing a gas stream over the surface of the medium, and allowing gas exchange to take place passively by gas exchange between the gas stream and the medium. External loops involve circulating the medium from the reactor around a loop; in the loop are placed some type of filter to retain the cells within the reactor and a gas exchange system to aerate the medium prior to it being recirculated back into the bioreactor. Bubble-free gas exchange involves passing the gas stream through a network of tubing or membranes permeating the bioreactor; the walls of the tubing are gas-permeable/liquid-impermeable, so that gas exchange occurs without formation of bubbles.

9.6 If cells could be genetically engineered so that receptors were permanently activated (without regulatory growth factors being present) or so that growth factors were permanently produced, then the need for serum addition to medium could be overcome. Alternatively, if the growth promoting factors and corresponding genes could be identified it may be possible to produce such factors relatively cheaply by fermentation processes (using genetically engineered microorganisms); the factors would then be available for inclusion in tissue culture medium.

9.7 The chemostat is a continuous culture system in which the inflowing medium contains one nutrient at concentration low enough to limit the growth rate of the cells. A steady state develops in which cell concentration, growth rate and residual nutrient concentration remain constant. This system is described in Section 8.6.4.

9.8 Immobilised culture with cells immobilised in beds or matrices act as plug-flow systems with medium flowing in a given direction through the reactor. In such systems gradients occur, as nutrients are progressively used up and metabolites progressively accumulate. Unless the system is thoroughly mixed, as is usually the case with suspension culture, then conditions are unlikely to be homogeneous within the reactor.

9.9 Within an aggregate of cells there is likely to be concentration gradients of nutrients and gases, with cells at the exterior of aggregates being subjected to concentrations of nutrients, oxygen and CO_2 identical to those in the bulk medium and with cells at the interior of aggregates being subjected to lower concentrations of nutrients and oxygen and higher concentrations of CO_2 than the concentrations in the bulk medium.

9.10 In continuous-culture systems without biomass recycle/retention, the cells (biomass) are(is) washed out of the reactor in the outflowing medium stream. Thus the number of cells available for synthesis of the product is reduced. If biomass can be retained within the bioreactor, or separated from the outflow stream and returned to the reactor, then the concentration of cells within the reactor will be greater than the corresponding system without retention/recycle, and thus the productivity will be greater. This was discussed in Section 8.6.4 — see Assignment 8.26.

9.11 Lab-scale tissue culture flasks are batch systems which are not amenable to control systems on fermenters (pH and gas concentrations) and thus the growth environment of cells within them is difficult to control and optimise. Also the production of large quantities of product requires the use of many flasks, which increases labour costs. For anchorage-dependent cells the area for cell attachment within a flask is limited, limiting the quantity of product produced by such a system.

9.12 Immobilisation is the term used to describe the act of attaching cells or enzymes to a solid inert support or trapping cells or enzymes within some sort of porous gel or matrix (Chapter 7). The advantage of doing this to mammalian cells is that cells can be immobilised on or in a support/ matrix which is packed into or retained within a bioreactor. The medium can be run continuously through the reactor in such a way as to retain the support/matrix and cells within the reactor, giving rise to a cell-free product stream.

9.13 The fixed-bed reactor for cell growth is analogous to a packed bed reactor for immobilised enzymes (Chapter 7). As there is no back mixing, and the medium flows in one direction through the reactor, the concentration of nutrients and oxygen would decrease gradually with distance through the bed, and concentration of waste metabolites (including carbon dioxide) and product(s) would increase. Thus concentration gradients of all items listed would occur.

9.14 If cells are growing in the extracapillary space, aerated medium can be passed into the centre of each hollow fibre. Molecules in the medium pass out through the wall of the fibre to reach cells growing outwith the fibre in the extracapillary space. If the pore size of the fibres is such that high molecular weight molecules cannot pass through, the product may be retained in the extracapillary space and serum can be introduced in this location without its loss in the product stream. In some hollow fibre systems gas-permeable/liquid-impermeable fibres are used with the gas stream passing into the centre of each hollow fibre, and with medium perfused into the extracapillary space where the cells are growing.

9.15 In a fluidised bed system the solid particles (cells or immobilised cell masses) are suspended in an upward flow of medium, so that the upward flow of medium counteracts the sedimentation of the particles. The solid particles move freely in the liquid stream and behave in the manner of molecules in a fluid — hence the name fluidised bed.

9.16 Extraneous proteins may have enzymic properties which may denature the product during storage, or they may induce allergic responses in recipients. Extraneous DNA may be of viral origin and pose a threat of disease transmission (for example, Kreutzfeldt Jacob's syndrome). Endotoxin is pyrogenic, inducing fever and other responses in recipients.

9.17 Microbial growth in the product stream can in turn contaminate the product stream with protein, DNA and endotoxin (see Assignment 9.16). Microbial growth may cause the pH of the medium to rise or fall, creating conditions in which the product is unstable. The microbial biomass may clog up processing equipment, making processing difficult.

9.10 Responses to self assessment exercises

9.1 The correct response is as follows. — ii, iv, vi and vii.

9.2 The correct matching is as follows. — (i) with (d), (ii) with (e), (iii) with (b), (iv) with a).

9.3 If you cannot think of an answer to this exercise, re-read Section 9.3.2. High glycolytic rates in cultured cells tend to result in lactic acid production, lowering the pH of the medium. Glutamine metabolism can lead to ammonia producton, raising the pH of the medium.

9.4 The proteins in serum, being amphoteric (capable of ionising to produce net positive or negative charge), have the capacity to neutralise acids and alkalis, and so add to the buffering capacity of the medium. This means that the proteins in serum can neutralise lactic acid produced by cells, stabilising the pH without the need for alkali addition. In serum-free medium the lactic acid causes the pH of the medium to fall to a greater extent.

9.5 By measuring the ammonia concentration in the medium, when ammonia levels start to rise the glutamine feed rate can be reduced. When ammonia levels are low glutamine feed rates can be increased. By this means the levels in the medium of growth promoting glutamine can be maintained at levels as high as possible without producing inhibition of growth due to ammonia accumulation.

9.6 The correct responses are as follows:
(i) True. (ii) True. (iii) False — the opposite is the case. (iv) True.

9.7 The correct matching is as follows. (i) with (c), (ii) with (d), (iii) with (a), (iv) with (b).

9.8 Cell death in any system has the undesirable effect of lowering productivity as the cell will have used up nutrients as it grows, if it then dies it makes no (further) contribution to product formation. Cell death may also release proteins and nucleic acids into the medium, which may make product separation and purification more difficult.

In suspension cultures cell death releases DNA which stimulates aggregation and hence more cell death. In a ceramic-matrix system, which operates as a plug-flow system, cell death leads to detachment of dead cells, the debris of which can clog channels. The restriction of medium flow because of channel clogging deprives cells downstream of clogged areas from nutrients causing their death. The situation in a packed bed reactor (which also operates in the plug-flow mode) is the same.

9.9 A ceramic-matrix culture acts as a plug-flow system with medium perfusing in directional flow through the sponge-like matrix on which the cells are growing. In such a system gassing by conventional methods (involving introduction of gas [air/CO_2] and medium agitation) is not applicable (medium cannot be agitated in this way in a matrix). Thus medium is gassed/degassed in a separate reactor prior to entering the cell reactor (see Fig. 9.12).

9.10 The correct responses are as follows:
- (i) True — suspension cultures would contain clumps within which the environment would not be homogeneous.
- (ii) False — in a packed bed reactor (which acts as a plug-flow) there will be a concentration gradient through the reactor.
- (iii) True — in a ceramic-matrix reactor concentration gradients occur.
- (iv) False — in a hollow fibre reactor concentration gradients will occur.

9.11. Cell recycle presents considerable technical difficulties (Table 9.4), and may cause cell damage which itself causes problems (as outlined in 9.7, above). External systems may be difficult to operate aseptically, rendering the system prone to contamination. Immobilised reactors with biomass fixed in position do not experience these difficulties, though they suffer from lack of environmental homogeneity, difficulty of control and tendency to clog.

10 Immunotechnology

Chapter outline

10.1 The immune system

10.1.1 Defences and immunity

The immune system of vertebrates is specific, in that it is capable of recognising and discriminating between the shapes of particular molecules. It is reactive, in the sense that it is capable of carrying out a variety of reactions against a target molecule or organism (antigen), and it has memory, in that it can remember previous encounters with an antigen. The immune system consists of a complex network of cells and molecules that interact to eliminate infectious organisms from the body. However, cells of the immune system do not 'see' foreign organisms as a whole; they recognise and react against potentially any molecule, so long as it is foreign or 'non-self'. The immune system therefore serves to discriminate between 'self' and 'non-self' and to eliminate 'non-self'.

The immune system can be thought of as the body's third line of defence against the entry of foreign substances. The first line of defence is formed by physical barriers such as the skin while the second consists of a variety of non-specific mechanisms that include low pH conditions, ciliary motions, sneezing, and secretion of mucus. These mechanisms are non-specific in that they seek only to prevent the entry of a pathogen into the body. Together they constitute what is known as innate immunity. The immune system, the third line of defence, is specific both in its acquired recognition of targets and its action. Phagocytic cells are key players in both specific and non-specific types of defence. In blood circulation, various kinds of phagocytic cells are found, such as polymorphonuclear leukocytes, large granular lymphocytes and monocytes. Circulating monocytes enter tissues and develop into cells called macrophages, which exhibit increased phagocytic capacity. Their role includes that of filtering out foreign material from the blood and the tissues. Macrophages begin to secrete factors, called cytokines (Section 10.3), after they have phagocytosed bacteria, and these cytokines are molecules of great significance in both non-specific and specific immune responses. Among molecules that constitute the non-specific effector molecules of the immune system are interferons α and β which defend cells from viral infections, and the proteins of the complement system that, along with antibodies, lyse bacteria and also make them more 'attractive' to phagocytic cells.

Non-specific, innate immunity is not inducible by 'foreign' antigens. Specific immune responses, on the other hand, are acquired: they are induced by foreign stimuli and generally react only against the stimulating molecule: the antigen or immunogen. An antigen or immunogen is defined as any molecule that induces a specific immune response. The cells that participate in specific reactions are called lymphocytes and are found in the blood and in lymphoid tissues such as the thymus, spleen and lymph nodes. Lymphocytes function by recognising the shape of small parts of antigens, known as epitopes, by means of receptors on the lymphocyte cell surface. Lymphocytes are broadly classified into T-cells and B-cells, depending on the site in the body in which the cells mature, their function and their profile of surface antigens. T-cells are so called because they mature in the thymus.

Bacteria, viruses, yeasts, parasites and filamentous fungi all exhibit different antigens, and each antigen may be composed of several different epitopes. Tumour cells are recognised as 'non-self' by the immune system, because they often express antigens that are not present on the surface of normal cells.

Self assessment exercise 10.1

The vertebrate body's first lines of defence against infection take the form of physical barriers: the horny layer of skin, the mucous membranes, cilia, acidic and bactericidal secretions. Are such defence systems innate or acquired? Give reasons for your choice.

Self assessment exercise 10.2

Explain the difference between an antigen and an epitope.

10.1.2 Cells of the immune system

All lymphocytes arise from stem cells in the bone marrow, but different development patterns separate the two main populations: the T- and the B-lymphocytes. B-lymphocytes, when stimulated by antigen and factors produced by T-helper cells (see below), proliferate and develop into antibody (immunoglobulin) secreting cells.

An immunoglobulin (Ig) molecule is Y-shaped with a four-chain structure — 2 light (L) and 2 heavy (H) chains — joined together by disulphide bonds (Fig.10.1). The arms (F_{ab}) consist of constant and variable regions.

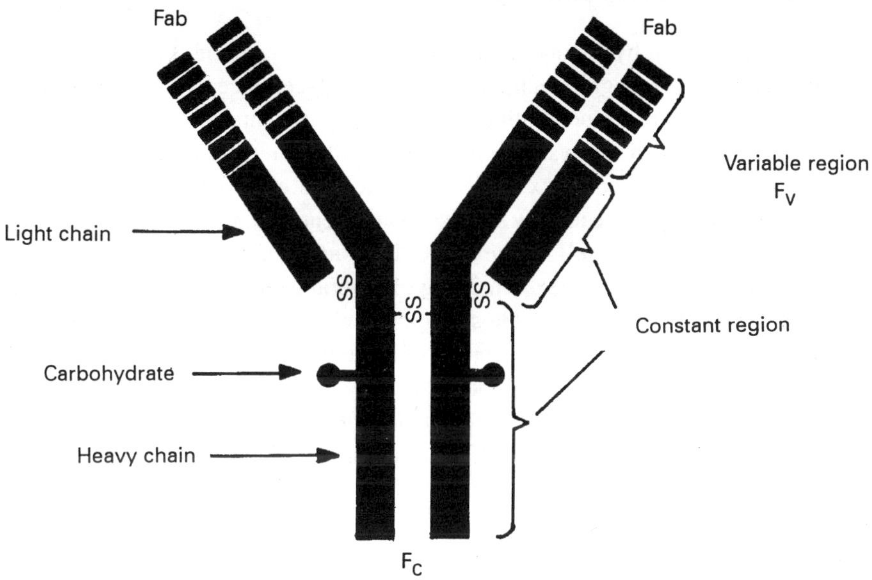

Fig. 10.1 *Structure of an immunoglobulin (Ig) molecule.*
Ig molecules consist of two heavy chains and two light chains; the two chains form two 'antigen-binding' pockets.

The inner (N-terminal) segment of Fab fragments has an amino acid sequence that is largely unaltered from molecule to molecule and hence is termed the constant region. The C-terminal end shows considerable variation in its amino acid sequences and is responsible for the high diversity exhibited by immunoglobulin molecules. It is the variable region of the molecule that forms the binding site to the antigen (or epitope). The remaining portion, the stem of the antibody molecule (F_C), performs functions which include complement fixation, transplacental passage of antibodies, and binding to various cell types bearing F_C receptors. The smallest antigen-binding unit of the antibody is the F_V fragment which consists of light and heavy chain variable regions. Based on the structure of their heavy chain constant regions, immunoglobulins have been classified into five major classes, termed IgG, IgA, IgM, IgD and IgE. Antibody-mediated immunity usually involves the binding to and elimination of antigen by specific antibodies, and is also referred to as humoral immunity.

The T-lymphocytes are responsible for cell-mediated immunity. The T-helper cells secrete stimulatory factors that facilitate the proliferation and differentiation of cytotoxic T-cells and B-cells. Cytotoxic T-cells recognise endogenous foreign antigens on the surface of other cells, and destroy these cells by lysis. Such foreign antigens may arise on the surface of tumour cells, or on the surface of virus-infected cells.

Another set of T-cells, known as T_{DTH} (Delayed Type Hypersensitivity), are stimulated to proliferate and lay down a barrier around the invading pathogen by causing local inflammatory responses. During an inflammatory response, the blood supply to an infected area is increased along with blood capilliary permeability. Antibodies and lymphocytes are therefore able to migrate more easily towards the site of infection.

A further subset of T-lymphocytes, which may overlap with other subsets, is the suppressor T-lymphocyte population (T_S). Their function may be to regulate immune responses by suppressing them.

10.1.3 The major histocompatibility complex

B-cells recognise the stimulating antigen by means of the immunoglobulin (antibody) that is attached to their own surface. Similarly, T-cells recognise antigen by means of T-cell receptors that they carry on their surface. However, while B-cell receptors (antibodies) can recognise antigen by themselves, T-cell receptors can do so only if the antigen is physically associated with molecules on the cell surface encoded by the major histocompatibility complex (MHC). Molecules encoded by the MHC are cell-surface antigens on all cells and they serve as very sophisticated 'identity cards' which the immune system uses to distinguish self from non-self. The MHC in man is called the HLA (human leukocyte antigen) complex, with the gene complex consisting of several genetic loci termed HLA-A, HLA-B, HLA-C and HLA-DP, -DQ and -DR. These loci code for glycoproteins which are extremely polymorphic. This means that they exist in a variety of different forms within a population of individuals. HLA-A, for example, is represented in the population by several types such as HLA-A1, HLA-A2, HLA-A3 and so on. The polymorphism at the different loci of HLA is manifested by differences in the HLA proteins' amino acid sequences, and hence in differences in protein structure. These differences are sufficient to lead to incompatability of tissues from different individuals, which is why transplantation of a kidney from one individual to another usually results in tissue rejection (unless the donor and recipient happen to share the same HLA surface antigens). Given the fact that there are several HLA loci, and that each one of the loci is polymorphic, this means that there must be an enormous diversity of cell surface antigen within a population. Any two individuals chosen at random are likely to differ from one other in at least one of their HLA loci unless they are genetically identical twins. In summary, MHC molecules are polymorphic glycoproteins that can be likened to 'identity cards'. Antigenic epitopes must associate with MHC molecules before T-cells can recognise them and respond to them by proliferating and carrying out their specific tasks. The MHC proteins are also known as transplantation antigens, since failure to match the MHC proteins of tissue transplanted from a donor to a recipient will result in the recipient's immune system recognising it as non-self and rejecting it.

MHC molecules have been experimentally classified into two main types: Class I and Class II. The two classes differ in their general structure, their expression and their modes of action. Class I MHC proteins are expressed by most cells of the body, whereas only cells of the immune system (B-cells, macrophages and certain T-cells) express class II MHC molecules. While both classes of MHC function by binding antigenic epitopes and then presenting them to T-cells, Class I molecules generally present epitopes to cytotoxic T-cells and Class II molecules present epitopes to T-helper cells. MHC molecules therefore play crucial roles during immune responses: they act to restrict responses and thereby govern both the extent and the nature of responses generated. MHC molecules dictate not only whether an individual will respond to a given antigen, they also govern the course of an immune response.

Self assessment exercise 10.3

Match each lymphocyte type given in Column A with the activity appropriate to it from Column B.

	Column A		Column B
(i)	T-helper lymphocyte	(a)	Causes of lowering in intensity of the immune response.
(ii)	T-suppressor lymphocyte	(b)	Secretes a factor which stimulates formation of other types of lymphocyte.
(iii)	B-lymphocyte	(c)	Causes local inflammatory responses.
		(d)	Develops into an antibody-forming cell.
(iv)	T-cytotoxic lymphocyte	(e)	Recognises non-self antigens on other cells and lyses them.

10.1.4 The immune response

Let us briefly follow the fate of a pathogen or antigen after it enters the body and encounters the immune system. When infection takes place or when an antigen enters the body the invading antigen is first trapped by macrophages in lymphoid organs. Antigens are swept into the lymphatic vessels and brought either to a lymph node or via the blood to the spleen. Phagocytic cells are constantly patrolling the body for the presence of any foreign substance and, whatever they find, they engulf and digest. Some of the partially digested antigen, the so called processed antigen, returns to the cell surface as peptide molecules so that the antigenic epitopes can be presented to antigen-specific lymphocytes. The majority of antigens fail to stimulate T-helper cells unless these antigens are presented by antigen-presenting cells (APC) which are primarily macrophages and B-cells. MHC molecules bind the antigenic fragment(s) and then direct the subsequent events of the response. In summary, T-helper cells interact with antigen presented by antigen-presenting cells, and this interaction results in the proliferation of antigen-specific T-helper cells, whose primary task is to stimulate B-cells and cytotoxic T-cells to proliferate.

Antigen-specific T-helper cells interact with specific B-cells and induce or activate them to mature into plasma cells that secrete antibodies specific to the antigen. Antibodies carry out their appointed tasks of which there are many. If an antigen is a toxin, antibodies are able to bind to and neutralise it. If the antigen is a bacterium, antibodies can bind to it and bring about cell lysis with the help of molecules of the complement system. This form of immunity is called *humoral* immunity (from the Latin word *humor*, meaning fluid, since antibodies are found in the fluids of the body such as blood and mucosal secretions).

Other antigen-specific T-helper cells stimulate the proliferation and differentiation of effector cytotoxic T-cells. Effector cytotoxic T-cells are basically killer cells; they can lyse specific antigen-bearing cells such as tumour cells or virus-infected cells. Thus T-and B-cells proliferate in response to a primary antigenic stimulus and within just a few days of the the original encounter with antigen in a lymphoid organ, effector T-cells are generated and B-cells start producing antibodies. While the initial battle is won by the immune system, the enemy is not forgotten. An army of antigen-specific memory T-and B-cells is left behind after an encounter, and these cells are able to elicit a secondary response; should the same antigen be encountered again, the antigen-specific memory cells multiply faster than they would have done in the absence of a primary encounter.

Assignment 10.1

Give an example of how humans make practical use of the 'memory' of specific immunity to particular antigens.

Through vaccination it is potentially possible to stimulate the immune system to provide immunity from any kind of pathogen, and we shall look at this in detail later on in Section 10.5.

> ***Self assessment exercise 10.4***
>
> Explain how each of the following types of lymphocyte is presented with antigen:
> (i) B-cells (ii) T-helper cells (iii) cytotoxic T-cells.

10.2 Dysfunctions of the immune system and their modulation

We have seen that the immune system must mount an appropriate immune response against antigens that are foreign to the body, and not against self antigens. Early in life, the immune system is programmed to differentiate between self and non-self: it undergoes a period of instruction in the thymus termed thymic education, in which lymphocytes are selected on the basis of their lack of reactivity to self. In spite of this education, however, it is still possible for aberrant immune responses to be generated. These abnormal responses constitute dysfunctions of the immune system.

Immune dysfunctions can be divided broadly into two categories: (i) those immune responses that are directed at self antigens, and (ii) disproportionate immune response that are directed at foreign or non-self antigens. The latter may be an exaggerated immune response leading to dysfunction, or a lowered response resulting in poor functioning of the immune system as a whole.

10.2.1 Immune response to self antigens

The consequences of the immune system mounting a response against self are serious. Responses which have evolved for the elimination of pathogens, if invoked against self tissues, results in tissue damage, illness and even death. Thymic education is not always foolproof and there are many instances known where antibodies directed at self molecules such as DNA, hormones and membrane components are found circulating in the host. If they are found in sufficiently high concentrations, such antibodies are capable of adversely affecting the normal functioning of the host, and result in autoimmune disease.

> ***Assignment 10.2***
>
> Name an autoimmune disease.

Both T- and B-cells are capable of reacting abnormally and causing autoimmune problems, and experimental animal model systems are useful in analysing problems associated with such dysfunctions. Information gained from these animal models is useful in unravelling the processes involved in human immune diseases. The models also provide insights into the design of appropriate treatments to overcome immunological defects. Some examples of animal model systems are described in Section 10.2.3.

10.2.2 Disproportionate immune responses against foreign antigens

Disproportionate responses may be either hyper- (excessive) or hypo- (below optimal) responses. Both kinds of dysfunctions are observed clinically, and they may either be genetically inherited or acquired during one's lifetime. When an immune response is exaggerated or inappropriate it often results in tissue damage, and this phenomenon is called hypersensitivity.

Hypersensitivity reactions are caused by an over expression of the immune response, and have been classified into four different types, of which three are antibody-mediated and the fourth cell mediated. In Type I hypersensitivity an IgE antibody is produced in response to an antigen; the antibody then binds to a special type of cell called a mast cell, and induces an inflammatory response such as that seen in asthma and rhinitis. Type I hypersensitivity is also seen in allergic skin reactions to agents such as pollen, house dust and cat fur. Type II hypersensitivity results when an antibody directed at a cell-surface self antigen binds to cells and initiates phagocytosis or complement-mediated lysis. Examples of Type II reactions are seen after blood transfusions and in drug-induced reactions to blood components. In Type III hypersensitivity, antigen–antibody complexes initiate local damage after being deposited in tissues. Sometimes such complexes cannot be cleared rapidly from the circulation and serum sickness results. Type IV hypersensitivity is cell-mediated and is also known as delayed type hypersensitivity (DTH). It is initiated when cytokines (Section 10.3) are secreted by antigen-sensitised T-cells when the T-cells encounter the antigen a second time. DTH-mediated effects are seen in rejection of foreign tissue grafts and in allergic dermatitis caused by chemicals such as those found in poison oak, poison ivy and other substances.

Hypo-functioning of the immune system (immunodeficiency) was identified when patients presented with repeating bacterial or viral infections. Owing to the partial or complete absence of one or more components of the immune system, patients would often develop infections which were difficult to control, and which were often fatal unless antibiotics were administered. In many cases of immune deficiency the defect is genetically inherited. Examples of genetic immunodeficiencies include agammaglobulinaemia (low concentrations of IgG antibodies), severe combined immunodeficiency (SCID), and Bare Lymphocyte Syndrome (low levels of MHC Class II expression resulting in lowered immune responses).

Some forms of acquired immunodeficiency arise as a side-effect of exposure to gamma radiation and anti-cancer treatment with cytotoxic drugs such as methotrexate, vinca alkaloids and corticoids. These drugs often lead to a depression in the number of progenitor cells in the bone marrow, and this results in a generalised decrease in the red and/or white cell components of the blood. In the last decade, acquired immunodeficiency syndrome (AIDS) which results from infection with human immundeficiency virus (HIV) has spread alarmingly throughout the world.

Assignment 10.3

Explain one of the ways in which HIV infection results in immunodeficiency.

The mechanisms involved in HIV-induced immunodeficiency are quite different to those induced by other states of acquired and inherited immune disorders; in HIV infection, mature immunocytes are primarily affected whereas in most of the other conditions listed above, cells in the developing stage are affected.

10.2.3 *Use of animal models in the study of immune dysfunction*

For ethical reasons it is difficult to study immune dysfunctions and assess treatment regimes in humans. Experimental animal models offer the best approach. Mice, rats and guinea pigs are commonly used as convenient laboratory animals for most experimental work. Inbred strains of mice (which are genetically identical) have proved to be particularly useful: Lpr/1pr mice for the study of systemic lupus erythomatosus, and NOD mice for the study of insulin-dependent diabetes mellitus, are two examples. Some of these studies have shown that there are often clear associations between MHC types and certain autoimmune diseases such as insulin-dependent diabetes. In the case of experimental encephalomyelitis, where the disease pathology is due to an immune response directed at a self myelin basic protein, it has been possible to prevent the generation of the relevant T-cell receptors. Some autoimmune diseases are clearly mediated by T-cells and not by B-cells, a good example being insulin-dependent diabetes.

In other types of dysfunction of the immune system, similar approaches to those described above are being used to study the intricacies of the immune response. For example, a mouse strain that suffers from severe combined immunodeficiency has been generated to help both in the analysis of developmental defects and in the modification of the results of such defects. There are also animal models for the study of hypersensitivity reactions which have helped to elucidate the role of antigen-presenting cells, B-cells, cytokines, and the immunoglobulins IgG and IgE, in allergic responses.

10.2.4 Approaches for correcting immune dysfunctions

Until the underlying mechanisms of immune dysfunction are clearly understood at both the molecular and the cellular levels it will not be possible to design rational immune modulators that can be used for correcting it in human patients. However, given our current state of knowledge, some approaches can be made empirically: approaches which have withstood the test of time, and others which are based on information gained from experimental work.

It now looks feasible to modulate immune responses for the treatment of immune deficiencies or hyper-responsiveness. Vaccination with TB vaccine, Bacillus Calmette Guerine (BCG), has been shown to boost cell-mediated immune responses in immunosuppressed individuals by activating macrophages and cytolytic cells called Natural Killer (NK) cells. Bone marrow dysfunction can be remedied by transplantation of appropriate functional bone marrow cells. Corticoids and other drugs can be used to decrease the intensity of allergic responses. Treatment for autoimmune disorders is less well established than that for hyper- and hypo-responsive states. Hypersensitivity is routinely treated by injecting increasing doses of the allergy-causing antigen (or allergen). This results in an increase in levels of allergen-specific IgG and/or suppressor T-cell activity and a reduction in the amount of circulating IgE. Specific peptide therapy has been successfully employed in the treatment of auto reactive experimental encephalomyelitis in the mouse, but research has not reached the stage at which patients with multiple sclerosis may be similarly treated.

Self assessment exercise 10.5

Answer true or false to each of the following:

(i) Type II hypersensitivity reactions could be a consequence of ineffective thymic education.

(ii) Type IV hypersensitivity reactions are caused when cytokines are secreted by antigen-sensitised B-cells.

(iii) Hypersensitivity to pollen (hay fever) can be cured by repeated injection (vaccination) with pollen in order to boost levels of IgE in the circulation.

(iv) An AIDS sufferer could be protected from the diseases which occur in AIDS by vaccinating him/her against those diseases.

10.3 Cytokines

10.3.1 What are cytokines?

Cells of the immune system communicate in two ways: one way is mediated by direct contact between cells, and the other by soluble messenger molecules that carry signals from one cell to another. These soluble messengers are called cytokines, and they play crucial roles in the initiation and maintenance of immune responses. Many of the messenger proteins are growth factors for other cells, some are stimulators of activity in effector cells, and some are both growth factors and cell-stimulators. Cytokines promote activities aimed at (i) the elimination of invading micro organisms and parasites, and (ii) the repair of damaged tissues. They can act either as short-range messengers between cells, or as long-range messengers by circulating in the blood and affecting cells at remote sites, just as hormones do. Several cytokines are named interleukins, abbreviated to IL.

Table 10.1 Biological activity demonstrated by cytokines in a variety of cells

	Primarily produced by	Principal function
IL-1	Monocytes, macrophages	Stimulates T-cells; inflammatory, tumouricidal
IL-2	T-helper cells	Proliferation and differentiation of T-and B-cells; activates NK cells
IL-3	T-helper cells	Stimulates growth of stem cells and mast cells
IL-4	T-helper cells	Stimulates growth of T and B-cells and mast cells; stimulates IgE production
IL-5	T-helper cells	Eosinophil and B-cell growth and differentiation factor
IL-6	T- and B-cells, monoctyes, endothelial cells	Stimulatory and differentiation factor for B-cells
IL-7	Bone marrow stromal cells	Production of B-cell precursors in bone marrow; induces proliferation of thymocytes
IL-8	Mononuclear leukocytes	Neutrophil chemotactic factor
IL-9	T-Helper cells	T -helper cell growth factor
IL-10	T-helper cells	Suppresses cytokine synthesis
Il-11	Bone marrow stromal cells	Stimulates precursor cell growth
IL-12	Transformed B-cells	Stimulates NK cells
IFN-γ	T-cells and NK cells	Anti-viral; activates macrophages and Nk cell
TNF	macrophages and lymphocytes	Inflammatory, anti-tumour effects
CSF	T-cells	GM—CSF-Stimulates granulocytes and macrophages G-CSF—Stimulates neutrophils and granulocytes

IL = Interleukin IFN = Interferon
TNF = Tumour necrosis factor CSF = Colony stimulating factor

The term cytokine also applies to lymphokines (which are produced by lymphocytes), and monokines (which are produced by monocytes). The different names given to these factors has made this subject area somewhat confusing at times, with the same molecule being referred to by different names by different laboratories. Furthermore, substances referred to as lymphokines are also produced by non-lymphocytes and can act on non-lymphocytes. For reasons of simplicity we will use the term cytokine to describe the whole group of intercellular messengers in general.

Cytokines, as messengers between interacting cells, act on target cells by binding to specific receptors on the target cell surface. The binding of a cytokine to its receptor is followed by signal transduction across the cell membrane and into the cytoplasm. Smaller molecules or ions, which diffuse within the cytoplasm, in turn, carry the signal to other cell proteins or enzymes and cause activation of intracellular processes. Interestingly, the same lymphokine may have quite different effects on different target cells, which indicates that the end result is a programmed property of the target cell. Thus, the main function of cytokines is to amplify immune responses and inflammatory responses, by inducing cell proliferation selectively while at the same time keeping responses under control.

Close to twenty cytokines have now been characterised and analysed for their molecular nature, function, source, and target. As Table 10.1 shows, cytokines are produced by a variety of cells and many of them have several functions.

10.3.2 Production of cytokines in vitro

Four types of cell secrete cytokines *in vitro*.

(a) Stimulated T-cells
Lymphocytes isolated from peripheral blood (in humans) or from spleen (in rodents) and cultured in appropriate growth medium, when stimulated with mitogen or with specific antigen to undergo mitosis, become activated and release cytokines into the culture medium. The stimulation can be followed by cytokine purification and identification of the cytokines released.

(b) T-cell hybridomas
Mitogen-activated or antigen-activated T-cells can be cloned or fused with appropriate cells to form hybridomas (see Section 10.4). The advantage of these hybridomas is that the cytokines produced from them in culture medium are more clearly defined compared to the material produced by bulk culturing of blood cells.

(c) Constitutive or inducible cell lines
These are established cell lines (leukaemic cell lines for instance) that constitutively produce certain cytokines in culture medium, or can be induced to do so.

(d) Genetically engineered cells containing genes coding for cytokines
Large amounts of pure cytokines can be produced by recombinant DNA methodologies, and is accomplished by the expression of the appropriate cytokine gene in an appropriate vector.

10.3.3 Therapeutic uses of cytokines

The cytokine that has attracted most attention in terms of potential applications in clinical medicine is interleukin-2 (IL-2). IL-2 is of central importance in the cascade of immune responses, and has proved to be extremely useful in experimental work on T-cells *in vitro*.

A great deal of interest has been generated by the use of IL-2 in cancer immunotherapy and in the treatment of immunodeficiency disease. In addition to its role in inducing the proliferation and differentiation of T- and B-cells, IL-2 is able to increase the cytotoxic capacity of natural killer (NK) cells. NK cells can spontaneously kill some tumour cells and virus-infected cells.

IL-2 also promotes the differentiation of lymphokine-activated killer cells (LAK): following exposure to IL-2 *in vitro*, leukocytes from cancer patients often exhibit strong cytotoxic reactions against a variety of tumours but have no effect on normal tissues. While high doses of IL-2 by themselves appear to be effective in causing tumour regression, the administration of IL-2 with LAK cells seems most effective. The injection of LAK cells directly into the tumour tissue in melanomas and breast cancer results in partial regression of the lesions. There is hope that the use of tumour-infiltrating lymphocytes may be substantially more effective than the use of peripheral blood leukocytes, and that the use of cocktails of cytokines may also be beneficial.

The related cytokine IL-3 plays important roles as it promotes the production and differentiation of a variety of blood cells such as granulocytes, monocytes, erythrocytes and platelets. Granulocyte-macrophage colony-stimulating factor (GM-CSF) stimulates the production of neutrophils, eosinophils, monocytes and macrophages; these features may find applications affecting the differentiation of stem cells into the various lymphoid cells. In patients receiving radiation or chemotherapy for cancer (which results in the depletion of stem cells and an inability to combat infections) IL-3 therapy may be beneficial in promoting stem cell differentiation, while GM-CSF has been tested in diseases where white blood cell counts are low.

Interferons comprise a group of proteins that are important in combating viral infection. Gamma interferon (IFN-γ) exhibits a variety of effects ranging from the inhibition of viral replication, inhibition of cell growth, increase in MHC expression, to an increase in phagocytosis and cytotoxicity in phagocytes. IFN-γ activates macrophages which then exhibit increased phagocytosis and increased cytotoxic effects. IFN-γ also stimulates NK-mediated cytotoxicity, which is another feature to be considered in favour of IFN-γ as an immuno-modulator.

There are several clinical conditions for which IFN-γ treatment may be beneficial. For example, the treatment of HIV-positive patients with IFN-γ may result in improvements in combating the opportunistic infections that are common in AIDS. IFN-γ has also proved useful in treating a particular type of cancer called hairy cell leukemia. However, the danger of the occurrence of possible deleterious side-effects, such as the induction of autoimmune reactions, has to be kept in mind.

Tumour necrosis factor, TNF, is another multipotent factor, with its name implying only one of its capabilities. TNF is a cytokine by definition, even though it does not have an IL designation; the TNF produced by macrophages/monocytes is called TNF-α and the TNF from T-lymphocytes is termed TNF-β. The tumoricidal properties of TNF have prompted investigations into its possible use in eliminating tumours; a combination of TNF and IFN-γ has proved to be effective in inhibiting the growth of tumours, as has a mixture of TNF and IL-1. Again, as with some other cytokines, optimism has to be balanced by clinical observations that TNF induces septic shock and weight loss.

In summary, while actual treatment regimens must be carefully selected because of problems with possible side-effects, it is envisaged that cytokines will become important immunotherapeutic tools in modern medicine.

Self assessment exercise 10.6

Answer true or false to each of the following:

(i) It is possible for a single messenger molecule to be correctly termed a cytokinin, a lymphokine and an interleukin.

(ii) TNF cannot be correctly termed a cytokinin, a lymphokine and an interleukin.
(iii) IL-2, TNF and IFN-α can all cause anti-tumour activity by stimulating NK cells.

10.4 Monoclonal antibody technology

10.4.1 Production of monoclonal antibodies

Molecular genetic studies on antibody synthesis by B-cells indicate that as a group they have the potential capacity to respond to at least ten billion different antibody specificities, which would be sufficient to recognise any antigen. The clonal selection theory recognises that each B-cell is able to produce an antibody response to only a single antigenic determinant/ epitope. A typical antigen is a complex molecule with several antigenic determinants which therefore induces in an animal a polyclonal response, in which antibodies of differing specificity, affinity and avidity are produced by a number of different lymphocyte clones. The response differs from animal to animal, has a finite life span, and requires induction by the presence of the antigen in the host.

In the disease called multiple myeloma, which is a clonal tumour of B-cells, all malignant cells generally produce one type of Ig (of a single specificity). Production of Ig by myeloma cells, which have an infinite life span, does not require induction by the presence of an antigen. Most of the myelomas are secretory, but some non-secretory myeloma cell lines are also available, especially in rodents.

The antibody producing cell, the B-cell, has a definite life span, but it can be immortalised artificially by fusing it with a myeloma cell which, by virtue of its cancerous nature, can multiply indefinitely *in vitro*. The hybridisation of these two cell types was achieved for the first time by Georges Kohler and Cesar Milstein in 1975. The basic procedure involves the fusion of B-cells from an immunised animal (each making antibody against a given epitope) with myeloma cells which do not secrete immunoglobulins and which lack functional hypoxanthine guanine phosphoribosyl transferase (HGPRT, a key enzyme in the salvage pathway of DNA synthesis, Chapter 3). Fusion of the two cell types to produce a hybridoma is carried out by stimulating fusion of the cell membranes using polyethylene glycol. Hybrid cells are selected from the myeloma cells in a medium containing aminopterin (which blocks the *de novo* pathway of DNA synthesis), thymidine and hypoxanthine.

Assignment 10.4

Cell fusion in hybridoma production is a random event and hence the resulting cell mixture contains unfused lymphocytes and myeloma cells, cells formed by fusion of lymphocytes with lymphocytes, cells formed by fusion of myeloma cells with myeloma cells, as well as the required hybridomas formed by fusion of lymphocyte with a myeloma cell. Cells need to synthesise nucleic acids in order to grow. Nucleic acids may be synthesised *de novo* using appropriate simple precursors, or by the salvage pathway using hypoxanthine (as a substrate for HGPRT for synthesis of purines) and thymidine for synthesis of pyrimidines. Explain how HAT (hypoxanthine: aminopterin:thymidine) medium selects for hybridomas.

In this technique B-lymphocytes are immortalised randomly, whether they secrete antibody against the required epitope or another epitope in the immunogen used to vaccinate the animal from which the B-cells were taken. There is thus a need to segregate clones of hybridoma which produce antibody against the required epitope. This can be achieved by using a limiting dilution technique in which the cell suspension is diluted in tissue culture medium so that only one cell occurs in aliquots added to tissue culture dishes. After growth, the contents of each dish are screened for the presence of the required antibody. By repeating this procedure a population (clone) of cells producing antibody of required specificity (reacting against one epitope only) can be obtained. The antibody secreted by a hybridoma clone is referred to as a monoclonal antibody (MAb). Figure 10.2 summarises the various steps in the production of a hybridoma clone.

Hybridomas can be grown up to produce MAbs in the laboratory on a small scale, or on a larger scale in a production facility in air-lift fermenters or other bioreactors (Chapter 9). Alternatively, mouse hybridoma clones can be grown as ascites after injection into the peritoneal cavity of the same strain of mice from which the progenitor cells were obtained. This procedure produces high yield of antibody without the need for sophisticated tissue culture facilities. However, the mice need to be maintained in a germ-free environment so that they are not producing antibody against other antigens.

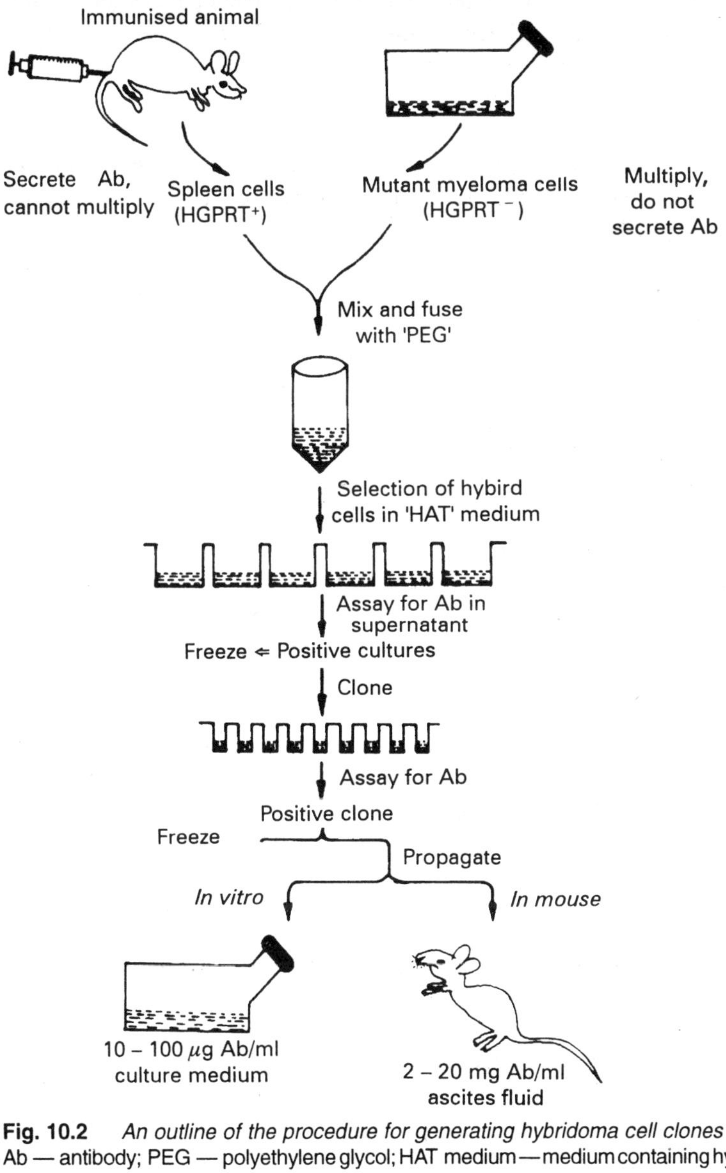

Fig. 10.2 *An outline of the procedure for generating hybridoma cell clones producing monoclonal antibodies.* Ab — antibody; PEG — polyethylene glycol; HAT medium — medium containing hypoxanthine, aminopterin and thymidine.

10.4.2 Strategies for the production of human monoclonal antibodies

While mouse monoclonal antibodies have revolutionised biology in general and immunology in particular, there are limitations when it comes to using them for injection into humans (the reasons for wanting to do this are discussed in Chapter 11).

> ***Assignment 10.5***
>
> Explain why monoclonal antibodies from mice are of limited use for administration to humans.

To circumvent this problem attempts have been made to produce monoclonals from human cells. Despite the difficulty that this technology poses, the potential of this treatment to human therapeutic and prophylactic use has stimulated a great deal of research and development in this area.

To generate human MAbs the following strategies have been used:

(a) Human–human hybrids

Attempts have been made to fuse sensitised human B-cells (from an individual exposed, naturally or by vaccination, to an epitope) with human plastocytoma or lymphoblastoid cell lines (immortal cells from tumours of lymphocytic cells).

Progress in the development of such human–human hybridomas has been slow owing to the lack of suitable non-antibody-secreting human plasmacytomas or lymphoblastoid cell lines with which to fuse human B-cells. However, the first human MAb against an antigen has been produced by fusing B-lymphocytes with a human plasma cell line.

(b) Interspecific hybrids

Attempts have been made to fuse human B-cells with non-secretory mouse or rat myeloma cells (the antibody produced by such interspecific hybridomas would be of human type as its production is coded for by the human B-cell). However, such interspecific hybrid cells are found preferentially to eliminate human chromosomes. Human B lymphocytes isolated from peripheral blood, bone marrow, spleen, tonsil or lymph node have been used for fusion. The number of B-cells secreting specific antibody is quite small in peripheral blood, but their number can be increased by stimulating growth of these cells *in vitro* with pokeweed mitogen or antigen, or a combination of both.

(c) EBV-transformation

The Epstein–Barr virus (EBV) transforms cells and results in uncontrolled growth of the transformed cells. EBV transformation of sensitised B-cells has been used as a means of producing an immortalised line of B-cells. However, EBV-transformed cell lines often secrete low amounts of antibody and are unstable.

> ***Assignment 10.6***
>
> What potential hazards would there be of injecting monoclonals produced by EBV-transformed B-cell lines into humans?

(d) Fusion of EBV-transformed human B-cells with rodent myelomas

In order to overcome the instability of EBV transformed B-cells, attempts have been made to fuse antibody-secreting EBV-transformed human cell lines with either mouse myeloma or human myeloma in order to produce stable hybrid cell lines secreting human antibody.

(e) Growth of human B-cells in immunodeficient mice
Human lymphocytes have recently been grown by injection into mice with severe combined immunodeficiency (SCID), after which the animals have been immunised to make human antibodies.

> ***Assignment 10.7***
>
> Explain why the human B-cells need to be grown in an immunodeficient mouse; why cannot a normal mouse be used?

(f) Humanising mouse monoclonals by means of genetic engineering
Using genetic engineering (recombinant DNA technology) it is possible to produce chimeric antibodies. The term 'chimeric' is derived from Greek mythology, in which the chimera was a monster with the head of a lion, the body of a goat and the tail of a dragon. A chimeric molecule (such as a monoclonal antibody) is thus one in which diverse segments are coded for by genes which are not normally associated with each other. These genes may be interspecific (from different species) or intraspecific (from the same species).

> ***Assignment 10.8***
>
> The F_C region of a mouse MAb is primarily responsible for the immunogenicity of this molecule in the human body. How could genetic engineering be used to produce a chimeric antibody to overcome this problem?

Recently transgenic mice have been made with genes coding for human immunoglobulin heavy chain regions, which will allow human antibodies to be produced directly from hyperimmunised mice.

> ***Self assessment exercise 10.7***
>
> Answer true or false to each of the following:
> (i) Antibody produced from a clone of hybridoma cells derived from fusion of one sensitised B-lymphocyte and one cell of a secretory myeloma can be called a monoclonal antibody;
> (ii) antibody produced from a clone of hybridoma cells derived from one sensitised human B-lymphocyte and one cell of a mouse non-secretory myeloma can be called an intraspecific chimeric monoclonal antibody;
> (iii) antibody produced from a clone of hybridoma cells derived from one sensitised mouse B-lymphocyte (in which the gene coding for the F_C fragment of immunoglobulin has been replaced by the equivalent human gene) and one cell of a mouse non-secretory myeloma can be called an interspecific chimeric monoclonal antibody.

10.4.3 Bispecific monoclonal antibodies

Bispecific antibodies are those in which the F_{ab} (antigen-binding) region of one arm has specificity for (is directed against) one epitope and the F_{ab} region of the other arm has specificity for another (different) epitope (Fig. 10.3). By definition, these antibodies are chimeric. They may be formed by genetic engineering or by fusing cells of two different hybridoma lines (so producing a hybrid hybridoma). Such antibodies allow refinements in and extentions of uses of monospecific antibodies (see Chapter 11).

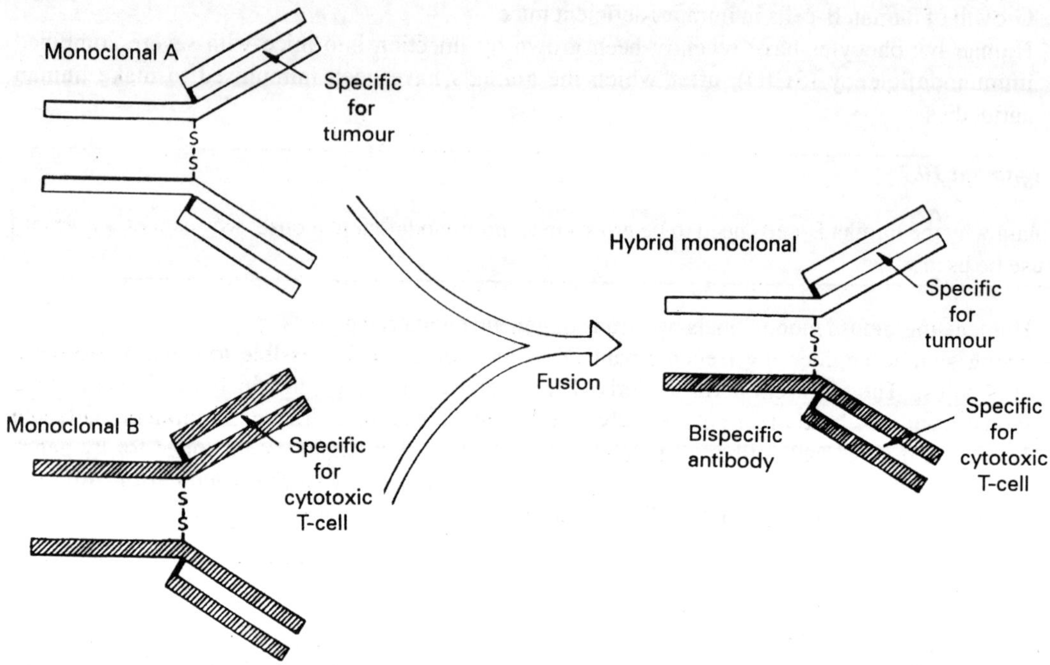

Fig. 10.3 *Bispecific antibody.*

10.4.4 Conjugated monoclonal antibodies

Immunoglobulins, including monoclonals, can be conjugated to other molecules using a variety of techniques to produce stable bonding between the two molecules. MAbs can be linked to enzymes such as alkaline phosphatase (for use in enzyme linked immunosorbent assay — ELISA), to enzymes such as peroxidase (for use in histological analysis), to radionuclides (for radioimmunoassay — RIA), and to cytotoxic molecules and drugs (for use in therapy). These applications will be discussed in more detail in Chapter 11.

10.4.5 Cloning and expression of immunoglobulin genes

Initially, antibody genes were isolated from hybridomas, cloned into plasmid vectors, and expressed as complete antibodies in mammalian cells or as fragments in bacteria. More recently a new approach was proposed which bypasses the need for hybridoma formation. Antibody genes are cloned directly from the lymphocytes of the immunised animals, expressed in bacteria and the antibody products then screened for their capacity to bind to antigen. Alternatively, it is possible (by using molecular biology approaches) to express in *E. coli* the DNA coding for the variable domains of immunoglobulin chains, obtained from libraries of DNA from genomes of spleen cells.

Looking ahead, it may become possible to bypass animal systems completely and still make new antibodies *in vitro*. This will involve the preparation of naive repertoires of antibody genes, expressing a gene library from bacteriophage in bacteria, screening for antigen-binding activity, and following this with enhancement of affinity by introducing point mutations. It has been possible to express functional antibody fragments on the surface of bacteriophages. Research in the next few years should tell us whether we will be able to imitate *in vitro* the production of high affinity antibodies which an animal accomplishes so efficiently *in vivo*. It may even be possible to produce functional antibodies from genetically engineered plants (Chapter 12).

10.5 Vaccinology

Vaccines have been very successful in the prevention of communicable diseases such as diphtheria and smallpox. Vaccines act by stimulating host immunity and rendering the body inhospitable for the growth of pathogens.

10.5.1 *Vaccines in the expanded programme of immunisation (EPI)*

Vaccines are the backbone of the "Health for All" programme of the World Health Organisation. Those that are part of the Expanded Programme of Immunisation undertaken by the governments of several countries are the triple vaccine DPT (consisting of antigens against organisms causing diphtheria, whooping cough and tetanus), and vaccines against poliomyelitis and measles. BCG vaccination against tuberculosis is not mandatory in some countries, since BCG vaccination has resulted in a fairly low efficacy against tuberculosis in extensive trials carried out in Myanmar (Burma), the United States and southern India. On the other hand, BCG has demonstrated high efficacy in Uganda and the United Kingdom. It continues to be included in the EPI Programme in India in the absence of an alternative.

The Sabin oral vaccine against polio (using live attenuated polio virus) is in use in many developing countries. It requires 3 to 5 repeat doses before the desired level of protection is attained and the vaccine must be kept refrigerated if its activity is to be maintained. An alternative polio vaccine, developed by Jonas Salk, employs inactivated virulent polio virus and produces good immunity. Furthermore, the Salk vaccine is not as sensitive to storage temperature and is in use in several countries. Unfortunately it costs more to produce than the Sabin version.

Assignment 10.9

Live vaccines such as BCG (for TB) and smallpox require only one vaccination (in the arm) for effective immunity to be induced as the live organism grows in the host, thereby effectively amplifying the effective dose of vaccine. However, live polio vaccine, given by mouth, requires a minimum of 3 doses of vaccine for effective immunity and may need as many as 5. Explain this.

10.5.2 *Vaccines undergoing improvement*

The present pertussis (whooping cough) vaccine in the triple DPT vaccine is a crude preparation of whole cells and is not entirely free of side-effects. Purified sub-unit vaccines have beeen developed in Japan and USA and are presently undergoing clinical trials, and are likely to replace the present pertussis vaccine in the near future. The traditional whole-cell vaccine against cholera (given by injection) is also a fairly crude preparation and poses problems such as the induction of poor and short-lived immunity and of side-effects such as malaise. A new oral vaccine for cholera, which consists of the B chain of cholera toxin along with cell wall preparations, is currently undergoing clinical trials in Bangladesh. This vaccine is reported to have an efficacy of approximately 85%, much greater than that of conventional vaccine.

The traditional typhoid vaccine (given by injection) confers only short-term immunity. A better oral vaccine, consisting of a self-limiting double mutant of *Salmonella*, is being developed for typhoid. In recent field trials conducted in Egypt, the vaccine registered 96% protection. In Chile, however, the efficacy was of a lower order, and trials are being repeated. An injectable vaccine consisting of the *Vi* capsular polysaccharide of *Salmonella typhi* is also being tested in Nepal; it has demonstrated protective efficacy of 70%. Improved immunogenicity and efficacy for prevention of typhoid has been demonstrated in animal models with the *Vi* capsular polysaccharide linked to either cholera toxin or diphtheria toxoid as carrier.

10.5.3 Progress in vaccinology

Vaccinology today is an interdisciplinary area of research involving immunologists, protein chemists, recombinant DNA technologists, crystallography experts and computer graphics specialists. First generation vaccines were suspensions of killed pathogens and which contained more material than was necessary to confer immunity. Some of the components were pyrogenic (fever-inducing), which is why transient fever usually accompanied immunisation. These vaccines were replaced with purified proteins. Good examples here are tetanus and diphtheria vaccines, where the respective purified toxins are employed as vaccines after they are converted into non-toxic toxoid by treatment with formaldehyde. The idea of targeting the secreted toxin rather than the bacterium itself was based on the crucial observation that the pathogenic action of the infective organisms, *Clostridium tetani* and *Corynebacterium diphtheriae* occurred at a site distant from the site of colonisation of the bacteria, suggesting that the pathology was mediated by toxins secreted by the pathogens. Toxoids remain invaluable effective vaccines with little accompanying side-effect. These vaccines may undergo further modification in future in view of recent observations that repeated immunisation often results in immunosuppression in both experimental animals and humans. Suppressor epitopes have been described in some molecules such as lysozyme and galactosidase and such epitopes are thought to induce T-suppressor cells that may inhibit T-helper cells and/or B-cell activities. In line with other developments in this field, future toxoid vaccines may consist of sub-molecular domains which are experimentally selected on the basis of their immunogenicity and lack of suppression potential.

In the course of the development of many vaccines the following strategy is used. First, suspensions of killed or inactivated pathogens may be used to raise an immune response. Second, proteins from the pathogen are isolated and characterised and the immune system's response to them analysed in animal models. Third, particular domains of the proteins that are responsible for eliciting a protective response are characterised. A good example of this progression in vaccine technology is provided by the vaccine against the Foot and Mouth Disease Virus (FMDV). The initial vaccine was an inactivated virus suspension. Research indicated that one of the coat proteins, VP1, was mainly involved in protection. Further research has shown that a 21 amino acid segment of VP1, amino acids 140–160, is immunodominant (the dominant epitope) and has receptors on both B and T-cells. Whereas the vaccine consisting of the whole inactivated virus would only immunise against one particular serotype of the virus, peptide 140–160 is common to all strains of FMDV and can induce protection against all serotypes.

10.5.4 Recombinant vaccines

The first vaccine prepared by recombinant DNA techniques is already in public use. The vaccine is based on the surface protein of the hepatitis B virus. The American version of this vaccine utilises the gene product expressed in yeast, while the vaccine developed in France uses the gene product expressed in animal cells. The recombinant vaccine is presently more expensive than the previous vaccine employing aggregates of the hepatitis B surface protein purified from the blood of human carriers. While the purified aggregates are more immunogenic than the recombinant vaccine, concern regarding the possible presence of contaminating pathogens such as HIV make the recombinant vaccine preferable from the safety point of view. Interestingly, this vaccine immunises the recipient not only against hepatitis but also against the subsequent development of hepatocellular carcinoma (a consequence of hepatitis B infection in a number of individuals infected by the virus).

Even though the recombinant vaccine against hepatitis B was introduced only a few years ago, it was realised that better immunity could be attained by using not only the hepatitis B surface protein but also the viral proteins pre-S2 and pre-S1. (The protein pre-S1 has domains which associate with the liver cell surface and enable the virus to enter hepatocytes.)

> ***Assignment 10.10***
>
> By what mechanism would pre-S1 in a vaccine be likely to confer resistance to hepatitis B virus?

The core protein of hepatitis B virus is highly immunogenic and important from the perspective of protection, because it is known that IgG antibody responses against the core antigen are found in patients who have successfully overcome the infection. These findings should lead to further improvements in the present vaccine.

Another area in which there has been considerable progress is the research to develop synthetic peptides as vaccines for hepatitis B. It has been shown that the two loops created by linkages of cysteine within the hepatitis B surface protein are immunodominant and constitute the "a region". These loops, between amino acids 124 and 137 and between 139 and 147 of the surface protein, project from the membranes of infected cells. A synthetic candidate peptide vaccine based on these has been developed at the International Centre for Genetic Engineering and Biotechnology, New Delhi. Another exciting aspect of this peptide construct is its ability to self-aggregate by forming inter-peptide disulphide linkages, since peptide aggregates are more immunogenic than the monomers. While the self-aggregating loop sequences of hepatitis B are attractive candidates for a hepatitis B vaccine, mixed aggregates of hepatitis B and HIV peptides have also been made which exploit the interpeptide disulphide linkages in the chosen sequences of the two viruses.

A live recombinant vaccine against rabies in wild animals has been approved for use in Europe. The vaccine consists of a genetically-engineered vaccinia virus into which the gene coding for rabies virus glycoprotein has been inserted. This vaccine is highly effective: a single intradermal inoculation confers protection on dogs and raccoons. The vaccine can also be given orally in bait. In New Delhi, scientists have expressed the rabies glycoprotein in the vaccinia employing the glycoprotein gene from a different strain of rabies virus. A single intradermal immunisation with this vaccine confers 100% protection of mice challenged with a virulent rabies virus. This vaccine is intended for use in dogs.

10.5.5 Vector-borne recombinant vaccines

The incorporation of foreign genes into viral and bacterial vectors is now beginning to find useful applications. The example of vaccinia as a vaccine vector against rabies described above is a good example. Vaccinia is an excellent inducer of both cellular and humoral immunity, has a wide spectrum of infectivity in animals and humans, and vaccinia-HIV constructs are currently undergoing clinical trials. For greater safety, avian pox viruses are being explored as vectors in place of vaccinia. A rabies vaccine using the canary pox virus as vector has entered clinical trials in humans, and the fowl pox virus is also under study as a potential vector for a birth control vaccine in New Delhi. In view of their affordability and other desirable characteristics, research is in progress with possible alternative vectors such as BCG, adenoviruses and *Salmonella*. However, none of these have yet been developed to the stage of clinical testing.

Assignment 10.11
What are the advantages and disadvantages of a live recombinant vaccine (compared to non-live vaccine) to a developing country?

10.5.6 Synthetic vaccines

The use of peptides is also being explored in the quest for an anti-malarial vaccine. Spf 66, a synthetic hybrid molecule, is an example of a potential peptide vaccine and has been developed in Colombia. The vaccine affords protection levels of approximately 50%. Synthetic or peptide vaccines based on immunodominant epitopes of *Plasmodium falciparum* circumsporozoite protein have also been produced in the USA and are undergoing clinical trials. Peptides are generally poor immunogens. They require conjugation to a carrier peptide or protein and often require use of strong adjuvants in order to elicit a good immune response. It should also be borne in mind that, with a limited number of epitopes in a given peptide, genetic (MHC) restriction often prevails (Section 10.1.3), and this prevents universal immune responses to the immunogen. Furthermore, the antibodies induced by peptide immunisation may not be of sufficiently high affinity for immunity to be effective. Synthetic peptide vaccines that employ several relevant epitopes may eventually emerge as being the most effective.

A semi-synthetic vaccine, directed against the hormone gonadotropin releasing-hormone (GnRH), has reached Phase I clinical trials after successful toxicological studies and experimentation in laboratory animals, including sub-human primates. The decapeptide hormone, common to both males and females, regulates the secretion of the pituitary gonadotropins — luteinising hormone (LH) and follicle stimulating hormone (FSH) — which are responsible for the generation of gametes (spermatozoa in the male and eggs in the female) as well as for the production of steroid sex hormones. Immunisation against the decapeptide leads to a block in the functions controlled by GnRH, spermatogenesis in the male and ovulation in the female. The synthesis of testosterone in the male and that of estradiol and progesterone in the female are also impaired. The vaccine has therapeutic potential against several clinical disorders, such as precocious puberty and sex hormone-dependent cancers. It is also a much cheaper alternative to GnRH analogues used currently. A GnRH candidate vaccine developed in New Delhi has been approved for trials in patients in India and Austria with carcinoma of the prostate. A large percentage of such cancers are androgen-dependent and the ability of the vaccine to inhibit synthesis of androgens is being exploited as a means of treatment. The vaccine causes a marked atrophy of the prostate in rats and monkeys. In early clinical trials, vaccination improved the clinical status of many, but not all, immunised patients. The size of the prostatic tissue was reduced as seen by ultrasonography and confirmed by a decline in the levels of the prostate-specific antigen.

10.5.7 Birth control vaccines

Vaccines traditionally mobilise the body's immune system to react against external antigens, but an interesting variation on this theme is the induction of immunity against self antigens in order to intercept and control a normal process. One example is the GnRH vaccine, described above. Its role is to inhibit reproduction by inducing antibodies against a hormone crucial to the reproductive system. The vaccine against GnRH can block fertility in both males and females but, since it also impairs the secretion of sex steroid hormones, it may not be an acceptable form of contraception unless complemented with hormone supplements.

Currently the most advanced birth control vaccine is the hCG vaccine, which induces antibodies against human chorionic gonadotropin (hCG). The appearance of hCG is an early signal of pregnancy and it plays a vital role in its establishment and maintenance.

> ***Assignment 10.12***
>
> How would vaccination wmith hCG act in birth control?

The strategy of targeting hCG does not interfere with a woman's normal pituitary and ovarian functions: she ovulates normally and has regular menstrual cycles. The interception occurs only when she conceives. A drawback with the most commonly used steroidal contraceptives is that they block the pituitary – ovarian axis; the user does not make her normal hormones and these therefore have to be replaced by synthetic substitutes in the contraceptive pill. This is not the case with the hCG vaccine.

hCG is normally made in appreciable amounts only in pregnancy (which is why diagnosis of pregnancy is based on testing for the presence of hCG in the urine). It is, however, also made by a variety of cancers. An observation was made that a certain type of human lung cancer makes hCG (and its constituent subunits, the α- and β-subunits). This acts as an autocrine growth factor for the tumour cells. It was noted that anti-hCG antibodies can block the proliferation of lung tumour cells *in vitro* and *in vivo* in nude mice. Based on this observation, the hCG vaccine developed originally for the control of fertility alone, has entered clinical trials in Mexico in patients with such lung cancers. This is an unexpected but interesting additional application of the hCG vaccine, particularly because there is presently no effective chemotherapy for this type of cancer.

A vaccine was initially devised in which the β-subunit of hCG was linked to tetanus toxoid in defined proportions. This structured conjugate vaccine overcame immunological tolerance to hCG (or its subunits) that can develop in women. An improved formulation of the vaccine employs the β-subunit of hCG associated with a heterospecific (non-human) α subunit. The heterospecific dimer (HSD) so created, is a tailored analogue of hCG which has the correct conformation to recognise tissue receptors (whereas the dissociated α- and β-subunits fail to do so). The heterospecific α-subunit, being foreign to humans, enhances immunogenicity of the vaccine without provoking cross reactions with human follicle stimulating hormone and human thyroid-stimulating hormone, since the heterospecific α-subunit is not immunologically cross reactive with the human α-subunit. Moreoever, the bio-neutralising capacity of the antibodies induced by the HSD vaccine is superior.

The HSD–hCG vaccine was approved by the Drug Regulatory Authorities and the Institutional Ethics Committees for Phase II efficacy trials which have recently concluded in three major medical institutions in India. Sexually-active women of proven fertility received the immunisation. The protocol required observations over 750 menstrual cycles to determine whether antibody levels above 50 ng/ml were protective. These trials have demonstrated for the first time the feasibility of preventing pregnancy by a birth control vaccine. The reversibility of the vaccine (ability to regain fertility once the antibody levels decline in the absence of a booster injection) has also been verified. In earlier Phase I clinical trials, the safety of the vaccine and the lack of side-effects were demonstrated. This HSD–hCG vaccine is currently the most advanced fertility control vaccine.

10.5.8 *Follicle stimulating hormone (FSH) vaccine for the control of male fertility*

Scientists in India have developed a vaccine that induces antibodies against follicle stimulating hormone (FSH), a hormone required for spermatogenesis in primates. The vaccine completed experimental studies in animals and pre-clinical toxicological studies, and is expected to enter Phase I clinical trials in the near future.

10.5.9 Other birth control vaccines under development

Scientists at the Indian Institute of Science in Bangalore are developing a vaccine against the chicken riboflavin-binding protein. The antibodies are observed to prevent pregnancy in rodents and primates. Scientists at the Institute for Research in Reproduction in Bombay are developing a vaccine against a prostatic peptide which has inhibin-like activity. Antibodies to this peptide agglutinate sperm.

Self assessment exercise 10.8

Answer true or false to each of the following:

(i) Whole-cell vaccines have the drawback that as well as containing antigens they may contain substances producing unwanted reactions to vaccination and substances which depress immune response to the vaccine.
(ii) toxoid vaccines are purified toxins which have been treated with formaldehyde in order to reduce their immunogenicity.
(iii) peptides may make poor vaccines because they contain so few epitopes that there may not be MHC sites for these epitopes to bind to on T-helper cells and thus an immune response is not induced.
(iv) vaccination against gonadotropin-releasing hormone and human chorionic gonadotropin can be effective birth control vaccines in both males and females.

10.5.10 New vaccines for the neglected communicable diseases

Research is in progress to develop vaccines against malaria and schistosomiasis. In malaria research, antigens expressed in three different stages of parasite development are being explored: the sporozoite, the merozoite and the gamete. Possible vaccines have reached the stage of clinical evaluation. Peptides denoting epitopes of interest have been made by scientists in Colombia. Scientists of the International Centre for Genetic Engineering and Biotechnology at New Delhi have proposed selected peptides for eliciting immunity against the malarial parasites *Plasmodium falciparum* and *Plasmodium vivax*. Scientists in France have developed candidate vaccines against schistosomiasis. The field of vaccinology is very active and it is expected that a number of new vaccines against parasitic infections will be developed in the years to come.

Three anti-leprosy vaccines are currently undergoing clinical trials. Scientists in Venezuela have developed a vaccine consisting of killed *Mycobacterium leprae* and live BCG. This vaccine confers protection similar to that given by BCG, which is known to vary from country to country. In Uganda and the UK protection conferred by BCG is high, while in countries such as India, Myanmar and in the southern regions of the USA, it is low. Two vaccines for leprosy have emerged from research conducted in India. One is based on the killed ICRC bacillus (belonging to *Mycobacterium avium*) primarily investigated by scientists at the Cancer Research Institute in Bombay. It is currently under comparative large scale immunoprophylactic trials in southern India with the WHO vaccine (developed in Venezuela) and a *Mycobacterium* vaccine developed in India. The results of these trials will begin to accumulate within a few years, but the final conclusions will have to await the early part of the next century.

10.5.11 Vaccine delivery systems

Several vaccines, such as DPT require the administration of multiple doses, and the lack of compliance with the proper regimen often results in incomplete immunisation and therefore lack of protection. Research is progressing to make biodegradable microspheres (of polylactic and polyglycolic acids) in which multiple doses of the vaccine could be encapsulated. Such systems would demand only one contact point with the vaccinees for delivery of the full vaccine complement. It is evident that these delivery systems would offer important logistical advantages for immunisation.

10.6 Concluding remarks

The ability of the immune system to mount specific immune responses to specific antigens (or epitopes) enables higher animals to resist infection by invading pathogens. Our increased understanding of the mechanisms involved in such processes enables us to develop improved vaccines and overcome dysfunctions of the immune system from which certain individuals suffer. The specificity of immunoglobulins for particular epitopes can be made use of in diagnosis, cytology and drug delivery.

In the last section of this article we have reviewed progress with vaccines (currently the backbone of the "Health for All" programme) whose use has resulted in decreased infant mortality in developing countries. We have also provided a review of the improvements being made to established vaccines, as well as the way in which we see the field progressing from crude beginnings to sophisticated vaccine designs. We have described a new category of vaccines which are designed to control fertility. There are vaccines for only a few communicable diseases, and even here numerous problems are encountered. We have seen that BCG does not protect efficiently against tuberculosis in all countries. In India, for example, there are two million new cases and half a million deaths every year due to this disease alone. The global picture is equally alarming. We are in urgent need of a better vaccine against tuberculosis, and the same may be said for a dozen or more other communicable diseases. As our knowledge of the intricate workings of the immune system becomes more sophisticated our ability to combat the spread of infectious diseases through vaccination programmes will increase. Unfortunately, the major successes of vaccination programmes against smallpox and poliomyelitis have yet to be repeated in the fight against malaria and tuberculosis and HIV. We still have a lot to learn.

10.7 Suggested further reading

1. Biotol Series (1992). — *Cellular interactions and immunology.*
   — *Defence mechanisms.*
   — *Technological applications of immunochemicals.*
   Butterworth-Heineman, Oxford.
2. Cosman, D. (1988). Colony-stimulating factors *in vivo* and *in vitro*. *Immunol. Today*, **9**, 97.
3. Dawson, M.M. (1991). *Cytokines and interleukins*. Open University Press, Buckingham.
4. Feinman, R., Henriksen–Destefano, D., Tsujimoto, M. and Vilcek, J. (1987). Tumour necrosis factor is an important mediator of tumour killing by human monocytes. *J. Immunol.*, **138**, 635.
5. Feldmann, M., June, C.H., McMichael, A., Maini, R., Simpson, E. and Woody, J.N. (1992). T cell-targeted immunotherapy. *Immunol. Today*, **13**, 84.
6. Hood, L.E., Weissman, I., Wood. W.B. and Wilson, J.H. (1989). *Immunology*. Benjamin/ Cummings, Menlo Park, California.

7. McMichael, A.J. and Fabre, J.W. (Eds.) (1982). *Monoclonal antibodies in clinical medicine.* Academic Press, London.
8. Paul, W.E. (1990). *Fundamental immunology.* Raven Press, New York.
9. Roitt, I., Brostoff, J. and Male, D. (1990). *Immunology*, Gower Medical Publishing, London.
10. Rosenberg, S.A. (1985). Lymphokine-activated killer cells; a new approach to immunotherapy of cancer. *J. Natl. Cancer Inst.*, **75**, 595.
11. Rosenberg, S.A. (1988). Immunotherapy of cancer using interleukin 2: current status and future prospects. *Immunol. Today* **9**, 58.
12. Sinha, A.A., Lopez, M.T. and McDevitt, H.O. (1990). Autoimmune diseases: failure of self-tolerance. *Science*, **248**, 1380.
13. Springer, T.A. (Ed.) (1985). *Hybridoma technology in the biosciences and medicine.* Plenum Press, New York and London.
14. Winter, G. and Milstein, C. (1991). Man-made antibodies. *Nature*, **349**, 293.

10.8 Responses to assignments

10.1 Immunisation by vaccination is such an example. Vaccination involves exposing a person to the antigen(s) of a given pathogen. In this way memory T-and B-lymphocytes are created. On subsequent exposure to the pathogen itself, these memory cells produce a secondary immune response more rapidly and more intensely than would be the case without vaccination (or previous exposure to the pathogen).

10.2 Some common autoimmune diseases are: systemic lupus erythematosus (where anti-DNA antibodies are produced), rheumatoid arthritis (which is caused by antibodies to immunoglobulins: complexes form with immunoglobulins and are deposited in joints), and autoimmune thyroiditis (a disease characterised by antibodies to thyroid antigens which causes tissue damage).

10.3 There are several theories as to how HIV infection brings about destruction of the human immune system. However, we do know that HIV specifically infects and destroys T-helper cells. Remember that T-helper cells boost the immune response by stimulating proliferation of B-cells and cytotoxic T-cells in response to an antigen. If T-helper cells are destroyed the immune response fails.

10.4 Lymphocytes have a finite lifespan, so unfused and fused lymphocytes will not grow in culture. Fused and unfused myeloma cells cannot grow in the provided culture medium, since they cannot synthesise the precursors needed for DNA synthesis (the *de novo* pathway is blocked by the presence of aminopterin and the salvage pathway is blocked by the lack of HGPRT enzyme). Hybridoma cells will grow, as the hybrid cell inherits a gene coding for HGPRT from the B lymphocyte and so can synthesise DNA using hypoxanthine (as substrate for HGPRT in the salvage pathway) and thymidine as precursors.

10.5 Murine monoclonals are recognised as foreign by the human immune system and one injection can result in anti-mouse antibodies being induced. These antibodies react with murine monoclonal antibody molecules and reduce their effectiveness. The interaction of anti-mouse antibody and monoclonal may also lead to anaphylaxis (shock reactions).

10.6 EBV causes transformation of cells resulting in uncontrolled growth; this is essentially a neoplastic (tumour-like) transformation. Such an event would be unacceptable in a human recipient of monoclonals contaminated with EBV so any monoclonal produced by EBV transformed cells would have to be suitably processed and rigorously tested to ensure it was free of EBV or EBV nucleic acid.

10.7 A normal mouse would recognise the human B-cells as foreign (non-self) and would mount an immune response which would eliminate the human cells. By using a severe immunodeficient mouse (which cannot mount an immune response to human cells) this problem is avoided and the human B-cells are not destroyed.

10.8 By isolating the human gene coding for the F_C region of Ig and substituting this for the gene coding for the F_{ab} region of Ig in the mouse myeloma cell, the mouse cell will produce chimeric MAbs with mouse F_{ab} regions and human F_C regions. This should result in an immunoglobulin with reduced immunogenicity in humans. Such is the case, although there is still sufficient residual immunogenicity in the mouse F_{ab} regions of these chimeric Ig molecules to render them of limited use in humans.

10.9 TB and smallpox organisms exist in only one immunological type (serotype) so the one vaccine protected against all forms of the organism. The live vaccine grows in the tissues after injection causing intense stimulation of immune responses. Poliovirus exists in three serotypes and as cross protection between types is low, immunity against each of the three types must be induced. All three types cannot be administered orally at once as only one virus can cause infection of the gut at one time (due to interference, in which infection by one virus prevents infection by other viruses). Each serotype must therefore be administered separately and at spaced intervals. Where enteric viral infections are rife, infection of the gut by an enteric virus may interfere with the oral vaccine, leading to failure of vaccination, so that more than three doses are needed. Salk vaccine (killed virulent virus) can be polyvalent (all three serotypes are included in it) as intereference does not affect the response to vaccines of this type.

10.10 Pre-S1 antigen is involved in attachement of the virus to the hepatocyte in which it replicates. Vaccination with pre-S1 will induce IgG antibodies (in the blood stream) which should react with pre-S1on the virus surface and thereby interfere with (block) attachment of the virus to the hepatocyte cell surface.

10.11 Live vaccines (natural and recombinant) grow in the host, and may thus induce effective immunity with just one vaccinaton; non-live vaccines take at least two, usually three, vaccinations. Live vaccines induce longer immunity, and the production cost of a recombinant vaccine (once the genetic engineering has been performed) is relatively low. The major disadvantage of a live vaccine is that the vaccine must be prepared and/or stored in such a way as to preserve the organisms contained in it in a viable state; this can add to the cost of a vaccination programme.

10.12 hCG plays a vital role in the establishment of pregnancy. Vaccination to induce antibodies against hCG in the circulation would mean that any hCG produced would become neutralised by the binding of antibodies, and effectively removed from the system. In this way the establishment of pregnancy is prevented.

10.9 Responses to self assessment exercises

10.1 The correct answer is innate. This is because an innate defence mechanism is non-specific and does not depend on prior exposure to the antigen. Acquired defence mechanisms are specific to an antigen and their magnitude is influenced by prior exposure to the antigen ('memory' is involved). All the factors mentioned are of the former type.

10.2 An antigen is an entity which induces antibody formation. As an example, an enzyme protein from a mouse will induce antibody when injected into a rabbit. The enzyme molecule is thus an antigen. However, if the enzyme (polypeptide) is cleaved into smaller oligopeptides, several of these may also each induce an antibody response. Each oligopeptide which induces antibody is thus an epitope. The enzyme molecule will thus induce several types of antibody, whereas an epitope will only induce one type of antibody.

10.3 The correct matching is as follows.
(i) ... (b), (ii) ... (a), (iii) ... (d), (iv) ... (e).

10.4 The correct answers are as follows.
- (i) B-cells — antibody on surface of that cell recognises antigen (free or on MHC) sites of another cell and also class II MHC molecules.
- (ii) T-helper cells — recognises epitope when epitope is in conjunction with MHC molecules of class II on APC.
- (iii) Cytotoxic T-cells — recognises epitope on surface of largest cell when in conjunction with MHC molecules of class I.

10.5 The correct answers are as follows.
- (i) True — successful thymic education would eliminate all immune dysfunctions.
- (ii) False — this would be true if 'B-cells' were replaced by 'T-cells' in the question.
- (iii) False — IgE is the cause of hypersensitivity to pollen. Vaccination with pollen results in production of IgG (which binds to pollen antigens in the bloodstream and leads to their removal from the system) or suppressor cells (which inhibit IgE production).
- (iv) False — infections characteristic of AIDS arise due to destruction of T-helper cells and associated failure of the immune response. Vaccination (which relies on an effective immune response to be mounted in response to injection of the antigens in the vaccine) is thus useless for AIDS sufferers.

10.6 The correct answers are as follows.
- (i) True — many cytokines are produced by lymphocytes (see Table 10.1) and are thus lymphokines and can also be classed as interleukins.
- (ii) True — TNF (see Table 10.1) is produced by macrophages (which are not lymphocytes) and thus cannot be termed a lymphokine. It is, in fact, a monokine.
- (iii) False — TNF has anti-tumour (nectrotising) activity in its own right; its activity is not mediated through NK cells.

10.7 The correct answers are as follows.
- (i) False — a secretory myeloma would produce one or more type of immunoglobulin, so the immunoglobulin formed from a hybridoma made from such a cell and a B-cell would contain at least two types of immunoglobulin (one from the B-cell, at least one from the myleloma); it would not therefore be a monoclonal antibody.
- (ii) False — it could not be called an intraspecific monoclonal as it is formed by fusion of a human and mouse cell; it is thus interspecific and not intraspecific. The antibody formed will be human (coded for by the human genes from the B lymphocyte); it would not therefore be a chimeric antibody.
- (iii) True.

10.8 The correct answers are as follows.
- (i) True.
- (ii) False — this would be true if 'immunogenicity' was replaced by 'toxigenicity' or 'toxicity'.
- (iii) False — it is true that peptides may not have enough epitopes for them to be recognised by MHC sites on cells of the immune system, but T-helper cells do not in themselves bind such epitopes — they only recognise such epitopes when the epitopes are bound to the MHC receptors of other cells.
- (iv) False — the statement is true for gonadotropin-releasing hormone, but for human chorionic gonadotropin it is only true in females.

11 Biotechnology as a new frontier in health

Chapter outline

11.1 Introduction

Biotechnology has been used in health sciences for more than fifty years. Antibiotics, mostly obtained from fungi and the bacterium *Streptomyces* have completely transformed disease patterns, especially in developed countries, where bacterial infections have largely been controlled, explaining in part why age-related disorders such as cancer have become the predominant cause of death. The availability of antibiotics has facilitated large scale animal cell culture used in production of viruses for preparation of vaccines. Hormones such as insulin have been isolated from animal tissues and used extensively in replacement therapy. In a limited way, immune sera have also been used in passive immunisation and for diagnostic purposes. However, in these technologies, which are cell culture based, there is inherent batch to batch variation and strict quality control is difficult.

During the last two decades hybridoma technology and genetic engineering have revolutionised disease management. The former gives an unlimited source of highly specific monoclonal antibodies (MoAbs). Using recombinant DNA technology, it is now possible to clone normal and novel genes, and obtain, in bulk quantities, quality products of defined specificity. A number of such products have already been marketed. Nucleic acid probes are now widely used in pre- and post-natal diagnosis of genetic disorders, as well aş in the study of genetic epidemiology to determine susceptibility to environmental factors. Correction of a missing or defective gene through gene therapy is no longer a dream, and a number of clinical trials are already in progress. The human genome project, aimed at molecular characterisation of the complete genome, marks the beginning of a new era in the identification of novel genetic disorders and elucidation of the role of genes in a wide spectrum of acquired diseases (infections, autoimmunity, degenerative and age-related disorders).

This chapter examines the development of techniques of hybridoma technology and gene technology as applied to health care. The technology associated with hybridoma production and gene technology have been presented in Chapters 10 and 5 respectively.

11.2 Hybridoma technology

11.2.1 Applications of MoAbs in diagnosis

(a) Biochemical analysis

In the last two decades, many varieties of monoclonal antibody have been produced. They have varied applications in medicine, including laboratory diagnosis. MoAbs are routinely used in radioimmuno assays (RIA) and enzyme-linked immunosorbent assays (ELISA) to measure circulating levels of enzymes, hormones and tissue and cell products. They are also used for blood group typing, and for HLA matching in organ transplantation. A number of diagnostic kits, using MoAbs, are now commercially available.

Immunoassays rely on the quantitative binding of MoAb to epitopes, and the quantitation of bound MoAb by enzyme activity (in ELISA) or level of radioactivity (in RIA). A technique such as ELISA can be performed in a variety of ways, and an example (the assay of parathyroid hormone in blood) is given below.

(i) The antigen (given volume of serum sample) is immobilised on (bound to) a solid surface (usually the well of a microtitre dish). Anti-parathyroid hormone MoAb may be used for this (the surface of the well may first be coated by the MoAb, so that when serum is added the parathyroid hormone is specifically bound by the monoclonal in the well).

(ii) The well is washed in buffer.

(iii) Anti-parathyroid hormone: alkaline phosphatase enzyme conjugate is added to the well (MoAb binds specifically to bound parathyroid hormone).

(iv) The well is washed in buffer.

(v) Enzyme substrate (para-nitrophenyl phosphate) is added and the mixture incubated.

(vi) The amount of product formed by the enzyme (para-nitrophenol) can be measured using a spectrophotometer. The amount of product formed is proportional to amount of conjugate bound, which is proportional to the amount of parathyroid hormone.

Assignment 11.1

Would a bispecific antibody, with specificity for both parathyroid hormone and alkaline phosphatase enzyme be of use in an ELISA assay for parathyroid hormone?

(b) Infectious diseases

Two approaches have been used — (i) to detect and quantify levels of antigens circulating (in blood) specific to an infectious agent and (ii) to identify infectious agents. Detection of antigens is indicative of recently acquired infection, whereas detection of antibodies indicates past episodes of infections. Circulating antigens can be assayed using ELISA or RIA, as outlined above. An example is show in Fig. 11.1. The presence of a pathogenic organism in the body can be diagnosed by the direct testing of clinical specimens for antigens using fluorescent labelled MoAbs.

> ***Assignment 11.2***
>
> Outline how a fluorescent labelled antibody against HIV antigen could be used to demonstrate the presence of HIV in cells in a histological section of a lymph node.

In the past, this approach was seldom used because of high levels of background fluorescence associated with polyclonal antiserum. With the help of fluorescent labelled MoAbs, herpes simplex virus has been demonstrated in swab specimens, *Chlamydia trichomatis* in urethral and cervical

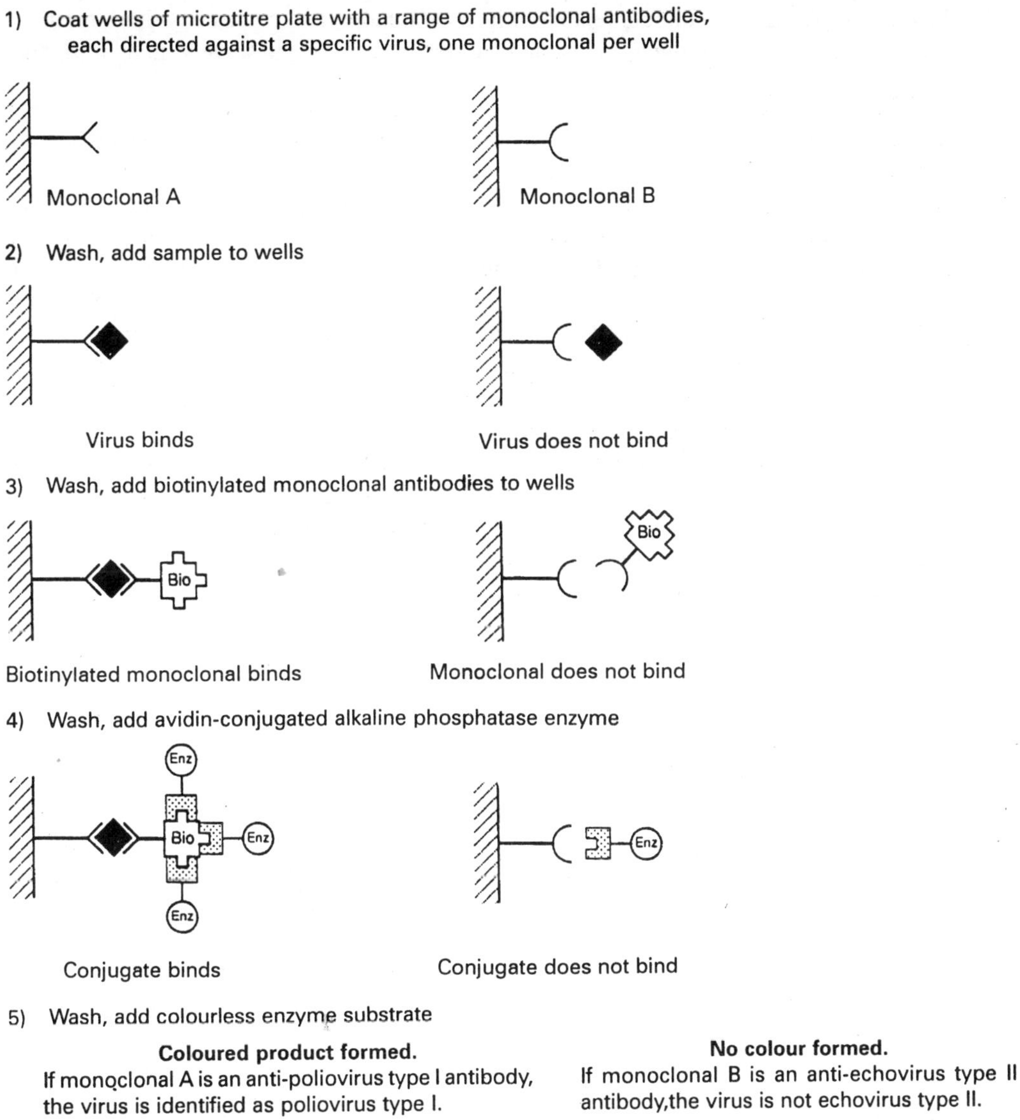

Fig. 11.1 *Use of ELISA in the detection/quantitative estimation of a viral antigen (epitope).*
The technique can be used qualitatively to demonstrate the presence or absence of viral antigen in the sample. The technique can also be used quantitatively, as the concentration of coloured product formed in a well is directly proportional to the concentration of antigen in the sample. Automated testing using ELISA is possible.

smears and cytomegalovirus in lung biopsies. Association of Epstein–Barr virus (EBV) with two different tumours — Burkitt lymphoma and nasopharyngeal carcinoma, has been demonstrated with MoAbs to the viral antigen. MoAbs are also used to type and sub-type bacteria and viruses; a panel of MoAbs that recognise different epitopes on the pathogen can be used in concert.

(c) Cancer

Although polyclonal antisera (produced in animals) have been used in diagnosis of cancers they are not very reliable since they cross-react with antigenic determinants on normal tissues. MoAbs have helped overcome these problems. They are used in three different ways in cancer diagnosis — (i) to detect tumour related antigens or tumour markers in body fluids or on cells, (ii) to demonstrate tumour antigens in histological sections and (iii) to serve as imaging agents.

(i) Tumour markers

Blood levels of certain tumour products are used as tumour markers in the diagnosis and prognosis of cancers. Alterations in the levels of the markers are variable and inconsistent in early stages, but are often raised in advanced cancers. When the tumour load is reduced, the blood level of the tumour marker is also reduced considerably. Hence this can be used to monitor the tumour burden and judge the efficacy of therapy. Carcinoembryonic antigen is a tumour marker elevated in 91% to 100% of persons with colorectal cancer. CA-125, another marker, is elevated in 82% of women with ovarian cancer. Prostate-specific antigen is a marker for late stage prostate cancer. Circulating levels of alpha–feto protein are raised in hepatoma and germ cell tumours, and, in the latter it is associated with concomitant rise in the hormone, human chorionic gonadotropin. Levels of such compounds in body fluids can be measured using ELISA or RIA.

MoAbs raised against a variety of tumours have helped in the definition and characterisation of tumour specific antigens, their correct 'typing' and diagnosis. Accurate diagnosis of the precise tumour type is essential since subsequent therapy depends on the type of tumour. MoAbs have made possible the accurate classification of lymphocytic and granulocytic leukaemias which require different optimal therapies. In certain cases of neoplastic meningitis, results of cytological analysis of cerebrospinal fluid (CSF) are often ambiguous, and repeated examinations often prove to be non-diagnostic. Fortunately, monoclonal antibody-based tests have greatly increased the accuracy with which malignant cells in CSF may be identified and characterised. With early and correct diagnosis many patients can be spared from tedious, extensive investigation or inappropriate treatments.

(ii) Diagnostic tumour pathology

The study of pathological changes has been performed by microscopical examination of tissue sections stained with MoAbs using immunohistochemical techniques. The sensitivity of such techniques is influenced by the stability of the antigen, the nature of the fixative and the methods used. A two-step immuno-peroxidase procedure, which gives a colour reaction at the end, is commonly used to demonstrate location of epitopes in a cell.

Immunohistopathology often helps in assessing the grade of malignancy of tumours. One such example is the use of MoAbs directed against keratins, which are cytoskeletal proteins expressed by squamous cell carcinomas; depending on the stage of differentiation, tumours express different sets of keratins. Such knowledge is useful in directing therapy. An example of such a reaction is shown in Fig. 11.2.

Assignment 11.3

Outline how a monoclonal antibody directed against a particular type of keratin could be used in an immuno-peroxidase system to demonstrate the presence of that keratin in a cell in a tissue section.

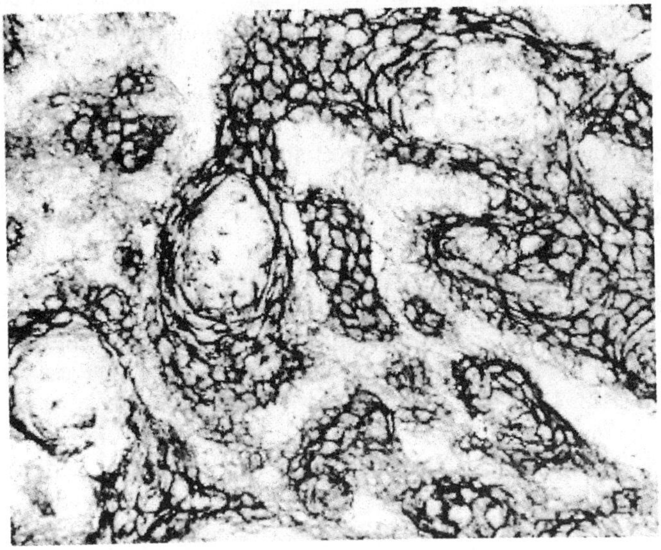

Fig. 11.2 *Immunohistochemical staining (using the peroxidase reaction) on a frozen secton of human squamous cell carcinoma of tongue.*
Mab 3F8E3 against a cell surface antigen of squamous cells has been used. The staining shows infiltrating cancer cells with positive staining on the cell membrane (x96).

MoAbs are widely used in diagnosis of haematopoietic malignancies. Haemopoietic stem cells in the bone marrow are the precursors of different blood cells including B- and T-lymphocytes, which undergo stepwise maturation. Lymphocytes carry cell surface markers characteristic of the stage of their maturation. A panel of MoAbs has been raised to the different cell surface markers, some of which are indicated in Table 11.1. During malignant

Table 11.1 Commonly used monoclonal antibodies to lymphocyte surface antigens

Monoclonal antibody	Antigen (ag)	Reactivity with cell type
1 OKT6, Leu6	T6 (T-cell ag)	Cortical thymocytes, epidermal Langerhan's cells
2 OKT3	T3, CD3 (T-cell ag)	Subset of thymocytes, all mature T-cells
3 OKT4	T4, CD4 (T-cell ag)	Cortical thymocytes, helper-inducer mature T-cells
4 OKT8	T8, CD8 (T-cell ag)	Cortical thymocytes, suppressor/cytotoxic mature T-cells
5 OKT11	T11, CD2 (T-cell ag)	All mature circulating T-cells
6 J-5, BA-3	CALLA, CD10 (Common acute lymphoblastic leukaemia)	Subset of normal B-cells, mature granulocytes kidney tubular epithelial cells
7 BA-1, B1	B-cell antigens	B-cells
8 SMIG	Surface membrane Ig	B-cells
9 anti-IL-2R/Tac	CD-25	Activated T- and B-cells and macrophages
10 CAMPATH-IH (chimeric)	—	Human lymphocytes
11 anti-T9	transferrin receptor	Activated T- and B-cells
12 anti-B4	CD-19	pre-B-cells and mature B-cells

transformation, the lymphocyte is arrested at a particular stage of maturation, and this can be determined with the help of a panel of stage-specific MoAbs. This is now routinely carried out in the characterisation of leukaemias and lymphomas. Sometimes it is very difficult to distinguish a poorly differentiated tumour of lymph nodes from a carcinoma; this problem can be overcome by using antibodies to lymphocytes (which recognise lymphoid cells) and to cytokeratin (which recognise most carcinomas). MoAbs to lymphoid and myeloid cell surface markers are extensively used to assess the level of immunodeficiency in primary and acquired immunodeficiency syndromes (AIDS).

(iii) Immunoscintigraphy
MoAbs with specificities to many types of human cancers are now available. These can be tagged with radioisotopes such as iodine-131 or technetium-99, and injected intravenously into the patients; the antibodies localise in the tumour, which can then be detected by imaging the radioactivity (Fig.11.3). This technique of imaging is known as immunoscintigraphy. Although imaging techniques such as CT scanning, ultrasound scanning and magnetic resonance imaging can detect small lesions, they are unable to differentiate between cancerous and non-cancerous growths. Being tumour specific, antibody imaging has therefore certain advantages. Small metastatic lesions, which remain undetected by CT scan, can be detected successfully, in many cases, with radiolabelled MoAbs. Since an antibody is a large molecule (150 kDa) and thus does not readily enter cells and tissues, attempts are being made to improve the efficiency of the procedure by using smaller Fab (12–15 kDa) or F_V fragments (Chapter 10).

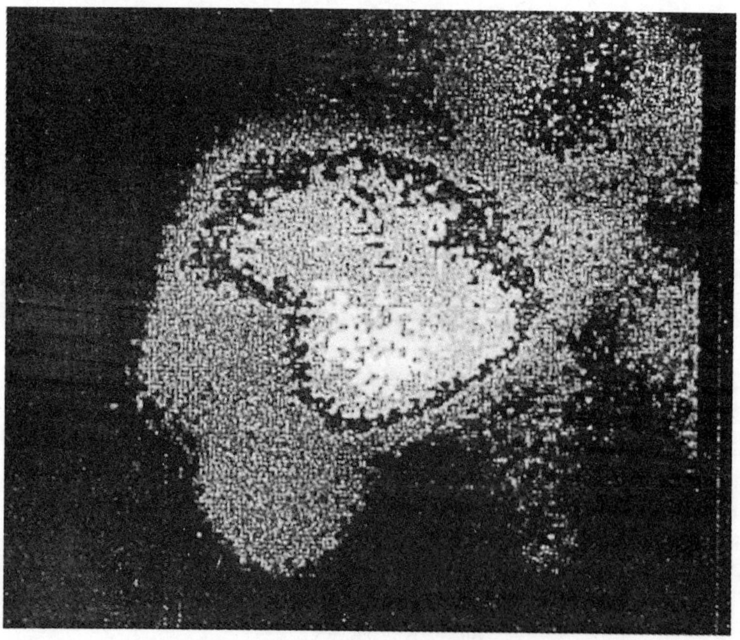

Fig. 11.3 *Immunoscintigram of the liver, 1 hour after administration of 99mTc labelled (Fab1)2 fragment of anti-CEA MoAb in a patient of colorectal carcinoma.*
Metastasis in the liver is demonstrated by the lighter area.

Self assessment exercise 11.1

A bispecific monoclonal with affinity for both a given cell-surface tumour marker and peroxidase enzyme has been developed. Would this monoclonal be of use in demonstrating the presence of the tumour marker in cells in a tissue section? Explain your answer.

Self assessment exercise 11.2

Answer true or false to each of the following statements:

(i) A monoclonal antibody having affinity to carcino-embryonic antigen and conjugated to alkaline phosphatase enzyme could be used to detect colorectal cancer markers in blood using ELISA.

(ii) A monoclonal antibody having affinity for a surface marker of a tumour cell type and conjugated to a radioisotope could be used to locate that type of tumour in the body using radioimmunoassay.

(iii) A monoclonal antibody having affinity for an HIV antigen and conjugated to a fluorochrome could be used to detect HIV infection of cells from the blood by immunofluorescence.

11.2.2 Applications of MoAbs in therapy

The highly specific antigen-binding properties of MoAbs make them ideal candidates for use in therapy. Two approaches have been considered — (a) direct use in therapy and (b) MoAbs as drug targeting agents.

(a) MoAbs as therapeutic agents

Mouse monoclonal antibodies directed against the particular organism have been found to protect chimpanzees against challenge with hepatitis-B virus and rodents against infections with the bacteria *Haemophilus influenzae, E. coli, Pseudomonas* sp. and *Streptococcus* sp. In the field of virology, it has been observed that some monoclonal antibodies which did not neutralise viral infectivity of host cells *in vitro* do confer protection *in vivo*. Protection *in vivo* is brought about principally by efficient opsonisation of the pathogen (coating it with antibody) which enhances phagocytosis. Monoclonal antibodies against antigens present on the surface of cancer cells can exert site-specific cytotoxic effects, brought about by mechanisms such as complement-mediated cytotoxicity, antibody-dependent cell-mediated cytotoxicity, phagocytosis of antibody-coated cells by cells of the reticuloendothelial system and direct antiproliferative effects.

Monoclonal antibodies produced in mice have been used with variable success in human patients suffering from leukaemia, lymphoma, melanoma and colorectal cancer and clinical trials have indicated several limiting factors such as:

(i) heterogeneity of tumour cells — not all malignant cells carry the relevant antigen;

(ii) circulating free antigen — the presence of circulating free antigen(s) can effectively block MoAbs from binding to the target cells by binding to the Fab sites on the antibody molecules;

(iii) antigenic modulation — antigen is modulated off the cell surface as a consequence of binding of MoAbs to the cancer cell (as in leukaemia cells). By employing monovalent MoAb, this problem can be circumvented to some extent;

(iv) immune responses to xenogeneic proteins — repeated administration of mouse MoAbs in humans leads to development of anti-mouse antibodies and hypersensitivity reactions.

In patients with leukaemia, there may be some residual leukaemic cells present in the bone marrow even during apparent remission. A monoclonal antibody specific to leukemic cells has been used to eradicate residual leukaemia cells without affecting hematopoietic progenitor cells. Furthermore, autologous bone marrow transplantation (reimplantation of the patient's own cells) is often necessary owing to the non-availability of matched donors, in which case MoAbs can be used *in vitro* for the removal of residual tumour cells prior to autologous bone marrow transplantation.

In attempts to overcome rejection by cell-mediated immune responses, patients undergo immunosuppressive therapy following organ transplantation. This therapy is in the form of immunosuppressive drugs such as cyclosporin plus prednisone, or patients are generally treated with glucocorticoid hormones in high doses for short periods or with the globulin fractions of xenogeneic antilymphocyte sera (sera from animals vaccinated with human lymphocytes). MoAbs specific for human T-cell surface antigens (and for specific subpopulations of T-cells) are now available. These MoAbs can be used selectively to eliminate T-cell populations responsible for allograft (transplant) rejection; for example, a monoclonal antibody reacting with mature intravascular T-cells has been found to be useful in subjects who have undergone renal transplantation.

One of the best examples of use of MoAbs in therapy is monoclonal OKT3 which is directed against CD3 antigen. CD3 is an antigen on the surface of T-lymphocytes, and CD3-activated T-lymphocytes can play a crucial role in allograft (organ transplant) rejection. OKT3 can be used to prevent acute renal allograft rejection in humans. The treatment has been found to be more effective than the administration of broad spectrum immunosuppressive agents or polyclonal anti-lymphocyte serum. OKT3 has also been used to purge donor bone marrow prior to its use in marrow transplantation, so as to reduce graft *vs.* host reaction. The chimeric 'humanised' antibody CAMPATH-1 H, which is cytotoxic to all lymphoid cells and monocytes but not to other cell types, has given promising results in the treatment of certain lymphomas.

In the search for effective immunotherapy against HIV infection, genetic engineering has been used to link the Fc portion of mouse monoclonal to human CD4 complex. CD4 is a cell surface receptor on T-helper cells and one which is the receptor site for HIV. HIV attaches to T-cells, prior to infecting them, by specifically attaching to CD4 receptors using the gp-120 receptors on its own surface. After infection of a T-cell, gp-120 antigens appear on the surface of that cell. If F_C:CD4 molecules are given to the sufferer, these bind to HIV infected cells (displaying gp-120) and the exposed mouse F_C fragment induces cell-mediated destruction of the infected cell (Fig. 11.4). This strategy may prove useful in preventing the spread of HIV throughout the T-cell population.

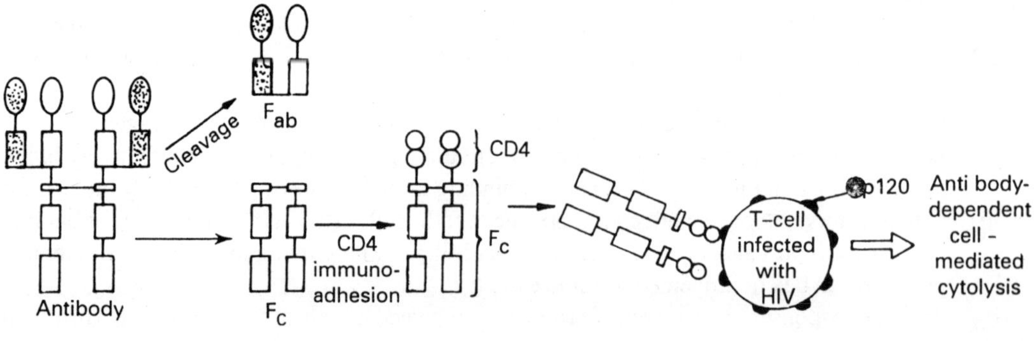

Fig. 11.4 *Modification of antibody molecules for therapy.*
The F_C, or the constant portion, of the antibody molecule is attached to a CD4 molecule which has an affinity for the gp-120 molecule expressed on the surface of HIV-infected cells. These modified antibodies will bind to virus-infected cells and direct the subsequent binding of special cells which recognise their F_C portion. These specialised cells then initiate antibody-dependent cell-mediated cytolysis.

(b) Drug targeting

Three approaches have been used for targeting drugs :

(i) Coating of drug containing liposomes with tissue specific MoAbs

Several attempts have been made to specifically target drugs using liposomes coated with tissue-specific MoAbs.

> ***Assignment 11.4***
>
> Explain how coating liposomes with tissue-specific MoAbs could be of assistance in drug targeting.

Such liposomes have shown some tissue specificity *in vitro*, but when administered *in vivo* most of them are retained by reticulo-endothelial cells of the liver and spleen and often do not reach the target tissue.

(ii) Chemical conjugation of MoAbs, including chimeric and bispecific MoAbs, with a toxin to create an immunotoxin

MoAbs can be used as vehicles for the delivery of drugs *in vivo*. Conjugates of antibodies with drugs are able to bind to an appropriate target cell and the antibody-linked drug is then introduced directly into that cell. This allows higher concentrations of the drug to reach the desired site and minimises systemic toxicity. Drugs such as daunamycin, vindesin, methotrexate and adriamycin can be coupled to MoAbs. Better survival rates are observed in experimental mice with lymphomas given daunamycin attached to anti-lymphoma MoAbs compared to those receiving the drug alone.

Anti-Tac monoclonal, reactive against IL2-R (T-cell growth factor receptor) has been chemically coupled with exotoxin from *Pseudomonas* sp. (PE). Resting T-cells lack IL2-R, which is expressed by abnormal T-cells in patients with certain lymphoid malignancies or autoimmune disorders and in individuals rejecting allografts. The immunotoxin has been used to purge malignant T-cells in patients with adult T-cell leukaemia. However, these chimeric antibodies produced by conventional chemical coupling techniques tend to be generally toxic and have several undesirable side-effects.

Plant-derived toxins such as ricin and gelonin can also be covalently coupled to MoAbs to form 'immunotoxins' which have been shown to be effective in the treatment of some forms of cancer. Ricin consists of a toxic polypeptide (the A-chain) which is attached to another polypeptide (the B-chain) which binds to cell membranes. Ricin binds to cells via the B-chain and this binding facilitates the entry of the ricin A-chain into the cytoplasm of the cell. Inside the cytoplasm, the A-chain inhibits protein synthesis.

> ***Self assessment exercise 11.3***
>
> Ricin is to be injected into a person to treat a tumour (type X). Which of the following would it be best to use for this purpose? Explain your answer.
>
> (i) A monospecific MoAb, specific for tumour X, conjugated with ricin (A- and B-chains).
>
> (ii) A bispecific MoAb, specific for tumour X, and the B-chain of ricin.
>
> (iii) A monospecific MoAb, specific for tumour X, and conjugated to the B-chain of ricin.
>
> (iv) A monospecific MoAb, specific for tumour X, and conjugated to the A-chain of ricin.

In another approach at drug targeting, a recombinant chimeric MoAb against fibrin coupled to tissue plasminogen activator (tPA) has been used in animal models. Because of its affinity for fibrin, the recombinant protein becomes concentrated in a clot, thereby locally raising the concentration of tPA. This helps in promoting an enhanced lysis of the clot. Such thrombolytic agents would have a crucial role to play in cardiovascular diseases. Conceptually, drug targeting is very attractive. In practice, however, there are several problems and, as a clinical method, it still has a long way to go.

(iii) Genetically engineered chimeric proteins containing F_{ab}/F_V fragments of MoAbs, and cytotoxins

In order to reduce toxicity of anti-Tac MoAbs, another approach has been conceived using genetic engineering. A hybrid gene construct, which encodes for the F_V fragment of anti-IL2-R MoAbs and PE has been expressed in *E.coli*. The recombinant protein so obtained contains the specificity of the F_V fragment and shows the toxicity of the PE.

Self assessment exercise 11.4

Which of the following, after injection into the bloodstream, might be expected to bring about death of abnormal T-cells by the action of a cytotoxic polypeptide, polypeptide A.

(i) A bispecific chimeric MoAb having affinity for surface markers specific to the abnormal T-cells and for polypeptide A.

(ii) A monospecific MoAb having affinity for surface markers specific to abnormal T-cells and conjugated to polypeptide A.

(iii) A chimeric bispecific MoAb having affinity for alkaline phosphatase enzyme and for polypeptide A and conjugated to polypeptideA.

(iv) A chimeric bispecific MoAb having affinity for surface markers specific to abnormal T-cells and for alkaline phosphatase enzyme and conjugated to polypeptide A.

(v) A recombinant protein, containing the F_C fragment of MoAb having affinity for the surface markers specific to abnormal T-cells linked to polypeptide A, and produced by a recombinant (genetically engineered) microorganism.

11.3 Recombinant DNA technology

The human genome is a massive storehouse of information. The haploid genome consists of 3 billion base pairs, of which only 3–5% is functional and capable of encoding 30,000 to 100,000 gene products. The DNA is packaged into 23 pairs of chromosomes (22 pairs of autosomes and 1 pair of sex chromosomes) and each chromatid in a pair is derived from one parent. The chromosomal location of many genes, which are inherited in accordance with Mendelian laws, is not known. The genome is highly polymorphic, which means that at a given position the DNA sequence can be either variable or relatively constant. This results in extreme variability, in which each copy of the genome differs from others in millions of sequence-specific ways. Variations in the DNA sequence result in a very wide range of biological effects, from insignificant to drastic. Generally, such variations are greater in the non-functional part of the genome and much less in protein encoding sequences or in sequences with regulatory functions. Variations, sometimes even a single nucleotide substitution, in the coding or regulatory region of the gene can bring about phenotypic change through a change in the encoded protein. Phenotypic change can occur as a consequence of deletion of the gene, abnormality in its regulatory mechanisms, or wide range of molecular alterations including point mutations, frameshifts or nucleotide substitutions (Chapter 4).

Because of the enormous complexity of the genome, the diagnosis of human genetic abnormalities has been, until recently, indirect and often depended on demonstration of an abnormal protein or an associated phenotypic change. For example, sickle cell anaemia, which is the result of a single nucleotide change resulting in abnormal haemoglobin, was diagnosed by observing either the sickling of the patient's erythrocytes under reduced oxygen tension or electrophoretic abnormalities of the sickle cell globin. Although these methods are still used, the advent of recombinant DNA technology has ushered in a revolution in laboratory diagnosis. So sensitive and specific are the techniques that genetic defects can be diagnosed with ease and precision, even prenatally. Also carrier states (heterozygous individuals, who have a normal gene allele and a defective gene allele) can be identified, thereby facilitating genetic counselling.

Major developments which have facilitated laboratory diagnosis of genetic disorders are: (i) the discovery of restriction endonucleases enabling cloning of genes, (ii) the ability to sequence DNA, (iii) the ability to synthesise oligonucleotide probes which hybridise with complementary DNA sequences, (iv) the ability to amplify DNA by the polymerase chain reaction and (v) the ability to identify genes or their products in tissue sections or smears by *in situ* hybridisation. These techniques have been described in Chapter 5. Several of these techniques have now been automated. The technology is also being used in diagnosis of infectious diseases as well as in genetic epidemiology. Cloning of normal and novel genes has also opened up the possibility of gene therapy as well as the production of therapeutically important recombinant proteins.

11.3.1 In situ hybridisation (ISH) and FISH

ISH is a technique which allows the detection and location of given nucleic acid sequences directly in cell smears and tissue sections. The slides are incubated in the hybridisation buffer at elevated temperature so as to denature the cellular DNA (separate the two strands). This is followed by hybridisation with a radio-labelled probe (Chapter 5) for the given sequence and autoradiography. The advantage of this technique is that it gives information about morphological localisation of DNA sequences in intact cells. The technique could also be used to study the expression of a gene within a cell by detecting the mRNA, using complementary probes to particular mRNA.

Another application of ISH is in determining the location of genes on chromosomes. A recently introduced modification is FISH or fluorescent *in situ* hybridisation, in which the specific

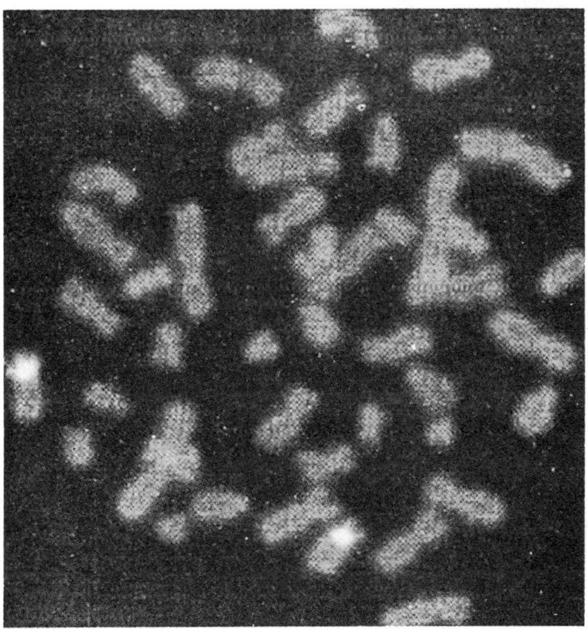

Fig. 11.5 *FISH on a normal human metaphase plate using biotinylated-satellite DNA probe (pSP12-1) which hybridises only with the centromeres of the two chromosome 12s.* Hybridised probe is shown by the light spots.

DNA/RNA probes are tagged with a fluorescent dye and hybridised with denatured chromosomal DNA. An illustration of this technique is given in Figure 11.5.

> ***Assignment 11.5***
>
> Explain how ISH or FISH would demonstrate the location of a nucleotide sequence within a cell or chromosome.

Chromosomal abnormalities, in which the number or relative position of a specific gene has become changed, can be detected by FISH. Several human chromosome-specific probes are now available. Human satellite DNA (alphoid) probes, which are specific for centromeric regions of individual chromosomes, are now used widely in FISH to detect chromosomal abnormalities, such as trisomy, even in interphase cells (Fig. 11.6).

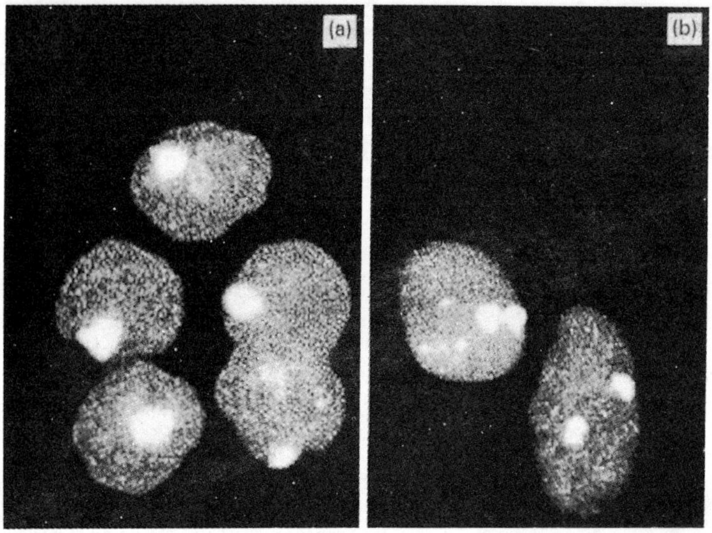

Fig. 11.6 *FISH using human Y-specific probe, pY431A. (a) normal male lymphocytes showing only one fluorescent spot in each nucleus. (b) XYY human amniocytes exhibiting two spots.*

One of the major advantages of FISH is that, unlike Southern blotting, it permits quantitation of gene dosage and hence can be readily used in the diagnosis of carriers, for example Duchenne muscular dystrophy, which is a sex-linked disorder affecting males and resulting from a deletion in the dystrophin gene. Female carriers, in which there is a deletion in the gene in one of the alleles, can be identified by FISH using interphase or metaphase cells, since it results in one labelled and one unlabelled chromosome. A marker known as Philadelphia chromosome (Ph1), which is usually a translocation between chromosome 9 and 22, is present in over 90% of the patients of chronic myeloid leukaemia. Probes specific for Ph1 have been used with FISH in interphase cells to demonstrate this abnormality.

11.3.2 Polymerase chain reaction (PCR)

The polymerase chain reaction (PCR) (Chapter 5) is a powerful and sensitive technique to study genetic diseases as well as infections. Trace amounts of DNA can be amplified if the nucleotide sequences flanking the DNA segment of interest are known. There are two basic limitations to PCR. As the result of contamination of the specimen with DNA, false positives may be obtained. Therefore, with every reaction, appropriate controls have to be included. Secondly, it is not possible to correlate PCR results with a specific cell in a tissue. To overcome this problem, another technique, called *in situ* PCR, has recently been developed. *In situ* PCR is a combination of PCR amplification and *in situ* hybridisation, in which DNA is amplified in intact cells in cyto-smears or histological tissue sections. If the tissues are appropriately fixed, the PCR reagents can diffuse inside the cells and amplify the target DNA. The amplified DNA is then detected by *in situ* hybridisation with a labelled probe. This technique has been used to demonstrate the presence of minute amounts of DNA/RNA from infectious agents in histological and cytological preparations.

11.3.3 Applications in medicine

Recombinant DNA technology has been utilised in medicine in six major ways: (a) for the diagnosis of genetic disorders associated with chromosomal alterations; (b) identification of infectious agents such as viruses and bacteria; (c) tumour biology; (d) forensic medicine; (e) the production of recombinant proteins; (f) gene therapy. Some aspects of these applications of gene technology have already been given in Chapter 5.

(a) Diagnosis of genetic diseases
Genetic diseases occur due to a variety of alterations in gene structure and function. One of the most studied examples of molecular heterogeneity is that of the globin chain (of haemoglobin of the blood), in which hundreds of single-base changes in the gene cause amino acid substitutions in haemoglobin. Molecular abnormalities in β-thalassaemia, an inherited disorder of β globin synthesis, arise from mutations causing non-sense codons, frameshifts, abnormal splicing, abnormal transcription and impaired RNA cleavage and polyadenylation. However, it is now possible to precisely identify these molecular lesions in a routine diagnostic laboratory using Southern blotting with nucleic acid hybridisation probes and PCR (Chapter 5). Some laboratories also have the

Table 11.2 DNA diagnostic techniques for some common disorders

	Disease	Affected gene	Type of defect	Detection technique/s
1	Duchenne muscular dystrophy	Dystrophin gene	Large deletion (x-linked)	1) Southern blot analysis, RFLP 2) PCR wiith ASO
2	Lesch Nyhann's syndrome	HGPRT gene	Point mutation	1) PCR with ASO 2) RFLP
3	Sickle cell anaemia	β-globin gene	Point mutation	1) RFLP analysis 2) PCR with ASO and DNA sequencing
4	β-thalassaemia	β-globin gene	Point mutation	1) RFLP 2) PCR 3) Sequencing
5	Chromin leukaemia	bcr-abl gene	Chromosomal translocation	1) FISH 2) Southern blotting 3) PCR
6	Dwarfism - familial growth hormone deficiency	Growth hormone gene	Deletion	1) RFLP analysis 2) PCR
7	Retinnoblastoma	Rb locus	Deletion	RFLP
8	Cystic fibrosis	CF gene	Point mutation	RFLP

capabilities of sequencing the PCR-amplified DNA. These techniques are ideal for the detection of diseases due to point mutations or deletion of functional genes (Table 11.2). Where the basis for the genetic disorder is understood, diagnosis by these procedures is possible, whereas if the basis for the genetic order is not understood diagnosis by these procedures is not possible.

Deletion in genes, which is frequently observed in patients in Duchenne muscular dystrophy, β-thalassaemia and familial dwarfism, can be readily diagnosed by Southern blot analysis using specific probes. Figure 11.7 shows that even a carrier of the molecular pathology can be diagnosed.

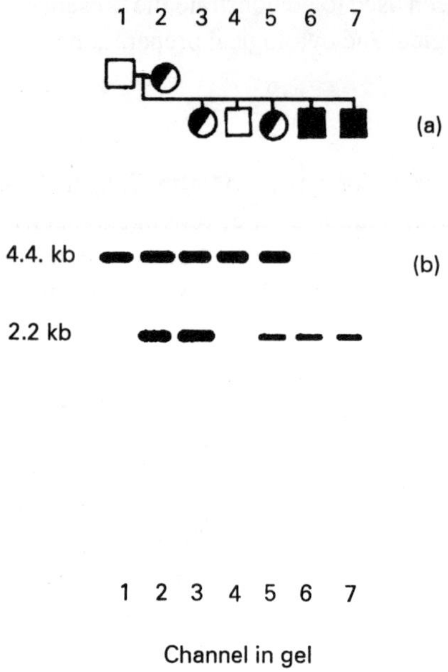

Fig. 11.7 *Southern blot analysis of the family of a patient of Duchenne muscular dystrophy (DMD), showing the deletion of a 2.2kb fragment. (a) The family tree, with 1 – father, 2–mother, 3 and 5 daughters and 4, 6 and 7 sons. (b) A diagrammatic representation of Southern blot analysis of the DNA of individuals in the family.*

In this technique the DNA of each individual was treated with the restriction enzyme Bst XI. The DNA fragments thus produced were separated by gel electrophoresis, channel 1 for the father, channel 2 for the mother, and so on. The separated DNA fragments were transferred to a membrane by Southern blotting, and denatured (unwound). The DNA was then exposed to radiolabelled probe pERT87-8, which binds with the DNA sequence comprising the DMD gene, and hybridisation of probe allowed. Autoradiography was then performed on the membrane, and diagram to represent the autoradiograph produced.

It can be seen that the father (channel 1) and one son (channel 4), who are both unaffected by the disease, exhibit the DMS locus only on a 4.4 kbp restriction fragment and a 2.2 kbp fragment (in which a 2.2 kbp deletion has occurred); they are thus heterozygous and carriers of the defective gene. Other sons (channels 6 and 7) exhibit the DMS locus on a 2.2 kbp fragment, and, having defective (deleted) allelic genes for DMS are sufferers of the disease.

Detection of a single base mutation, such as that causing sickle cell anaemia, can be carried out by Southern blotting or dot blot, in which the hybridisation probe is a synthetic allele-specific oligonucleotide (ASO) encompassing the sequence containing the mutation. Genomic DNA or PCR amplified DNA segments of the affected gene can be used (Figs. 11.8 and 11.9).

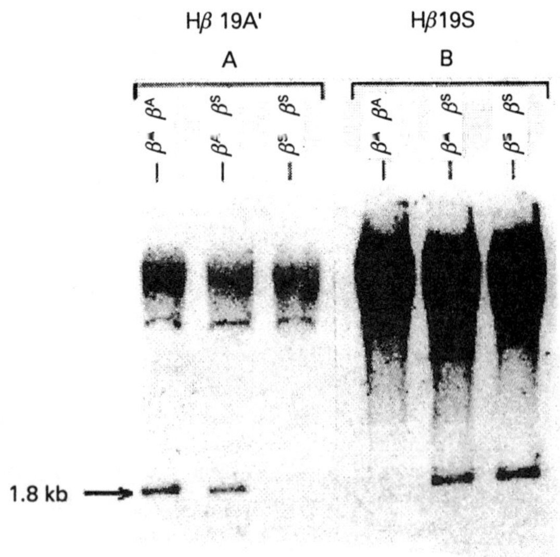

Fig. 11.8 *Southern blot analysis of Bam H1 digested genomic DNA from normal individual (A A) sickle anaemia patient (S S) and carrier (A S), hybridised with ASO.*

HB19A' is for normal globin gene (A), and HB19S for mutated globin B, A A and S S hybridise with HB19A' and HB19S respectively. DNA sample/s of the carrier hybridise with both the probes.

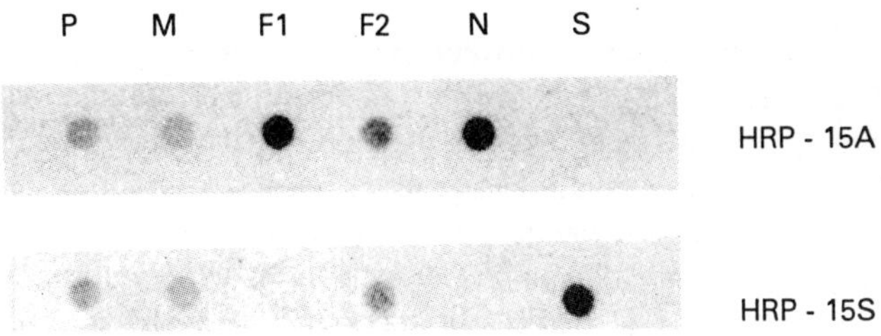

Fig. 11.9 *Diagnosis of sickle cell anaemia in a family with dizygotic twins, using dot blot analysis with probes specific for normal sequence (HRP-15A) and sickle cell mutation (HRP-15S).*
Both paternal (P) and maternal (M) samples were heterozygous A S, twin 1 (F1) was normal, A A; and twin 2 (F2) was heterozygous, A S. Control sample N was obtained from a normal subject homozygous for A A and control sample was from a sickle cell anaemia patient known to be homzygous for S S.

Another method of detecting the mutation is to use a restriction enzyme that will cut the wild type gene but not the mutated gene. This restriction digestion can be coupled with Southern blotting or with PCR for rapid detection. For example, in sickle cell anaemia, there is a point mutation which occurs in the 6th codon substituting T for A; this results in loss of the restriction site for the enzyme Mst II, and thus also results in restriction fragment length polymorphism (RFLP). An illustration of this is given in Fig. 11.10.

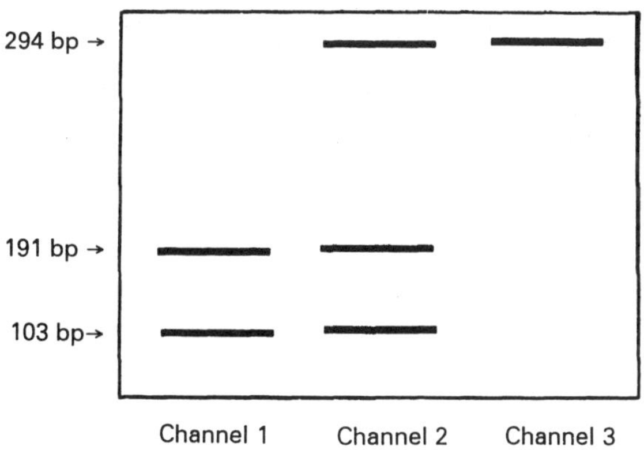

Fig.11.10 *Diagrammatic representation of RFLP.*
Diagnosis of sickle cell genotype using PCR amplified DNA, digested with Mst II which cuts the product from normal gene but not the mutant and fragments separated by gel electrophoresis. Due to the A and T substitution, the Mst II recognition site is lost (normal CCTGAGG, sickle cell haemoglobin CCTGTGG). In normal individuals (channel 1), homozygous for the unaffected gene, restriction enzyme cleaves the DNA at the recognition sites, resulting in fragments of 103 bp and 191 bp in length. In an affected individual, homozygous for the sickle cell gene, the recognition sites are lost and so a 294 bp fragment results (channel 3). In a heterozygous individual the normal allele is cleaved, and the sickle cell allele is not, resulting in 103 bp, 191 bp and 294 bp fragments (channel 2).

Assignment 11.6

Explain how restriction fragment length polymorphism by Mst II would demonstrate the presence of the sickle cell anaemia point mutation.

Genetic disorders, which cannot be diagnosed by direct methods for want of suitable probes, can be sometimes diagnosed by linkage analysis (Chapter 4) of the particular gene with its adjoining genes. The analysis in genetic linkage is based on the exchange of genetic material (recombination) between the maternally and paternally derived chromosome homologues during meiosis (in formation of germ cells). When two genes are on separate chromosomes, they segregate independently but when the genes are close together on the same chromosome they behave as if they are linked and seldom segregate during recombination in meiosis. Such linked genes serve as markers as they are coinherited. The genetic distance between two markers is defined as 1 centiMorgan (cM) when the markers have a probability of 1 in 100 of undergoing recombination events. This corresponds to an average of 10^6 base pairs between the markers (the total genetic length of the human genome is 3300 cM). Besides using linked genes, sequence polymorphism which occurs in the genome can also serve as valuable genetic linkage markers.

Most of the linkage markers are RFLPs, which are demonstrated using a highly polymorphic probe. Predictive testing in affected families can be carried out by following the inheritance of genetic markers linked to the mutant gene. Such an approach has been applied to diseases such as Duchenne muscular dystrophy, cystic fibrosis, adult polycystic kidney disease and Huntington's chorea. In Fig. 11.11 an example of analysis of a family affected with adult polycystic kidney disease is shown.

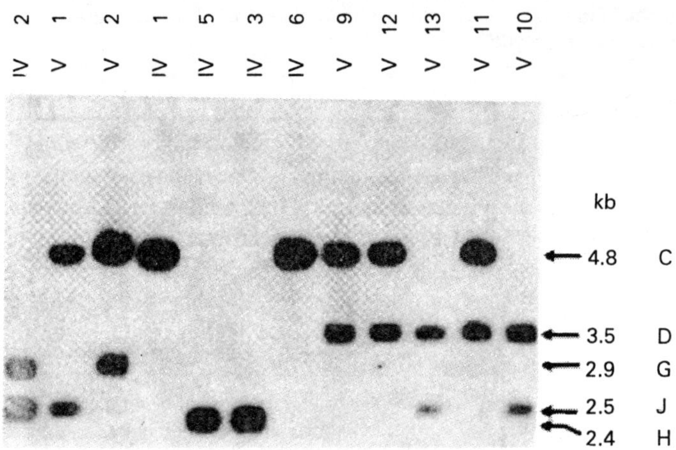

Fig. 11.11 *Southern blot analysis of a family affected with adult polycystic kidney disease.*
The alleles C, D, G, J, H are detected with a linked highly polymorphic probe. Every person with the C allele (4.8 kb band) has polycystic disease.

Prenatal diagnosis

During the last two decades, the diagnosis of genetic disorders in foetuses has become an integral part of obstetrical care in Western countries. The number of disorders that can be diagnosed has increased dramatically, particularly with the advent of allele-specific oligonucleotide probes. Foetal cells have conventionally been obtained by amniocentesis (removal of amniotic fluid), although this procedure is carried out only after the 16th week of gestation, which is rather late to abort a foetus, if necessary. To overcome this problem, chorionic villus sampling is being carried out at an earlier stage of pregnancy, with biopsy of the placenta undertaken with the guidance of ultrasonography. However, efforts are underway to develop reliable methods for foetal testing by less invasive procedures. One of the ways being actively pursued is to subject trophoblasts from maternal circulation to PCR analysis. These cells of foetal origin are frequently present in maternal circulation although in very small numbers.

(b) Diagnosis of infectious diseases

Although monoclonal antibodies are widely used in clinics for diagnosis of viral as well as bacterial infections, the use of DNA probes is proving to be more sensitive and reliable. Presence of viruses such as HIV can be detected by using PCR and probes at a very early stage. Blood cells from infected individuals are lysed in hypotonic solution and HIV-specific mRNA is converted to cDNA, with the help of reverse transcriptase (RT) enzyme and then amplified by PCR using suitable primers (RT-PCR). Gel electrophoresis can be performed on the sample to indicate whether or not amplification of DNA has occurred. The diagnosis can be confirmed by Southern blotting of

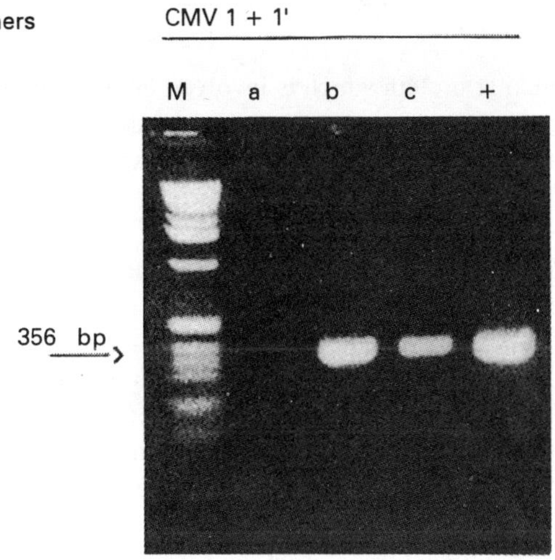

the amplified sample using HIV-specific oligonucleotide probes. Important human pathogens such as herpes virus, cytomegalovirus, EBV and varicella-zoster have been detected in this way in urine, blood and, in some cases, cerebrospinal fluid or tissue biopsies. An example of the technique for detection of cytomegalovirus is provided in Fig. 11.12.

Fig. 11.12 *Detection of human cytomegalovirus (CMV) by PCR using CMV specific primers in blood, urine and sputum of a patient.*
The ethidium bromide stained gel shows amplification in urine and sputum but not in blood. A positive control is also included. a: blood; b : urine; c : sputum : + control.

Mycobacterial infections are common in many parts of the world. In addition to the most common pathogen, *Mycobacterium tuberculosis*, infections with atypical mycobacteria of the avium complex are becoming more and more prevalent, especially in immuno-suppressed (AIDS) patients. Routine diagnostic procedures, such as cultivation of the pathogen, are very time consuming. Excellent correlation has been shown between diagnosis by PCR and conventional culture procedures. PCR has also been used successfully to detect *Chlamydia trichomatis*, an obligate intracellular parasite, which is a common cause of sexually transmitted diseases. The diagnosis of this disease is currently cumbersome and expensive as cell cultures must be used.

(c) Tumour biology

Cancer is basically a disorder of growth. Normally, growth is regulated through the action of proto-oncogenes and suppressor genes, which are positive and negative regulators, respectively, of cell growth. Oncogenes, which are the activated counterparts of proto-oncogenes, play a major role in cancer. There are more than 100 oncogenes which can induce tumours in experimental animals. Anti-oncogenes or tumour suppressor genes are the recessive-acting genes. Loss of both copies of anti-oncogenes is seen in some tumours.

Oncogenes may be activated through mechanisms such as amplification, point mutations or rearrangements of their regulatory regions. Point mutations have been extensively studied in the *ras*-family of oncogenes. The gene termed K-*ras* is mutated in a majority of pancreatic carcinomas and some colorectal carcinomas. Likewise, about one third of patients with tobacco-associated oral cancers show point mutations in the H-*ras* oncogene, particularly at codons 12/13 and/or 61. In chronic myelogenous leukaemia, a portion of chromosome 9 (long arm) is translocated to the long arm of chromosome 22 resulting in activation of the oncogene *bcr-abl*. In neuroblastomas, activation of the N-*myc* gene occurs by the amplification of the gene to almost 300 copies. This amplification has been correlated with poor prognosis. In the familial form of retinoblastoma, one of the mutant gene copies of the suppressor gene is inherited at birth. Loss of the second copy, the mechanism for which is not yet fully understood, results in malignancy. Although a number of cancers show abnormalities of both oncogenes and suppressor genes, these alterations are not consistent and therefore cannot be used in tumour diagnosis or as prognostic markers to assess effects of therapy.

> **Assignment 11.7**
>
> A cancer-causing gene (oncogene) codes for a particular protein which is involved in the malignant transformation of cells. Explain how incorporating an antisense gene to the cancer-causing gene into cells might prevent expression of the cancer-causing gene and reverse/prevent malignant transformation of those cells.

One disadvantage of treating cancers with cytotoxic drugs or radiation is that normal cells are also destroyed. The search for effective tumour-specific therapeutic agents has eluded us so far. However, the knowledge that some cancers show alterations in proto-oncogenes has raised the possibility of using the very lesion itself as a tumour-specific target for treatment. In several types of tumour, alterations in the sequence of a given gene lead to the production of an altered protein which plays a role in the malignant transformation process. Suggestions have been made that the production of such proteins could be specifically blocked by inserting into cells an 'antisense gene' (the complementary sequence to the gene on the chromosome in reverse orientation).

Antisense oligonucleotides to c-*myc*, which is activated in Burkitt lymphoma, a cancer of the B-lymphocytes, have been recently used to arrest cancer growth. This novel approach is still in the experimental stage.

(d) Forensic medicine

In the genome of humans there are particular oligonucleotide sequences which are duplicated (tandem repeats). The tandem repeats are themselves repeated and the number of such repeats is variable from individual to individual but specific to an individual. The occurrence of variable number tandemly repeated DNA sequences (VNTRS) means that restriction fragment length polymorphism occurs (with restriction enzymes having cleavage sites within

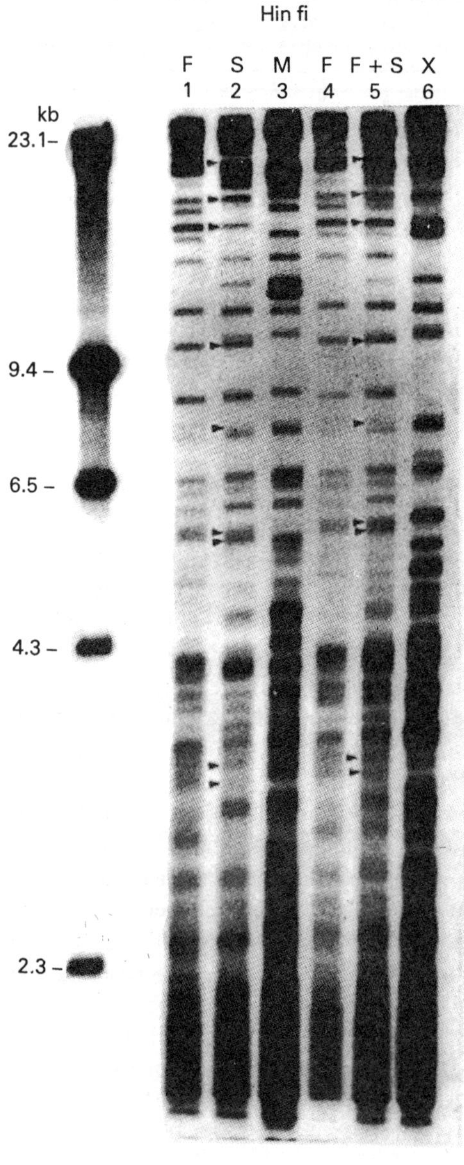

Fig. 11.13 *Establishing Paternity. Track 1 and 4 : DNA profile of the alleged father (F), track 2 : DNA profile of the son (S), track 3 : DNA profile of the mother (M), track 5 : DNA profile of equal amounts of DNA of the alleged father and son (F+S), track 6 : DNA profile of an unrelated individual (X). Arrows indicate that the DNA fragments inherited by the son are similar to those present in the DNA of the alleged father. F is therefore the biological father (kindly provided by Dr.Lalji Singh, Centre for Cellular & Molecular Biology, Hyderabad, India).*

VNTRS). VNTRS are distributed throughout the genome and they appear to be non-functional. The VNTR loci have been cloned to generate probes which when hybridised to genomic DNA can detect alleles from multiple related loci. When used in conjunction with restriction enzyme cleavage and gel electrophoresis, the method produces a pattern of bands which is specific to individuals. It has been accepted as evidence in paternity and other legal disputes in law courts in many countries. This technique is commonly known as DNA fingerprinting or genetic fingerprinting.

The VNTRs contain sequences which are found not only in the genomes of humans but also in other vertebrates and plants. One such sequence has been isolated from the female Indian banded krait and designated as Bkm (banded krait minor satellite DNA). The Bkm sequences have been used as a probe in paternity disputes and crime investigations. The probe can be used to detect RFLPs after digesting human DNA with restriction enzymes. The DNA fingerprint obtained from one family is shown in Fig.11.13.

(e) Recombinant proteins and vaccines

One of the most exciting developments in recombinant DNA technology has been the ability to design and construct novel proteins and enzymes with modified functions. It is now possible to introduce one or more point mutations into genes coding for proteins by site-directed mutagenesis or even introduce genetic changes leading to substitution or deletion of polypeptide fragments. Small changes in the structure of the protein can sometimes lead to dramatic changes in the useful properties of the molecule. Thus protein engineering has been used to redesign antibodies, enzymes and vaccines (Chapter 6).

One of the latest developments in recombinant DNA technology is the production of transgenic animals. In this technique, a foreign gene (transgene) is introduced into the pronucleus of a single celled embryo. The embryo is then reimplanted into the oviduct of a pseudopregnant recipient female and allowed to develop further. The newborn animals are screened for presence of the transgene by Southern blot analysis of DNA extracted from cells. Transgenic animals have great potential for synthesising recombinant proteins for use in therapy for many diseases. Human Protein C, which has anti-coagulant activity, is a regulator of haemostasis. The cDNA for this protein has been inserted into the first exon of the gene coding for mouse whey acidic protein (a component of milk). This gene has then been inserted into swine embryos. High levels of the heterologous (human) protein were found in the milk of the transgenic swine, showing the successful introduction of the transgene. Similarly, human factor IX and human alpha-antitrypsin have been synthesised in the milk of transgenic sheep. The yield of recombinant protein is high in such systems. A number of recombinant proteins, which are now commercially available, are used in therapy (Table 11.3).

Table 11.3 Recombinant therapeutic proteins

	Recombinant proteins	Systems used	Applications
1	Somatostatin	Bacteria	Treatment of human growth disorders
2	Tumour Necrotic Factor (TNF)	Bacteria, mammalian cells	Cytokine, used as anti-cancer and anti-viral agent
3	Erythropoietin (EPO)	Bacteria, mammalian cells	Stimulates committed erythroid precursor cells
4	Interleukin-2 (ILO2)	Bacteria, yeast, mammalian cells, insect cells	Cytokine, used as an anti-cancer agent
5	Interferons (IFN)	Bacteria, yeast, mammalian cells, insect cells	Cytokine, anti-cancer and anti-viral agents
6	Granulocyte-Macrophage colony stimulation factor (GM-CSF)	Bacteria, yeast, mammalian cells	Cytokine, accelerates neutrophil recovery in neutropenia, anti-cancer agent

(Contd...)

Table 11.3 *(Contd...)*

Recombinant proteins	Systems used	Applications
7 Human Urokinase(tPA)	Bacteria, yeast, mammalian cells	Plasminogen activator used in vascular disorders
8 Human protein C	Mammalian cells, transgenic animals	Anti-coagulant, regulator of haemostasis
9 Human factor IX	Mammalian cells, transgenic animals	Clotting factor for haemophilia B
10 Human factor VIII	Mammalian cells	Clotting factor for haemophilia A
11 Hepatitis B surface antigens	Yeast, mammalian cells, transgenic plants	Vaccine against Hepatitis B
12 Attenuated pseudorabies virus antigens	Mammalian cells	Vaccine against rabies

The first successfully recorded attempts at vaccination, which was against smallpox, were carried out by Edward Jenner, and the scientific basis for vaccination was laid down by Louis Pasteur. He showed that when organisms are attenuated, they lose their virulence, but still retain their immunogenic component/s. Since then a number of vaccines have been prepared using this principle. Although they may contain attenuated organisms, many vaccines still produce unacceptable side-reactions. Moreover, the organisms may also contain antigens that evoke suppressor responses. Efforts are therefore underway to generate vaccines containing only the immunogenic 'subunits'. Peptide vaccines containing only the immunogenic epitope/s would be a further refinement.

Genetic engineering has been used to generate recombinant proteins in microorganisms for vaccine preparation. A vaccine based on recombinant hepatitis B surface antigen in a microorganism is now widely available.

Assignment 11.8

What advantages would there be in producing a vaccine against a virus such as HIV by producing the viral antigens in a recombinant microorganism such as yeast?

A number of vaccines, prepared from recombinant HIV envelope protein/s, are now ready for clinical trials. Live recombinant vaccines have also been prepared. In this approach, genes coding for one or more antigens of interest are introduced into the genome of vaccinia, a rather innocuous virus. The viral genome is large and can accommodate additional genes. As the recombinant virus grows, it would act as a good source of the immunogenic HIV protein. This approach, which is still in the experimental phase, has been tried for a number of viral vaccines including hepatitis B, influenza and herpes simplex.

Self assessment exercise 11.5

Answer true or false to each of the following:

(i) If a point mutation on a chromosome leads to a genetic disorder and the loss of activity of a restriction enzyme at that site, then the point mutation can be identified using PCR.

(ii) If given priming sequences (sequences flanking a given oligonucleotide), and the interlinking oligonucleotide sequence is specific to *Mycobacterium leprae*, then

amplification of DNA from a tissue biopsy by PCR using those given primers indicates that *Mycobacterium leprae* is present in the tissue sample.
(iii) If the sequence of a mutant gene which codes for a protein responsible for malignant transformation is known, then the malignant transformation in a cell may be prevented or reversed by incorporating into the cell copies of the mutant gene.
(iv) Genetic fingerprinting is based on the fact that the pattern of RFLP produced by given endonucleases with cleavage sites in VNTR sequences is characteristic to an individual.

(f) Gene therapy

The ease with which a given gene can be cloned has increased the feasibility of gene therapy, which is aimed at the correction of genetic defects through the introduction of the desired gene into humans. Strategies for therapy depend on the nature of the genetic defect. Normally, gene expression is tissue-specific and the chromosomal location of genes has physiological implications. Gene expression is not only self-regulated but also influenced by neighbouring genes. Thus, for optimal functioning, the therapeutic gene should integrate at the chromosomal site where its normal counterpart is located. Also the defective mutant gene should be removed from the chromosome if it encodes a product harmful to the host. At the time of writing this has not been possible, although efforts are being made to achieve this objective by homologous recombination. For gene transfer *in vivo*, appropriate cells of the host are withdrawn from the body, transfected with the desired therapeutic gene using retrovirus vector systems and reintroduced into the host. The cells generally used are those which can be easily obtained such as bone marrow cells and their descendants, fibroblasts, hepatocytes and myoblasts. However, these cells have a limited life span. Ideally, the transfected cells, when reintroduced in the body, should survive the life-span of the host, so that gene therapy becomes a one-off procedure.

At present, most effort is being directed towards correction of gene defects in somatic cells (Table 11.4).

Table 11.4 Gene therapy: on-going clinical trials

	Disease	Gene inserted	Cell type	Current status	Comments
1	ADA-deficient severe combined immune deficiency	ADA gene	1) peripheral T-lymphocytes 2) Bone marrow cells	Clinical trials are on-going since 1990 in USA, since 1992 in Italy	Repeated injections of gene-corrected cells have to be given
2	Duchenne's muscular dystrophy	Dystrophin gene	Myoblasts	Clinical trials initiated in USA	It is yet to be determined how long the gene remains active
3	Malignant melanoma	TNF gene	TIL, bone marrow cells	Clinical trials initiated in USA since 1991	Homing of the TNF carrying TIL to the tumour is not very satisfactory
4	Haemophilia B	Factor IX gene	Autologous skin fibroblasts	Clinical trials initiated in China since 1991	Duration of gene activity is not known
5	Brain tumour	Herpes simplex thymidine kinase	Mouse fibroblasts	Clinicial trials underway in USA since 1992	Will help in reducing the tumour load, effective only locally
6	Cystic fibrosis	CFTR gene		Clinical trials underway in Scotland since 1994	Gene delivered to lung cells via liposomes

The most amenable disorder for gene therapy is a monogenic deficiency disorder due to deletion of the gene, specially if the gene product has no tissue specificity and does not produce any adverse effects, even when its level in the circulation is high. Cells transfected with such a replacement gene are injected at a suitable site in the host. In such disorders, even if the gene expression is improperly regulated and excessive, it would have no adverse effects on the host. Enzyme deficiencies generally fall in this category. The first clinical trial has been conducted in severe combined immunodeficiency (SCID) of both B and T-cells. This genetic disease is due to a deficiency of the enzyme adenosine deaminase (ADA), caused by deletion of its gene. Lymphocytes transfected with the ADA gene have been reintroduced into patients, and the results have been very promising. Unfortunately, since lymphocytes have a finite life-span, the disease cannot be cured permanently by a single treatment; repeated injections of modified lymphocytes are needed. Trials are also underway in which the gene has been introduced into bone marrow stem cells in an attempt to overcome this problem. SCID was discussed in more detail in Chapter 5.

Recently myoblast cells genetically engineered with retroviruses have been used successfully in animal models to deliver recombinant proteins. Retroviruses carrying the growth hormone gene have been introduced into myoblasts which were then reintroduced into donor animals. Growth hormone produced by the introduced gene could be detected in the animals' circulation for at least 3 months. Similarly, in another recent clinical trial, patients with Duchenne muscular dystrophy were implanted with myoblasts producing the missing gene product, dystrophin. The myoblasts made dystrophin for at least 6 months after implantation. Thus myoblast-mediated gene transfer provides a novel form of drug delivery system which should prove useful in treatment of myopathies as well as non-muscular disorders such as haemophilia, heart diseases and cancer. Another cell type used in somatic gene therapy, especially in animals, is the hepatocyte.

A novel approach in gene therapy of cancer is the introduction of genes expressing proteins which assist in the destruction of tumour cells. It has been shown that certain lymphocytes isolated from tumours, termed 'tumour-infiltrating lymphocytes' (TILs), when reinjected into the patient 'home in' predominantly onto the tumour. TILs transfected with the gene coding for tumour necrosis factor have been injected into patients with melanoma, resulting in regression of tumours in some cases.

Another approach is to use the thymidine kinase (TK) gene of herpes simplex (HS) virus. Cells expressing HSTK are targets for anti-viral agents (such as ganciclovir). A replication-deficient recombinant retrovirus is constructed carrying the HSTK gene and is introduced via transfected fibroblasts, at the site of the tumour. The fibroblasts produce and release numerous copies of the retroviral nucleic acids which become integrated only into dividing cells, which are mainly the tumour cells. The infected cells produce TK and can be killed by administration of ganciclovir. In a rat brain tumour model, this strategy has resulted in the regression of tumours.

Self assessment exercise 11.6

For which of the following would use of hybridisation with a labelled DNA probe be appropriate?

(i) to demonstrate the location of the gene coding for alkaline phosphatase in yeast cells.

(ii) to demonstrate the intracellular location of alkaline phosphatase in yeast cells.

(iii) to demonstrate the presence of hepatitis B virus in a sample of liver tissue.

(iv) to show that the cells of a newborn transgenic sheep carry a gene coding for human Factor IX.

(v) to show that the cells of a newborn transgenic mouse carry an antisense gene for tumour-inducing protein.

(vi) to show the location of hepatitis B antigens on the surface of recombinant yeast cells.
(vii) to assay the levels of carcinoembryonic antigen in blood.

(g) The human genome project

With the advent of recombinant DNA technology, it has become possible to identify and map genes on human chromosomes. The phenomenal task of the complete mapping and sequencing of the human genome, involving some 3 billion base pairs, which would provide rational insight into the role of genetics in human health, is the aim of the human genome project. The hope is that it may lead to the identification of novel genes and their functions, leading to increased understanding of genetic disorders. This is important, especially in population groups where consanguinity (inbreeding) is common, since this favours the expression of defective recessive genes. Comprehensive genetic linkage maps would also permit novel approaches to diagnosis of genetic disorders and offer new tools with which to study genetic epidemiology, genetic susceptibility to noxious agents, and the permissive role of host genetics in the pathogenesis of complex disorders such as cancer, cardio-vascular diseases, auto-immunity and mental illnesses. The mapping and sequencing of the human genome constitutes a new human anatomy, albeit at the molecular level. Many see in the human genome project the dawn of a new era of predictive medicine based on the identification of susceptibility revealed in an individual's genome.

Several laboratories worldwide are now participating in this gigantic project which was launched in the USA in 1987. Because of the extreme polymorphism of the human genome, one of the major problems is to define what constitutes a normal gene, especially in nonfunctional regions. Over the last six years a few chromosomes have been mapped including the Y chromosome and chromosome 21. Both chromosomes are important, especially chromosome 21 which houses the genes involved in Down's syndrome, Alzheimer's and other neurological diseases. Previously no more than 40 kilobases could be cloned at one time. That meant tens of thousands of clones had to be analysed to map just one chromosome. In mapping the Y chromosome and chromosome 21, use was made of modified yeast chromosomes called YACs (yeast artificial chromosomes) into which stretches of DNA 650 kilobases long could be inserted. These mega-YACs made it possible to map the two chromosomes in less than 2 years.

11.4 Concluding remarks

Hybridoma technology and genetic engineering have pervaded every branch of bio-medicine. The enormous potential of the combination of thc two tcchnologics is obvious. It is now possible to design novel molecules to suit certain functions.

Chimaeric antibodies and chimaeric and recombinant proteins are just the beginning. Unlimited combinations can be clearly visualised. The synthesis of xenogenic proteins, which do not evoke strong adverse immunological reaction, have become a reality through transgenic modulations of genes. Novel prokaryotes have been constructed and tailor-made to serve certain functions, for example to produce viral antigens (recombinant vaccines).

The capacity to clone normal and novel genes has increased the feasibility of manipulating genes at will. Although gene therapy is currently being tried only in somatic cells, future efforts could be to attempt it in germ cells. There is an increasing recognition that genes play a major role in host response to infections and noxious agents, and in immune modulations and degenerative processes. A complete analysis of the human genome, as envisaged in the human genome project, will result in newer approaches to genetic disorders and may make it possible to predict, looking at the genes, individual host responses to noxious agents and predisposition to certain diseases.

11.5 Suggested further reading

1. Gusella, J.F. (1986). DNA polymorphism and human disease. *Ann. Rev. Biochem.*, **55**, 831–854.
2. Erlich, H.A. (ed.) (1989). *PCR technology: principles and applications for DNA amplification.* Stockton Press.
3. Winter, C. and Milstein, C. (1991). Man-made antibodies, (review article). *Nature,* **349**, 293–299.
4. Epenetos, A.A. (ed.) (1991). *Monoclonal antibodies: applications in clinical oncology*, Chapman and Hall Medical, London.
5. Rolfs, A., Schumacher, H.C. and Marix, P. (eds.) (1991). *PCR topics*. Springer-Verlag, Berlin.
6. Huber, B.E. and Lazo, J.S. (eds.) (1994). Gene therapy for neoplastic diseases. *Annals of New York Academy of Science*, **716**.
7. Martin, J.B. (1989). Molecular genetic studies in the neuropsychiatric disorders. *Trends Neurosci*, **12**, 130.

11.6 Responses to assignments

11.1 Yes. Instead of having to conjugate the enzyme with the monoclonal, the bispecific antibody coula be used to attach firstly to bound hormone in the test, and if enzyme is added afterwards the enzyme would bind to bound antibody molecules. This would overcome the need to conjugate the antibody and the enzyme. This assumes binding of the enzyme to the monoclonal does not interfere with enzyme activity.

11.2 The tissue section can be exposed to the fluorescent-labelled monoclonal, upon which the monoclonal will specifically bind to any HIV antigen on the cell surface or internal structure exposed in the section. After unbound monoclonal has been washed off, the section can be examined under a UV microscope. Control sections using HIV-free tissue and/or fluorescent labelled antibody against a virus other than HIV should also be prepared and examined. If the tissue section in question shows fluorescence when control sections do not, presence of HIV has been demonstrated.

11.3 The tissue section can be exposed to the monoclonal conjugated to peroxidase enzyme, in which case the conjugated monoclonal binds specifically to any keratin present. After unbound monoclonal has been washed off, a substrate for peroxidase which results in formation of a coloured product is added. After washing, the section can be examined under a light (white light-illuminated) microscope and areas of any cell containing the keratin will be stained by the coloured product formed as a result of conjugated peroxidase activity.

11.4 Liposomes are lipid or phospholipid vesicles which fuse with cell membranes (which have a similar structure) and thus deliver any drug they contain into the cells. This fusion is, however, non-specific (liposomes may fuse with any cell). By coating liposomes with MoAbs directed against surface receptors (antigens) specific to a given type of cell, then the coated liposomes will be directed to, and preferentially fuse with, that given type of cell.

11.5 The specific oligonucleotide probe is radiolabelled (for ISH) or fluorescence labelled (for FISH). After suitable preparation of the specimen and hybridisation, the specific probes hybridise with complementary sequences in the DNA of the specimen. For ISH, the specimen is then exposed to sensitive film and examination of the exposed film shows location of radioactivity (i.e., probes); this can be related back to the original specimen, thus allowing location of probes to be determined. For FISH, the specimen can be viewed directly under a UV microscope after hybridisation; location of probes is demonstrated by their fluorescence. This latter technique is much more rapid than ISH, as it avoids the time consuming step of autoradiography.

11.6 The restriction enzyme *Mst*II cleaves normal DNA at its specific restriction sites, producing fragments of various lengths; these can be identified (characterised) by gel electrophoresis. Where the sickle cell anaemia mutation has occurred, the enzyme no longer cleaves DNA at that site (as the mutation leads to loss of enzyme action at the site) and thus an uncharacteristically long DNA fragment will arise; this can be identified by gel electrophoresis.

11.7 The oncogene is transcribed into mRNA which is translated into the particular protein. An antisense gene is the complementary sequence to the oncogene inserted into DNA in reverse orientation to the original gene, so that mRNA transcribed from the antisense gene is complementary to mRNA transcribed from the oncogene. The complementary RNAs may thus bind to each other (to form double stranded mRNA) and translation will thereby be prevented (translation of double-stranded mRNA cannot take place, only single-stranded mRNA can be translated). This may prevent or reverse the malignant transformation in affected cells. Normal cells (without an active oncogene) will not be affected.

11.8 Vaccines against viral diseases such as rabies and poliomyelitis are conventionally prepared by growing the virus in cell culture and preparing a vaccine from the cultivated virus. However, it would be difficult and dangerous to cultivate HIV in this way. Vaccines use either virulent virus which has been killed by chemical treatment, or live avirulent (attenuated) virus. However, with HIV there are fears associated with vaccinating humans with the virulent virus (even after chemical inactivation) and fears of using live virus (which may revert to virulence). All these problems may be overcome by using a recombinant microorganism expressing HIV antigens; yeasts are easy to cultivate and most are non-pathogenic, and as the recombinant organism contains only those genes from HIV which code for antigens there is no possibility of HIV infection arising from vaccination.

11.7 Responses to self assessment exercises

11.1 The answer is yes. The monoclonal could be used as follows:
(i) the tissue section would be exposed to the monoclonal (which would bind to any tumour-marker antigens present); (ii) unbound monoclonal would be washed off; (iii) peroxidase enzyme would be added (which would bind to the monoclonal bound to the tumour-marker antigens); (iv) unbound peroxidase would be washed off; (v) colorigenic substrate for peroxidase would be added; (vi) after incubation unreacted substrate would be washed off and the section examined for tissue coloration under a microscope.

The advantage of this technique over using peroxidase-labelled antibody is that the conjugation (labelling) process can be technically difficult and the conjugated antibody relatively unstable on storage.

11.2 (i) True.
(ii) False - this would be true if 'radio immunoassay' was replaced by 'immunoscintigraphy'.
(iii) True.

11.3 The best MoAb to use would be (iv) and this is specific for tumour X and has the A-chain toxic moiety of ricin, but lacks the B-chain of ricin which causes non-specific binding to cells.

MoAbs (ii) and (iii) do not contain the toxic A-chain moiety of ricin and would therefore not be cytoxic.

MoAb (i) would produce cytotoxicity in the cells of tumour X but as it is conjugated to the B-chain of ricin (as well as the A-chain), the B-chain may cause non-specific attachment of the MoAb to cells other than those of tumour X.

11.4 The answer is (ii) and (iv), as both antibodies would have specificity for T-cell markers and are conjugated to polypeptide A — they are both immunotoxins.

MoAb (i) would not cause cytotoxicity to T-cells as it does not carry a cytotoxic molecule, it has only affinity for peptide A.

MoAb (iii) would not bind to T-cells as it does not have affinity for T-cell receptors.

MoAb (v) would not bind to T-cells, as F_C fragments do not contain antigen-binding sites. If the F_C fragment was substituted for F_{ab} or F_V fragments, then the molecule would have affinity for T-cells and might be cytotoxic.

11.5 (i) False — this would be true if 'PCR' was replaced by 'RFLP'. PCR is a means of replicating a given DNA sequence.

(ii) True.

(iii) False — this would be true if '... copies of the mutant gene' was replaced by 'copies of the mutant gene in reverse orientation'.

(iv) True.

11.6 Labelled gene probes would be useful in (i), (iii), (iv) and (v). All these tests are genetic tests and a specific probe for the specific gene in question would demonstrate, in each case, the presence of that specific gene, after a suitable hybridisation process.

In (ii), (vi) and (vii), the tests are not genetic tests, rather they are tests for the products of specific genes. Gene hybridisation is not therefore applicable to these tests. Immunoassay would be useful for (vii), and immunofluorescence for (ii) and (vi).

12 Plant biotechnology

Chapter outline

12.1 Introduction

The oldest of all biotechnologies is perhaps plant biotechnology, heralded by the introduction of crop rotation. This was essentially aimed at improving crop productivity. Classical plant biotechnology is limited to the introduction of required characteristics into plants by genetic crossing during sexual reproduction. Consequently, the breeding system of a plant determines whether this approach could be useful or not. Further, this classical approach is useful if the desired genetic improvement is encoded by a single gene. If a trait is determined by several genes, it may be extremely difficult to introduce all the genes responsible for the crop improvement using conventional breeding techniques. Perhaps the most serious limitation of classical plant breeding technology is the time needed to develop a usable plant variety; it is estimated that it takes about 15 years to introduce a new variety using this method.

Agriculture has always been under pressure to cope with the needs of the ever expanding human population. World grain production has had to increase at a rate of 50 billion tonnes a year to match population growth. There is a real danger that even the so-called Green Revolution that occurred over the past few decades may be falling short of the goal of meeting the needs of the human race. It is in this context that the newly emerging area of biotechnology, *viz.* genetic engineering, assumes greater significance. Genetic engineering has provided the technology with which plants can be genetically modified by manipulating a plant's own genes and also by introducing genes from taxonomically unrelated plants and other organisms such as bacteria, fungi, viruses and even animals.

Plant biotechnology has affected all avenues of human life. Besides being used as food, plants serve as raw materials for the production of fabrics, medicines, chemicals, oils, dyes, paper and timber. With the rapid depletion of fossil fuels such as oil, coal and natural gas, greater attention is now being paid to the possibilities of utilising plants or plant-derived products as substitutes for coal and oil for the generation of energy and production of organic chemicals. Genetic engineering will pave the way for increased food production and also allow modification of plants for industrial needs. In this chapter we are going to examine how methods for manipulating plant cells at the cellular and genetic levels are being applied to the benefit of humankind.

12.2 Plant tissue culture

Plant tissue culture is a process by which parts of plants and also single cells can be grown *in vitro* in sterile culture in a medium containing a carbon source, minerals, growth factors and growth regulators such as hormones. The totipotency of plant cells, which confers ability to differentiate into shoots and roots, has greatly contributed to progress in this area. Tissue culture is of great value, particularly in speeding up conventional breeding and propagation procedures. Let us examine the major applications of plant tissue culture.

12.2.1 Micropropagation

The term propagation, as applied to plants, means the generation (reproduction) of plants by asexual means. Methods of conventional propagation involve taking portions of a plant and inducing each portion to form roots and/or shoots, thus forming a new plant.

Assignment 12.1

What are the advantages and disadvantages of using propagation to produce plants as seedlings (for planting by farmers), as opposed to using seeds?

Micropropagation is the term used for propagation in the laboratory (it is propagation in controlled and aseptic conditions, rather than in a greenhouse or nursery).

Compared to conventional propagation, micropropagation has the advantage of allowing rapid propagation in a limited space, it allows complete control of growth conditions, and the plants produced are microbiologically sterile (guaranteed disease-free) and thus exempt from quarantine regulations that may apply to export/import of normal plants.

(a) Methods in micropropagation

Micropropagation can be carried out by several means, as outlined below:

(i) Shoot tip culture

Most micropropagation methods use apical or lateral shoot tips containing meristems. Each of these shoot tips has the shoot apex and several leaf primordia, at the axils of which axillary shoots could potentially develop. Excised shoots can be cultured on an agar-solidified culture medium with high concentration of a cytokinin hormone. This allows a large number of axillary shoots to develop. Each of these axillary shoots can be separated and the same process repeated to rapidly multiply the shoots in an exponential manner. The shoots are then planted in an agar culture medium containing root-inducing hormone so that a complete plantlet is obtained from each (Fig. 12.1).

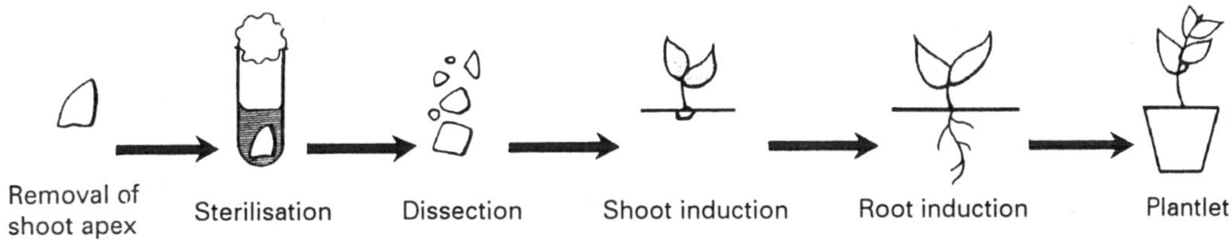

Fig. 12.1 *Micropropagation of plants by shoot tip culture.*
Shoot tips are excised, sterilised (in ethanol or mercuric chloride solution), dissected, and each explant induced to form shoots and roots. In this way a clone of germ-free plants can be produced.

(ii) Culture of nodes

In certain plants it is not easy to generate multiple shoots using cytokinin hormones. Serial subculture of stem nodes is the method of choice in such plants. Nodes comprising axillary buds are excised from the plant, placed on agar-solidified culture medium and the axillary buds allowed to grow into shoots. Each of the nodes of this shoot are again subcultured. Shoots can then be rooted, as described above, to form complete plantlets. This serial subculturing could lead to the generation of as many as 2000 plants from a single node in a period of three months.

(iii) Micropropagation by organogenesis and somatic embryogenesis

Tissues from leaves, stems, roots and inflorescences can be cultured to promote direct morphogenesis (without going through a callus or suspension stage) or indirect morphogenesis (after going through callus and/or suspension cultures) to obtain either direct shoot generation (organogenesis) or somatic embryos (somatic embryogenesis) (Fig. 12.2). Micropropagation through somatic embryogenesis is the most widely used method. Somatic (or asexual) embryogenesis is the formation of embryo-like structures from somatic cells. Somatic embryos are structures similar to zygotic (sexually formed) embryos and they can be excised from parent tissues and induced to germinate in tissue culture media. In the case of direct somatic embryogenesis, embryos develop directly on the excised plant tissues without an intervening callus phase, thus avoiding the possibility of introducing what is referred to as 'somaclonalvariations' in the propagated plants. In the case of indirect embryogenesis, callus is first generated from the excised plant tissues from which suspension cultures are raised and somatic embryogenesis is induced in cells in these suspension cultures. Indirect somatic embryogenesis is commercially very attractive because very rapid scaling up is possible and potentially vast numbers of somatic embryos can be generated in small volumes of culture media in a synchronous and reproducible manner. For instance, as many as 10^5 somatic embryos can be obained from 1 g of tissue. An added advantage of this method is the feasibility of automating some steps.

Somatic embryogenesis is generally induced by auxins such as 2,4-dichlorophenoxyacetic acid (2,4-D). Somatic embryos are allowed to mature by removing 2,4-D. These embryos are germinated in a medium with reduced amounts of macro and micronutrients and lacking hormones. Methods are now available to make artificial seeds by encapsulation of somatic embryos: somatic embryos coated with sodium alginate and nutrient solution are dipped into a 50 mM $CaCl_2$ solution, whereupon calcium ions induce rapid cross-linking of sodium alginate to form small gel beads each containing an encapsulated embryo. These encapsulated embryos can be maintained in a viable state till the planting season.

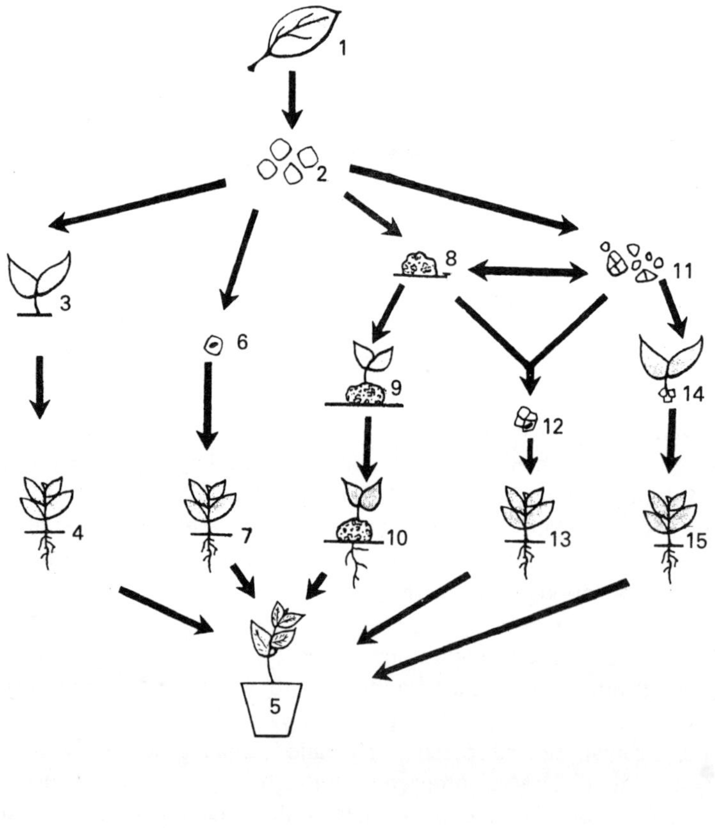

Fig. 12.2 *Micropropagation of plants by organogenesis and somatic embryogenesis.*
Tissue such as a leaf (1) is sterilised and dissected to form explants (2). Explants can be grown on medium and induced by appropriate hormone treatments to form shoots (3) and roots (4) and thus a complete plantlet (5). This is direct organogenesis. Explants can be induced to somatic embryos (6), which can be induced to develop (germinate) with formation of roots and shoots (7). This is direct embryogenesis. Explants can be transformed into callus (8) or suspension cultures (11). Cells from callus or suspension culture can be induced to form shoots (9) and (14), and roots (10) and (15). This is indirect organogenesis. Alternatively, cells from callus or suspension culture can be induced to form somatic embryos (12), which can be induced to form shoots and roots (13). This is indirect embryogenesis.

Plantlets produced in this way from a plant should all be clones of that original plant. The exception may be plants generated through a callus stage, as genetic and somaclonal variations sometimes arise in plants generated this way.

(b) Commercial aspects of micropropagation

A vast number of ornamentals (such as chrysanthemum, orchids and ferns), several woody plants (such as *Eucalyptus, Populus, Pinus, Sequoia, Tectona* and *Rosa*) and several crop plants (such as *Asparagus*, celery, carrot, banana, cardamom and cassava) are examples of plants in which micropropagation is already a commercial success. The Netherlands, a country that has a leading tissue culture/glass house industry, has more than ten companies that produce in the range of 10 million micropropagated plants a year and about 50 companies serving smaller markets. A micropropagation industry has now emerged in the UK, USA, France, Germany, Israel, Japan, China, Singapore, India and Malaysia. As this industry is labour intensive and needs a large number of trained workers with expertise in aseptic handling, developing countries such as India with a lot of trained manpower have potential to develop this industry.

(c) Virus and pathogen elimination

In natural conditions, plants are susceptible to a variety of pathogens such as viruses, bacteria, fungi, insects, mycoplasma and nematodes. Infected plants are not suitable as a source material for tissue culture propagation in the production of plantlet stocks (for farmers/planters) as infected tissues grow slowly in tissue culture conditions and those infected with viruses fail to grow, and in several vegetatively propagated plants, such as potato and cassava, the yield is very low if the farmers/planters use infected plants as stocks. Therefore the plantlets provided to farmers/planters should be made disease-free using tissue culture methods.

Viruses spread systemically in plants and many of them are seed-borne. As there are no control measures by which virus infection can be eliminated from an infected plant, the only means of avoiding losses due to viral disease is to start with seeds or plantlets which are virus-free. Early observations that, even in a systemically infected plant, distribution of virus particles is uneven and especially that there are fewer virus particles in apical meristems, led to the use of meristem cultures to obtain virus-free plants. Apical shoot tips of about 0.1 mm in diameter and 0.25 to 0.33 mm length are collected and surface sterilised.The apical meristem consisting of a dome of undifferentiated, rapidly dividing cells and adjacent leaf primordia are dissected out and cultured on a suitable medium to first obtain callus and subsequently to regenerate whole plants. Although most plants obtained this way are free from viruses, it is very important that the absence of virus particles in these plants is confirmed, either by ELISA using virus-specific antibody or by nucleic acid hybridisation using viral nucleic acid probes (Chapters 5 and 11). Once virus-free plants are obtained, numbers could be scaled up by micropropagation and maintained free from virus infection.

Another useful discovery for virus elimination is the fact that viruses are sensitive to elevated temperatures. Infected plants can thus be incubated at 30 to 35°C for 10 to 25 days before excision of explants for meristem culture. This procedure increases the efficiency of obtaining virus-free plants from infected plants.

The programme of virus elimination by tissue culture is very critical for those plants which are propagated by vegetative methods. As many as 136 cultivars of potato have been freed from viruses using the above methods. A large number of cassava cultivars in Africa have been freed from African Cassava Mosaic Virus and all the freed stock plants in this programme are certified as virus-free by the Scottish Crop Research Institute, UK, before these plants are used for micropropagation to scale up the number of virus-free plantlets.

Virus-free plants germinated by tissue culture should not be confused with virus-resistant plants obtained by genetic engineering. Virus-free plants obtained by tissue culture may be susceptible to virus infection. However, these virus-free plants offer a great boon to the planter who is assured of starting with an uninfected seedling.

12.2.2 Embryo rescue

Sexual crosses between varieties of the same species generally yield viable hybrids. If sexual crosses are made between plants that belong to different species of the same genus or between plants that belong to different genera, hybrid production is difficult due to barriers at the level of pollination, fertilisation and different stages of development of zygotic embryos. In quite a few cases, pollination and fertilisation may be successful but embryos may not develop beyond certain stages either due to their inherent deficiencies or due to incompatibility between the developing embryo and endosperm. Such embryos can be made to grow by 'embryo rescue' in which immature zygotic embryos are dissected out from the developing fruits (seeds) and grown in a suitable medium that either allows the embryos to grow directly into plantlets or allows differentiation of embryo tissue into shoots and roots.

Besides the application in 'wide hybridisation' as discussed above, 'embryo rescue' also helps to complete the growth of embryos in plants in which embryo development is not completed due to defects in seed development. Another application is in the potential for breaking the dormancy of certain seeds in which the period of dormancy is very long; such dormancy can be easily broken after excising the embryos and growing them in a tissue culture medium that favours rapid germination.

12.2.3 Production of haploid plants

Haploid plants serve two important purposes. Direct screening for recessive mutations in plants is not possible since plant cells are diploid or polyploid. Generation of haploid plants or cell lines would be particularly useful for screening for recessive mutations. The second important use of haploid plants is the scope it offers to develop homozygous diploid plants following chromosome doubling of haploid plant cells.

> ***Assignment 12.2***
>
> Homozygous diploid plants are extremely useful as parental types for seed production. Explain why this is so.

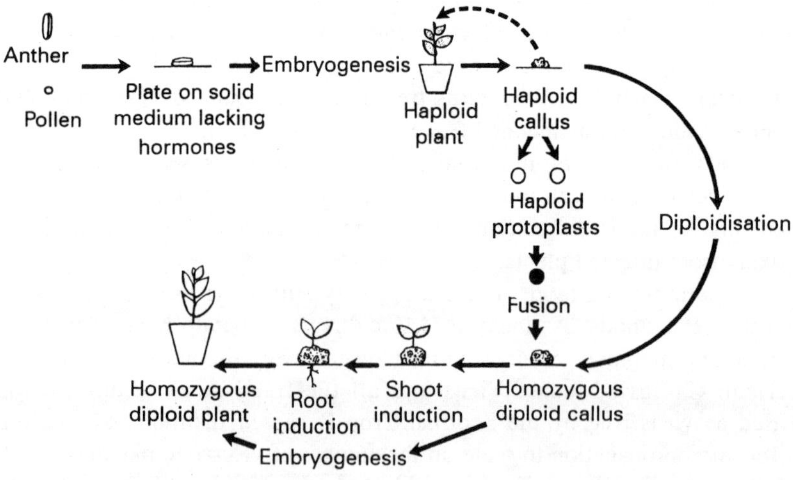

Fig. 12.3 *The production of haploid plants.*
Haploid cells are induced to form somatic embryos which develop into haploid plants. From haploid plants callus can be developed. Cells from this can be diploidised (the chromosome complement doubled), and from this a diploid plant can be developed (as also shown previously in Fig. 12.2). Such a diploid plant will be homozygous for all genes.

Haploid plants have now been generated in over 50 species. Most *in vitro* work involves the use of developing anthers. Immature pollen grains (microspores) and isolated ovules have also been used as sources of haploid plants. Haploid plants obtained through anther cultures or ovule culture are first confirmed by counting their chromosome numbers. Homozygous diploid lines are then established by inducing doubling of chromosomes spontaneously or by exposure to colchicine (Fig. 12.3).

Homozygous diploid lines obtained through haploids are very widely used in China; the most widely grown wheat variety in China is a doubled haploid. Successful use of anther culture in rice has enabled Chinese scientists to reduce the time taken from initial culture to yield trials of a new variety to 18–24 months, compared to four to six years by conventional breeding.

12.2.4 Phytochemicals from plant cell cultures

Many phytochemicals (plant metabolites) such as alkaloids, terpenoids, steroids, anthocyanins, anthraquinones and polyphenols are used as drugs, food flavours, pigments, perfumes and agrochemicals and could potentially be obtained from cell cultures (Table 12.1).

A unique feature of most of these chemicals is that they are secondary metabolites of plants and unique to specific groups of plants. In many cases they are highly tissue-specific, for instance, crocin and picrocrocin of saffron are synthesised only in the stigmas of the plant *Crocus sativus*. Many of these chemicals have a complex structure and stereospecificity and are difficult to synthesise chemically. Efforts have been taken in the past 35 years to explore the possibility of using cultured cells as the source of these phytochemicals.

> ***Assignment 12.3***
>
> What might be the advantages and disadvantages to a country of producing a given phytochemical by tissue culture rather than extracting it from the plant?

The cultured plant cells are utilised in two ways in production of phytochemicals.

(a) Biotransformation

In biotransformation one or more specific biosynthetic steps in plant cells are used to carry out stereospecific conversions which are very difficult to achieve using organic synthesis. Precursors involved in these reactions are provided in the medium. An example of such biotransformation is the conversion of digoxin by C-12 hydroxylation to digitoxin, a heart drug, by cultured cells of *Digitalis lanata.*

(b) Synthesis *de novo*

In synthesis *de novo,* the product is synthesised from basic media components such as sucrose, salts and vitamins. Sometimes the required chemical is synthesised by plant cells from a relatively distant precursor added to the medium. Examples of complex synthesis *de novo* are synthesis of atropine by *Atropa belladonna* cells and vinblastine from the roots of *Catharanthus roseus.*

The initial excitement over prospects of using plant cells for the production of phytochemicals soon subsided once scientists understood the limitations of this approach. The first limitation is the fact that several chemicals are made in a tissue-specific manner and are not made in substantial quantities in cell cultures. The second limitation is the lower yield of the phytochemicals often occurring in cell cultures: cell cultures often produce only small amounts of the secondary metabolites of interest. Because of these limitations only a few cell cultures have been adopted for large-scale production of useful secondary metabolites.

Table 12.1 Plant secondary metabolites and their applications

Metabolite	Application	Plant
	Drugs	
Atropine		*Atropa belladonna*
Hyoscyamine	Muscle relaxants	*Hyoscyamus* spp
Hyoscine		*Datura* spp
Quinine	Anti-malarial	*Cinchona* spp
Morphine		
Codeine	Analgesics	*Papaver* spp
Thebaine		
Vinblastine	Leukaemia	*Catharanthus roseus*
Vincristine		
Digoxin	Cardiatonic	*Digitalis* spp
	Flavours	
Crocin, Picrocrocin	Saffron	*Crocus sativus*
Glycyrrhizin	Liquorice	*Glycyrrhiza*
Capsaicin	Chilli	*Capsicum frutescens*
Vanillin	Vanilla	*Vanilla* spp
	Pigments	
Anthocyanins	red/blue	Various plants
Xanthophylls	yellow	Various plants
Shikonin	red	*Lithospermum erythrorhizon*
Anthraquinones	red	*Morinda citrifolia*
	Perfumes	
Various complex mixtures of oils and terpenoids	Cosmetics	Various plants.
	Agrochemicals	
Various pyrtethroids		*Chrysanthemum* spp
Nicotine	Insecticides	*Nicotiana tabacum*
Piperine		*Piper nigrum*

Two strategies have been used for the large-scale production of secondary metabolites. Firstly, fermenter systems (Chapter 8) have been used to grow up suspension cultures in one or more steps to a stationary phase. The cultures are then harvested to extract the product. Secondly, immobilised cell systems have been used in which the cells are embedded or entrapped in an inert polymeric matrix such as a gel or a cartridge of hollow fibres (Chapters 7 and 9), through which the medium with the precursor is passed at a steady flow rate, and the outflow is collected and processed for the purification of the desired chemical.

In a recent approach, attempts have been made to induce root tumours in plants using the bacterium *Agrobacterium rhizogenes* (the cause of 'hairy root' disease). Root tumours obtained in this way grow very rapidly in culture and have been found to be very useful for the production of root-specific phytochemicals.

One of the best examples of industrial production of secondary plant products from cell cultures is the manufacture of the red pigment 'shikonin' by cultured cells of *Lithospermum erythrorhizon*. In this two-stage production process, the cells are grown up in bulk in the first stage in growth medium and then transferred to a production medium in a second stage in which shikonin biosynthesis occurs (while the cells are in the stationary phase).

Phytochemicals have a large world market. Many of the chemicals are highly useful and very expensive. The best example is vinblastine, worth about $20,000 per gram. Plant-derived drugs have a market of about US $20 billion ($2 \times 10^{10}$) annually in the United States alone. It is also estimated that only 5 to 15% of potential useful plants have so far been systematically explored for useful chemicals. Therefore, there is great potential for using plant cultures for the production of many highly valuable chemicals.

12.2.5 *Storage of germplasm in vitro*

Constant world demand for increased food production has forced scientists and farmers to resort to a narrow range of high yielding crops, thus narrowing the genetic base of crop plants.

Assignment 12.4
Explain the disadvantages of having crop plants with a narrow genetic base (with limited genetic diversity).

A possible consequence of this is the outbreak of major diseases, as has been the case with Southern Corn Leaf Blight in maize in the USA. Therefore, it is very important to preserve as much germplasm as possible for future needs. Additionally, for purposes such as micropropagation and maintenance of pathogen-free plants and true breeding lines, it will be highly advantageous to preserve the stocks in the laboratory rather than maintaining them in fields (where it is difficult to maintain their purity).

The most economical form of preserving germplasm of seed-propagated plants is as seeds. However, some crop plants do not produce viable seeds, some seeds have a very short storage life, some seeds deteriorate rapidly due to endogenous pathogens and some may be heterozygous and are unsuitable for maintaining as true breeding line. Germplasm storage *in vitro* is therefore crucial for the future development and safety of agriculture. Two techniques, 'slow growth technique' and 'cryopreservation' are the main approach for germplasm storage *in vitro*.

(a) Slow-growth technique
Prior to the development of modern methods of germplasm preservation, stocks were maintained by continuous growth of shoot cultures under normal, optimal conditions. This method offers the advantage of enabling monitoring of the health of stocks. However, the disadvantages are the need for greater manpower for frequent subculturing (4–6 weeks) and the risk of accidental loss of the stocks due to contamination, errors of media preparation, etc., which are more likely to happen with frequent subculturing. With the objective of reducing the frequency of subculturing, various methods have been developed to slow down growth in culture, such as growth under conditions of reduced temperature, light or nutrient concentration.

(b) Cryopreservation
Storage of cultures in liquid nitrogen (–196°C) will ensure zero metabolic activity and an almost indefinite period of storage. Plant cultures which normally have large vacuoles are highly prone to freezing damage. Therefore procedures have been developed to prevent ice crystal formation during freezing and thawing of plant cultures.

The ability to generate intact plants from tissues, callus or suspenson cultures, means that such tissues can be conveniently stored by cryopreservation, and intact plants regenerated from stored material as required.

12.2.6 Plant gene transfer by protoplast fusion

In classical breeding, sexual hybridisation between (interbreeding [or crossing]) two compatible plants is performed to achieve desirable exchange of genes (hybrid formation). This mode of genetic exchange is normally possible only between two cultivars within a species or very rarely between very closely related species. The development of viable embryos in most interspecific or intergeneric crosses is not achieved because of specific inhibition or prevention of key steps in pollination, pollen tube growth, fertilisation and embryo or endosperm development. These mating barriers limit the usefulness of breeding programmes. The development of technologies for the isolation, purification and culture of viable protoplasts and the possibility of fusion of protoplasts of genetically different lines and regeneration of plants from them has opened the possibility of overcoming such mating barriers.

(a) Formation and fusion of protoplasts
Protoplasts are formed from cell suspensions, which have been prepared either by treatment of plant tissues with pectinase enzymes or from suspension cultures formed from callus material (callus material is friable, and when incubated in shake-flask culture can disintegrate and form cell suspensions). To form protoplasts, cell suspensions are incubated in hypertonic mannitol solution (to cause plasmolysis — shrinkage of the cytoplasmic membrane away from the cell wall) and then exposed to hydrolytic enzymes which degrade the cell walls.

Assignment 12.5

Which enzymes would be needed to degrade the walls of plant cells, and how could you tell when this had been achieved?

Protoplasts can be induced to fuse by exposure to polyethylene gycol (PEG) followed by perturbation of ionic composition of the medium, or by electrofusion in a fusion cell (a low voltage between two electrodes causes protoplasts to clump together and a short pulse of higher voltage stimulates fusion).

(b) Formation of somatic hybrids and cybrids
Upon fusion, there is always a coalescence of cytoplasms of the two protoplasts (Fig. 12.4). The nuclei of the fused protoplasts may fuse together or may remain separate. Cells formed by the fusion of protoplasts of the same genotype are called homokaryocytes. Fused cells containing non-identical nuclei are referred to as heterokaryocytes and the fusion of two such nuclei results in the formation of a true hybrid protoplast or 'synkaryocyte'.

After fusion of a mixture of protoplasts of two different genotypes, the resultant cell mixture contains unfused protoplasts, cells formed by the fusion of protoplasts of the same genotype (homokaryocytes) and cells formed by the fusion of protoplasts of different genotypes (heterokaryocytes). It is only synkaryocytes formed from heterokaryocytes which are of interest, as only plants regenerated from them will be of the desired (hybrid) genotype. Thus, after fusion the cell mixture is plated out in a medium which allows only heterokaryocytes (and synkaryocytes) to grow. Compare this to the situation which exists after the fusion of lymphocytes and myeloma cells in the production of hybridomas (Chapter 10).

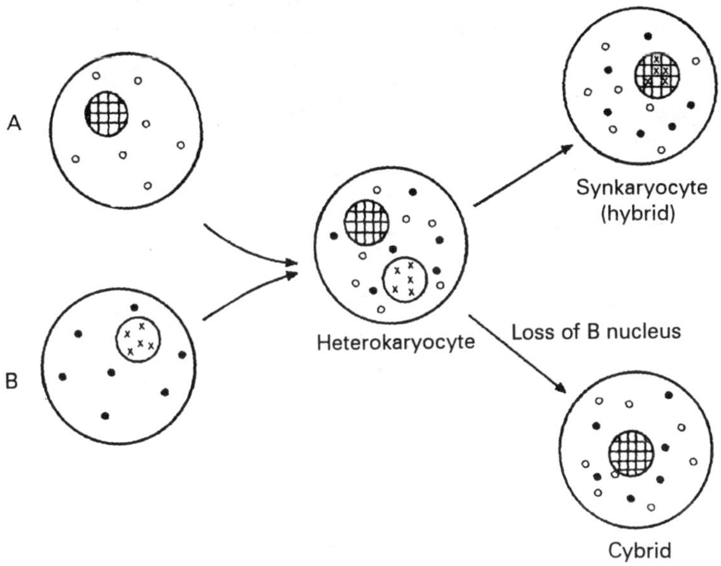

Fig. 12.4 *Hybrid/cybrid production by protoplast fusion.* A fusion product that contains nuclear and cytoplasmic genomes of both parental protoplasts is a 'hybrid'. A fusion product in which both cytoplasmic genomes and one nuclear genome are maintained but the nuclear genome of one of the parents is completely lost is called a 'cybrid'.

Assignment 12.6

How could heterokaryocytes/synkaryocytes be selected for in a medium after fusion of protoplast mixtures?

Heterokaryons/synkaryons that develop in a selective medium can be isolated (and cloned) by dilution plating or micromanipulation (Fig. 12.5). From each colony of cells developed from one heterokaryon/synkaryon, a plantlet or clone of plantlets can be regenerated, as described previously. Thus, using these techniques, a clone of hybrid plants can be produced through protoplast fusion.

Frequently, there is a tendency for selective elimination of nuclear genetic material from one of the parents in synkaryocytes. Instability of the fused nuclear genome is the major problem in somatic hybrids. In extreme situations the nuclear genetic material of one of the parents may be completely lost, while the cytoplasms from both the parents are still retained. Such fusion products are called 'cybrids'. Several methods have been developed to obtain cybrids. One successful method is to irradiate one of the protoplast donors which will damage the nuclear genome but the damage of the organelle genome is minimal due to their multiplicity; irradiated protoplasts are then fused with protoplasts with intact organelle and nuclear genomes. The fusion product will be a cybrid.

(c) Importance of somatic fusion in agriculture

Somatic hybrids have been obtained in several plants and the technique holds potential for the production of true hybrids of sexually incomaptible species (called amphidiploid hybrids) and for the production of heterozygous diploids within a plant species which does not normally reproduce sexually. Yet, commercial application of somatic hybrids has not occurred, basically because somatic hybrids are unstable and almost all of them are sterile. The generation of cybrids has, however, led to important applications such as the introduction of the mitochondrial male-sterility trait of a wild variety of rice into cultivated varieties of rice; plants regenerated from such cybrids have the nuclear and cytoplasmic genes of the commercial cultivar and cytoplasmic genes (with cms) of the wild variety of rice. Cybrid technology in rice is a great success in China. Such plants are very useful in producing hybrid seeds without the need for emasculation. Cybrid technology has also been successfully applied to carrot, *Brassica* spp., sugarbeet, tobacco and citrus.

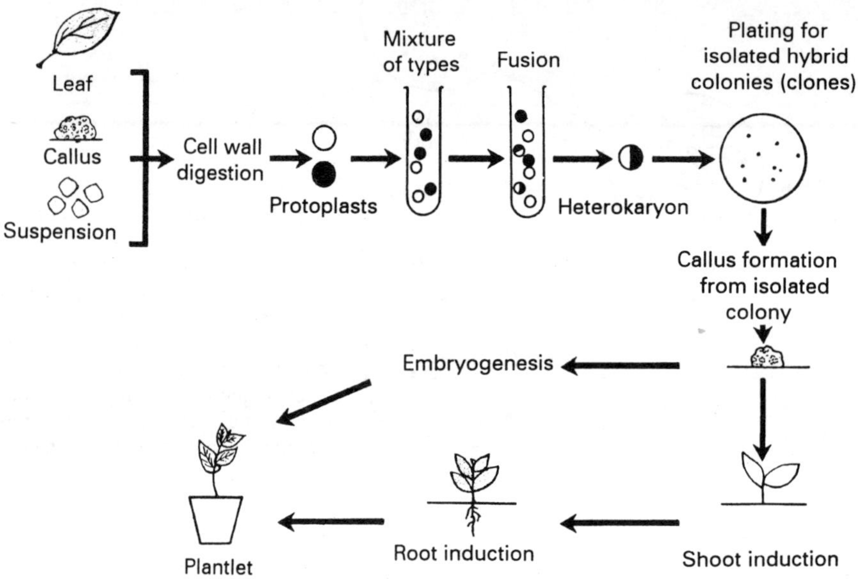

Fig. 12.5 *Protoplast fusion, somatic hybridisation and plant regeneration.*
Protoplasts are prepared from leaves, suspension cultures or callus. Protoplasts from cells of differing genotypes are mixed and fused. Protoplasts are then diluted and mixed with a selective medium in molten agar. The medium is poured into a Petri dish and allowed to solidify. After incubation a colony (microcallus) develops from each synkaryon. From this an intact plant can be produced (as also shown previously in Fig. 12.2).

12.3 Features of plants relevant to their genetic manipulation

Among living things, plants are unique in several respects. Amongst these, totipotency is of special value to plant biotechnologists: every cell, when separated from the plant and incubated under appropriate conditions, can give rise to a complete new plant. Totipotency is somehow lost in animals, perhaps because of the extreme developmental differentiation of cells during embryo development in animals. Plant cells, in that respect, do not undergo extreme differentiation and consequent irreversible inactivation of certain genes.

> ***Assignment 12.7***
>
> A single plant cell can be induced to develop into a complete plant. How could this be useful to plant genetic engineers?

Plants are photoautotrophic (they utilise solar energy and inorganic carbon [CO_2]), converting CO_2 to carbohydrates using the process of photosynthesis. This process takes place in chloroplasts which are organelles within plant cells. A chloroplast has its own genome that codes for about half of the proteins located within the chloroplast, as well as its own transcriptional and translational machinery. There is thus potential for the selective alteration of the genetic organisation of chloroplasts. Interestingly, several proteins present in the chloroplast are encoded by nuclear genes, necessitating a complex interaction between the chloroplast and nuclear genomes. The photoautotrophic capabilities of plants makes them useful life forms with which to study this property. Explorations of this particular property could have great potential in the development of novel plant varieties with increased yield.

Molecular analyses of plants has always lagged behind that of animals and microorganisms, largely because biochemical analysis of plants is difficult; plant cells have relatively low concentrations of enzymes, they have several compounds that inhibit or inactivate enzymes, and extraction and purification of nucleic acids from plant cells is difficult. Plants also have very large genomes necessitating the construction and screening of large genomic libraries in order to clone the genes of interest.

As in other organisms, plants also have several accessory genetic elements including transposons. These are also unique and differ in their behaviour from their counterparts in bacteria, fungi and animals. Transposons are useful in genetic engineering of plants.

Plants exhibit several symbiotic relationships with microorganisms. Symbiotic interactions between *Rhizobium* and legumes and *Azolla* and *Anabaena* lead to fixation of molecular nitrogen. Association between plant roots and mycorrhiza and between algae and fungi in lichens are other examples of symbiotic interactions. Such symbiotic relations may be extended to a wider range of plant species if their genetic bases can be understood and manipulated.

An important feature of plants relevant to genetic engineering is the relative ease with which genes can be introduced into plant cells. Genetic engineering in plants is generally achieved using the bacterium *Agrobacterium tumefaciens* which has evolved a mechanism of introducing genes into the nuclear genomes of plants. The genes of interest are first introduced into *Agrobacterium* which then introduces the genes efficiently into plants. More recently, other methods such as electroporation and particle gun bombardment have been developed which help in the introduction of genes into plants which are not susceptible to *Agrobacterium*. The ease of performing genetic engineering in plants has led to the generation of a steady stream of transgenic plants in the past few years. Plant genetic engineering is being used for introducing a variety of useful genes into plants as well as to gain more insight into plant gene function. These techniques and their applications are described in the following sections.

12.4 Architecture of the plant genome

12.4.1 Structure of plant genes

The architecture of plant chromosomes and the structural organisation of plant genes resembles that of other eukaryotes. The structure of plant genes could be described under two headings, *viz.* protein-coding genes and RNA-coding genes. Figure 12.6 shows the structure of the intermediates in the expression of a typical protein-coding gene.

In all cases transcription begins 40–80 nucleotides upstream of the ATG initiation codon. A typical gene carries one or more intervening sequences (introns) and these are removed at the pre-mRNA level to form the mature mRNA. More about this process will be described later. The genes have controlling sequences at the DNA level at the 5' and 3' ends of the genes. The 5' region has several regulatory elements and some are common and some are unique for specific genes. Two common regulatory elements

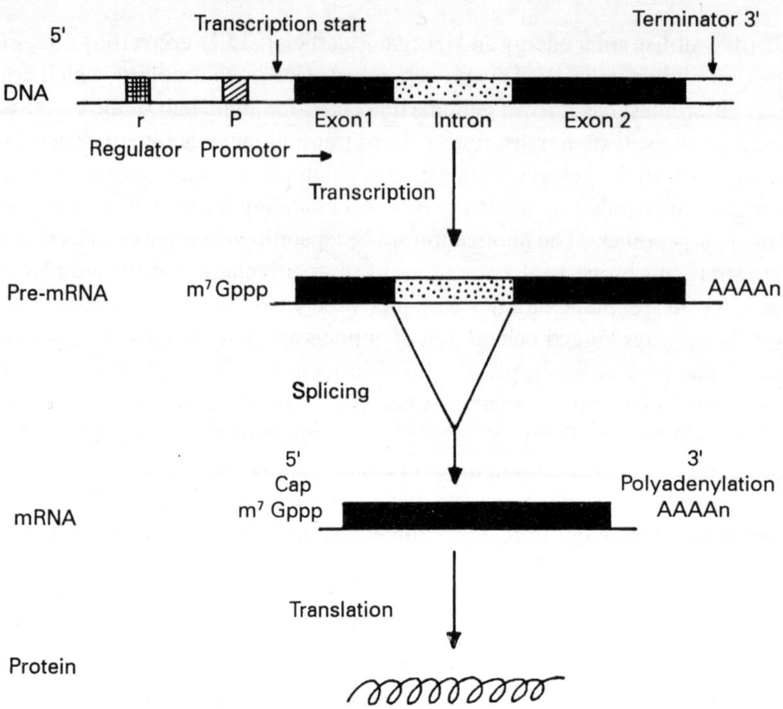

Fig. 12.6 *Structural organisation and expression of a typical plant gene.*

seen in all plant genes are called TATA and CAAT boxes. The TATA box is located 25–40 nucleotides upstream of the transcription initiation site. RNA polymerase II requires this site for accurate initiation of transcription. The consensus sequence of the TATA box is TCACTATATATAG. The CAAT box located further upstream of the TATA box is essential for regulating transcription initiation. In some plants the CAAT box is replaced by the AGGA box. At the 3' end, a sequence needed for polyadenylation of mRNA is present along with the sequence needed for the termination of transcription.

The coding region of the mRNA begins with an AUG codon, which is part of the conserved sequence AACA<u>AUG</u>GCT. The 40S ribosomal subunit scanning the mRNA sequence from the 5' end selects this conserved sequence for initiation of translation. The 3' end of the mRNA will have one of the three termination codons where translation stops. The 3' end also has a less well defined nucleotide sequence where the pre-mRNA dissociates from the DNA template during transcription. The 3' end of the pre-mRNA is then processed and a stretch of adenosines are added at the site where the conserved sequence AATAAA is located.

12.4.2 Organisation of plant chromatin

Chromatin accounts for 10% of the total nuclear mass. In order to fit into the limited space the chromatin must be folded. This compaction at the same time must not interfere with ordered gene expression.

As in other eukaryotes, DNA is coiled around histones to form nucleosomes. The nucleosomes are further condensed into a solenoid structure of 30 nm diameter. In this highly condensed chromatin, active genes are located in regions which assume a more relaxed state.

Recent studies show that the chromatin is attached to the nuclear envelope and also to a proteinaceous scaffold. The regions of DNA attached to the scaffold are called scaffold attachment regions (SAR). The loop-like structures (Fig. 12.7) are sequences bordered by SARs. What is surprising is that the SARs are regions which do not have functional genes.

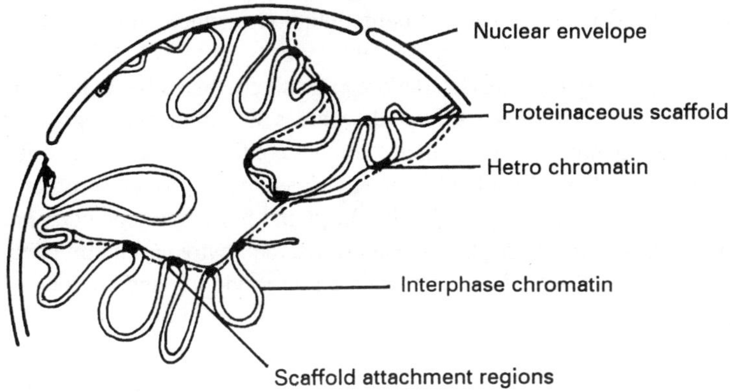

Fig. 12.7 *Organisation of chromatin inside the nucleus.*

Consequently the regions covered by two SARs have one gene or several coexpressed genes. These genes are protected from the influence of genes outside the SAR. This is called genomic compartmentation. By using SAR fragments as border sequences, expression units can be isolated. This approach is very useful in constructing transgenic plants.

Apart from this compartmentation, the active single copy genes and the repetitive sequences are near the centromere, at least in cereals. It should be appreciated that such architectural aspects of the genome may influence the expression of the genes.

12.4.3 C-value paradox

The haploid genome content of organisms, defined as the C-value, increases with their biological complexity. For instance, mammalian cells have a larger genome (higher C-value) than the fruit fly *Drosophila* which in turn has a genome larger than the bacterium *E. coli.* However, in plants one witnesses 'C-value paradox' since very often no correlation is seen between biological complexity and genomic complexity (C-value). For instance, angiosperms which are more evolved than gymnosperms have lower C-value. Plants that belong to the same family may have a 10-fold difference in their haploid genome content. There are even examples where two different species of plants belonging to the same genus may exhibit an order of difference in their genome sizes.

12.5 Regulation of gene expression

The regulation of expression of a plant gene can be controlled at least at three different levels, as outlined below.

12.5.1 Transcriptional regulation

The predominant mode of transcriptional regulation is by the modulation of the rate of initiation. The rate of initiation is controlled by the promoter as well as other *cis-* and *trans-*factors called enhancers and silencers. Enhancers increase transcription and silencers reduce transcription. The enhancers are sequences located well away from the gene which they control and they could exert their effect in either orientation. There are also other gene-specific control sequences which regulate the expression of specific genes under specified conditions. For example, storage protein genes are expressed in a tissue-specific manner. In the case of wheat, rye and barley a unique conserved sequence and corresponding transcription factors regulate the expression of storage proteins. Similarly genes which respond to light also have conserved sequence motifs and corresponding transcription factors.

In addition to the *cis*-acting elements and *trans*-acting proteins, gene expression in plants is also regulated by physiological processes. It has been shown that demethylation of nucleotides leads to enhanced transcription in plants. The sequences which carry 5-methylcytosine are CpG and CpXpG (where X is any other nucleotide). Plant genomes are methylated more extensively than other eukaryotic genomes. For example, 1–8% of the vertebrate genome is methylated, whereas 30% of the wheat genome carries methylated sequences. In mammalian genomes cytosine methylation is confined to the dinucleotide 'CpG' and 80% of these are methylated. These CpG dinucleotides are confined to 5' regions or promotor regions of the genes in animals and are hence called CpG islands. In plants 5-methylcytosine is not confined to CpG dinucleotides alone, as the trinucleotides CpXpG also have their cytosines methylated. Surprisingly, in plants CpG and CpXpG nucleotides are also distributed non-randomly and are located predominantly in the promotor regions of genes. Therefore methylation plays an important role in regulation of gene expression in plants.

12.5.2 Post-transcriptional regulation

The involvement of post-transcriptional regulation became evident when the discrepancy between rate of transcription and the steady-state level of the mRNA was observed. This regulation may involve one or more of the mRNA processing steps described below.

(a) mRNA capping
The addition of a guanosine residue (cap) through a 5'-5' triphosphate linkage to the 5' end of mRNA is a characteristic feature of plant mRNAs. This residue is also methylated at the N7 position. The cap protects the 5' end of the mRNA from degradation by exonucleases and presumably has other functions in mRNA splicing and initiation of translation.

(b) Polyadenylation
The processing of the 3' end of the mRNA has been demonstrated conclusively in several plant genes. This is the first process to occur, even before mRNA splicing. Details regarding termination, cleavage of the poly A addition site by endonucleases and the process of adenylation are not well understood. In plant genes the DNA sequence AATAAA is conserved at a location 10–30 nucleotides upstream of the poly A site. Immediately downstream of the poly A site another sequence YGTGTTYCY (where Y is any other nucleotide) also is conserved in many cases. Plants, unlike other eukaryotes, also have other factors involved in the selection of the poly A site. The poly A sequence has been shown to be involved in the regulation of gene expression by stabilising the mRNA, however its postulated role in translation initiation has not been proved *in vivo*.

(c) Pre-mRNA splicing
A protein-coding gene is transcribed normally into a pre-mRNA which contains non-coding intervening sequences (introns). In order to create an open reading frame, the introns must be removed and the exons must be ligated accurately. This process, called splicing, must be specific and accurate. The process is complex involving the mRNA, U-type small ribonucleoprotein particles (Usn RNPs) and other less well defined proteins. The splicing process is initiated by assembling all these components into complexes called spliceosomes. This process is not well understood at present.

Introns are not totally without function, since some introns influence gene expression. Further, alternative splicing using different sequences as splice sites can produce different proteins. For example, in the case of rubisco activase from spinach and *Arabidopsis thaliana*, pre-mRNA utilises either of two splice sites separated by 22 nucleotides at the 3' splice site; this results in formation of different proteins.

> ***Assignment 12.8***
>
> Explain how alternative splicing within an intron can lead to the formation of different proteins.

12.5.3 *Translation and its control*

The ribosome scanning model applies to plants as well as animals. The 40S subunit of the ribosome recognises the cap structure and other protein factors and scans the 5' untranslated region of mRNA until it encounters an AUG codon. Not all AUG codons are potential translation initiation sites. The nucleotide sequences surrounding the AUG play an important role. In plants, the consensus nucleotide sequence which allows initiation is AACA<u>AUG</u>GCT. This sequence is different from the sequences found in animals and yeast.

In contrast to bacteria, eukaryotic mRNAs exist as nucleoprotein complexes. In the nucleus these are called hnRNPs. Once transported into the cytoplasm they form complexes with separate cytoplasmic proteins and form cytoplasmic RNPs (cmRNPs) and polysomal messenger ribonucleoprotein particles (pmRNPs). The pmRNPs are thought to be stablised forms of mRNA, although the mechanism of conversion of inactive cmRNP to pmRNP has not been demonstrated.

Nuclear encoded mRNAs are translated on cytoplasmic ribosomes. The translated proteins may remain in the cytoplasm or may be transported into other cell compartments such as the nucleus, chloroplasts, mitochondria, endoplasmic reticulum, Golgi apparatus, lysosomes, storage bodies or cell wall. The information needed to target a protein to its respective compartment is found in the primary structure of a protein and consequently in the coding sequence of its gene. Precursor proteins with an additional N-terminal peptide called the 'signal peptide' first enter the endoplasmic reticulum. While the precursor protein passes through the endoplasmic reticulum, the signal peptide is cleaved. The proteins destined to be transported into chloroplasts, mitochondria and the nucleus are synthesised on cytoplasmic ribosomes as larger precursor proteins with an additional N-terminal peptide called the 'transit peptide'. The transit peptide, specific to each organelle, is cleaved from the mature protein after the precursor protein is transported into the organelles. The primary sequence of the precursor protein and the corresponding coding sequence of its gene have the information to target a protein into a specific organelle.

The signal peptide and transit peptide sequences are very important in plant genetic engineering since an engineered gene should be designed in such a way that its final product, namely the protein, is targeted to the organelle where it is expected to function.

Glycosylation (addition of sugar residues) of proteins is carried out in the endoplasmic reticulum before their transport into various compartments.

> ***Self assessment exercise 12.1***
>
> Which of the following promotes (enhances) the expression of a gene (the formation of a protein from a gene by transcription and translation)?
>
> (i) A silencer
> (ii) An enhancer
> (iii) A promoter
> (iv) Demethylation of nucletotides
> (v) Polyadenylation of mRNA

(vi) Pre-mRNA splicing
(vii) mRNA capping
(viii) A 'CAAT' box
(ix) An AUG codon
(x) A signal peptide
(xi) A transit peptide.

12.6 Transposable elements (transposons)

Transposable elements (or transposons), unlike other DNA sequences, move from one location to another within the genome. This intragenomic mobility is mainly restricted to the movement to a different location in the same chromosome or to sister chromatids.

Molecular analysis of transposable elements has revealed the complexity of their mechanism of transposition. Besides being present universally in almost all organisms, what is becoming increasingly evident is the biological relevance of transposons. They are elements which actively promote evolutionary divergence. This aspect is clearly illustrated by plant transposable elements in general and maize transposons in particular. Furthermore, in the case of maize elements, it has been shown that these elements can be differentially expressed at specific intervals in certain parts of the plant. Such developments could be utilised in using these elements for plant genetic engineering.

As early as 1973, it was noted that transposon-vacated donor sites do not return to the wild type sequences. Although precise excision should theoretically restore the gene function, in the case of some transposable elements, an 8 base target sequence becomes duplicated and upon excision the element is removed but not all of the duplicated sequences are excised. This results in the addition of some nucleotides (called footprints) at the original site.

Assigment 12.9

What would be the effect of insertion of extra bases (footprints) into DNA during imprecise excision of a transposon?

Some transposable elements increase the spontaneous mutation rate considerably. Transposition events may also lead to a more drastic alteration in the chromosome, such as breakage or inversion. Such events may be harmful or lethal to the plant.

12.6.1 Transposons as tools in plant genetic engineering

An active transposon is a mutagen. Once the sequence of a transposon is known, its location on the chromosome can be identified using Southern blot analysis. Extension of this approach could allow the sequences flanking the transposon to be cloned and these sequences could be used once again to clone the original gene. This is shown in Fig. 12.8. This procedure is known as transposon tagging.

In summary, transposons are genetic elements which are capable of moving location within the genome. Transposition can lead to mutation, which can be recognised by a change in phenotype in the plant. If the DNA sequence of a transposon is known, its location on the genome of the mutant plant can be detected (using labelled transposon sequences as probes). The sequences surrounding the transposon will be the

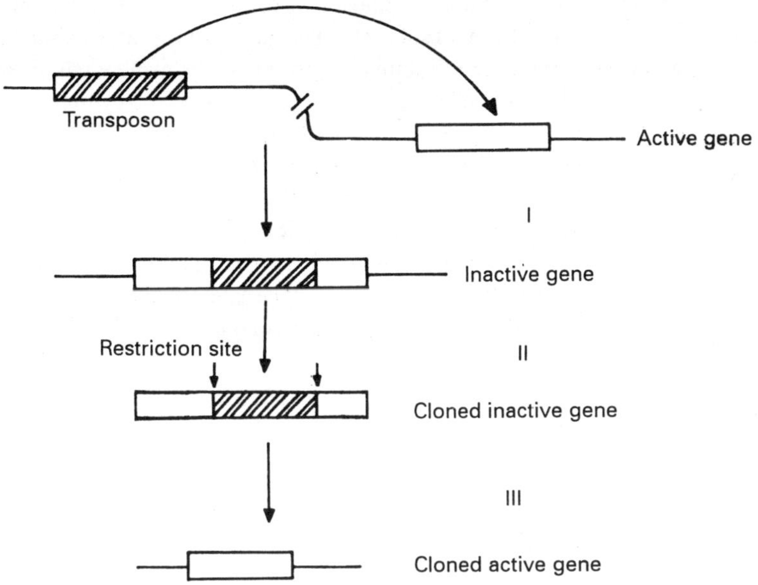

Fig. 12.8 *Use of transposons in isolating plant genes.* A transposon is introduced into a gene and this results in a mutant phenotype which could be detected. In the second step a restriction enzyme is used for cleaving the DNA which is cloned on a plasmid. The recombinant plasmid could be isolated using the transposon as probe. In the third step the transposon is deleted and the wild-type gene could be cloned from a non-mutant host and the sequences flanking the transposon could be used as gene-specific probes.

gene which has become mutated by the transposition event. In this way the gene coding for the trait which has been mutated can be isolated. This gene can be cloned in a process called transposon tagging. The isolation and cloning of a particular gene is a prerequisite for its use in genetic engineering (see Section 12.10). Transposons are thus important tools in the genetic engineering of plants; they permit the isolation and cloning of particular plant genes. Other methods used to characterise and isolate plant genes are RFLP (restriction fragment length polymorphism — Chapter 11) and a method known as chromosome walking.

> ***Assignment 12.10***
>
> Explain why it is important to be able to identify and clone the gene or genes responsible for a given plant phenotype (characteristic).

12.7 The chloroplast genome

Chloroplasts have their own genome (called the plastome), transcription system and translation machinery. The organisation of the chloroplast genome, the regulatory elements that control transcription and the translation apparatus (e.g., ribosomes) typically resemble systems in prokaryotes. Translation on chloroplast ribosomes is also inhibited by inhibitors of bacterial protein synthesis.

Chloroplast genomes of tobacco, rice and the liverwort *Marchantia* have been completely sequenced. Many chloroplast genomes have a set of inverted repeats which carry the genes for the 23S, 16S, 5S and 4.5S rRNA (Fig. 12.9). Based upon the analysis of protein synthesis in isolated chloroplasts and DNA sequence information, a chloroplast genome is expected to encode 120 genes of which about 100 could be protein-coding sequences.

Interestingly, although chloroplasts have their own genomes and protein synthetic apparatus, these do not code for all proteins that are present in the chloroplast: nuclear genes may also be involved in coding for proteins found within chloroplasts. An example is ribulose bisphosphate carboxylase, a key enzyme in CO_2 fixation, which has eight large subunits and eight small subunits. The genes for the large subunits are present in the chloroplast genome and genes for the small subunit are present in the nuclear genome. The small subunits are synthesised on the 80S cytoplasmic ribosomes as precursors with an N-terminal 'transit peptide' and are targeted by the transit peptide to the chloroplasts. Once the small subunit reaches the stroma of the chloroplast, the transit peptide is cleaved and the large and small subunits are associated into a holoenzyme with the help of chaperonin-type assembly proteins which are coded for by the nuclear genome.

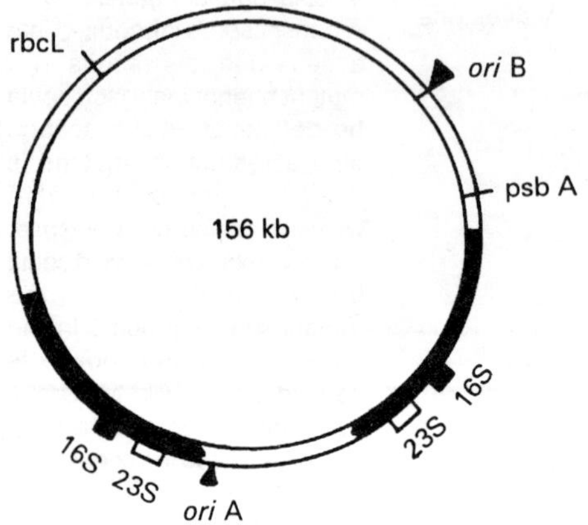

Fig. 12.9 *Chloroplast genome of tobacco (Nicotiana tabacum).* The genome has a pair of inverted repeat sequences comprising of genes for 16S and 23S rRNAs. The location of the genes of the large subunit of ribulose bisphosphate carboxylase (*rbc*L) QB protein (*psb*A), and origins of replication (*ori*A and *ori*B) are marked.

Protein targeting is even more interesting in the case of thylakoid proteins such as chlorophyll *a/b* binding protein and thylakoid lumen proteins such as plastocyanin (Fig. 12.10). Both these chloroplast proteins are coded for by the nuclear genome. Chlorophyll *a/b* binding protein is synthesised in the cytoplasm as a precursor with an N-terminal transit peptide sequence which targets it to the chloroplast

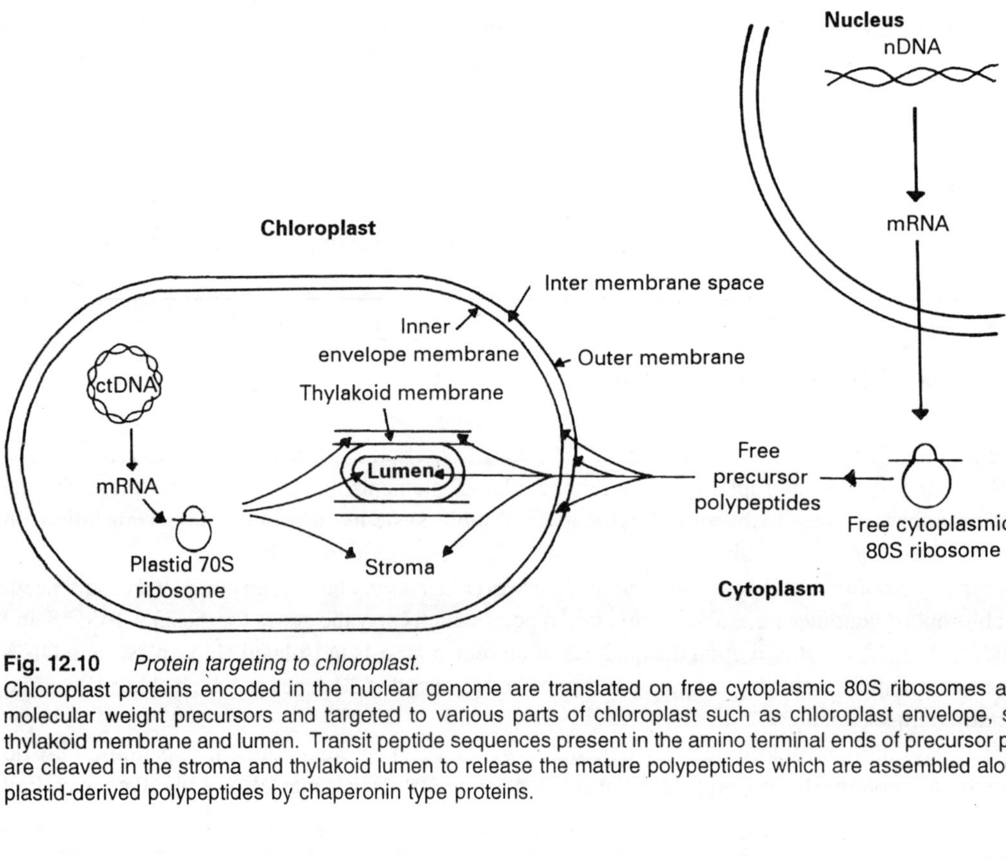

Fig. 12.10 *Protein targeting to chloroplast.*
Chloroplast proteins encoded in the nuclear genome are translated on free cytoplasmic 80S ribosomes as large molecular weight precursors and targeted to various parts of chloroplast such as chloroplast envelope, stroma, thylakoid membrane and lumen. Transit peptide sequences present in the amino terminal ends of precursor proteins are cleaved in the stroma and thylakoid lumen to release the mature polypeptides which are assembled along with plastid-derived polypeptides by chaperonin type proteins.

and the mature protein that is released there has the structural motif that targets the protein to the thylakoid membrane. Plastocyanin is translated on cytoplasmic ribosomes as a precursor with two N-terminal targeting signals. The first N-terminal signal is the transit peptide that targets plastocyanin to the chloroplast stroma where the transit peptide is cleaved. The rest of the precursor protein is targeted to the thylakoid by the second signal sequence where it also becomes cleaved. These examples highlight the complex interaction between the nuclear genome and the chloroplast genome.

12.7.1 Genetic engineering of chloroplasts

Besides housing the entire photosynthetic machinery, chloroplasts also carry out several other metabolic processes such as biosynthesis of amino acids, nucleotides, lipids, starch and several secondary metabolites. As we saw earlier, some of the genes involved are encoded in the chloroplast genome and some in the nuclear genome. Genetic engineering aiming at modifying any of the chloroplast functioning may thus involve modification of chloroplast genes and nuclear genes. The feasibility of manipulation of nuclear genes for chloroplast function has been established in several cases.

While many methods of plant transformation have failed to incorporate foreign genes into the chloroplast genome, particle gun bombardment (see Section 12.9.2) has been found to be successful in introducing genes stably into chloroplasts. The foreign gene was cloned between sequences homologous to the chloroplast genome. The regenerated plants containing the modified plastome are designated as transplastomic plants. This technology is still in a developing stage and has not become as routine as transformation of the nuclear genomes of plants.

Assignment 12.11

Particle gun bombardment involves coating minute projectiles with copies of a gene and then firing the projectiles into plant cells (and chloroplasts). Explain what advantage there would be in sandwiching the gene between sequences homologous to (the same as) those on the host cell genome.

Protein targeting is an important component of plant genetic engineering. If a pathway to be manipulated by genetic engineering is located within the chloroplast, then the foreign gene should be constructed such that it will encode the transit sequence that will be necessary to target the protein to the chloroplast (for nuclear genes) and to the correct compartment within the chloroplast.

12.8 The mitochondrial genome

12.8.1 Genome organisation

The sizes of plant mitochondrial genomes and their complex organisation are not well understood. The sizes of mitochondrial genomes in other organisms are as follows: humans 16 kb; *Tetrahymena* (protist) 55 kb; *Agaricus* (fungus) 176 kb. In higher plants the size of the mitochondrial genome ranges from 208 kb in *Brassica* to 2400 kb in muskmelon. Strikingly, the coding potential of mitochondrial genomes of all the above mentioned organisms is approximately the same. This means that most of the plant mitochondrial DNA is non-coding and redundant. It is not clear as to why plant mitochondrial genomes are large and why their sizes are highly variable.

Plant mitochondrial genomes harbour direct and inverted repeats. Direct repeats lead to the generation of subgenomic circles of mitochondrial DNA. For instance, the 218 kb *Brassica campestris* mitochondrial genome has one pair of direct repeats that allow the genome to recombine and generate two subgenomic circles of 135 and 83 kb. Consequently, *B. campestris* mitochondria will have three mitochondrial DNA circles, a master circle of 218 kb and a pair of subgenomic circles of 135 and 83 kb. The organisation is even more complex in the case of maize mitochondrion which has 570 kb mitochondrial DNA with four pairs of direct repeats and one pair of inverted repeats. A large number of possible recombinations between the pairs of direct repeats lead to the generation of a multitude of subgenomic circles. The presence of a pair of inverted repeats leads to changes in the orientation of the sequences between the repeats. The presence of repeats in the mitochondrial genome and recombination between these repeats makes the mitochondrial genome organisation exceedingly complex.

12.8.2 Mitochondrial plasmids

In addition to the mitochondrial genomic DNA, a large number of circular and linear plasmids have been detected in mitochondria. These plasmids do not share homology with genomic sequences. The role of the circular plasmids is not clear. Double-stranded linear plasmids called S1 and S2 have been reported from maize mitochondria. These plasmids are associated with cytoplasmic male sterility type S (Section 12.7.4). The plants are male sterile when the S1 and S2 plasmids remain free and restoration of fertility in these plants is often associated with the integration of S1 and S linear plasmids into the mitochondrial genome.

12.8.3 Mitochondrial genes, transcription and translation

In spite of the relatively large size of the mitochondrial genome, its coding potential is relatively small. Conserved regions among the different plant mitochondrial genomes code for mitochondrial 26S, 18S and 5S ribosomal RNA, a few tRNAs and about 18 polypeptides, most of which are involved in electron transport reactions. Mitochondria rely on the nuclear genome for most of the tRNAs and mitochondrial proteins.

A comparison of nucleotide sequences of protein-coding genes of plant mitochondria with the corresponding genes in other systems led previously to a suggestion that codon usage in plant mitochondria may slightly deviate from the universal genetic code. However, recent findings indicate that plant mitochondria and chloroplasts have an 'RNA editing' mechanism by which many of the 'C' residues are converted to 'U' residues on the mRNA after transcription. RNA editing corrects the possible anomalies between the genetic code in mitochondria and the universal genetic code. For example, in a place expected to have the tryptophan codon UGG, CGG was present in the mitochondrial gene, which as per the universal codon is for arginine. The editing of CGG to UGG by RNA editing corrects the anomaly and allows tryptophan to be added by the translation machinery although the codon in the gene is for arginine. The significance of RNA editing in plant organelles is not yet understood.

12.8.4 Cytoplasmic male sterility and mitochondria

Male sterility in plants is inherited in both a biparental (nuclear) and a maternal (cytoplasmic) manner. Cytoplasmic male sterility (cms) is invariably due to defect(s) in mitochondrial functions and can be traced to the mitochondrial genome. Plants with cms, which are otherwise normal, do not make fertile pollen grains; generally, cms is associated with defective functioning of the tapetum of the anthers that supply nutrients to the developing pollen grains. The female components of these plants maintain fertility. Cytoplasmic male sterility has so far been reported in over 40 plant species.

Cytoplasmic male sterility has been studied in detail in maize in which there are three types of cytoplasmic male sterility: cms-T (Texas cytoplasm), cms-S (USDA) and cms-C (Charrua). These cms types were originally distinguished on the basis of specific nuclear genes, designated as fertility-restoring genes (Rf) which restore pollen fertility upon crossing. For instance if the stigmas of cms-T are pollinated with the pollen from a restorer maize with Rf1 and Rf2 and normal mitochondrial genes then the seeds obtained through this cross give rise to plants which are male fertile. Therefore a defect in the mitochondrial function could be complemented (overcome) by a product encoded in the nuclear genome. Remember that a fertilised ovum inherits mitochondria only from the maternal cell; on fertilisation by pollen, nuclear genes from the pollen enter the ovum but not mitochondria from the pollen cell.

Cytoplasmic male sterility in cms-T is attributed to the presence of an additional DNA sequence in mitochondria designated as T-*urf*13, that codes for a 13,000 MW polypeptide. The same polypeptide is also responsible for the susceptibility of the cms-T plants to the toxins of the fungi *Biopolaris maydis* race T (renamed *Helminthosporium maydis*) and *Phyllostica maydis*. Restoration of fertility by crossing with a nuclear restorer with Rf1 or Rf2 leads to cessation of the synthesis of the 13,000 MW polypeptide, thus establishing a strong link between cms-T and T-*urf*13. Cytoplasmic male sterility has become a major focus of agricultural biotechnology because of its usefulness in developing hybrid seeds.

Hybrid vigour or 'heterosis' is a phenomenon in which the F1 generation out of a cross (sexual breeding of two parental types) produces the best yield; the yield reduces rapidly in subsequent generations. To ensure a perfect cross, the anthers of the 'female' (hybrid-seed producing) plant should be removed before they mature.

Assignment 12.12

In the production of F1 hybrid seeds, the anthers of the hybrid-seed producing parental plants must be removed before they (the anthers) mature. Explain why.

Emasculation (anther removal) is a labour intensive process and is impractical in several plants. The cms lines are 'genetically emasculated' in the sense that they are not capable of producing viable pollen, thus obviating the need for emasculation and thereby helping to make the production of hybrid seeds commercially viable. A general scheme of hybrid seed production is shown in Fig. 12.11.

In a typical hybrid seed production programme involving cms, three lines of plants are maintained. The A-line (the hybrid seed-producing parent) has cms cytoplasm (defective mitochondrial genes) and recessive nuclear restorer genes. In order to maintain this line, the seed parent of the A-line should be crossed with pollen parents of B-line which are called 'maintainers'. The maintainers have normal (N) cytoplasm effective

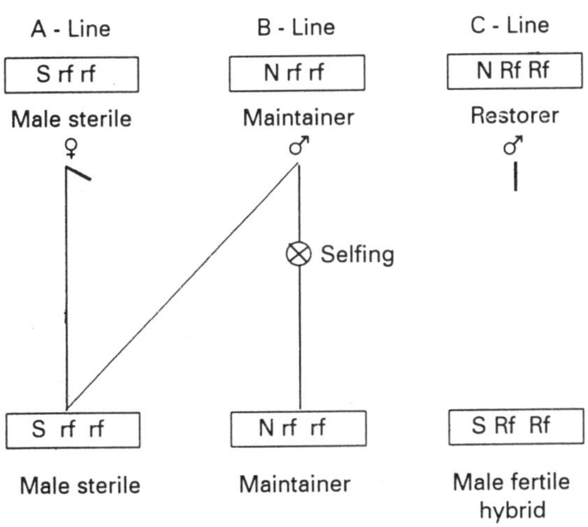

Fig. 12.11 *Strategy of hybrid seed production.* Male sterile, maintainer and restorer lines are called as A-B-and C-lines, respectively. Male sterile line is maintained by dusting the pollen of maintainer on the stigmas of male sterile plants. The progeny remain male sterile. Whenever hybrid seeds have to be produced, the pollen from the restorer plant is dusted on the stigma of male sterile plant. Male fertility is restored in the F1 progeny that exhibits 'heterosis' or hybrid vigour.

mitochondrial genes but have only recessive restorer (rf) genes. The hybrids between A-line and B-line will continue to be male sterile since they have the cms cytoplasm and no restorer. This process is important because it ensures that male sterile line is maintained. The third is the C-line which is called 'restorer' with the normal (N) cytoplasm and dominant nuclear restorer (Rf) genes. Using cms (A-line) as the hybrid seed- producing plant and restorer (C-line) as the pollen parent plant, a hybrid seed is obtained in which the male fertility is restored due to Rf genes. These hybrid seeds sold to the farmer give rise to male fertile plants that exhibit hybrid vigour.

Self assessment exercise 12.2

Answer true or false to each of the following:

(i) Cms is a phenomenon in which defects in mitochondrial genes cause defects in pollen grain formation leading to ineffective production of pollen grains (i.e., male sterility).

(ii) Cms can be overcome by pollinating a fertility restoring (Rf) plant with pollen from a cms plant.

(iii) To maintain a seed-producing line of plants with cms (A-line), A-line plants need to be pollinated with pollen from plants of a maintainer line (B-line) which has normal (non-cms) mitochondrial genes and recessive restorer genes (rf).

(iv) To produce hybrid seeds from a cms A-line, plants of the A-line must be pollinated with pollen from plants of a restorer line (C-line) which has cms mitochondrial genes and dominant (Rf) restorer genes.

Genetic engineering of male infertility is discussed in Section 12.10.2.

12.9 Methods of genetic engineering in plant cells

12.9.1 Agrobacterium Ti plasmid vectors and their use in plant genetic engineering

Agrobacterium tumefaciens is a soil bacterium that incites tumour development in wounded plants and causes crown gall disease by a naturally evolved 'genetic engineering' process. The causative agent of tumorigenesis is an exceptionally large (200 kbp) plasmid designated as tumour-inducing (Ti) plasmid (Fig. 12.12). A part of the plasmid called transferred DNA (T-DNA) is transferred from *Agrobacterium* into plant cells and becomes integrated into the plant genome. The T-DNA genes, when expressed in plant cells, code for (amongst other things) the production of: an auxin and a cytokinin hormone (which cause the development of a tumour by promoting uncontrolled growth of plant cells); an opine (such as octopine and nopaline) which acts as a growth-promoting nutrient for the infecting strain of *Agrobacterium*.

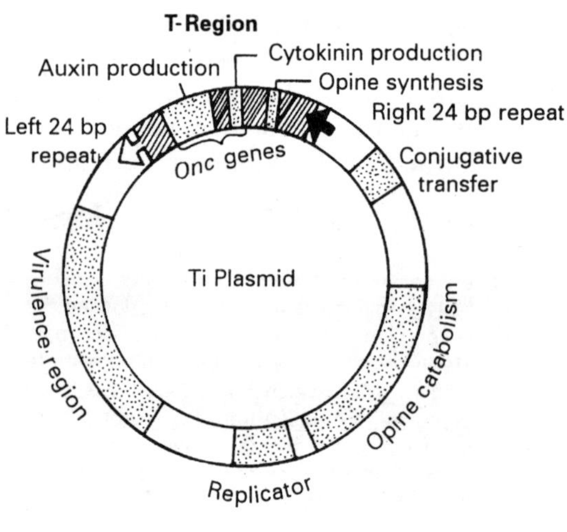

Fig. 12.12 *Genetic organisation of Ti plasmid.* Ti plasmid has three important regions: (1) T-region with genes for hormone and opine biosynthesis and flanked by 24 bp direct repeat sequences is transferred to the plant cells; (2) Virulence region codes for proteins involved in T-DNA transfer; (3) Opine catabolism region codes for proteins involved in the uptake and metabolism of opines.

The ecological advantage for *Agrobacterium* in inciting tumours on plant tissues is the availability of opines in the tumours (opines are made by plant cells but can be utilised only by *Agrobacterium*) and the expression of hormone biosynthetic genes in the tumour cells which causes the tumour cells to proliferate at a rate faster than the normal cells (thereby increasing the population of cells that could make and secrete opines for *Agrobacterium*). Since *Agrobacterium* exploits plant cells by inducing them to produce compounds that are of specific use only to itself and does this by way of genetic engineering, this process is referred to as 'genetic colonisation'. This is a unique example of natural genetic engineering, and is depicted in Fig. 12.13.

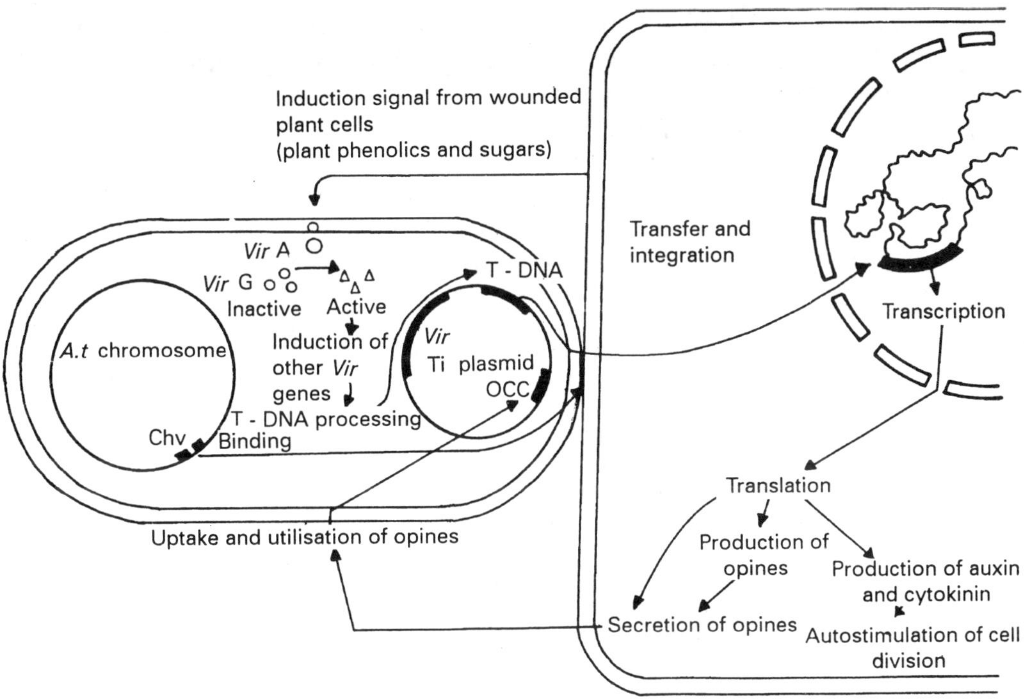

Fig. 12.13 *Biology of Agrobacterium tumorigenesis.*
Agrobacterium remains in the intercellular space of crown gall cells. Opines, synthesised and secreted by the transformed plant cells, are taken up and utilised by *Agrobacterium.* Opine-biosynthetic genes in the T-DNA region are transferred to the plant cells. Genes coding for opine uptake and catabolism are present in the *occ* region of Ti plasmid.

(a) Mechanism of T-DNA transfer

Whole cells of *Agrobacterium* are necessary for Ti plasmid transfer to plant cells. The mechanisms involve chemotaxis of the bacterial cell to the plant cell and firm attachment of the bacterial cell to the plant cell; both these events are influenced by genes on the bacterial cell chromosome. The T-DNA portion of the Ti plasmid is selectively processed into T-DNA transfer intermediates and transferred to a plant cell where it becomes integrated into the plant genome.

The major components of the Ti plasmid involved in the T-DNA transfer process are; (i) T-DNA borders, and (ii) Ti plasmid virulence region.

(i) T-DNA borders

A set of 24 bp, direct repeat sequences flank the T-DNA region. Since the DNA sequences between the 24 bp repeats are invariably transferred to the plant cells, these sequences are referred to as T-

DNA 'borders' The left repeat is termed the left border and the right repeat is termed the right border. All the T-DNA borders of various Ti plasmids exhibit a high degree of homology. Genetic analysis reveals that the right border is more critical for T-DNA transfer and tumorigenesis (Fig. 12.14). The orientation of the borders is also critical; if the native orientation of the border is changed, the tumour-inducing genes are not transferred to the plant cells. Subsequent studies have revealed that the T-DNA borders are the only structural requirements for transfer of a T-DNA region; all T-DNA sequences between the borders can be replaced with any DNA sequence without affecting the T-DNA transfer process. In fact, any DNA sequence placed between two T-DNA borders is treated as T-DNA by *Agrobacterium.*

Nucleotide sequences flanking the right borders of octopine-coding Ti plasmids were found to be very critical for efficient tumorigenesis (Fig. 12.14). When a synthetic right border repeat was placed at the right end of the T-DNA, tumorigenesis was drastically reduced. Almost complete tumorigenesis was restored by adding the right border-flanking sequences. A 24 bp consensus was characterised in this region and termed as 'overdrive'. Overdrive was found to be responsible for the quantitative increase in the T-DNA transmission that was brought about by the right border-flanking sequences. Therefore, the overdrive sequence is termed the quantitative T-DNA transmission enhancer. Overdrive is present to the right of octopine-coding T-DNA right borders but not to the right of nopaline-coding T-DNA right borders.

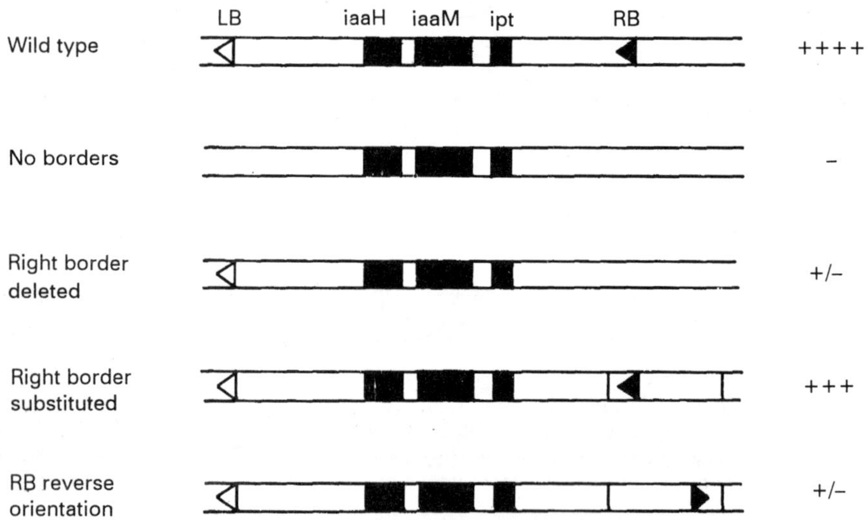

Fig. 12.14 *Role of T-DNA borders in tumorigenesis.*
LB and RB represent left and right T-DNA borders. Tumour genes encoding tryptophan monooxygenase (*iaa*M), indoleacetamide hydrolase (*iaa*H) and isopentenyl-transferase (ipt) are indicated. Approximate quantitation of tumorigenesis on carrot root discs is indicated on the right: (++++) wild-type tumorigenesis, (–) no tumorigenesis, (+/–) very low or rare tumorigenesis.

(ii) Ti plasmid virulence (*vir*) genes.

A 40 kbp region of the Ti plasmid encompasses about 24 *vir* genes that code for a large number of proteins that function in *trans* and mobilise T-DNA into plant cells. All the *vir* genes are organised into 8 operons termed *vir*A to *vir*H, which are coordinately expressed and thus constitute a *vir* regulon. Genetic studies by transposon mutagenesis revealed that *vir* A, *vir* B, *vir* G, *vir* D and *vir* E are essential for T-DNA transfer to all plants while *vir*C, *vir*F and *vir*H are dispensable in a number of plants.

The expression of *vir* genes is highly regulated and is the 'master switch' in the transformation process. Acetosyringone and several other phenolic compounds synthesised and secreted by the wounded plant cells serve as plant signal molecules and induce the expression of all *vir* genes in the *vir* regulon. A variety of monosaccharides such as D-glucuronic acid and xylose and opines such as octopine and nopaline also increase the *vir* gene expression in the presence of acetosyringone.

(b) *Agrobacterium* Ti plasmid-based vectors

The ability of *Agrobacterium* to introduce genes into plants has been successfully exploited to develop Ti plasmid-based vectors for introducing desirable foreign genes into plants.

Self assessment exercise 12.3

You are constructing a vector, based on a Ti plasmid, to carry a given cloned gene into plant cells. Which of the following regions of the Ti plasmid do you think it would be desirable to include in the vector, and which regions of the Ti plasmid do you think it would be desirable to exclude from the vector?

(i) 'Overdrive' region
(ii) Border regions
(iii) DNA between border regions
(iv) Auxin hormone-coding gene
(v) Cytokinin hormone-coding gene
(vi) *Vir* genes
(vii) Opine-coding gene

A major limitation of Ti plasmids is their large size (about 200 kbp), making purification of intact plasmids very difficult. In addition, Ti plasmids have multiple sites for almost all common nuclease enzymes thus rendering them unsuitable for cloning. Efforts were thus made to disarm the Ti plasmid by deleting the hormone biosynthetic genes and to develop smaller Ti plasmid derivatives with conveniently-placed restriction sites for inserting genes of interest. Two types of Ti plasmid vectors, namely, (i) cointegrate vectors and (ii) binary vectors, are currently used in plant transformations.

(i) Cointegrate vectors

A derivative of the nopaline-coding Ti plasmid *pTiC*58 was constructed by retaining the left and right border regions but deleting the entire intervening DNA except for the nopaline synthase gene. In this process the T-DNA genes for hormone biosynthesis were removed (the Ti plasmid was 'disarmed'). In place of the deleted DNA a *pBR*322 (bacterial plasmid) sequence was placed. The plasmid thus formed, called 'receptor plasmid', has all the structures of the T-DNA necessary to transfer to plant cells (a right border and a left border) but it lacks a selection marker needed in plant formation (Fig. 12.15).

An intermediate vector was constructed having:

(i) a *pBR*322 sequence homologous to that found in the receptor Ti plasmid (which provides sequence homology so that the intermediate vector can integrate [by recombination] into the receptor plasmid);

(ii) a plant transformation marker, such as a gene coding for neomycin phosphotransferase II (*npt* II) (which confers kanamycin resistance on recipient plant cells, and thus permits their selective isolation [on kanamycin-containing medium]);

(iii) a bacterial transformation marker, such as a gene coding for spectinomycin resistance (which confers spectinomycin resistance on recipient bacterial cells, and thus permits their selective isolation [on spectinomycin-containing medium]);
(iv) a multiple cloning site (which allows insertion of foreign genes);
(v) a *ColE*1 origin of replication (which allows replication of the plasmid in *E. coli* but not in *Agrobacterium*);
(v) an *ori* T sequence with a *bom* (basis of mobilisation) site (for mobilisation [tranfer] of the intermediate vector from *E. coli* into *Agrobacterium*).

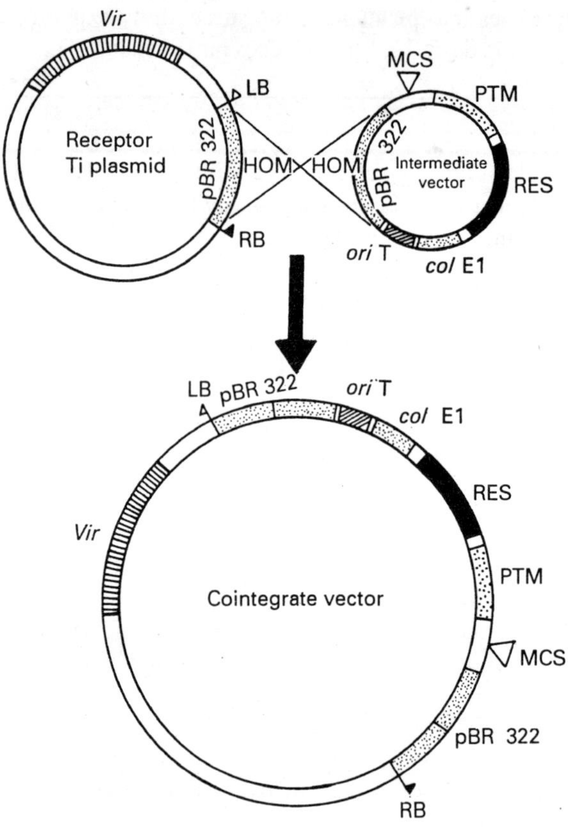

Fig. 12.15 *Cointegrate vector system.*
Receptor Ti plasmid is a disarmed derivative of Ti plasmid retaining the right and left T-DNA borders. *VIR* – Ti plasmid virulence region; HOM – homologous regions within which recombination leading to cointegration occurs; LB and RB – left and right T-DNA borders; MCS – multiple cloning site; PTM – plant transformation marker; RES – bacterial resistance marker; *ori*T – origin of transfer and *bom* site for conjugative plasmid mobilisation; *colE*1 origin a of replication from *colE*1 plasmid.

A gene of interest is first cloned in the multiple cloning site of the intermediate vector. The cloning is performed in *E. coli* in which the process is most efficient. The intermediate vector with the foreign gene is mobilised into *Agrobacterium* with the receptor Ti plasmid by a process called triparental mating. Selection for transformed *Agrobacterium* is made on a minimal medium containing spectinomycin to select *Agrobacterium* with the receptor Ti plasmid and intermediate vector; *E. coli* does not grow on a minimal medium in which *Agrobacterium* grows.

The intermediate vector carrying spectinomycin resistance cannot thus replicate in *E. coli*. The only way the spectinomycin marker could be maintained in *Agrobacterium* is by the integration of the intermediate plasmid into the receptor Ti plasmid. Such a cointegration event will place the entire intermediate plasmid between the right and left T-DNA border sequences of the receptor Ti plasmid. In this resulting cointegrate plasmid, the plant transformation marker (e.g., *npt*II gene) and the cloned foreign gene are inserted between the T-DNA borders and will be transferred to the plant cells as a part of the T-DNA. The transformed plant cells can be selected on a medium containing kanamycin after cells of the plant and *Agrobacterium* have been incubated together, or after plant cells have been transformed by exposure to isolated plasmids by another means (see Section 12.9.2).

The advantage of cointegrate vectors is that the foreign genes are easily cloned in the relatively small intermediate plasmid which has a number of single-restriction sites. After performing all the cloning in *E. coli* the intermediate plasmid can be mobilised into *Agrobacterium* and placed between the T-DNA borders, for transfer to plant cells.

(ii) Binary vector

Studies on the T-DNA transfer mechanism revealed that the T-DNA region between the T-DNA borders could be transferred to plants even if the T-DNA is not physically linked to the Ti plasmid virulence region. In other words, *vir* genes could mobilise T-DNA in *trans*. This finding led to the development of more convenient Ti plasmid vectors called 'binary vectors'. A binary vector system consists of an *Agrobacterium* strain with a disarmed Ti plasmid called *vir* helper plasmid (in which the entire T-DNA region inclusive of borders is deleted but the *vir* gene is present) and a binary vector with T-DNA that replicate in *E. coli* and *Agrobacterium* (Fig. 12.16).

A typical binary vector has the following sequences: left and right border repeats that delimit the T-DNA region of the binary vector; a plant transformation marker such as *npt*II gene that confers kanamycin resistance in transformed plant cells; a multiple cloning site for introducing desirable foreign genes; an *E. coli* origin of replication such as *ColE*1 for replication in *E. coli*; bacterial resistance marker such as tetracycline resistance gene for selecting the binary-vector clones in *E. coli* and *Agrobacterium*; a broad host-range origin of replication such as *RK*2 that allows the replication of the binary vector in *Agrobacterium*; and an *ori*T sequence for conjugal mobilisation of the binary vector from *E. coli* to *Agrobacterium*.

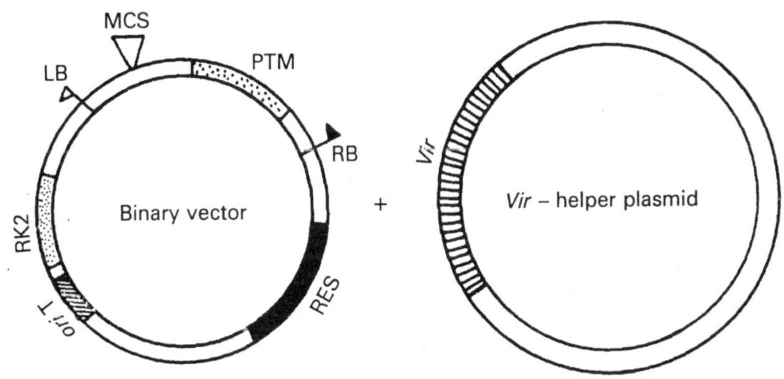

Fig. 12.16 *Binary vector system.*
Vir helper plasmid is a non-oncogenic derivative of Ti plasmid in which the entire T-DNA, inclusive of T-DNA borders, is deleted. Binary vector replicates in *E. coli* and *Agrobacterium*. LB and RB, left and right T-DNA borders; MCS – multiple-cloning site; PTM – plant transformation marker; RES – bacterial resistance marker; *ori*T – origin of transfer and *bom* site for conjugative plasmid mobilisation. *RK*2 – wide host range origin of replication from plasmid *RK*2.

A foreign gene of interest is inserted into the multiple cloning site of the binary vector, thus placing it between the right and left border repeats, and cloned in *E. coli*. The binary vector is then mobilised from *E. coli* to *Agrobacterium* (*vir* helper strain) by triparental mating. The *vir* gene-encoded proteins act on the T-DNA borders in *trans* and mobilise the T-DNA of the binary vector into the plant cells.

Whereas the cointegrate system involves the transfer of the intermediate plasmid into *Agrobacterium* and its subsequent integration, the binary system involves only transfer of the binary plasmid to *Agrobacterium* and does not involve integration. Therefore, binary vectors are more convenient than the cointegrate vectors and are most commonly used in plant transformation.

Self assessment exercise 12.4

Choose the correct completion to the following statement:
The difference between a cointegrate vector and a binary vector is that

(i) a cointegrate vector carries a gene for a bacterial transformation marker which allows cells of *Agrobacterium* into which the binary vector has been introduced to be selected for.

(ii) a cointegrate vector has to integrate with another vector after it has been introduced into *Agrobacterium*.

(iii) a cointegrate vector carries a gene for a plant transformation marker which allows cells of a plant into which the binary vector has been introduced to be selected for.

(iv) a cointegrate vector needs expression of *vir* genes in order for T-DNA to be transferred into a plant cell.

12.9.2 Direct transformation of plants by electroporation and particle gun bombardment

Most dicot plants can be efficiently transformed by *Agrobacterium*. However, many important dicot crops such as pulses and the majority of monocot crops, particularly cereals, are highly recalcitrant to *Agrobacterium*-mediated transformation, and so direct transformation methods have been developed to transform such plants. Examples of such direct transformation methods are: calcium chloride and PEG-mediated transformation; liposome-mediated transformation; electroporation; and transformation using microprojectiles. While the first three methods are applicable only to protoplasts, the last method, involving microprojectile bombardment, can be used on intact tissues. Electroporation and microprojectile bombardment methods have been found to be most effective in direct transformation and will be discussed here.

(a) Electroporation

Protoplasts of a large number of monocots and dicots have been successfully transformed by electroporation. Electroporation is a process whereby electrical pulses of high field strength are used to reversibly permeabilise cell membranes to facilitate uptake of large molecules, including DNA. Electroporation has been successfully used for transforming plants in which efficient regeneration of plants from protoplasts is possible. Direct transformation vectors are different from the Ti plasmid vectors in that they do not have border repeat sequences. Vectors for direct transformation generally have a plant transformation marker and a reporter gene such as luciferase, chloramphenicol acetyl transferase (CAT) or β-glucuronidase. The advantages of electroporation are its convenience, low toxicity and equal efficiency in dicots and monocots. The disadvantage is that it is effective only in protoplasts, from which intact plantlets need to be regenerated; this is technically extremely difficult in many crop plant species.

(b) Particle-gun bombardment

The direct transformation methods such as electroporation, PEG-mediated transformation and liposome fusion demand isolation of protoplasts and regeneration of plants from protoplasts. This is technically difficult and a major limitation in several crop plants. A second major problem is the risk of introducing somaclonal variations when the plants are regenerated from protoplasts through microcalli. In this regard it is advantageous if intact plant tissues can be transformed and plants regenerated from transformed tissues. Bombardment of DNA-coated microprojectiles using particle guns allows the transformation of intact cells of embryos, shoot apices and other tissues.

The technology (Fig. 12.17) basically involves coating tiny tungsten or gold spheres (microprojectiles) with DNA (gene(s) to be introduced) and spreading the particles on the surface of a mobile plate or plastic or nylon bullet called the 'macroprojectile'. Then under partial vacuum, the macroprojectile is fired against a retaining plate or mesh either by an explosive discharge (ballistic device) or by using shock waves initiated by a high voltage electric discharge (electrostatic device). The macroprojectiles are retained by the retaining plate or mesh, while the microprojectiles, normally 1—3μm diameter, pass through due to their small size and higher density and penetrate the target plant tissue. A large number of reports have been published where microprojectile bombardment of calli, embryos, pollen, epidermis and shoot apices led to transient expression of the introduced reporter genes such as β-glucuronidase. Significantly, a number of plants have been stably transformed by microprojectile bombardment and the transformed characteristic demonstrated in the subsequent generation.

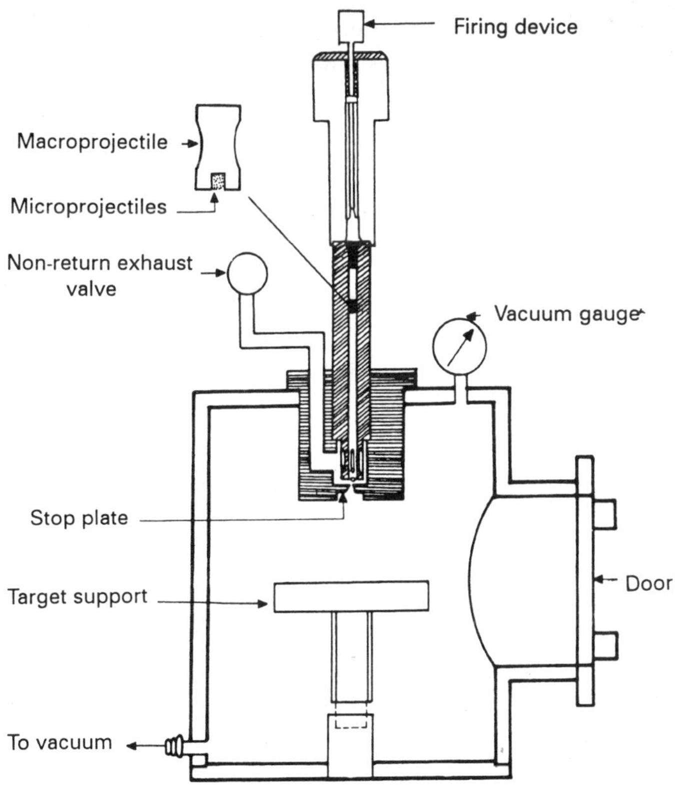

Fig. 12.17 *Schematic diagram of a particle gun.* Microprojectiles of tungsten or gold are coated with DNA and placed on a nylon macroprojectile. The macroprojectiles are propelled by a firing device. While the macro projectile is stopped by the stop plate, the accelerated microprojectiles pass through the stop plate and enter the target tissue. The complete system is maintained under vaccum to prevent any deceleration of microprojectiles.

The major advantage of microprojectile bombardment is its ability to introduce genes into intact tissues thereby obviating the need for protoplast isolation and plantlet regeneration from protoplasts. The limitations of this method are the high cost of the equipment, the high level of variations in transformation efficiencies and the low frequency of achieving stable transformation.

As previously discussed, plants have a tripartite genome organisation: the chloroplast genome codes for a large proportion of its proteins and the mitochondrial genome codes for a limited number of its proteins. However, organelle transformation turned out to be very difficult; none of the tissues transformed by *Agrobacterium* and direct methods such as electroporation and PEG-mediated transformation exhibited any organelle transformation either because of the impermeability of the organelle double membranes to DNA or from inability to detect expression of a single introduced gene copy in a cell having multiple copies of the organelles. Therefore, manipulation of certain chloroplast proteins was achieved by introducing their genes into the nuclear genome along with the transit peptide-coding sequence so that the proteins translated in the cytoplasm could be targeted into the chloroplast. This approach has been successful in a few cases, such as the glyphosate resistance gene. However, certain chloroplast proteins are not properly targeted into the chloroplast even if they are expressed with the transit peptide. These limitations have now been overcome using microprojectile-mediated gene transfer. Successful chloroplast transformation was first achieved in the green alga *Chlamydomonas,* which has the advantage of having a single large cup-shaped chloroplast that completely surrounds the nucleus. Subsequently, an excellent non-lethal selection system based on spectinomycin was developed to achieve tobacco chloroplast transformation; plant cells with a spectinomycin-resistance gene in the plastome appear green whereas those cells without the spectinomycin resistance gene appear white. Since this is a non-lethal selection, sufficient time is given for the transplastomic cells to amplify the transformed plastids. Chloroplast transformation by microprojectile bombardment occurs at a very low frequency when compared with the transformation of nuclear genomes by this same method. Transformation of mitochondrial genomes is also being attempted using this technique.

One interesting outcome from all attempts of chloroplast transformation is the finding that all observed integrations are by homologous recombination. Therefore, it is possible to direct a particular gene into a particular location on the chloroplast genome.

Self assessment exercise 12.5

Explain how a particular gene could be directed into a particular location on the chloroplast genome using homologous recombination.

Self assessment exercise 12.6

In which of the following could micropropagation be usefully applied?

(i) Generation of plants from a 'rescued' embryo.
(ii) Generation of plants free from particular virus diseases.
(iii) Generation of plants from fused protoplasts.
(iv) Generation of haploid plants from haploid cells.
(v) Generation of homozygous diploid plants after diploidisation of haploid cells.
(vi) Generation of plants from stored germ plasm.
(vii) Generation of plants from cells transformed by Ti plasmids.
(viii) Generation of plants from tissue transformed by particle gun bombardment.

12.10 The manipulation of phenotypic traits in plants using genetic engineering

12.10.1 Herbicide resistance

The feasibility of using genetic engineering for crop improvement was first demonstrated by developing plants with resistance to commercial herbicides. A chemical is considered as a good herbicide if it has the following features: the chemical should selectively kill weeds without affecting the crop plant; it should not have toxicity towards animals and microorganisms; it should be rapidly translocated within the target plant such that the rapidly growing parts accumulate high concentrations of the herbicide with lethal effect; the chemical should be rapidly degraded in the soil so that it does not cause any environmental damage. None of the commercially used herbicides fulfill all the above criteria. The main limitation is the specificity of the herbicide; there is no known herbicide that is toxic to weeds that is not also toxic to crop plants to some degree. To circumvent this problem, a herbicide can be chosen which would fulfill the other conditions and then genetic modification of the targeted crop plant can be carried out to produce resistance against the herbicide. The approaches which have been taken to introduce herbicide resistance have been to: identify genes coding for modified proteins that are no longer sensitive to (not inhibited by) the herbicide and introduce such genes into crop plants to confer herbicide resistance; characterise a gene that codes for an enzyme that modifies the herbicide and renders it inactive and then introduce the gene into crop plants of interest. Both approaches have been successfully used for genetically engineering plants with resistance to broad-spectrum herbicides such as glyphosate, sulfonylurea and bialophos. For example, glyphosate, known commercially as Round Up, is a broad-spectrum herbicide with low toxicity to animals. It has a very short half life in soil because it is rapidly degraded by microorganisms. Glyphosate is rapidly translocated to the growing points of plants and kills the plants rapidly at low applied concentrations. Glyphosate is a competitive inhibitor of EPSP synthase (5-enolpyruvylshikimic acid-3-phosphate synthase), which catalyses the synthesis of 5-enolpyruvylshikimic acid-3-phosphate from phosphoenolpyruvate and shikimate 3-phosphate (Fig. 12.18). This is a key step in the shikimic acid pathway that leads to the synthesis of aromatic amino acids, phenols and other secondary metabolites. Glyphosate is toxic to plants and microorganisms which possess the shikimate pathway.

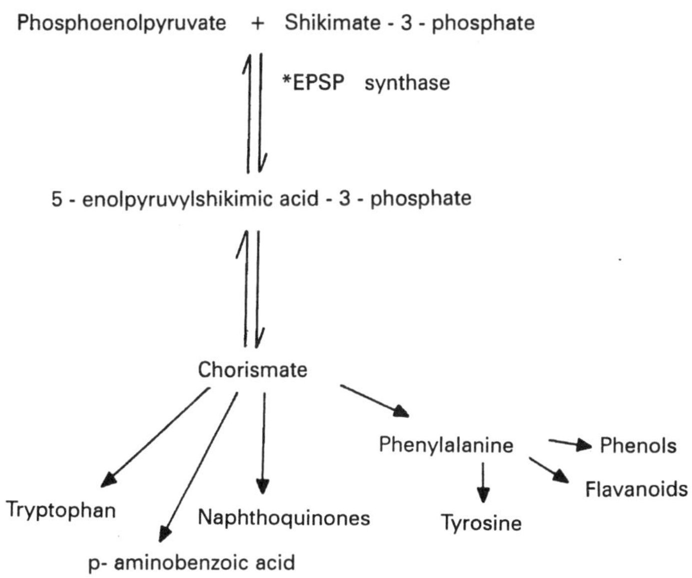

Fig. 12.18 *Key role of EPSP synthase in the biosynthesis of aromatic amino acids and other aromatic compounds.*

Assignment 12.13
Animals (including humans) do not possess the shikimate pathway; instead of synthesising aromatic amino acids via this pathway animals obtain them from their diet. In view of this, predict whether glyphosate would be toxic to animals.

Assignment 12.14
Predict what changes would need to be induced in EPSP synthase in order for resistance to glyphosate to be conferred.

EPSP synthase in plants is coded by a nuclear gene. EPSP synthase with its transit peptide is translated on the 80S ribosomes and then targeted to the chloroplast where it is functional. An EPSP synthase gene which conferred resistance to glyphosate was first found in a mutant of the bacterium *Salmonella* that had a mutation in its *aro*A locus that encodes EPSP synthase; a single base substitution (C to T) causing a change from proline to serine in EPSP synthase abolished the binding of glyphosate to the enzyme. A chimaeric gene was made with an octopine synthase (OCS) promoter, mutant *aro*A coding sequences and OCS polyA signal. The chimaeric gene was introduced into tobacco plants using *Agrobacterium Ti* plasmid vectors. Although transgenic tobacco plants expressed high level of active bacterial EPSP synthase resistant to glyphosate, the plants exhibited only marginal glyphosate tolerance. It then became apparent that the shikimate pathway takes place in chloroplasts but the glyphosate-resistant EPSP synthase was expressed only in cytoplasm. The need for targeting the proteins to the plastids was realised. A chloroplast-specific transit peptide sequence was added to the coding sequence of the mutant EPSP synthase gene to target the protein to the chloroplasts. Expression of an *E. coli* mutant gene with chloroplast-transit peptide directed the glyphosate resistant EPSP synthase to chloroplasts and conferred a much higher level of glyphosate resistance.

In a parallel approach, an over-expressing EPSP synthase gene of a *Petunia* (plant) variant was found to confer glyphosate tolerance. Introduction of this over-expressing gene into other plants led to an increase in the level of glyphosate tolerance in the transgenic plants.

Several economically important plants such as tobacco, tomato, oilseed rape, flax, sugarbeet, alfalfa, soybean and cotton have been engineered by glyphosate resistance and are at different stages of field trial.

12.10.2 Male infertility

We have seen in Section 12.8.4 that male sterility of the seed parent and restoration of fertility in the hybrid seed are key factors in using cms plants for hybrid seed production. Is it possible to introduce these traits in a plant which does not have male sterile lines by using genetic engineering? Since the tapetum is an important tissue in pollen development, any interference in tapetum development or functioning affects pollen development and can lead to male sterility. This was achieved by introducing a ribonuclease gene from *Streptomyces* into tobacco and oil seed rape plants. A hybrid gene was constructed with a ribonuclease-coding sequence and a tapetum-specific promoter. The ribonuclease gene was expressed only in cells of the tapetum and the ribonuclease thus produced degraded the RNA molecules in tapetal cells arresting their development and leading to male sterility in the transgenic plants. This remarkable achievement was made possible by earlier work in which genes that were expressed only in the tapetum were characterised. If the expression of ribonuclease gene was not strictly tissue-specific, other tissues would have been affected by ribonuclease. Along with the excitement, the following limitation of this approach was also realised; the above mentioned transgenic plants were always male sterile; male fertility could not be restored in them by crossing with restorer (C-line) plants.

Assignment 12.15

What type of plant would be needed as 'restorer' for seed parent plants which were male sterile due to expression of a transgenic ribonuclease gene?

It was subsequently reported that male fertility in these transgenic plants could be restored by crossing them with a second set of transgenic plants which expressed a chimaeric ribonuclease inhibitor gene under the control of tapetum-specific promoter. The inhibitor protein as well as ribonuclease are expressed in the tapetal tissue, thus causing blocking of ribonuclease activity and restoration of normal functioning of the tapetum and male fertility. This approach allows the introduction of a restorable male sterility trait into other crop plants in which hybrid seed technology is viable.

12.10.3 Virus resistance

Infection of crop plants by viruses can lead to reduced yield, lowered product quality and sometimes to complete crop failure. Chemical methods that are available to control other types of plant pathogens cannot be used to control plant viruses since they (the viruses) are intracellular, obligate parasites. Several safe agricultural practices currently used to control or limit virus diseases are: using virus-free seeds; controlling insect vectors that spread plant viruses; controlling weed species that serve as alternate hosts for viruses; using cultivars bred for virus resistance.

Engineering virus resistance into crop plants is a prime area in crop biotechnology. Several approaches have been taken to introduce virus resistance into desired crop plants and we will examine these below:

(a) Host-encoded resistance
A number of plant-encoded virus-resistance genes have been characterised by genetic analysis. There is a possibility of isolating these resistance genes from resistant plants and introducing them into susceptible crop plants. The major limitation in this approach is the fact that all the resistance genes have so far only been characterised genetically and not at the molecular level. Isolation and sequencing of these resistance genes should permit genetic engineering of susceptible plants for virus resistance.

(b) Resistance using virus-encoded genes
Cross protection of plants against severe virus infections by pre-infecting the plants with mild virus strains has been successfully used in the protection of tomatoes against tobacco mosaic virus and citrus trees against triesteza virus.

Assignment 12.16

What would be the disadvantages of using a mild strain of virus to infect plants in order to protect them from virulent viruses?

One solution to these problems is to generate transgenic plants which produce components of the virus that confer cross protection without causing viral disease.

(i) Virus coat protein
It has been known for some time that tobacco mosaic virus (TMV) coat protein severely inhibited the translation of TMV RNA *in vitro*. Taking a clue from this, a chimaeric gene was constructed with a TMV coat protein-coding sequence with a CaMV 35S promoter and nopaline synthase polyA signal and this construct was introduced into tobacco plants using *Agrobacterium*. The CaMV 35S promoter is a promoter from the genome of cauliflower mosaic virus. This promoter is used as it is constitutive — that is it is expressed all the time and not down regulated by factors within the plant cell. This is of obvious advantage to the virus. The transgenic tobacco plants expressing the TMV coat protein gene showed high levels of resistance to TMV infection. Following this remarkable success, coat protein-mediated protection has been successfully engineered for over 20 different viruses. Some examples are alfalfa mosaic virus (A1MV), potato virus X (PVX),

potato virus Y (PVY), tobacco streak virus (TSV) and tobacco rattle virus. Coat protein-mediated protection is by far the most successful approach for genetically engineering virus resistance in plants. This approach is very successful for viruses with single-stranded RNA genomes. This approach is not successful in the case of double-stranded and single-stranded DNA viruses.

In spite of the remarkable success of coat protein-mediated protection, the molecular basis of it is not yet clearly understood. Recent experiments indicate that engineering of coat protein genes with deletions or mutations that affect coat protein function also confer resistance. Therefore, functional coat protein is not essential for conferring protection.

(ii) Replicase

Expression of a non-structural 54 kDa protein, representing a part of the replicase enzyme of TMV, conferred a very high level of resistance in tobacco against TMV. The protein product has not so far been detected in TMV infected plants and its function is not clearly known. The sub-genomic RNA corresponding to the 54 kDa protein has however been detected. Transgenic plants engineered with the 54 kDa protein gene make the transcripts but do not make detectable levels of the protein. It is not clear at this stage whether the protection seen in the transgenic plants is due to the 54 kDa protein or its transcript.

(iii) Movement protein

Systemic infection by a virus results from rapid cell to cell spread and long distance spread through vascular tissues. A 30 kDa movement protein of TMV binds to TMV-RNA and some host proteins and enables the ribonucleoprotein complex to pass through the plasmodesmata (intercellular junctions) thus mediating cell to cell spread of virus. Transgenic tobacco plants expressing a mutated 30 kDa movement protein showed a reduction in the final yield of infective TMV particles in infected plants. It was proposed that the defective 30 kDa protein of the transgenic plants competes with the wild-type TMV-encoded 30 kDa protein and thus reduces the spread of the virus. In another experiment, the 32 kDa movement protein of brome mosaic virus (BMV) expressed in transgenic tobacco effectively lowered the spread of TMV. It was proposed that the constitutively expressed BMV 32 kDa protein acts as a molecular decoy to compete with the TMV-encoded 30 kDa movement protein. The BMV protein probably binds only to the movement-related host protein or to TMV-RNA but not to both. Binding of movement protein to both the host protein and RNA may be necessary for virus spread. Fortunately, the movement protein strategy seems to work for the single-stranded DNA viruses (gemini viruses) as well. A recombinant movement protein having parts from tomato golden mosaic virus and African cassava mosaic virus has been shown to severely interfere with the spread of both viruses.

(iv) Transmission protein

There is a high level of specificity in the interaction between plant viruses and the insect vectors that spread them from one plant to another. In the case of aphid-transmitted cauliflower mosaic virus (CaMV), a bifunctional virus-encoded protein recognises the viral coat protein and also a 'helper component' of the aphid, referred to as 'aphid transmission factor'. Mixing a mutated (defective) 'helper component' of the aphid, referred to as 'aphid transmission component', with the virus severely interfered with the transmission of the virus by the aphid. The defective 'helper component' apparently competed with the aphid protein in binding to the 'bifunctional' insect spread protein of the virus. Thus it may be possible to prevent the spread of insect-transmitted viruses by engineering crops to express a defective virus-transmission protein.

(v) Disease attenuation with satellite RNA

Satellite RNAs infect plants only with the help of helper viruses. Satellite RNAs are encapsulated together with the respective helper viruses. Satellite RNAs either increase the severity of the symptoms caused by the helper virus or at times attenuate the symptoms caused by the helper virus.

The latter property of satellite RNAs has been used in biologic control of the spread of certain viruses. For example, when cucumber mosaic virus (CMV) infects pepper plants, severe symptoms appear. However, when CMV is coinoculated with a satellite RNA the disease symptoms are attenuated and the yield of the plants is higher. This approach does have the drawback that the first infection of the satellite RNA with a mild virus causes a penalty in the yield. Additionally a mutation in the first helper virus may convert a mild strain to a virulent strain and may lead to spread of the virus. One way to avoid the need to use a helper virus is to introduce the DNA sequence corresponding to satellite RNA into the plants such that the symptom-attenuating satellite RNA is expressed in transgenic plants. DNA copies of cucumber mosaic virus-satellite RNA have been introduced into tobacco plants. These transgenic plants expressed the satellite RNA and this conferred protection against cucumber mosaic virus and tomato aspermy virus (TAV). On the basis of the fact that CMV and TAV are helper viruses for the same satellite RNA, it was proposed that the satellite RNA expressed in the transgenic plants competed with the helper virus for the limiting quantities of replicase and thereby reduced the replication of the infective virus.

The satellite RNA approach has several drawbacks: satellite RNAs that attenuate symptoms in one crop may cause severe disease in another crop plant, for example CMV-D satellite RNA which attenuated CMV symptoms in tobacco caused a lethal infection in tomato; satellite RNAs mutate very rapidly, for example a single nucleotide change can convert ameliorative satellite into a highly virulent agent; recombinations between satellite RNAs have been observed. Because of these limitations researchers are highly guarded in using satellite RNA sequences for controlling the helper viruses.

(vi) Defective interfering DNA

Defective interfering (DI) DNAs and RNAs are very rare in plants. Like satellite RNAs, DI nucleic acids can intensify or ameliorate the symptoms of their respective parental viruses. This approach has been successfully tested in the case of African cassava mosaic virus (ACMV), a geminivirus with two SS-DNA genomes (A and B). A subgenomic B component of ACMV was engineered into tobacco plants which are susceptible to ACMV. The DI of ACMV interfered with the replication of both A- and B-DNAs in the transgenic tobacco plant and ameliorated the symptoms of virus infection. Interestingly, the ACMV defective DNA did not confer resistance against another geminivirus, tomato gold mosaic virus. Attempts are being made to extend this approach to other groups of plant viruses.

(vii) Antisense RNA

Using genetic engineering the complementary DNA strand of a gene sequence can be inserted in reverse orientation ($3' \rightarrow 5'$ as opposed to $5' \rightarrow 3'$) into a construct under control of a suitable promoter. Such a complementary sequence in reverse orientation is termed an antisense gene. If such an antisense gene is expressed in a cell, the mRNA transcribed from it will also be antisense and will be complementary to the mRNA transcribed from the normal gene. The complementary normal mRNA and antisense mRNA may hybridise (bind together) and the translation of the normal mRNA may be blocked (ribosomes cannot translate mRNA in hybridised form). Thus engineering an antisense gene into a cell may result in formation of antisense mRNA and blocking of gene expression.

Viral antisense genes can be introduced into plants such that mRNA complementary to the viral coding sequences or sequences involved in replication of the virus will be transcribed. Such antisense mRNAs may block replication of the viruses. The antisense RNA approach was initially attempted for single-stranded RNA viruses.

Assignment 12.17

To make an antisense gene (DNA) of an RNA virus gene (RNA), a DNA copy must be made of the RNA gene. How could this be achieved?

Transgenic plants expressing antisense mRNA complementing different regions of the viral RNA genome conferred very low level of protection against low concentrations of challenging viruses. The reasons attributed to this failure are: an exclusively cytoplasmic replication cycle in RNA viruses; high level of genome-sense mRNA thus necessitating much higher concentrations of antisense mRNA; association of sense mRNA with proteins at all stages.

The antisense RNA approach seems to hold promise for DNA viruses, for which replication and transcription takes place in the nucleus. Tomato golden mosaic virus (TGMV) replicase-coding sequence was cloned in its antisense orientation under the control of a CaMV 35S promoter and introduced into tobacco plants using *Agrobacterium*. Transgenic plants that expressed the antisense RNA of TGMV replicase showed resistance to TGMV when infection with the virus was performed.

(viii) Ribozymes

Ribozymes are small RNA molecules, derived from the satellite RNA of tobacco ringspot virus (TRSV), or certain viroids and viroid-like satellite RNAs, which promote highly specific catalytic cleavage of RNA. Ribozyme-mediated catalytic activity is generally intramolecular. However, by appropriate designing of the catalytic domain with flanking antisense arms it is possible to achieve intermolecular cleavage of target viral RNA molecules. Such 'ribozyme' sequences can be incorporated into the genomes of mild viruses such that 'ribozyme'-containing sub-genomic RNA active against a severe virus can be expressed. This approach would be an effective extension of the classical "cross protection" with the added advantage of amplifying ribozymes in the host plants. The feasibility of the ribozyme approach, though very attractive, remains to be tested.

(c) Resistance using animal genes

Genes for mouse monoclonal antibodies engineered into plants were expressed well and functional antibodies were shown to assemble within plant cells. The biological significance of such antibodies (plantibodies) has recently been demonstrated using monoclonal antibodies against phytochrome; phytochrome-related functions were inhibited in plants expressing phytochrome plantibodies. Attempts are being made to introduce genes for plantibodies against important viral proteins such as coat proteins and replicase so that multiplication of viruses within plant cell can be limited. The usefulness of such an approach in the control of viral diseases is currently in the testing stage. This is, in effect, using animal immune mechanisms to confer immunity onto plants.

In summary, the variety of approaches discussed so far highlight the importance of genetic engineering to introducing virus resistance into crop plants. In the next phase of development, emphasis will be towards checking whether the success of laboratory and greenhouse experiments will be reflected in field conditions.

12.10.4 Pest resistance

A global survey in 1988 revealed that the annual cost of use of chemical pesticides is in the range of 6075 million US dollars, with cotton and rice cultivation accounting for almost half of the total. Scientists have constantly been looking for alternate methods of pest control because: chemical pesticides are highly inefficient — about 98% of the sprayed chemical is washed away from the plant surface and ends up in soil; chemical pesticides are not degraded efficiently in the soil and their residues therefore build up and cause environmental pollution; chemical pesticides have a broad spectrum of activity and are toxic to several non-target organisms — toxicity to animals and humans is a major concern; it is difficult to deliver chemical pesticides to highly vulnerable parts of plants such as roots or internal regions of stems and fruits.

Scientists have discovered two biotechnological alternatives to chemical pesticides: insecticidal crystal proteins from the bacterium *Bacillus thuringiensis* that are toxic to specific groups of insects at low concentrations; proteinase inhibitors which are harmful to insects when present in the diet at high concentrations. Tremendous progress has taken place in the genetic engineering of plants using both these approaches.

(a) *Bacillus thuringiensis* toxins

Bacillus thuringiensis (*B.t.*), a Gram-positive soil bacterium, produces upon sporulation a parasporal crystalline proteinaceous toxin with insecticidal activity. This protein is referred to as insecticidal crystal protein (ICP). When the larvae of susceptible insects ingest the ICP, a combination of alkaline pH and proteinases in the midgut solubilises the protein and converts the relatively harmless protoxin to the active toxin. The effect of the toxin is very rapid; within minutes of ingestion, the binding of toxin molecules to specific receptors in the midgut membranes disrupts ion-transport across the membrane and paralyses the midgut function. The larvae stop feeding almost instantly and eventually die. The toxin is harmless to higher animals, including humans.

A number of *B.t.* strains with insecticidal properties have been known for over 30 years. Biopesticides based on *B.t.* are produced by growing the organism in liquid media in fermenters, allowing the organism to sporulate and form protoxin on depletion of nutrients, then drying the product to a powder. The powder can be dusted onto plants or sprayed onto them after the powder is emulsified in water.

B.t.-based biopesticides have applications in environmentally-sensitive applications and home-garden use. Despite their remarkable safety and efficacy, they have not been a commercial success, basically due to the high cost of production and low persistence and stability of the product on the surface of plants.

Assignment 12.18

Predict how genetic engineering could be used to overcome the problems associated with the use of *B.t.* insecticides.

Scientists from Plant Genetic Systems, a Belgian Biotechnology Company, reported the first successful generation of transgenic plants with *B.t.* toxin. From a full length protoxin gene coding for 1155 amino acids, a truncated version inclusive of the N-terminal 29 to 607 amino acids was cloned under the control of the mannopine synthase promoter of Ti plasmid T-DNA. Tobacco plants were transformed and selected using kanamycin; since neomycin phosphotransferase that confers kanamycin resistance is translationally fused to the C-terminal end of the truncated *B.t.* gene, selection on kanamycin not only confirms T-DNA transfer but also ensures that the *B.t.* gene has been transcribed and translated. Coincidentally, the mannopine synthase promoter also turned out to be a wound-inducible promoter.

Assignment 12.19

What advantage would there be in having the *B.t.* toxin gene under the control of a wound-inducible promoter?

Since then a number of reports on genetic engineering of tomato, tobacco, potato and cotton have appeared. A general conclusion drawn from all these studies is that the level of expression of *B.t.* genes in transgenic plants is low and not sufficient enough for commercial exploitation. To overcome this problem, regions of localised A=T richness, regions resembling plant introns, potential polyA signals and ATTTA sequences were deleted from the *B.t.* gene producing a truncated gene without, however, altering the amino acid sequences coded for A 100- to 500-fold increase in the expression of the engineering (truncated) gene over the wild-type gene was observed. Thus a major limitation in the expression of *B.t.* proteins in transgenic plants was overcome. Cotton plants transformed with the engineered *B.t.* gene expressed the gene at high levels and exhibited impressive levels of insect resistance in greenhouse and field conditions.

There are two particular advantages in using transgenic plants with *B.t.* genes for pest resistance. Firstly, *B.t.* genes in a transgenic plant could be expressed in all parts of the plant including roots and internal regions of stems and fruits: these are the parts of the plant that cannot be readily reached by chemical or biological pesticide spray. Secondly, the toxin proteins are produced only within the plants. This is in contrast to chemical pesticide sprays which would be present in the environment generally. While chemical pesticides may affect other organisms apart from target insects, *B.t.* toxins made in the plants would kill only (target) organisms that feed on the transgenic plant. *B.t.* protein made in the plants could also be rapidly degraded in the environment. These characteristics of *B.t.* expression in transgenic plants promise that the widespread use of these insecticides should be relatively safe and compatible with environmental protection.

B.t. expression in plants is not without disadvantages. As *B.t.* toxin is a protein, and the receptor on the insect gut that mediates the toxic response is also a protein, there is fear that *B.t.*-resistant insects would evolve rapidly. This would be of importance if *B.t.* toxin was continuously expressed in all parts of the transgenic plant, as this would exert a great deal of selection pressure on the insect to evolve *B.t.* resistance. The fear has turned out to be correct: evidence is accumulating both in laboratory and in the field that widespread use of *B.t.*-protected crops may select for resistant insects. A number of approaches are being tested to avoid the development of such resistance in insect populations. An important approach being actively tested is to introduce two different *B.t.* toxins for the same target insect into a plant or to introduce two types of insect-resistance genes, such as a *B.t.* gene and a proteinase inhibitor gene, into a plant.

(b) Proteinase inhibitors

Many plants, including many used as food by humans, have evolved a natural mechanism of defence against herbivorous insects by accumulating proteinase inhibitors (proteins which inhibit the activity of proteinase enzymes) at concentrations that will cause metabolic inhibition upon ingestion by insects. Compared to *B.t.* toxins, proteinase inhibitors have a broader spectrum of metabolic inhibition and so can be used to control many different types of insects. An attractive approach to pest control is to introduce and express proteinase inhibitor genes into crop plants that normally do not make these proteins.

Proteinase inhibitors are classified into four classes: serine-, thiol-, metallo- and aspartyl-proteinase inhibitors. Serine-proteinase inhibitors and thiol-proteinase inhibitors are important with respect to pest control as different types of pests use either serine proteinases or thiol proteinases or in some cases both. Proteinase inhibitors offer several advantages as insect control agents: many insects not affected by *B.t.* can be controlled by proteinase inhibitors; proteinase inhibitors can be used as a second mechanism to reinforce the resistance already achieved through *B.t.* genes and simultaneous expression of both *B.t.* toxin and proteinase inhibitor will reduce the chance of evolution of insects resistant to either *B.t.* or proteinase inhibitor; proteinase inhibitors become inactivated by cooking. Limitations of the proteinase inhibitor approach are: unlike *B.t.* toxin that acts at low concentrations, high levels of proteinase inhibitors are needed to kill insects; expression of proteinase inhibitors should be reduced in parts of plants used as human food and increased in parts of the plants used as food by insect pests.

Transgenic tobacco plants expressing high levels of trypsin-inhibitor protein encoded by a gene of an African variety of cowpea showed decreased insect damage by *Heliothis virescences*, tobacco bud worm, a pest of tobacco. It was also demonstrated that cowpea trypsin inhibitor killed a wide variety of lepidopteran and coleopteran insects, including army worm (*Spodopteran littoralis*), corn earworm (*Heliothis zea*) and corn root worm (*Diabrotica undecimpunctats*).

In summary, there are now three different approaches for the biocontrol of insects: microbial pesticides derived from *Bacillus thuringiensis* or other bacteria expressing *B.t.* genes; genetic engineering of plants with *B.t.* genes; genetic engineering of proteinase inhibitor genes into plants. Though the success is reassuring, it should be appreciated that insects with their great diversity and large population could evolve very rapidly to gain resistance against a given biopesticide. In anticipation of this there is a constant search for newer strains with variations in the known pesticide and improved mechanisms for pest control.

12.10.5 Resistance towards fungal pathogens

Plants exhibit resistance to fungal infections in a number of ways. In a passive mechanism, plant cell walls can be strengthened to provide a physical barrier to limit penetration by fungal hyphae. In the active defence mechanism called 'hypersensitive reaction', a resistant plant synthesises toxic antimicrobial compounds called 'phytoalexins'. Hypersensitive reaction also leads to the induction and accumulation of proteinase inhibitors and lytic enzymes (against components of fungal cell walls) such as chitinase and β-1,3-glucanase. Hypersensitive reaction is a specific process that occurs as a result of interaction between a product of a dominant resistance-gene of the host and a product of a dominant avirulence-gene of the fungal pathogen. This kind of interaction is known as an incompatibility reaction, and it leads to resistance in the host. In spite of two decades of research, the precise mechanism that results in hypersensitive reaction is not known. Many genes conferring resistance against fungi have been characterised at the genetic level but none has been cloned. Attempts are underway to map and clone these genes. Genetic engineering for fungus resistance using host genes will only become feasible once this has been done.

An alternate approach is to introduce a heterologous (foreign) chitinase gene into plants in order to confer resistance to fungi. Chitinase catalyses the hydrolysis of chitin, a β -1,4-linked polymer of N-acetylglucosamine, that is a major component of cell walls of most filamentous fungi, except the Oomycetes. Purified chitinase inhibits growth of phytopathogenic fungi *in vitro*. Chitinase is generally found in low levels in healthy plants but its expression is increased during pathogen attack.

A hybrid chitinase gene was constructed by replacing the 5' regulatory sequences of a bean chitinase gene with the CaMV 35S promoter to achieve constitutive (continuous) expression. After introduction of the hybrid gene into tobacco plants, bean chitinase was expressed in the transgenic plants and the precursor enzyme was correctly processed post-translationally. Transgenic tobacco plants that expressed bean chitinase showed marginally higher resistance to the fungus *Rhizoctonia solani* as compared to normal tobacco plants: while 53% of control plants became infected by *R. solani* only 22 to 37% of the seedlings of transgenic tobacco plants suffered infection. Chitinase-mediated resistance was very specific, since the same transgenic tobacco plants did not differ from the control plants in their susceptibility to a second fungus, *Pythium aphanidermatum*, which does not have chitin in its cell wall; overexpression of chitinase confers resistance only against fungi that have chitin in their cell wall. *Brassica napus* plants engineered with bean chitinase showed significant resistance to *R. solani*, which causes a severe disease in this plant.

Assignment 12.20

Predict whether plants which overexpress chitinase are suitable as food for humans. Explain your answer.

12.10.6 Genetic engineering of plant oils

Plant oils represent a vast renewable source of oils. Two thirds of plant oil produced are used as food and one third is used for a variety of industrial purposes. Annual plant oil production in the world amounts to 60×10^6 tonnes, with a value of about 20×10^9 US dollars. Considering the high value of plant oils, and their vast market, enormous efforts are being made to increase the yield of oil producing crops and to alter or improve the quality of oils by structural modification, by both conventional plant breeding and by using modern tools of genetic engineering. Some of the goals towards which scientists in this area are working are outlined in Table 12.2.

Table 12.2 Genetic engineering of plant oils

	Goal	Purpose
1	Decrease linoleic acid in soybean oil	Increased stability, less tendency to spoil
2	Develop a temperate-zone, annual crop as source of lauric acid	Stabilise supplies, grow lauric acid crop in temperate zones, lower prices (soap and detergent industry)
3	Develop high erusic acid plant oils – particularly *Brassica*	Make industrial use of plant oils economically viable, reduce prices
4	Develop high palmitic lines of rape seed	Allow increased use of canola oil in margarines
5	Decrease levels of unsaturated fatty acids in *Cocoa seeds*	Make chocolates that remain solid in warm weather
6	Increase oil yield	Make vegetable oils competitive as diesel oil substitutes

Plant oils are synthesised in various compartments of plant cells and stored in lipid bodies. The cell compartments involved in oil biosynthesis are the proplastid, cytoplasm and endoplasmic reticulum (Fig. 12.19).

Acetate taken up by plastids is converted to acetyl-CoA and malonyl-CoA. These two molecules serve as oil precursors. An enzyme complex, fatty acid synthase, catalyses the synthesis of laurate (C12), myristate (C14), palmitate (C16), stearate (C18) and oleate (C18:1) as their ACP (acyl carrier protein) derivatives. Fatty acid moieties can then either be incorporated into complex plastid lipids or be exported to the cytoplasm, following hydrolysis to their free fatty acids by acyl-ACP thioesterases. Fatty acid modifications such as elongation, desaturation (introduction of double bonds) and hydroxylation occur in cytoplasm. Fatty acids are subsequently transported to the endoplasmic reticulum, where they combine with glycerol-3-phosphate to form phosphatidic acid and subsequently triacylglycerides (TAG) which are transported to lipid bodies and stored.

(a) Modification of fatty acid composition by plant breeding

The feasibility of modifying fatty acid composition of seeds was first realised by the success of Canadian plant breeders in breeding a rape seed (*Brassica rapa*) variety with reduced levels of erusic acid. This paved the way for marketing rape seed oil as edible "Canola oil". Conventional breeding has also resulted in the development of sunflower with a high content of oleic acid (up to 90% of the seed oil), which is used for industrial purposes.

Conventional breeding for modification of plant oils seems to have reached its limits. For example, desirable traits such as high erusic acid content (greater than 90% for industrial purposes) and the ability to synthesise medium chain fatty acids (for the soap and detergent industry) are found in plants that cannot be crossed with rape seed. Therefore, the next logical step is to overcome the

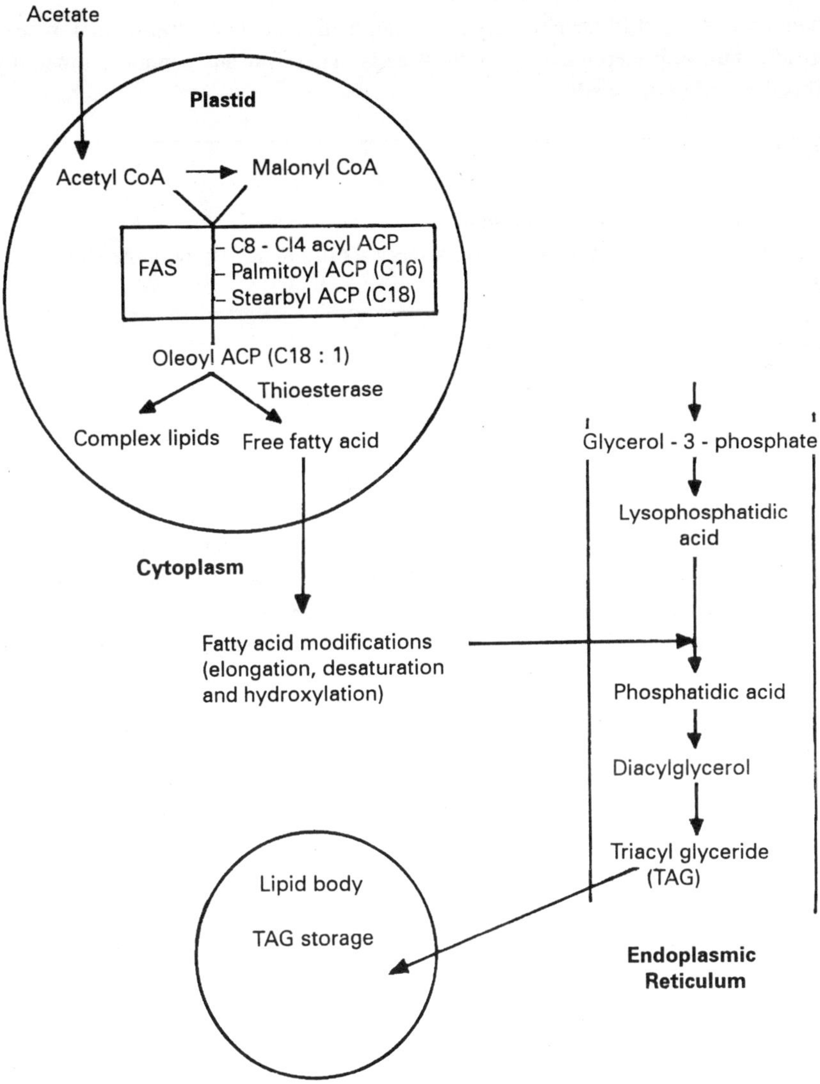

Fig. 12.19 *Biosynthesis of triacylglycerides in plants.*
Four cellular compartments are involved in triacylglyceride biosynthesis ans storage (plastid, cytoplasm, endoplasmic reticulum and lipd body). Abbreviations: FAS – fatty acyl synthase; ACP – acyl carrier protein; TAG – triacylglyceride.

limitation of conventional plant breeding by manipulating plant oil biosynthesis by genetic engineering.

(b) Modification of plant oil biosynthesis by genetic engineering
Because of the complex pathway of plant oil biosynthesis the following aspects have to be considered when genetic engineering of plant oils is attempted: the engineered gene should be expressed only in the seeds where oils are made and stored; the engineered gene for a specific enzyme should be designed such that the enzyme protein, when made, is targeted to the appropriate compartment in the cell where its reactions normally occur. It should be remembered that all plant cells need lipids

as constituents of membranes; a disturbance in fatty acid composition or concentration in non-seed cells may drastically affect the normal physiology of these cells. Therefore any genetic engineering of plant oils should be limited to the seeds.

Self assessment exercise 12.7

Choose the correct completion to the following statement.
In order to ensure that an engineered gene (transgene) is expressed only in seeds, the gene should be placed under control of :

(i) a wound-specific promoter.
(ii) a Ca 35S promoter.
(iii) a seed-specific promoter.
(iv) a leaf-specific promoter.

Medium chain fatty acids (C8 to C14) such as lauric acid (C12) are in high demand in the detergent industry; approximately 640,000 tonnes is needed for the USA alone. Natural sources of these fatty acids are the oils of palm kernel and coconut, both of which are tropical crops. Efforts have been undertaken to elucidate the pathway for the synthesis of medium chain fatty acids and introduce the appropriate gene(s) into a temperate oil crop such as *Brassica.* A gene responsible for synthesis of medium-chain fatty acids has been isolated from an unconventional oil plant, California bay (*Umbellularia californica*), that accumulated high levels of caproate (C10) and laurate (C12); the gene for a C-12 acyl carrier protein thioesterase (BTE) also coded for an N-terminal transit peptide sequence for plastid targeting. To ensure seed-specific expression, the DNA construct containing the BTE gene was fused to a *Brassica rapa* seed storage protein (napin) promoter and introduced into *Arabidopsis* using *Agrobacterium* vectors. *Arabidopsis* was chosen as a model temperate plant. The transgenic *Arabidopsis* plants expressed the introduced gene and showed BTE activity. Whereas the untransformed control plants accumulated predominantly C18 fatty acids, the transgenic plants accumulated high levels of medium chain fatty acids. Thus, an era has begun in which genetic engineering may permit scientists to alter the fatty acid chain length in conventional temperate oil crops such as *Brassica.*

A second approach has been reported recently in which a qualitative change in fatty acid synthesis in *Brassica rapa* and *B. napus* was achieved by blocking a step in the fatty acid biosynthesis pathway using the antisense RNA approach. The objective was to reduce the conversion of stearate (C18:0) to unsaturated fatty acids such as oleic acid (C18:1), linoleic acid (C18:2) and linolenic acid (C18:3). The first desaturation step of conversion from stearoyl-ACP to oleoyl-ACP is catalysed by the enzyme stearoyl-ACP desaturase. A gene for stearoyl-ACP desaturase was isolated from *B. rapa* and the complementary sequence cloned in reverse orientation under the control of two seed-specific promoters, namely napin and acyl carrier protein (ACP) of *Brassica.* The two seed-specific promoters were used to enhance antisense RNA transcription and thereby ensure the presence of antisense RNA throughout seed development. The antisense RNA transcribed in the seeds is thought to hybridise with the sense RNA coding for stearoyl-ACP desaturase and thereby prevent RNA translation and synthesis of this protein.

Self assessment exercise 12.8

Choose the correct completion to the following statement.
The effect of expression of an antisense stearoyl-ACP desaturase gene in a seed would be to

(i) increase the proportion of linolenic acid;
(ii) increase the proportion of linoleic acid;
(iii) increase the proportion of oleic acid;
(iv) increase the proportion of stearic acid.

Antisense genes to stearoyl-ACP desaturase have been successfully introduced into *B. rapa* and *B. napus* using *Agrobacterium*. While the control plants had about 1.3% stearic acid (of the total fatty acids), many of the transgenic plants with antisense genes had about 20 to 25% stearic acid, with concomitant decrease in the activity of stearoyl-ACP desaturase. These findings highlight the tremendous scope for application of genetic engineering to manipulating fatty acid composition in plant oils.

Genetic manipulation of lipids has also been shown to influence very complex traits such as 'chilling sensitivity' of plants. It had previously been postulated that plants that tolerate chilling (e.g., spinach, *Arabidopsis*) had high levels of *cis*-unsaturated fatty acids in the chloroplast-localised phosphatidylglycerol. Chill-sensitive plants such as squash have been shown to have lower levels of *cis*-unsaturated fatty acids. The chloroplast enzyme glycerol-3-phosphate acyltransferase was considered to be important in promoting phosphatidyl glycerol fatty acid unsaturation.

Self assessment exercise 12.9

Which of the following would be of benefit in a DNA construct to be used to confer chill tolerance on a chill-sensitive plant?

(i) a gene for glycerol-3-phosphate acyltransferase;
(ii) an antisense gene for glycerol-3-phosphate acyltransferase;
(iii) a seed-specific promoter;
(iv) a CaMV 35S promoter;
(v) a chloroplast-specific transit peptide sequence;
(vi) an endoplasmic reticulum-specific transit peptide sequence.

The gene for glycerol-3-phosphate acyltransferase of *Arabidopsis* with its chloroplast-specific transit peptide sequence has been introduced into chill-sensitive tobacco plants. The transgenic tobacco plants exhibited high levels of chilling tolerance, suggesting that a single enzyme such as glycerol-3-phosphate acyltransferase could control a complex trait such as chilling tolerance.

12.10.7 Genetic engineering for extended shelf-life of fruits

Delaying fruit ripening has become an important commercial aspect in plant genetic engineering. Delay in ripening extends the shelf-life and keeping quality of fruits thus enabling long distance shipping of fruits, with minimum damage, and it enables fruit to be ripened on the plant, thus improving flavour. Genetic engineering work has been carried out in tomatoes, basically because a great deal of physiological, biochemical and molecular biology work has already been carried out in this system.

Tomato is an example of a climacteric fruit in which the onset of ripening is accompanied by an increase in respiration. The respiratory climacteric is stimulated by the phytohormone ethylene, and a peak of ethylene biosynthesis by the fruit accompanies the respiratory climacteric. Ethylene level is thus a key factor in fruit ripening. Addition of exogenous ethylene promotes the rate of ripening, whereas inhibition of ethylene biosynthesis reduces the rate of ripening. Other molecular changes such as induction of the pectinase enzymes polygalacturanase and pectin esterase are very typical of ripening tomatoes. Genes involved in carotenoid biosynthesis are also induced in a ripening fruit.

Attempts have been made to manipulate all these processes by genetic engineering. All attempts have involved the 'antisense RNA' approach. As previously described, the synthesis of a specific protein is reduced by genetically engineering a plant with a construct in which complementary sequence of the targeted gene is placed in a reverse orientation under a fruit-specific and climacteric-specific promoter. The sense mRNA expressed in the fruit of the transgenic plant would hybridise with the complementary antisense mRNA made in the same fruit thus blocking its post-transcriptional processing and translation, so reducing production of the encoded protein.

The protein to be targeted first was polygalacturanase. Ripening of tomato fruit is closely associated with the increase in polygalacturanase activity leading to pectin breakdown and softening of the cell wall. Therefore a reduction in the levels of polygalacturanase was expected to lead to delay in fruit ripening. Expression of antisense RNA-gene constructs led to a marked decrease in the levels of polygalacturanase mRNA and enzyme. However, contrary to expectations, there was no delay in ripening. Subsequently the effect of expression of antisense mRNA directed against pectin esterase was tested. Though there was a reduction in the level of pectin esterase activity (the removal of methyl groups from pectin), this did not lead to any delay in the ripening of tomato fruits. These studies conclusively proved that polygalacturanase and pectin esterase are not the key determinants of tomato ripening.

Further studies targeted ethylene biosynthesis, since physiological studies revealed that ethylene is the key signal in the ripening of climacteric fruits. Ethylene is synthesised from S-adenosylmethionine via the intermediate 1-aminocyclopropane-1-carboxylic acid (ACC). This reaction is catalysed by an enzyme ACC synthase. ACC is converted into ethylene by ACC oxidase. Introduction of the antisense gene of ACC oxidase into tomato plants inhibited ethylene production from ACC by 97% in transgenic plants. This resulted in a significant reduction in the rate of ripening. Surprisingly, polygalacturanase level in these fruits was not affected. Antisense RNA against ACC synthase was also expressed in tomato plants and this inhibited ethylene biosynthesis by about 99.5% and markedly delayed fruit ripening. The gene of a bacterial enzyme ACC deaminase, which renders ACC unsuitable as a substrate for ACC oxidase, was expressed as a chimaeric gene with a plant promoter in tomato plants, and these transgenic tomato plants exhibited almost 90% inhibition of ethylene biosynthesis and fruit ripening was delayed by more than six weeks.

Genetically engineered tomatoes have become a reality. A commercial company, Calgene, is set to market genetically engineered "Flavr–Savr" tomatoes, developed using antisense RNA technology. These tomatoes have an extended shelf-life and increased resistance to bruising. This technology should soon be extended to other climacteric fruits.

12.10.8 Manipulation of starch biosynthesis

Starch biosynthesis determines the yield of all cereal crops and tuber crops such as potato and cassava. Genetic engineering has been used to manipulate both quality and quantity of starch in potato. Starch is composed of two types of polysaccharides, amylose and amylopectin. The first attempt at manipulation of starch biosynthesis was directed towards generating potato plants that synthesised only amylopectin. This was achieved by reducing the levels of starch synthase enzyme (responsible for amylose synthesis) using the antisense RNA approach: the complementary sequence to the starch synthase gene was cloned in a reverse orientation under the control of a CaMV 35S promoter and introduced into potato plants. The tubers of transgenic potato plants were devoid of amylose and contained only amylopectin. In this particular case the aerial parts of the plant also made only amylopectin since the antisense RNA was expressed throughout the plant using the constitutive CaMV 35S promoter.

Increasing the quantity of starch made by potato plants has been achieved recently. Starch biosynthesis in plants is highly regulated at a key step catalysed by the enzyme ADP-glucose pyrophosphorylase, which is allosterically regulated (feed-back inhibited) by several metabolites, including phosphate.

> ***Assignment 12.21***
>
> Predict what changes would need to be induced in ADP-glucose pyrophosphorylase in order to increase the quantity of starch synthesised.

Escherichia coli ADP-glucose pyrophosphorylase was mutated to alter its allosteric properties and was expressed in potato plants with a transit peptide sequence that would permit targeting of the *E. coli* enzyme to the stroma of amyloplasts (where starch biosynthesis occurs). These transgenic plants displayed increased starch level and a higher dry matter content in their tubers: on average, the tubers of transgenic plants contained 25% higher level of dry matter compared to control plants.

A detailed understanding of the regulatory steps in the biosynthesis of starch and other polysaccharides, and knowledge of the source-sink relationship that regulates the flow of nutrients, particularly sucrose, between plant tissues should be helpful in the manipulation of highly complex traits, such as yield, of crop plants.

12.10.9 *Improvement of quality of seed storage proteins*

Seed storage proteins, which constitute a major source of proteins in the human diet and animal feed, suffer from the disadvantage of having very low levels of essential amino acids. Proteins of pulses are deficient in the sulphur-containing amino acids, cysteine and methionine, and cereal grain proteins are deficient in lysine and tryptophan. Several attempts have been made to express cereal seed storage proteins in pulses and to make substitutions in pulse seed storage proteins by introducing genes coding for cysteine and methionine.

Seed storage proteins have evolved several unique structural and functional features. These proteins accumulate during seed development. They maintain their structure when water is almost completely removed during seed maturation. They become rapidly broken down when the seed absorbs water. The majority of genes coding for seed storage proteins have signal sequences for protein targeting to the endoplasmic reticulum, for glycosylation and for transport to protein bodies either directly from the endoplasmic reticulum or through the Golgi apparatus. Many seed storage proteins undergo post-translational processing and have appropriate sites for such processing. The structure of seed storage proteins should permit very compact folding so that large amounts of protein can be packed in the small volume of a seed. Genetic engineering strategies to improve the quality of seed storage proteins should take into consideration all these unique features of seed storage proteins.

Genetic engineering approaches have been limited to the identification of seed storage proteins that have higher levels of essential amino acids and then engineering their expression in target plants. In one instance a chimaeric gene with a phaseolin (seed storage protein) gene promoter and the coding region of a methionine-rich protein (2S albumin) from Brazil nut was constructed and introduced into tobacco. The transgenic tobacco plants expressed the 2S protein in the seed, resulting in a 30% increase in the levels of methionine. In another instance parts of the seed-specific 2S albumin genes of *Arabidopsis thaliana* were replaced by sequences coding for methionine-rich portions of the Brazil nut 2S albumin and introduced into *A. thaliana, Brassica napus* and tobacco plants. The chimaeric gene was expressed in these transgenic plants albeit at low levels. These experiments demonstrate the feasibility of improving the quality of seed storage proteins.

Lysine is synthesised in higher plants from aspartate via the aspartate family biosynthetic pathway. This pathway also leads to the biosynthesis of isoleucine, methionine and threonine. Aspartate kinase and dihydropicolinate synthase (DHPS) are the key regulatory enzymes in this pathway. DHPS is subject to feed-back inhibition by lysine. While plant DHPS is highly sensitive to lysine inhibition, bacterial DHPS is less sensitive. Reduced feed-back inhibition of DHPS is expected to lead to increased levels of all four amino acids. A chimaeric gene consisting of the coding region of the DHPS gene from *E. coli,* a chloroplast transit peptide sequence and CaMV35S promoter was constructed and introduced into tobacco plants. The transgenic plants contained 25% higher levels of DHPS and, as expected, the enzyme was less sensitive to lysine inhibition. These plants also exhibited up to 15-fold higher levels of free lysine. Thus, a dramatic change in amino acid composition of the plant was achieved by introducing a gene for a single enzyme with reduced allosteric regulation.

Self assessment exercise 12.10

Answer true or false to each of the following:

(i) It is reasonable to expect that, as glyphosate acts by competitive inhibition of EPSP synthase, glyphosate resistance could be induced in a plant by the expression of an antisense EPSP synthase gene.

(ii) It is reasonable to expect that male infertility conferred on a seed parent plant by inclusion of a tapetum-specific ribonuclease gene could be restored (to male fertility) by cross breeding with a restorer line which expresses a nuclear tapetum-specific antisense ribonuclease gene.

(iii) It is reasonable to expect that, as a TMV-coded movement protein mediates spread of TMV virus throughout a plant, TMV infection would be ameliorated in a plant expressing an antisense TMV-movement protein gene.

(iv) It is reasonable to expect that, as resistance to fungal pathogens is conferred by synthesis of particular phytoalexins, fungal infection would be ameliorated in a plant expressing antisense phytoalexin genes.

(v) It is reasonable to expect that, as starch biosynthesis is catalysed by ADP-glucose pyrophorylase, starch synthesis would be depressed in a plant expressing an antisense ADP-glucose pyrophosphorylase gene.

(vi) It is reasonable to expect that, as dihydropicolinate synthase (DHPS) is inhibited by the end-product of the pathway (lysine), lysine would be stimulated in a plant expressing an antisense DHPS gene.

12.11 Concluding remarks

Plants are fundamental to human welfare, providing food, feed, fibres, timber and phytochemicals. In the future, when fossil fuels have become depleted, plant biomass will also be required as feedstock for the production of organic chemicals and biofuels. The need to provide food for an increasing population, and biomass to replace fossil fuels, means that immense pressure will be put on agriculture (including forestry) to meet these needs. In addition, we may be on the verge of drastic global climatic changes that many scientists predict will result from global warming, as well as local changes in the environment occurring naturally or resulting from human intervention. For these reasons there is a pressing need to rapidly improve the quality of existing crop plants, develop new varieties and extend the range over which a crop plant can be grown.

Many useful agronomic traits of plants have been manipulated and gene banks are being established to preserve existing genetic resources for future use. The limitation of natural mating barriers can be overcome by protoplast fusion and somatic hybridisation in which nuclei of different genotypes are fused together in the hope that a useful hybrid genotype will arise. Genetic engineering can also be applied in a specific manner but it is limited to relatively simple genetic systems. Not only can the genotype of plants be manipulated to advantage, but also the genotype of microorganisms which form associations with plants.

For most crops, seeds are used for planting by farmers. F1 hybrid seeds with particular characteristics are used to ensure high yields, and biotechnology can be applied to their production in manipulation of male sterility and in production of homozygous diploid parental types (from diploidised haploid plants). For crops in which asexually propagated organs or plantlets are used for planting by farmers/planters, biotechnology can be applied to efficient production of clones of disease-free propagules.

The genetic manipulation of plants is not without controversy. There are fears amongst sections of the community about the ethics of genetic engineering, the release of genetically modified organisms into the environment and their safety as food for humans. Much of the development of high yielding pest resistant/disease resistant hybrids has been carried out by multinational agrochemical companies (the same companies who produce the pesticides, herbicides and fertilisers on which the high yield is dependent), giving rise to the accusation from users in developing countries of 'agricultural imperialism'. These issues are discussed in more detail in Chapter 17.

With the development of methods for large scale cultivation of plant cells in bioreactors, there is potential for production of phytochemicals in processes akin to microbial fermentation processes.

In all these ways, biotechnological processes involving the manipulation of plant cells are maximising human exploitation of plants. Some spectacular advances have already been made, and the potential for advances in the future is enormous.

12.12 Suggested further reading

1. Balcels, L., Swinburns, J. and Coupland, G. (1991). Transposons as tools for the isolation of plant genes. *Trends in Biotechnology*, **9**, 31–36.
2. BIOTOL Series (1992) — *Crop physiology.*
   — *Crop productivity.*
   — *In vitro cultivation of plant cells.*
   Butterworth-Heinemann, Oxford.
3 Breyne, P., Van Montagu, M. and Gheysen, G. (1993). Higher order organisation of plant chromatin and the regulation of gene expression. *Ag. Biotech. News and Information*, **5**, 183N–188N.
4. Fedoroff, N.V. (1989). *Maize transposable elements*. Cold Spring Harbor Laboratory, 375–411.
5. Gibson, S. and Somerville, C. (1993). Isolating plant genes. *Trends in Biotechnology*, **11**, 306–13.
6. Gierl, A. and Saedler, H. (1992). Plant transposable elements and gene tagging. *Plant Molecular Biology*, **19**, 39–49.
7. Grierson, D. and Covey, S. (1988). *Plant molecular biology* (II Edition). Blackie, London.
8. Grierson, D. (Ed.) (1992). *Plant genetic engineering*. Blackie, Glasgow.
9. Hooykaas, P.J.J. and Schilperoort, R.A. (1992). *Plant molecular biology*, **19**, 15–38.
10. Jones, M.G.K. and Lindsey, K. (1993). Plant biotechnology. In: Walker, J.M. and Gingold, E.B. (Eds.). *Molecular biology and biotechnology* (III Edition). The Royal Society of Chemistry, Cambridge, 123–154.
11. Lindsey, K. and Jones, M.G.K. (1990). *Plant biotechnology in agriculture*. Prentice Hall, New Jersey.
12. Ma, J.K.C. and Hein, M.B. (1995). Immunotherapeutic potential of antibodies produced in plants. *Trends in Biotechnology*, **13**, 522–7.

12.13 Responses to assignments

12.1 Seeds are convenient to use (storage and sowing) and are useful for crops (such as cereals) for which large numbers of plants are required. However, seeds are produced by sexual reproduction, and so are thus hybrid in genotype. Producing plants by propagation usually results in clones (all the plants so produced are of identical genotype to the parent plant), so that useful genotypes can be preserved intact. However, propagation is labour intensive (i.e., expensive) and cannot be applied on a vast scale (such as would be needed for cereals). For this reason it is limited to plants which are valuable in their own right and which are not required on a vast scale, or it is applied to producing plants of certain genotypes which can then be used in seed production.

12.2 When plants are sexually crossed to form seeds, a chromosomal gene will generally assort randomly, according to Mendelian law. Thus, if both parent plants are heterozygous for gene A (that is they are type Aa), then the progeny (seeds) will be type AA, Aa and aa. If Aa and/or aa genotype confers disadvantage to the plant (such as lower yield or decreased pest resistance), then a proportion of the seeds produced will be of the undesirable genotype. If, however, both parental types are homozygous (of AA genotype), then all progeny of the cross (seeds) will have the desirable AA genotype. In simple terms, sexual crosses using heterozygous parents can 'dilute out' desirable genes; sexual crosses using homozygous parents conserves desirable genes.

12.3 If a country cannot grow the particular plant and has to import it (or the phytochemical), then there may be economic, political and strategic benefit from producing the phytochemical from within. Even if a country produces the crop plant, economic benefits may still arise from producing the phytochemical in cell culture (the product from cell culture may be cheaper), and production in culture will be free from disruption of supply as a result of crop disease, crop pests or adverse environmental conditions (such as season, flood or drought). However, adoption of production in cell culture does require the necessary technical know-how and infrastructure and it may have adverse economic repercussions on farmers who previously grew the plant as a cash crop.

12.4 Important characteristics of crop plants (such as flavour, nutritional value, yield, resistance to pests, resistance to disease, resistance to drought) are coded for in the genes of the plant. If a crop such as maize is reduced to just a few major high-yielding varieties, then local varieties may fall into disuse and become extinct (and the genes they carry become extinct also). If in the future the genetic makeup of the major high yielding varieties needs to be expanded (to cope with problems caused by changing climatic conditions, new pests or new diseases), then natural genes from local varieties that could have been bred into these high yielding varieties to overcome the problem may not be available (they may have become extinct). For these reasons it is important to preserve germplasm of a wide variety of existing crop plants.

12.5 The structural components of plant cell walls are cellulose, hemicelluloses and pectins. Thus, to degrade cell walls a mixture of cellulases, hemicellulases and pectinases could be needed, depending on the actual composition of the walls of the cells being digested.

After digestion of the cell walls the wall-less protoplasts form spherical membrane bound structures. These can be easily recognised using a light microscope.

12.6 Suppose one type of protoplast was formed from plant cells which were resistant to herbicide/antibiotic X, and the other type of protoplast was formed from plant cells which were resistant to herbicide/antibiotic Y. If the cell mixture after fusion was plated on a medium containing X and Y, then only heterokaryocytes/synkaryocytes could grow, as only these cells have inherited genes for X resistance and Y resistance. If plant cells are not naturally resistant to such agents, then they could be made so by genetic engineering (see Section 12.10.1).

12.7 As a single plant cell can be regenerated into a whole plant, it is theoretically possible to regenerate many plants from a small piece of plant tissue; such regenerated plants should be clones (genetically identical) as each cell from which each plant was formed would be genetically identical. It is also easier to genetically engineer single cells than it is to genetically engineer whole plants, so genetic engineering of single cells followed by formation of intact plants makes a suitable technique for developing clones of transgenic plants.

12.8 The pre-mRNA contains functional regions (exons) interspersed with introns. During splicing, introns are excised (cut out) by endonucleases and the exons are ligated (joined up) by ligases, so that mRNA is formed. The mRNA has a particular reading frame (where codons of three nucleotides are translated into amino acids in sequential order). If an intron can be excised at two different points, then two different reading frames can result (mRNA formed can have two different reading frames) and thus two different proteins can be formed.

12.9 We saw in Chapter 4, how the genetic code comprises codons (triplets of bases) which are read in a given direction along DNA (the reading frame) and translated into specific amino acids. Inserting extra bases alters the reading frame, leading to different amino acids in the protein, and perhaps a truncated polypeptide (if a termination codon is created). Insertion of footprints during transposon excision has the same effects.

12.10 In order to alter the trait (phenotype) of a plant, the gene(s) responsible for the trait and the manner in which they are controlled must be understood. For genetic engineering purposes copies of the particular gene(s) must be available (for transfer into organisms to be engineered). For these reasons it is important to be able to identify and clone particular genes.

12.11 For a given gene to be expressed after insertion into the chloroplast, it would need to become inserted into the chloroplast genome, and at a site under control of a suitable promoter (which would initiate transcription of the inserted gene). If short sequences of DNA which are homologous to genome sequence at a given location on the chloroplast genome, are linked to each end of the given gene, then the likelihood of recombination is increased (identical DNA sequences swap over [recombine] with relatively high frequency during DNA replication). In other words, the (homologous) sequences linked to the ends of the given gene swap over with (homologous) sequences on the chloroplast genome during genome replication. As the given gene is interposed between these sequences, it thus becomes integrated into the chloroplast genome.

12.12 In order to produce seeds which have a required F1 hybrid phenotype and genotype, a plant of selected genotype which is to produce those seeds must be cross fertilised by pollen from a plant having different (selected) phenotype. If the seed-producing plant self-pollinates (fertilises itself with its own pollen) then the seeds produced from this self-fertilisation will not be the required hybrids. Thus, anthers from seed producing parental plants must be removed in order to prevent self-fertilisation.

12.13 Glyphosate would be expected to be non-toxic to animals as they do not possess the enzyme EPSP on which glyphosate acts. This is the case in practice.

12.14 Glyphosate competitively inhibits EPSP synthase; it binds to EPSP synthase enzyme which cannot then bind its substrates phosphoenolpyruvate and shikimate-3-phosphate. To become resistant to glyphosate action, the structure of EPSP would need to be altered so that glyphosate no longer binds to the enzyme. However, the EPSP molecule would still need to bind its substrates; any change in configuration should not affect this binding or the active site of the enzyme.

12.15 The 'restorer' line would need to have a nuclear gene which codes for a product which prevents ribonuclease formation, prevents its activity, or causes its degradation.

12.16 Infection of a plant with a mild strain of virus may in itself result in a reduction of cropyield. The inoculated mild virus may mutate or recombine with another virus present in the plant to produce a more virulent strain, and so cause an even greater reduction in yield. The mild virus could be transmitted to non-target crops in which it causes unwanted or more virulent infection.

12.17 The enzyme reverse transcriptase could be used to produce a DNA transcript of an RNA sequence.

12.18 If the gene coding for *B.t.* toxin was isolated and inserted into and expressed in plant cells, the plant cells would produce *B.t.* toxin. An insect larva ingesting such an engineered plant would succumb to the effects of the toxin. The toxin would be persistent on the plants (unlike the insecticide powder or spray it would not be washed off by rain).

12.19 By putting the *B.t.* toxin gene under the control of a wound-inducible promoter, the toxin gene would not be expressed in normal plants, lessening any resistance there might be amongst consumers to eating such engineered plants. Only upon wounding (attack by insect larvae) would toxin be produced in a plant.

12.20 Chitin is not a component of the human/higher animal body; it is found not only in fungal cell walls but in exoskeletons of insects and crustacea. As chitin is not a component of the intestinal tract of higher animals/humans, chitinase should have no enzymatic effect on it and thus transgenic plants expressing chitinase should be suitable as food and animal feed.

12.21 If a change could be introduced in the ADP-glucose pyrophosphorylase gene such that the enzyme coded for had an altered allosteric site (which is less sensitive to feed-back inhibition by metabolites such as phosphate), then the biosynthesis of starch would be stimulated.

12.14 Responses to self assessment exercises

12.1 Gene expression is promoted as follows:

(ii) enhancers increase transcription rates;
(iii) promoters initiate transcription;
(iv) demethylation of nucleotides enhances transcription;
(v) polyadenylation stabilises mRNA;
(vi) pre-mRNA splicing is necessary to remove introns from mRNA prior to translation,
(vii) mRNA capping prevents degradation of mRNA by exonucleases;
(viii) 'CAAT' boxes are required for the initiation of transcription;
(ix) AUG codons are necessary for initiation of translation.

(i), (x) and (xi) do not promote gene expression; (i) decreases gene expression, and (x) and (xi) are related to the behaviour of proteins after their synthesis (and are not involved in regulating that synthesis).

12.2 The correct answers are as follows:

(i) True.
(ii) False — pollen from a cms plant is infertile and would not fertilise an Rf plant; cms could be overcome by pollinating a cms plant with pollen from a Rf plant without cms.
(iii) False — A-line crossed with B-line would itself be a hybrid of A and B and not the true A-line; the statement would be true if 'A-line plants' was replaced by 'parental A-line plants' — that is, the B-line is used to pollinate the parents of the A-line, not the A-line itself.
(iv) False — a C-line plant with dominant nuclear Rf genes would still produce infertile pollen grains if it had cms mitochondrial genes; the restorer line must be male fertile (non-cms).

12.3 Of use in a vector could be (i) 'overdrive', (ii) border regions and (vii) *vir* genes, as all of these sequences promote or are essential for T-DNA transfer to plant cells.

Of little use would be (vii) as this gene does not influence T-DNA transfer. The same is true for (iii), unless you have assumed that it contains an 'overdrive' region (see above).

It would be undesirable to have (iv) or (v) in the vector, as these genes would induce tumour formation in the transformed plant. If you have listed (iii) as undesirable because you have assumed it contains (iv) and (v), then you are correct.

12.4 The correct completion is (ii).

12.5 For the answer to this question see the Response to assignment 12.11.

12.6 The correct answer is — all of them; it would be useful to micropropagate plants in all the situations listed.

12.7 The correct completion is (iii).

12.8 The correct completion is (iv). Antisense stearoyl-ACP desaturase gene would prevent synthesis and activity of stearoyl-ACP desaturase, thus preventing conversion of stearoyl-ACP to oleoyl-ACP and causing accumulation of stearate.

12.9 The components which would be of benefit in a DNA construct to confer chilling tolerance would be (i), (iv) and (v); the glycerol-3-phosphate acyltransferase gene is required to promote fatty acid unsaturation, the CaMV 35S promoter is required to ensure constitutive expression throughout the plant, and chloroplast-specific transit peptide is required to direct the enzyme protein to the chloroplast.

12.10 The correct answers are as follows:

(i) False — an antisense gene to EPSP synthase would depress synthesis of EPSP synthase and thus lower EPSP synthase activity — in other words it would have the same effect as glyphosate and kill the host plant.

(ii) True.

(iii) True — except that TMV is an RNA virus for which the antisense mRNA approach is not highly effective.

(iv) False — antisense-phytoalexin genes would prevent phytoalexin synthesis and exacerbate fungal infection.

(v) True.

(vi) False — an antisense-DHPS gene would depress synthesis and activity of DHPS which would depress synthesis of lysine.

13 Biotechnology in livestock production

Chapter outline

13.1 Introduction

The production of food from animals is expensive and exhausting. It is certainly less efficient than from plants, needing far more energy, nutrients and land to produce a similar amount of usable protein (Table 13.1). The hoofed animals that are farmed most commonly — cattle, sheep, goats and pigs — also require more intensive management than do plants. Their hooves degrade vegetation and soil, and their waste products can cause significant environ-mental and health problems. Despite these disadvantages, we continue to farm animals on a large scale for food, hides, fibre and by-products.

Table 13.1 Efficiency of land use in protein production systems

Protein source	kg protein per hectare per year
Algal ponds	30,000
Farmed fish	1,000
Potatoes	800
Rice	600
Peanuts	450
Wheat	360
Milk	120
Meat	80

Animal protein is an excellent source of dietary amino acids essential for human nutrition, but in addition to meat and hides we are able to continually harvest milk and fibre from live animals, and of course use them for labour. The contribution that animal protein products make to the human diet varies from region to region (Fig. 13.1), due to availability, cost, and cultural/ religious preference. There are also strong cultural reasons for continued animal produc-tion. Quite simply, humans like to work with animals. From the time our ancestors first dome-sticated them many thousands of years ago they have been our constant companions.

From that time, too, the animals themselves started to change. Captive males were no longer free to mate with females of their choosing. Natural selection ceased and the tamed animals were bred within a restricted gene pool. With the passage of time and continued isolation from their wild relatives, domestic animals started to display different characteristics. Some would have produced more or better milk than others, some would have grown faster or been more fecund, some would have been more vigorous, and some just looked different. Our ancestors undoubtedly noticed these variations and mated animals with similar desirable traits, but they did so without understanding the basis of heredity for we have only gained that knowledge within the last 90 years. Restricted breeding and selective mating gradually led to the many diverse breeds of livestock we have today, although most breeds of the dominant domestic cattle species, *Bos taurus*, arose in just the last couple of centuries. Herds of the ancestral species, *Bos primigenius*, still lived in Europe in the 18th century. Animal improvement is no longer a matter of interest — it is a matter of necessity. The pressures of continued population growth force us to improve the way in which we use our shrinking resources while improving

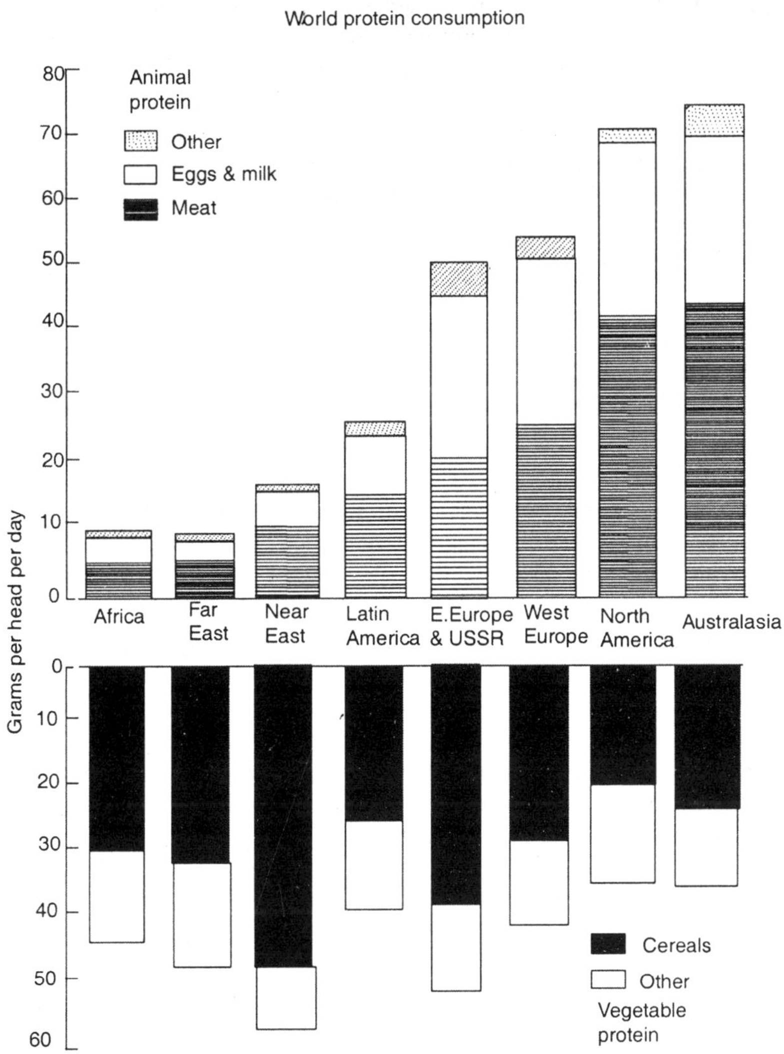

Fig.13.1 *Protein consumption in various world regions.*

the quality and quantity of food and fibre. We must use less natural resources and fewer chemicals to provide more and better products. Biotechnology gives us powerful tools that help us to do this. In this next 50 years, worldwide animal production will change radically as we learn to use these tools.

13.1.1 Animals are not plants

All living production systems have four intrinsic limitations: reproduction, genetics, health and nutrition. Animal production has to deal with constraints that do not apply to plant production, so the opportunities for improvement with biotechnology are different. Nutrition and environmental stress are common to both, as is disease; in fact animal production is totally dependent on the nutrition and health of pastures and fodder crops. So in a practical sense these production systems are interrelated, with plants dependent on inorganic nutrients to provide the organic intake for animals. The major practical difference between plant and animal production systems is in their reproduction. Plants produce large numbers of female gametes (ova) which develop into propagules ideal for farmers (i.e., seeds), whereas female mammals produce just a few gametes (ova or eggs). Indeed all common livestock species except pigs can produce just one fertilised egg each year. In addition, crop plants complete their life cycle within one year whereas livestocks require 3–4 years from a conception to maturity.

Assignment 13.1
Apart from their ability to produce multiple ova, what other characteristic do plants have that make reproduction on a large scale easier than that of animals?

The low fecundity of female animals has profound implications. In seeking to improve the intrinsic characteristics of our agricultural plants and animals, we must first identify, then manipulate and propagate desirable genetic traits. To identify the genetic basis of a desirable trait we set up matings between individuals with different phenotypes for that trait and determine its heritability by measurement of a large number of progeny. With crop plants this is not difficult since we can get thousands of F1 progeny within 12 months at little cost. It is even possible to identify minor quantitative trait loci (QTLs) without too much difficulty. With animals the task is impossible if we are restricted to using natural reproduction.

Manipulating genetics, too, is much easier with plants. In many cases we can produce fertile progeny from matings between different plant species so that genes from a wild relative can be introduced into a cultivated crop plant, and genes can be (and are) shuffled between different cultivated species by suitable crosses and back selection. The same cannot be done with mammals. The infinitely greater number of individuals of plant species also provides a much greater source of genetic variation to use in breeding selection.

Finally, of course, when a desirable plant has been identified or bred, large amounts of seed can be produced quickly and easily so the new variety can be propagated and used in agriculture. Natural reproduction does not allow us to do this with animals.

Quite simply, the most important role of biotechnology in animal production is to make the reproduction and genetics of livestock more like that of plants — to increase reproductive rate by increasing male and female fecundity, and to identify and use genes that determine desirable traits.

At present the application of biotechnology to animal production is limited because we do not yet know enough about animal development and function to allow us to make sensible and effective modifications. We are presently preparing the tools of biotechnology for future use. In this chapter we examine what those tools are and how they will be applied. Not all of the technologies will or should be applied to all animal industries. Some are suited only to intensive production, some are applicable only to certain species, while some may never be of economic value. Priorities for biotechnology research and application must be set by the needs of the production system. Technologies must also be developed in a way that allow them to be used in a rational production system. There is a world of difference between the controlled environment of a scientist's laboratory and the real world of farming.

We are concerned here with animal production, but we should point out that biotechnology may lead to the elimination of livestock as a major food source. As we learn how to manipulate plant functions (Chapter 12) there is no doubt that, in the near future, plants will be made to produce all of our essential foodstuffs.

13.2 Manipulation of reproduction in animals

Reproduction has two critical roles in production systems: it must be at a rate that ensures replenishment of stock is greater than its use, and reproduction has been, until the advent of genetic engineering, the only means by which the genetic quality of stock can be improved. Removal of the natural constraints of reproductive behaviour and fecundity has been responsible for large gains in productivity, most notably for intensively managed animals such as dairy cattle and domestic pigs.

The goals of reproductive technologies are to increase the number of progeny that a particular, selected male and a particular female can produce. All male mammals produce millions of gametes (sperm) daily and so in theory can inseminate many females at regular intervals to produce large numbers of offspring. Artificial insemination, the first reproductive technology and the first animal biotechnology, takes advantage of the male's excess capacity.

Female mammals, on the other hand, produce only limited numbers of gametes (eggs) — in all common livestock species except the pig just one at a time — at approximately monthly intervals. In addition, the female must carry her offspring within her reproductive tract while it develops from a fertilised egg to a stage where it can survive independently. She then continues to provide nourishment for her young through suckling. As a result, ruminant females can usually have no more than one offspring in a year. The strategy now used to overcome this limitation is multiple ovulation with embryo transfer (MOET); more advanced technologies are currently being developed.

We will not try to cover the details of reproductive technologies but will present the principles of artificial breeding systems and how they are being used as the basis for further applications of biotechnology.

13.2.1 Artificial insemination

Technologies affecting male reproduction aim to increase the number of females that can be inseminated by a male by decreasing the amount of semen needed to produce a pregnancy and increasing the length of time for which semen can be stored. The breeder must then take the place of the sire by inseminating the female artificially. Since fertilisation and pregnancy can result only if insemination occurs when the female ovulates and is receptive, the breeder must replace nature further by ensuring that females in a herd ovulate together at a convenient time. The economic value of artificial insemination is reduced if the breeder must wait for females to ovulate and then inseminate each one separately as she enters oestrous.

(a) Semen collection and storage

Artificial Insemination (AI) has been the single most effective factor in increasing the productivity of cattle. The semen ejaculate is collected, diluted ('extended'), and examined by microscope to determine the number and motility of sperm. The extended semen is dispensed into small plastic 'straws' that are plugged at each end. If diluted properly, it can be used fresh within a few days. For most applications a cryoprotectant is added to preserve the semen and the straws kept in liquid nitrogen for long term storage and transport.

Commercial producers now extend bull semen up to 100-fold and some 10 million motile sperm in 0.2 ml is used for insemination. For cattle, AI is done on standing animals without anaesthetic using a procedure known as "rectal palpation". Each ejaculate can be used, in theory, to inseminate as many as 500 cows. Prolonged storage of sperm in liquid nitrogen allows a single male to inseminate many thousands of females, even if the male is in a different country or has died. The genetic influence of a single superior male can be profound.

(b) Oestrous detection and synchronisation

Females are inseminated after ovulation, which is indicated by behavioural oestrous (a Greek word referring to the gadfly, since a mare in oestrous behaves as though stung). In many breeds of farm animals and in many management situations it is very difficult to detect oestrous, since it may last only a few hours and occurs most commonly at night. The alternative is to induce females to ovulate at a set time. If all females are induced similarly this has the added advantage of ensuring that all females can be inseminated at once so that management is simplified. In practice, it is never possible to achieve total synchrony of oestrous but some 80% of females usually respond as expected.

With ruminants, ovulation in females can be induced by administration of progesterone and/or prostaglandin (hormones which regulate the ovulation cycle).

Assignment 13.2

In the future, progesterone and prostaglandin hormones may be required in large quantities for use in ovulation control in intensively farmed animals. How might biotechnology assist in the production of such hormones?

13.2.2 Embryo transfer

Where AI is an obvious way of increasing the number of progeny that a selected male can sire, the solution to the problem of limited female fecundity is not so simple. We must change the animal's function, increasing the number of mature eggs that a selected female produces. After fertilisation those eggs must be implanted into less important foster mothers (recipients), one for each egg, since each ruminant female ideally carries just one pregnancy. The total cost of this embryo transfer procedure is very high, including drugs, labour and recipient females. Clearly, it should only be used for animals of high genetic or economic value.

The first pregnancy by embryo transfer was achieved using rabbits in 1890. By the 1930s it had been used on sheep and goats, while methods were developed for pigs and cattle by the early 1950s.

(a) Multiple ovulation with embryo transfer (MOET)

Just as studies of human contraception have led to oestrous control in livestock, so treatment of infertile women has led to our ability to 'superovulate' livestock.

In the reproductive cycle of a non-pregnant female one ovarian follicle matures and ruptures, releasing its fertile egg, every 21 days (cow, horse) or 16 days (sheep, goat). In fact some 20 or so ovarian follicles reach an early stage of maturity where oocyte maturation and ovulation can be triggered by a suitable dose of gonadotrophic hormone. In the normal course of events only one is so triggered by circulating hormones, but by increasing hormone dosage at the appropriate time a number of follicles can be induced to 'ripen' and ovulate. In practice, well managed domestic cattle yield an average of about eight eggs when superovulated, but the number can range from none to 60 depending on nutrition, health, environment, breed and previous reproductive experience.

Donor females are frequently injected with prostaglandin F2a (PGF2a) to induce a synchronised oestrous before treatment is started. Starting ten days after oestrous (i.e., during the luteal phase) they are injected with FSH (follicle stimulating hormone) over four days, followed by PGF2a to induce oestrous. Oestrous should occur within two days and the superovulated donors are then mated within 24 hours. Since the purpose of MOET is to increase the number of genetically superior progeny, the donor is usually mated by artificial insemination.

The fertilised embryos are recovered some 6–8 days after insemination for cattle (6 days for sheep and goats) when they have developed to the stage of a compacted cell mass (morula) or early blastocyst (Fig. 13.2) and are still in the oviduct. In cattle it is common to recover the embryos non-surgically by inserting a Foley catheter into the oviduct, using a procedure similar to AI. A saline solution is allowed to drain into the sealed oviduct and the embryos are flushed out through the catheter into a collection bottle. For sheep, with their smaller reproductive tract, surgical procedures are used (again similar to AI) to expose the oviduct and flush out the embryos by syringe with a small volume of saline.

After standing for some minutes to allow the embryos to settle, most of the saline solution is decanted. The small volume remaining is transferred to a petri dish and examined microscopically.

When the embryos have been identified and recovered they are immediately transferred into synchronised recipients, using a procedure identical to AI (except that the receipients are 6–8 days into the cycle to match the stage of development of the embryos). Alternatively, they can be frozen for indefinite storage and transport, or they can be subjected to manipulation as described below.

With good practice pregnancy rates of 50–60% can be achieved in cattle from transferred embryos, so that one superovulated female will give rise to five pregnancies on average.

(b) Embryo splitting

The most effective means of increasing pregnancy output rates is to produce identical twins by splitting embryos into two halves. In principle this will double the number of progeny from embryo transfer. The twins are also an excellent resource for genetic studies, providing a powerful approach to separating the effects of heredity and environment.

In recent years the procedure of embryo splitting has been refined. What was recently an experimental laboratory technique is now a routine production aid, requiring basic micromanipulation equipment and minimal training. It takes just a couple of minutes and is commonly used on farm during embryo transfer.

Embryos (usually at the blastocyst stage, Fig. 13.2) are transferred for a few minutes into standard cell culture medium (Chapter 8) that contains two additional components: hypertonic sucrose and BSA (bovine serum albumin). The medium penetrates the zona pellucida (the translucent shell membrane of the embryo) where its high osmotic strength causes the cells to contract. The BSA penetrates and adheres to the zona pellucida, giving it a net negative charge.

The embryo is then transferred into standard culture medium in a plastic Petri dish where it sinks to the bottom (due to the higher density of the sucrose medium within it) and sticks to it (due to electrostatic interaction between the negative charge on the embryo outer membrane and the positive charge of the plastic dish). The dish is placed on the stage of an inverted microscope, and a fine surgical blade controlled by a micromanipulator is used to orient the embryo and bisect it. It has been found that when cells have been shrunk by exposure to hypertonic sucrose the blade slices between them with minimal damage to individual cells.

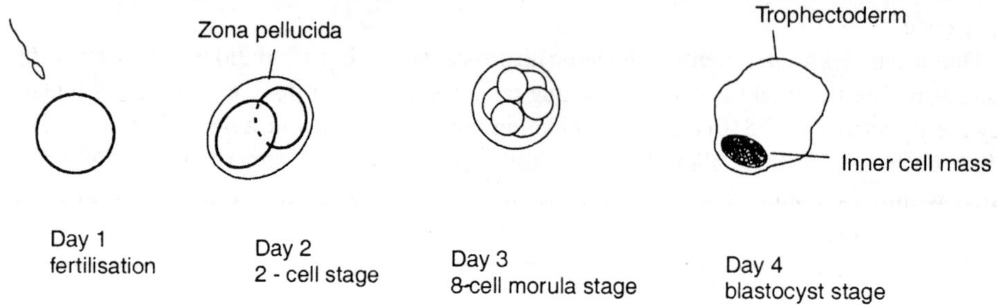

Fig. 13.2 *The early stages of embryo development.*

At the blastocyst stage, the embryo comprises two different cell types (Fig.13.2): the trophectoderm, a single layer of trophoblast cells lining the inside of the zona (that will become the foetal membranes of the placenta) surrounding a fluid-filled cavity (the blastocoel), and at one 'pole' of the blastocyst a mass of cells called the 'inner cell mass' (ICM) that will develop into the foetus. When an embryo is split to produce twin pregnancies it is essential that the ICM is split into roughly equal halves.

The half (or demi-) embryos do not need to be enclosed in a zona pellucida for successful implantation and pregnancy to occur. After splitting, the hemispherical cell masses re-form spheres, and they are simply transferred into the oviducts of synchronised recipients as for normal embryo transfer. The pregnancy rate with a demi-embryo is about 5–10% less than with an intact embryo; this means that the average number of progeny from a MOET procedure can be increased from five to nine.

In principle the splitting approach could be extended to give three or four identical progeny. In fact splitting the embryo into three appears to be a viable option for a careful technician, and four identical lambs have been produced by symmetrical splitting of a hatched (elongated) blastocyst and transferring the sections into recipients that have been synchronised to an earlier stage of the cycle. Implantation of an embryo in the wall of the uterus requires the uterus of the recipient female to know that an embryo is waiting. The embryo regulates this by sending chemical signals from the trophoblast, and they get stronger as the embryo grows. Transferring a section of embryo that contains too little trophoblast will not induce the uterus to prepare for implantation and the embryo will be aborted. Transferring too little inner cell mass will also fail, but we have no clear idea of how many cells of the ICM are needed for a viable pregnancy.

(c) Embryo biopsy

Development of a routine embryo splitting procedure has also allowed routine embryo biopsy — the removal of a small number of cells for genetic analysis. In this case a small section of trophoblast is removed. Cells of the incipient foetus, the ICM, are not needed since all cells in the embryo contain the same genetic information. In fact it is common to combine splitting with biopsy so that identical twins of known genotype can be produced.

At present the most common reason for taking a biopsy is to determine embryo sex, but it is becoming increasingly useful in breeding programs for detection of genetic diseases, preventing the wasteful transfer of embryos that are affected or are carriers. In the future we will see increasing applications to analysis of production traits.

> ***Self assessment exercise 13.1***
>
> What techniques would be used to screen the small numbers of cells taken at biopsy for the presence of genes coding for genetic diseases?

(d) Embryo sexing

Progeny sex is unique as a production trait since the knowledge is only useful before embryo implantation. The principle used to determine the sex of embryos is quite simple: since it is the Y chromosome that makes males different from females (refer above), the presence of DNA sequences unique to the Y chromosome in cells of an embryo biopsy tells us that the embryo is male; if no Y chromosomal DNA is present the embryo is female.

Because the assay is done on a very small sample of cells the assay we use must be very sensitive — ideally, able to detect a single Y chromosome from a single cell (although it is easier to remove 5–10 cells from the trophoblast). This can be done quite easily by using PCR (polymerase chain reaction, Chapter 11) to amplify a DNA sequence that is present only on the Y chromosome. In practice, the target sequence is usually a repeated motif; since each cell contains multiple copies the sensitivity of the assay is increased.

One word of caution is necessary: PCR amplification can detect single molecules of a DNA sequence. Contamination of samples is therefore a major problem. The greatest sources of contaminating male DNA are with airborne material present in every livestock breeding centre, and with sperm heads embedded in the zona of the embryo that may be recovered with the biopsy.

This technology is now in common use for cattle, sheep and goats — in fact livestock embryo sexing was the first commercial implementation of PCR.

13.2.3 *In vitro fertilisation (IVF)*

Transfer of embryos into prepared recipients is no more difficult or expensive than artificial insemination. The only significant difference is that more care is needed in thawing frozen embryos than frozen sperm. Yet the genetic advantages of embryo transfer are far greater than AI since complete diploid cells are transferred, not just a haploid gamete containing half a genome. The embryo develops in a foster mother that can be a poor quality animal since she makes no genetic contribution to the offspring she carries. The producer is not restricted to breeding from his existing high quality females.

Embryo transfer is not used widely because of its cost, technical difficulty and the limited supply of embryos available from superovulated donors. These limitations would be removed if we could fertilise *in vitro* the thousands of oocytes that are present in a female's ovaries. In doing so, domestic animals would indeed become similar to plants in the fecundity of both sexes. Breeding programmes could select for female genetics as easily as for male genetics. The genetic influence of a male would be increased further since the amount of semen required for fertilisation *in vitro* is a fraction of the amount needed for AI.

IVF is generally quite successful, resulting in about 70–80% of fertilised eggs. The practical difficulties arise in sourcing eggs for fertilisation and in the development of fertilised zygotes to term.

13.2.4 Embryo cloning

The procedures discussed so far increase the number of progeny that selected males and females can produce. The genetics of the progeny is not defined or known at conception since each parent contributes just half its genetic information to its progeny — which half is a matter of chance, resulting from recombination during meiosis.

It would be valuable to be able to increase the numbers of one particular embryo that has been identified as having desirable characteristics. In many plant species this is done by vegetative propagation. It is even possible to regenerate many plants from the somatic tissues of a single plant through *in vitro* culture techniques that are now used widely in commercial production. These techniques have been discussed in Chapter 12. Unfortunately, we cannot use these technologies with livestock animals, but we can produce quite a lot of animals from the cells of one embryo. This is referred to as embryo cloning. In principle, it results in a number of genetically identical animals, but we must still be cautious about possible effects of genomic imprinting and possible genetic damage during cell culture. We should also be aware of experience with the vegetative propagation of plants which tells us that genetic variation occurs in somatic cells. Not all plants arising from the culture of somatic cells of one plant are identical, a phenomenon known as 'somaclonal variation'.

There remains the problem, too, of knowing which embryos are worth cloning. There is no value in multiplying an individual that is not genetically superior. This problem can be addressed in two ways. The first is the only one available at present, but over the next two decades the second will become more viable.

1. The majority of embryos from a clone can be stored frozen and a small number transferred to recipients and allowed to grow to production maturity. Their performance can be evaluated in different environments and from this it can be determined if the remaining embryos in the clone should be put into production. This approach is slow and expensive. It is unlikely that production gains would be great enough to offset the time required for clonal testing, in comparison with alternative production technologies. It is not likely that any one clone will suit all management conditions, so many clones would need to be produced and tested. In addition, continued breeding selection would be expected to produce animals of higher merit during the testing programme. Nevertheless, this approach will find application with particularly valuable or rare genotypes.
2. When markers have been identified for all significant production traits under a range of environments, cloned embryos can be evaluated directly by DNA genotyping. Their genetic merit can be scored directly, and their phenotypic performance inferred from it. In other words, we will know in advance how well a particular embryo will perform under particular conditions. This detailed level of knowledge is a long time in the future — at least 20 years — but partial information will be used for this purpose over the next few years.

> ***Assignment 13.3***
>
> What techniques would be used to evaluate the genotype of a given clone of embryos?

A clear understanding of the molecular genetics of development and function will give us a sound basis for selecting embryos to be cloned, leading to broad adoption of cloned production animals. The technology of cloning is presently being developed in two ways: nuclear transfer and embryonic stem cells.

(a) Nuclear transfer

Most or all of the cells (blastomeres) in the early cleavage stage embryo of domestic ruminants are totipotent, i.e., the nucleus of each blastomere has all of the genetic potential required to produce a complete, normal animal. Differentiation of individual cells has begun by the morula stage (about day five in cattle) and has probably occurred to some extent in the late cleavage stages (64 cells) where the cells that are completely enclosed within the spherical cell mass are different from those on the surface (it seems that the enclosed cells are destined to become the inner cell mass, the embryo proper). The nuclei of differentiated cells are no longer capable of expressing all genes due to differential shut-down or inactivation of certain genes, a process that appears to be irreversible.

Up to the 32-cell stage, however, it appears that virtually all cells of the ruminant embryo remain totipotent. Nuclear transfer involves removing a single blastomere from a cleavage-stage embryo with a fine glass micropipette. The blastomere is then placed under the outer membrane of an unfertilised, mature oocyte whose haploid nucleus has been removed (with a micropipette) or destroyed (by ultraviolet light). The blastomere is compressed between the oocyte cell membrane and the inner surface of the outer membrane. Its membrane is then induced to fuse with the oocyte membrane. This is usually done by electrofusion, using special apparatus to control the shape and duration of electric field pulses and to align the cells between microelectrodes on a microscope slide for maximum efficiency.

Following fusion, the diploid blastomere nucleus is 'adopted' by the oocyte, ideally resulting in a normal zygote. The zygote is cultured *in vitro* to blastocyst stage, then transferred into a synchronised recipient. In principle the same can be done with every blastomere from a 32-cell embryo, giving 32 cloned individuals that are genetically identical. The procedure could even be repeated with the cloned embryos at cleavage stage, so that each blastomere from the original embryo could give rise to its own clone of 32 embryos, and so on.

In practice, though, we still have much to learn. The best reported result is eight progeny from a single embryo; more commonly only two or three are recovered, a result that offers no advantage (and is considerably more expensive and difficult) than embryo splitting. Second round cloning, where blastomeres of a cloned embryo are transferred into recipient oocytes, has been unsuccessful; the best published result is just two calves. A disturbing feature of calves produced by nuclear transfer is that many are oversize, requiring delivery by Caesarian section. The reason for this is unknown, and contrasts with the normal size of progeny produced from cloned embryos that are produced by splitting.

The future success of this technology clearly depends on a large number of factors that are presently unknown. Perhaps the most critical is the precise stage in the cell cycle of the rapidly dividing blastomeres that is best suited to occupying the oocyte cytoplasm. The precise stage of oocyte maturation is also likely to be important. Difficulties with embryo culture have already been mentioned and the physical manipulations must be done with great care. If the technology can be improved it may become useful, but unless it can be partially automated it will remain expensive and limited in application.

The approach would be viable if nuclei of cells from mature animals could be used, since this would allow cloning of animals with known performance characteristics and would provide an unlimited supply of cells. To date we do not know of any cells in the mature animal that are totipotent — whose nuclei have the complete potential for embryonic differentiation and development. It may be that the primordial germ cells of the testis are suitable, but this seems unlikely given the phenomenon of genomic imprinting. However, these cells have yet to be tested as nuclear donors.

(b) Embryonic stem (ES) cells

Scientists have been unable to clone mice by nuclear transfer, in contrast to ruminant animals.

The embryonic developmental programme of mice is greatly accelerated and it seems that as early as the first embryonic division the cells have begun their irreversible process of differentiation.

However, an alternative technology has been used with great success for cloning mice. When a mouse blastocyst is placed in appropriate culture conditions, the cells of the inner cell mass (those contributing to the future foetus) sometimes continue dividing and remain in an undifferentiated, totipotent state as 'embryonic stem' (ES) cells. It has been found that a peptide growth factor known as LIF (Leukaemia Inhibitory Factor) markedly increases the success of establishing and maintaining ES cell lines. Even so, few strains of mice and few blastocysts have been successful. Most experimental work is done with a few standard lines that are frozen and shipped around the world.

ES cells can be used for cloning in two ways: (i) they can donate their nuclei to enucleated oocytes for cloning by nuclear transfer (an approach that has not yet been tested since nuclear transfer does not work with mice); (ii) a small number of ES cells can be injected into the blastocoel space of a blastocyst, where they intermingle with the inner cell mass cells to create a chimaera. Ideally, all tissues of the mature chimaeric animal would be a mixture of the two cell genotypes and the results are often seen as patches of different coloured fur. If the germ cells are also chimaeric, a proportion of progeny will result from ES cells. Crossing male and female chimaeras allows selection of a homozygous strain of mice derived solely from the ES cells.

In many cases the ES cells do not contribute to the germ line of the mature chimaera. Indeed, ES cell lines differ markedly in their ability to populate the germ line of chimaeras. The sex of the cell line and of the recipient blastocysts is also an important consideration if the chimaeras are to be fertile.

Much research has been carried out on attempts to establish ES cells lines for domestic livestock species. No verifiable reports of success have yet been published, but we can be certain that such cell lines will become available very soon. They could provide an excellent source of genetic material for cloning by nuclear transfer, but the source and value of those genes will remain a major problem. Experience with mice has shown a great disparity in the ability of blastocysts from different animals to generate ES cell lines. The same will probably apply to livestock, but will be complicated further by the lack of inbred strains which might allow us to identify the basis of this variability. Cloning is only of value when applied to animals of exceptional genetic merit. To be useful, a cloning technology must work with any (or most) individuals. ES cell lines do, however, provide significant advantages in that their abundance allows detailed genotyping. Alleles of genes and linked markers for any desired trait can be determined so that the lines can be fully characterised.

It will be many years before this technology becomes a valuable production tool. Again, it will be introduced gradually in defined applications before its parameters have been completely defined. With time and knowledge the present problems will be resolved, with dramatic implications for farm management and product quality.

Before leaving this subject, it must be noted that the ES cell lines will undoubtedly be of great value in a related area — the production of defined transgenic animals. Since the cells are plentiful and are maintained in culture, they can be transformed with DNA very easily, and the resultant transformants can be fully characterised *in vitro*. Furthermore, their genome can be modified with precision by gene targeting using *in vitro* transformation and selection. This may be the area of greatest impact of livestock ES cells since, by definition, it produces cells of exceptional genetic merit that can be propagated by cloning.

13.2.5 Natural reproduction

Biotechnology is likely to also intervene to some extent in reproduction by natural mating, in those extensive industries in which economic considerations prevent artificial breeding. The clearest example is in modifying the progeny sex ratio. The results of continuing work on sperm and embryo antigens will allow females to be immunised against 'male' sperm or male embryos, giving rise to predominantly female progeny in broadacre farming.

In many parts of the world, harsh conditions at certain times of the year make it desirable to prevent pregnancies. Reversible, short-term immuno-contraception will aid in making the best use of limited resources. The techniques of immuno-contraception have been described in Chapter 11. As an aside, a similar approach that results in long-term sterility will be used to control feral animal populations.

Assignment 13.4

What would be the advantage, in extensive farming, of controlling feral animal populations?

13.3 Characterising and isolating animal genes

Best use of available genetic variation requires that we gain a detailed understanding of genes that are responsible for economic traits. The availability of markers of variants of these genes will allow us to use these markers in breeding selection instead of inferring genotype from the phenotype of animals and their progeny. This is true for plant breeding just as much as for animals. Breeding selection based on gene markers will allow genetic improvement much more accurately and rapidly.

This section deals with the technologies and applications used in developing markers for breeding selection. The first step is to identify DNA sequences associated with economic trait loci (either the genes themselves or tightly linked markers). These must then be incorporated into three maps: (i) a physical map of the chromosomes, in which the DNA sequences are assigned to specific sites on specific chromosomes; (ii) a linkage map, in which the linkage (coinheritance) of different genes and markers is assigned (linked genes and markers are in close proximity on the same chromosome); (iii) a genetic map, in which the inheritance of economic traits is correlated with the inheritance of genes and markers. The first two are essential but are merely steps along the way to a high-resolution genetic map. This is the map that will allow us to select on the basis of DNA markers.

13.3.1 Homologues in humans

How have molecular biologists identified genes that are responsible for genetic diseases of livestock? In most cases they have used the vast body of information that has been collected for human heritable diseases (Chapter 11), stored in the OMIM database that is accessible through computer networks (OMIM is an acronym for On-line database of Mendelian Inheritance in Man).

The genomes of all mammals are remarkably conserved in both their genetic content and their arrangement of genes, a reflection of their brief evolutionary history. We find that if a variant of a gene causes a heritable disease in one mammalian species, the homologous gene usually exists in similar variant forms in other species causing a similar disease. By far the most extensive information exists for humans. By searching the OMIM database we can usually find reference to defects similar to those affecting livestock. Increasingly, we find that the homologous human gene has been cloned and its DNA characterised.

> ***Assignment 13.5***
>
> How does the fact that a particular gene in humans has been cloned and sequenced help in locating (mapping) the homologous gene in an animal cell?

This is known as the 'candidate gene' approach, where we recover a gene that is a probable candidate for causing the trait of interest.This was the method used for characterising citrullinaemia, Pompe's disease and porcine stress syndrome (mentioned below), in addition to a number of other genetic diseases of animals.

A similar approach is frequently used to isolate genes in livestock species that are responsible for production traits. Bovine genes encoding milk proteins (caseins, α-lactalbumin, β-lactoglobulin) were isolated by means of their similarity with homologous rat genes; genes encoding growth hormone, many growth factors and hormone receptors as well as genes of the major histocompatibility complex have been isolated from livestock species by homology with their human counterparts. The Human Genome Project is proving to be of immense value to livestock producers.

13.3.2 Linked markers and linkage mapping

Not all genetic diseases have an unambiguous, identified homologue in humans or other mammals. In such cases one could use an indirect approach to identify polymorphic DNA markers that are linked to variants of the unknown gene. The defective genes that are responsible for the human diseases cystic fibrosis and Huntington's chorea, have only recently been identified. However, genetic screening has been available for quite some time, based on the diagnosis of DNA polymorphisms that were known to be tightly linked to the defective gene variants. Similar linked markers are starting to be identified for uncharacterised genetic traits of livestock.

Cattle have 30 pairs of chromosomes, each comprising an average of about 10^8 (100 million) base pairs of DNA. Genes and other DNA sequences that are a part of the same chromosome are said to be linked or coinherited. If we identify a specific variant (allele) of a DNA sequence that is known to be on the same chromosome as a defective gene allele, this should tell us that the defective gene is also present since the variant sequence and the defective gene are linked and hence will be coinherited. During meiosis, however, DNA is exchanged between homologous chromosome pairs by recombination. DNA sequences that are on the same chromosome as a gene variant in the somatic cells of a parent may be separated from that allele in a gamete, and hence in progeny that result from fertilisation with that gamete.

The nearer to the gene a particular sequence is, the less likely it is to be separated from it during meiosis and the more likely it is to be coinherited (Chapter 5). Linkage is defined by this coinheritance; tight linkage is used to describe DNA sequences or genes that recombine only very rarely, and hence are physically close to each other in the DNA sequence of the chromosome. In the absence of any knowledge of a disease-causing gene, we can use an identified polymorphism in an 'anonymous' DNA sequence that is tightly linked to the gene variant to tell us whether that variant is present. An anonymous DNA sequence is a sequence of unknown or no function; a polymorphism is a heritable variant of a gene or anonymous DNA sequence that occurs at a constant level in the population (approximately 1 in 300 bases in mammalian DNA is polymorphic).

Not more than 20% of the genome encodes genes. The remaining 80% is not subject to strong selective pressure and so accumulates a high number of mutations and DNA sequence variations — polymorphisms. The first type of anonymous DNA polymorphism that was identified is known as an RFLP (Restriction Fragment Length Polymorphism) (Chapter 11). These polymorphisms are due to mutations that cause loss or gain of a recognition sequence for a restriction enzyme; for example, near the normal allele of a gene there may be the sequence 5'-GAATTC-3' (comprising a recognition site for Eco R I) that we find has mutated to 5'-GGATTC-3' near a variant of the gene. The Eco R I RFLP and the gene variant are linked, and detection of the RFLP allows us to infer the presence of the variant gene (within the probable limits of recombination).

RFLPs can be detected by digestion of genomic DNA with the appropriate enzyme, followed by Southern blotting and hybridisation with a suitable labelled probe sequence (Chapter 11). RFLPs are not very informative since they are commonly found in just two variant forms; the chances of one of these being linked exclusively to a variant allele of a gene of interest are not very high.

(a) Microsatellites

The most informative markers are those which exist in a variety of different allelic forms. Of these, the class of anonymous DNA sequence that is proving most useful is known as microsatellites. These are short sequences comprising repeats of two, three, four or five bases. The best studied are the AC (or GT) microsatellites, which consist of sequences such as 5'-ACCACACACAC ACACACACAC-3', represented as $(AC)_n$ where 'n' may be as low as 1 or as high as 50. The human genome contains about 100,000 $(AC)_n$ microsatellite loci that appear to be dispersed evenly throughout the genome. Cattle, sheep and pigs seem to have a similar number.

Such sequences occur in allelic variants at defined genetic loci. The variants are inherited in Mendelian manner, as is any other trait encoded by nuclear chromosomes. The value of microsatellite loci stems from the fact that a specific locus may have a large number of different allelic variants — many dozen — which differ in the number of repeats. Thus, an individual may have two different alleles at a specific microsatellite locus that could be represented as 5'-$(AC)_5$-3' and 5'-$(AC)_7$-3', with one allele being inherited from its sire and one from its dam. Microsatellite loci such as these can be identified and isolated simply by screening a genomic DNA library with labelled synthetic $(AC)_n$ oligonucleotides (although technical refinements have been introduced to speed up the process). Each locus is defined by the unique DNA surrounding the repeat motif, which is different for all 100,000 loci but identical for all alleles at a single locus. PCR could be used to analyse microsatellite loci taking advantage of their unique flanking sequences to discriminate between different loci. Primers are used that are complementary to the unique sequences at each end of the repeat, so that PCR results in amplification of these microsatellites. Alleles are identified by the length of the product(s) which could be determined by gel electrophoresis.

Self assessment exercise 13.2

A cell contains a microsatellite locus with an allele $(AC)_5$ (inherited from the male parent) and $(AC)_7$ (inherited from the female parent). When the locus is amplified by PCR, unwound (DNA strands separated) and fragments separated by gel electrophoresis, what would be the result?

Microsatellites are abundant, dispersed, locus-specific, easy to characterise, and have many polymorphic variants that are very quick and easy to assay. They are thus an ideal class of sequence to use as linkage markers. Hundreds of microsatellite loci have now been characterised for most livestock species. Intensive work is now underway to determine the linkage of alleles at specific loci with production and disease traits.

(b) Quantitative Trait Loci (QTLs)

Some markers have been identified for genes of major effect, such as the poll (hornless) gene in cattle and the Booroola (FecB) gene in sheep. However, most production traits are not determined by single genes but result from the coordinated expression of many different genes.

The discipline of quantitative genetics seeks to unravel how many genes contribute to a trait, to what extent each contributes and where it is located, i.e., what is its linkage group (chromosome). The laborious task of correlating markers with Quantitative Trait Loci (QTLs) is just beginning. This depends on surveying herds of reference animals that have been bred for a broad range of phenotypic (and hence genotypic) variation and then using computer programs to examine the co-inheritance of specific microsatellite alleles with QTL variants. We can anticipate that within ten years we will have a detailed linkage map of most livestock species, allowing breeding selection with great speed and precision.

(c) Synteny

A second approach to identifying linked markers makes use of the extraordinary conservation of linkage groups in mammals. In extensive studies of many species, it has been shown that entire segments of chromosomes are conserved. Indeed, the conservation of linkage groups is so high that it can be used to track the evolutionary history of mammals. The most extensive data for a conserved linkage group in a livestock species are for chromosome 5 of cattle, where 15 identified gene loci are known to be linked. These 15 loci comprise what is known as a syntenic group.

We can use synteny to provide markers for an unmapped gene. Genetic studies may tell us that an unknown heritable defect of cattle is located on the same chromosome as two or three known genes but is not tightly linked to any of them. By comparing syntenic groups with the higher resolution human linkage map, we can identify markers that are tightly linked to our gene of interest. This approach will become increasingly important as the human gene mapping project nears completion. At that time, comparative examination of syntenic groups will suggest candidate genes rather than linked markers.

In the meantime, synteny is being used to speed the construction of linkage maps for all mammalian species. Certain loci on different chromosomes that are conserved in the genomes of all mammals have been nominated as 'anchor loci' to serve as fixed reference points for linkage analysis. In this way, comparative linkage groups can rapidly be assembled for all species.

(d) Chromosome microdissection

Knowledge of the broad chromosomal location of a gene (from loosely linked markers or synteny) allows us to physically cut out a piece of the chromosome that contains the gene and recover markers that are more tightly linked, or even the gene itself. This is done by identifying the chromosome microscopically in a metaphase spread and using a fine glass microknife to dissect the piece of the chromosome that we know contains the gene.

DNA in the minute piece of chromosome is amplified by PCR with random prlmers, then screened for microsatellite sequences (for tightly linked markers) or with mRNA from a tissue in which the gene is expressed (to identify candidate coding sequences). The procedure requires considerable technical skill, but is proving to be very useful.

13.3.3 *Bulked segregant analysis*

Classical genetic studies enable us to define genetic loci that are responsible for phenotypic traits. Many hundreds of traits have been identified in this way, making use of genetic variants at specific loci to segregate the variants by appropriate matings. Bulked segregant analysis provides an extremely powerful approach to recovering DNA marker sequences that are tightly linked to genetic loci, and hence to genes.

We start by mating two individuals that are homozygous for two different variants of a gene of interest. The first generation of offspring, known as F1, are all heterozygous. Two of these in turn are mated, giving rise to an F2 population in which one-half of the progeny are heterozygous, one-quarter is homozygous for one parental variant and one-quarter is homozygous for the other parental variant. The trait of interest is said to have segregated in the F2 population. If we now pool genomic DNA from all F2 individuals displaying one allele, and pool DNA from all F2 individuals with the second allele, we have two samples of DNA that are identical except for the gene of interest (and the chromosomal region surrounding it). These two samples of DNA are then subjected to PCR with a single random primer — an oligonucleotide of 8–10 bases — a procedure commonly referred to as RAPD PCR (Random Amplification of Polymorphic DNA). The products are resolved by gel electrophoresis. With most short random primers some 6–12 bands are seen. The products are examined for differences between the two bulked samples: bands that are present in one sample but not the other are due to differences in DNA between the samples and hence are linked tightly to the gene responsible for the segregating trait. Such bands are excised from the gel and cloned, providing a marker probe for the gene of interest.

This is an extremely powerful approach that is simple and fast. Usually, differential products are found after just 10–20 random primers have been used, a task that takes just one day. Markers for many genes have been identified in various plant species. Applications to livestock have been few, due primarily to the need for segregating F2 populations which take considerable time and resources to produce. However, alternative approaches to 'bulking' can often be used since its purpose is solely to provide identical genetic backgrounds to a single genetic difference. The major constraint is that variants of a gene of interest must be able to be reflected in variations in phenotype. There is no doubt that bulked segregant analysis will be a very important tool in dissecting the molecular genetics of production species.

13.3.4 *Identification of novel genes*

The tools of molecular genetics also provide direct approaches to the isolation of unknown genes that are responsible for a genetic disease or a production trait. Information about expression of the gene allows us to use a variety of methods to identify it: when and where the gene is expressed (at what stages of development or hormone status, in what tissues), or what is its product (protein sequence).

Biochemical analysis has revealed enzymes and structural proteins that are responsible for many traits and many have been purified. A large number of additional proteins have been identified by two-dimensional gel electrophoresis, simply as 'spots' that may be missing or enhanced in tissues that are known to be affected by a gene variant.Two approaches are commonly used to identify genes from purified proteins, whether the protein has been purified in bulk by chromatography or in trace amounts by electrophoresis.

(a) Identification from protein sequence

The first depends on knowing the sequence of amino acids that comprise at least a small part of the protein. This is usually the amino terminus since the chemistry of most protein sequencing methods depends on sequential degradation from a free amino group. Sensitive instruments are now capable of determining partial sequence from a few pg (picogram; 10^{-12} g) of protein, so even a minor spot in an electrophoresis gel can be eluted and sequenced.

Knowing the partial amino acid sequence of a protein, we can infer the corresponding nucleotide sequence of its gene (Chapter 4). Since each amino acid is encoded by a sequence of three nucleotides we can synthesise an oligonucleotide that encodes the partial protein sequence. In fact, most amino acids are encoded by more than one nucleotide triplet (referred to as redundancy of the genetic code) so rather than synthesising just one oligonucleotide that could encode the protein sequence, we synthesise all possible coding sequences.

To illustrate, consider the following N-terminal sequence which has been determined from an unknown, purified protein:

NH2-Val-Lys-Gly-Thr-Glu-Met-Phe-

The possible codons corresponding to these seven amino acids are:

GUU	AAA	GGU	ACU	GAA	AUG	UUU
GUC	AAG	GGC	ACC	GAG	–	UUC
GUA	–	GGA	ACA	–	–	–
GUG	–	GGG	ACG	–	–	–

We can represent all possible DNA sequences that could encode this part of the protein as shown, where each of the bases shown in a vertical array are used in the position shown: dashes represent bases shown above.

Synthesising a mixture of oligonucleotides that contains all of these possible coding sequences (512 in all) by automated synthesis is no more difficult or expensive than synthesising just one oligonucleotide, since four different nucleotides can be fed into the automated synthesis at any step as easily as can a single nucleotide.

The mixture of oligonucleotides is labelled (forming probes) and used to screen a cDNA library by hybridisation. One of the 512 oligonucleotides in the labelled probe mixture will correspond exactly to the sequence of a portion of the cDNA encoding our protein and that clone will hybridise to the labelled probe. In practice, the cDNA library is also screened with a second oligonucleotide mixture derived from the amino acid sequence of a different part of the protein. Only clones that hybridise with both oligonucleotide mixtures are purified and characterised, since only those clones that contain both sequences will contain cDNA corresponding to the desired protein sequence.

For this procedure to be used successfully, the oligonucleotides should comprise at least 20 bases. The reason is that the haploid animal genome contains approximately 3.2×10^9 bp of DNA, there are 4^{20} possible sequences of 20 bases, i.e., $1{,}100 \times 10^9$, so a single, defined oligonucleotide of 20 bases (a 20-mer) would not be expected to occur in the genome other than in a known sequence. Even with 500 different oligonucleotides in a mixture of 20-mers, we would expect to find very few occurring by chance in the genome, since the number of possible 20-mers is twice as large as the genome size.

This approach is extremely powerful and quite straightforward, now that technologies are widely available for microsequencing of proteins and synthesis of oligonucleotides. It has been used to isolate genes encoding a large number of proteins that have been implicated in phenotype variation.

(b) Identification with antibodies

Purified protein associated with a trait of interest can be used in an alternative approach to gene cloning. The protein is injected into a suitable animal (usually mouse, rabbit or chicken) to raise antibodies. Even if the protein has been identified only as a spot on an electrophoresis gel, the segment of gel containing the protein can be cut out, mashed up, and injected into the animal.

The antibodies will react specifically with the protein, so they can be used to detect its expression in a cDNA 'expression library'. Such a library is constructed by cloning the cDNAs of mRNAs from a tissue in which the protein is known to be expressed into a bacterial vector (phage or plasmid) that has a strong, inducible promoter allowing transcription and translation of the cloned sequences. Plaques (or colonies) containing the cDNA encoding the protein therefore express the protein, which can be detected with the antibodies using standard immuno-chemical techniques.

> ***Self assessment exercise 13.3***
>
> Would monoclonal antibodies be of use in identifying a particular protein produced from a given cloned DNA sequence? Explain your answer.

Again, this procedure has been used to clone DNA sequences encoding many proteins of interest.

(c) Subtractive hybridisation

In a large number of cases the protein(s) responsible for a particular phenotype is not known, although the tissue in which the gene has its major effect is known. It may be a gene (and phenotype) that is induced by a hormone or growth factor; it may be a gene that is expressed at a certain time in development, or it may be a gene that is normally expressed in a particular tissue but is not expressed in a homozygous variant. In these cases, it may be assumed that gene expression correlates with the presence of mRNA in the tissue, while lack of expression correlates with absence of mRNA. Isolation of mRNA from 'normal' and 'variant' tissues gives two populations of mRNA that are similar, except that one contains mRNA transcribed from the gene of interest whereas the other does not.

A variety of methods may be used to detect this single difference between the mRNA populations. One such method is known as 'hybridisation enrichment', or 'subtractive hybridisation'. In principle, the mixture of mRNAs containing the expressed gene product is transcribed into cDNAs and hybridised with the mRNAs from the depleted mRNA population.

> ***Self assessment exercise 13.4***
>
> Explain how the above procedure would allow the gene in queston to be isolated?

In practice, the procedure requires far more refinement to be fruitful. Genes that are transcribed at very low levels may not be recovered and it demands a high degree of technical skill. It has, however, been useful in recovering a variety of genes that are essential to development, hormone induction, and cancer.

(d) Differential display

The sensitivity of PCR allows it to detect DNA at extremely low concentrations. It has been applied to the identification of genes that are differentially expressed, as in the situations described above.

Two populations of mRNA, isolated from 'normal' and 'variant' tissues as above, are reverse transcribed into cDNA. Both populations of cDNA are subjected to PCR using primers of a particular type. One primer is a repeated sequence of the nucleotide T (usually 10) followed by two other bases, e.g., 5'-TTTTTTTTTTCC-3'; there are 64 possible primers of this type, referred to as 'anchor primers'. Each one anneals only at the 3' end of mRNA molecules that end with the complementary two-base sequence immediately adjacent (5') to the poly (A) tract. For instance, the primer illustrated above would anneal only to mRNA molecules that end with the sequence 5'-XXXXXGG$(A)_n$-3' (where X is any base and $(A)_n$ represents the poly (A) tract in which mRNA terminates).

The second primer is a random short sequence of DNA (usually 8–10 bases). Such a primer will be complementary to a large number of sequences in a population of different mRNA molecules. Of the 3.2×10^9 bp of DNA in the haploid genome approximately 10% is transcribed into RNA, i.e., approx. 3.2×10^8 nucleotides of mRNA sequence; there are 4^8 (65,500) possible 8-mers, so we would expect to find about 4,500 sites in mRNA at which an 8-mer would anneal (far less than this in most tissues since not all genes are expressed in all tissues). To be functional in PCR amplification, however, this second 'random' primer must anneal at a site on the complementary strand within a few hundred bases of the anchor primer, so the number of functional primer sites is far less — in practice, just half a dozen or so.

The two populations of cDNA (from 'normal' and 'variant' cells) are subjected to PCR with one anchor primer and one random primer — which ones are used is totally irrelevant. This usually yields a small number of amplification products of differing sizes. The products are separated by gel electrophoresis and the gels examined closely to see if any products from amplification of the 'normal' mRNA are not present in the products of the 'variant' mRNA (or vice versa) — differential display. The procedure can be repeated very quickly with various combinations of anchor primers and random primers until differences are found.

DNA bands that are present only in the amplification products of 'normal' mRNA are recovered from the electrophoresis gel and cloned. These should correspond to the (partial) products of genes that are expressed abnormally in the 'variant' tissue and hence are associated with the variant phenotype. The procedure is very sensitive and technically simple. It is a powerful approach to identifying genes of unknown function that are associated with known phenotypic effects.

(e) Livestock genome mapping projects

It is an overwhelming task to acquire and make sense of DNA sequence information, protein sequence information, data on gene structure and control, anonymous marker loci, physical chromosome maps, linkage maps, genetics maps and QTL correlations for a genome of up to 60 chromosomes containing 6.4×10^9 base pairs of DNA among which are 50,000 – 100,000 genes. International collaborative networks have been set up to coordinate genome research for each of the major livestock species. All groups with an active interest in a particular species share data by computer network with one acting as a central data repository. Participating groups conduct collaborative research programs to ensure that work is not duplicated and to allow the different tools of different laboratories to be applied to a single problem. Most importantly, the livestock consortia are intimately involved with the Human Genome Organisation, deriving benefits not only from DNA sequence homologies but from HUGO's sophisticated organisational networks, resources and experience.

13.4 Applications of molecular genetics

The most dramatic impact of biotechnology in animal production is yet to come. The development of detailed linkage and genetic maps, greater knowledge of gene function and control, and methods for large scale propagation will revolutionise animal production industries. However, gene technologies are already being applied in a growing number of situations.

13.4.1 Identification of animal products

DNA is remarkably stable. Fragments can survive for thousands of years in fossils and on prehistoric tools. DNA survives in decomposed carcasses and in many processed foodstuffs. With PCR we can conduct analyses on small traces of badly degraded DNA. By using suitable primers and probes we can determine the species of origin of many foods and animal products, and even the composition of many mixed foods. Tests of this type are becoming important in quarantine, health and nutrition, providing a high level of quality control. In another context, PCR and gene probe analysis of carcass remains aids the detection and prevention of slaughter of protected animals, and provides evidence of the identity (or parentage) of animals in cases of livestock theft.

13.4.2 *Breeding selected traits into livestock herds*

(a) Diagnosis and elimination of genetic diseases

The isolation of domestic animals in small breeding groups has tended to increase the homozygosity of recessive genes and increase the frequency of specific alleles at quantitative trait loci. The result is that most (probably all) breeds of domestic livestock now harbour variant genes that are deleterious or lethal in the homozygous stage — the frequency of such an allele in the breeding population may be as high as 15%, with 5% of all animals born being affected homozygotes. Common examples in Western livestock include citrullinaemia in Holstein–Friesian dairy cattle (due to a deficiency of the urea cycle enzyme argininosuccinate synthetase), porcine stress syndrome (PSS) in Landrace pigs (due to deficiency of the ryanodine calcium channel receptor in skeletal muscle) and the weaver trait in Brown Swiss dairy cattle (due to a neurological defect), while Brahman (Bos *indicus*) cattle are prone to Pompe's disease (due to a deficiency of the lysosomal enzyme α-mannosidase).

The continued propagation of these and many other genetic diseases in domestic herds is unpleasant for the affected animals and expensive for producers. Many variant genes that are responsible for genetic defects have now been identified and diagnostic tests have been developed that allow animals to be typed as normal, carrier (heterozygous: one normal allele and one variant allele) or affected (homozygous: two variant alleles). Such tests commonly make use of PCR to amplify the affected region of the gene. Variants are identified by the amplified DNA fragment being different in length or in base sequence. In the former, gel electrophoresis allows the variance to be distinguished. In the latter, a restriction endonuclease is used to cut the amplified fragments at a site that is present in one variant allele but not in the other (RFLP); the digestion products are then resolved by gel electrophoresis (Chapter 11).

PCR analysis provides a rapid and accurate way of screening large numbers of animals with automated instruments. A major advantage of PCR is that small samples of any tissue can be used for analysis, allowing assays to be done on a single, air-dried hair follicle rather than a fresh blood sample. The convenience and low cost of this unique method of sample collection is making genetic testing available to a large number of producers. However, the tests themselves are still relativelyexpensive (approx. US $50 per animal) so their economic value must be carefully assessed. Anticipated advances in PCR technology over the next decade should reduce the assay cost by two-to five-fold.

(b) Strategies for breeding with markers for genetic disease

Most genetic diseases result from a deficiency of an enzyme, or less commonly a hormone or structural protein. They are therefore classed as recessive; the phenotypic defect is observed only in homozygotes, when both alleles of the gene are defective. Heterozygotes (carriers) are usually not affected significantly. With the increasing availability of diagnostic genetic tests, economic management strategies can now be adopted that eliminate pregnancies of affected foetuses without production loss. The obvious starting point is to type all breeding animals so that carriers are identified. In most cases, carriers are phenotypically normal and should be retained, particularly if they are otherwise of high genetic merit. Indeed, no producer can afford to cull unaffected carriers.

> ***Self assessment exercise 13.5***
>
> A genetic disease is caused by a mutation in a particular gene, Y. The normal gene is designated as Y, the mutant variant as y. In a breeding programme, which of the following crosses give rise to affected (yy) offspring, and carriers (Yy), and at what frequencies?
> (i) Crossing two homozygous unaffected animals (YY) × (YY).
> (ii) Crossing a homozygous unaffected animal with an affected animal (YY) × (yy).
> (iii) Crossing two carriers (Yy) × (Yy).
> (iv) Crossing a homozygous unaffected animal with a carrier (YY) × (Yy).

If the producer wishes to eliminate the defect from a herd this can be done by using only normal (homozygous) animals for mating; however, the otherwise desirable genetic characteristics of carriers are then lost, and herd replacement would be slower (assuming 15% of animals are affected). An alternative approach is to use MOET with embryo biopsy and splitting on the best animals, regardless of whether they are normal or carriers. Embryos that are identified as affected or as carriers are not transferred; normal embryos are split and transferred. The long-term benefits of this approach are substantial, both in removing the defect from the herd and in genetic gain resulting from MOET with superior animals. However, it is costly so its potential benefits must be evaluated with great care. The tools of biotechnology are too expensive for them to be used simply because they are available.

(c) Inbreeding and introgression

Intensive breeding selection within certain livestock industries (dairy cattle, pig, poultry) has caused concern that some strains may be too inbred — too closely related. Common consequences of inbreeding are loss of productivity and poor reproductive fitness, but these are not of great concern in livestock industries since they are continually and strongly selected against. The major issues are a little more subtle.

First there is the danger of maintaining a high frequency of a deleterious gene allele within a partially inbred strain. A good example of this is the significant incidence of citrullinaemia in the Australian Holstein–Friesian dairy herd, introduced from the UK in the 1960s in one heterozygous bull of high genetic merit. A second example is the weaver trait in Brown Swiss dairy cattle: the trait could not be eliminated through conventional breeding because the defective gene is tightly linked to an allele of a gene that has a major effect on milk production; selection for milk production resulted in simultaneous selection for the linked weaver trait. As we have seen, genetic defects can now be precisely and easily eliminated from a breeding herd through DNA screening procedures, so they no longer are a necessary result of partial inbreeding.

The second major concern is loss of genetic diversity. One feared consequence is that the repertoire of immune defences will be weakened, so that a novel or particularly virulent pathogen may cause an epidemic with disastrous losses. There is a fear also that we may lose variants of genes whose importance we do not yet know that may be of great value in the future. The dangers inherent in loss of diversity were recognised some time ago, with the result that much effort has gone into preserving 'primitive' livestock species and breeds for their uniquely diverse genetic resources. The large number of distinct breeds of livestock also provides a buffer against loss of genetic variants. DNA technologies play an important role in this context by allowing us to determine the degree of relatedness of individuals (see below) and hence to maximise outbreeding in resource herds to maintain diversity.

In a similar vein, introgression of a trait (or a complex of traits) into a herd can be monitored accurately and easily by using chromosome-specific DNA markers to select animals for continued backcrossing.

(d) Parentage determination

Genetic improvement involves identification of animals of superior genetic merit, selective breeding from them to provide progeny of higher individual merit, and the continuing use of improved animals to provide production stock. It is critically important for genetic improvement that the parents of a desirable animal are identified.

In stud breeding the issue is simple: an animal cannot be registered unless its parentage is known. This is usually not difficult since stud matings are set up with a single, specific sire. However, mistakes do sometimes happen (it is estimated that as many as 15% of stud bulls may have incorrect pedigrees). If a mistake in parentage assignment is not detected, the genetic improvement programme is weakened since the genetics of the animal are different from what is assumed. If a mistake is detected, the animal is discarded, despite the fact that it may be of high merit.

The technique known as 'DNA fingerprinting' (Chapter 11) is proving of great value in these cases. DNA fingerprinting analyses anonymous DNA sequences at defined loci that have a large number of allelic variants. The probability of two animals having the same alleles at each of a number of loci is very small, so the total profile provides unique identification of an animal — a fingerprint (identical twins have the samc DNA fingerprint). It is now common practice to use PCR amplification of five or six different microsatellite loci (see above) to provide a DNA fingerprint. One of the pair of alleles at each locus is inherited from the sire and the other from the dam. Comparison of DNA fingerprints of a suspect animal, its dam, and all its possible sires thus allow assignment of the animal's sire. In many cases the dam does not need to be typed since only one of the probable sires could contribute one allele at each of the microsatellite loci tested. The use of DNA fingerprinting to assign parentage is now accepted by most livestock registered breed societies, with the result that the accuracy of genetic improvement is improving and animals need no longer be discarded due to unknown pedigree. Another benefit of parentage assignment by DNA fingerprinting is that it allows breeding selection for carcass traits. Breeding for the meat industries is based on characteristics that can only be measured accurately after slaughter when the carcass is being processed. Meat samples can now be removed from carcasses of high merit for DNA fingerprinting, allowing identification of the animal's sire and continued breeding selection.

Cost is an important element in extensive applications of DNA testing. Again, costs are reduced markedly by using samples of air-dried hair follicles for analysis. The DNA tests remain too expensive, but increasing automation will ultimately reduce the cost to \$10–\$15 per sample.

(e) Velogenesis

Velogenesis is a term coined by Genmark to define the future of livestock genetic improvement. It means, literally, rapid generation — accelerated genetic selection for any desired characters. Velogenesis combines the future outcomes of reproductive technologies and molecular genetics. At this stage it is a vision, but it is anticipated that it will be a reality by 2010.

By that time we will have identified and characterised all genes of major effect that are responsible for production, disease and disease-resistance traits. Tightly linked markers will be available for all significant QTL (Quantitative Trait Loci). Screening selected animals will allow identification of a sire(s) and dam(s) of the highest merit for a particular production purpose in a particular environment. These animals will be mated by MOET.

In early pregnancy (probably at about three months) the foetuses will be biopsied and screened for the desired traits. An ovary will be removed from female foetuses of the highest merit and its immature oocytes induced to mature *in vitro*. The oocytes will be fertilised *in vitro* and the cultured embryos transferred to foster dams, where at three months pregnancy, the cycle will be repeated. Within two years, eight cycles of intensive genotypic breeding selection will be completed, the most intensive and accurate selection imaginable. Animals will be produced to suit specific management practices, production environments, and market needs.

Ethical issues must be considered, but we should be aware that the debate has already been engaged. Already it is possible to produce mature oocytes from human foetuses and IVF has reached a high level of sophistication with fertilisation by microinjection of single sperm. Prenatal screening for genetic defects is routine, and the Human Genome Project is daily extending our knowledge of genome organisation and function. But could it be that the most valuable application of these extraordinary advances will be in the production of animal protein?

13.5 Concluding remarks

Humans depend on animals as a source of high protein food, fibre and leather. The scope for increasing and improving quality of production by natural means is time-consuming and limited by natural barriers. These natural barriers can be overcome, at least in part, by applicaton of modern technological advances in animal surgery, cell manipulation and culture, and molecular genetics and genetic engineering. It is not hard to envisage an intensive livestock rearing system based on genetically engineered animals fed on genetically engineered crops. Human welfare may also be improved by our understanding of certain diseases gained from the use of transgenic laboratory animals.

There are, however, important issues raised by artificial genetic selection and manipulation of animals (as indeed is also the case with plants). Are such procedures ethical or do they infringe the 'natural rights' of animals? Are such procedures safe? Who controls developments? Will developments benefit all countries, or just the technologically developed countries in which advanced biotechnological procedures are possible? Are transgenic animals patentable? Should transgenic animals be patentable?

13.6 Suggested further reading

1. BIOTOL series (1992). *Biotechnological innovations in animal productivity*. Butterworth-Heinemann, Oxford.
2. Church, R.B. (1987). Embryo manipulation and gene transfer in domestic animals. *Trends in Biotechnology*, **5**, 13–19.
3. Clark, A.J. *et al.* (1987). Pharmaceuticals from transgenic livestock. *Trends in Biotechnology*, **5**, 20–24.
4. Friedman, T (1987). HPRT-deficient mice: a useful new animal model for human disease. *Trends in Biotechnology*, **5**, 157–158.

5. Houdebeine, L.M. (Ed.) (1997). *Transgenic animals: generation and use.* Harwood Academic Publishers.
6. Puhler, A. (1993). *Genetic engineering of animals.* VCH.
7. Rogers, G.E. (1990). Improvement of wool production through genetic engineering. *Trends in Biotechnology*, **8**, 6–11.
8. Rexroad, C.E. Transgenic livestock in agriculture: New animals for a new world? *Outlook on Agriculture*, 24, 227.

13.7 Responses to assignments

13.1 We saw in Chapter 12 that, due to the totipotency of plant cells, individual plant cells or dissected plant tissues can be induced to undergo morphogenesis and produce an intact plant, and that this can be conveniently carried out in controlled conditions in the laboratory. It is thus relatively easy to create plants on a large scale by asexual means (propagation). This is not possible with animal cells.

13.2 We saw in Chapters 5 and 8 how microbes may be genetically engineered to produce animal proteins, such as hormones, and how such microbes can be efficiently grown in fermentation systems. We also saw in Chapter 9 how mammalian cells may be grown in culture, and engineered to yield high levels of a particular product. By such means progesterone and prostaglandin hormones could be efficiently produced on a large scale.

13.3 We saw in Chapter 11 how RFLP and gene probe technology can be used to identify particular gene loci in DNA. In these ways individual genes may be screened for. On a broader basis, RFLP (genetic fingerprinting) of larger genome fragments (Chapter 11) or sequencing of particular sections (Chapter 11) can also yield information as to whether a particular clone of embryos is of a particular genotype.

13.4 In extensive farming, animals are grazed on natural vegetation. Other feral animals may also graze on the same vegetation, and thus deplete the food source of the farmed animals. If populations of natural grazers are controlled, a greater quantity of vegetation becomes available to the farmed animals.

13.5 Once a gene has been isolated, it is possible to clone it and label it with radionuclides, enzymes or fluorochromes — in other words produce a probe for the complementary sequence (Chapter 11). Gene probes allow the position of genes on particular chromosomes to be determined, and with this informaton the sequencing of the relevant gene is made relatively straightforward (Chapter 5).

3.8 Responses to self assessment exercises

13.1 The amount of DNA present in the few cells taken at biopsy may not be enough in itself to allow analysis to be undertaken. If this is the case, then the cells must be grown in small scale tissue culture (Chapter 9). Alternatively PCR (Chapter 11) may be used to amplify regions of DNA in which defective genes are known to lie.

13.2 There would be two bands in the electrophoresis gel, as the products of PCR would differ by four nucleotides, one strand $(AC)_7$ having an ACAC sequence extra.

13.3 The answer is yes. A monoclonal antibody could be produced which has affinity towards a certain oligopeptide epitope on the protein (Chapter 10). This would make detection of the protein more specific than using polyclonal antibody (raised by immunising animals), as polyclonal antiserum may contain antibodies which cross react with peptide sequences (epitopes) shared by more than one protein.

13.4 When the mRNAs from the cell expressing the protein are cloned, each mRNA species is copied into the corresponding DNA olignucleotide. When these cDNAs are mixed with mRNAs from a cell which does not express the protein (and thus does not have mRNA coding for that protein) the only cDNA species not hybridising (with a mRNA which is complementary to it) will be that coding for the protein. This single-stranded DNA can be separated from the mixture in a suitable gel.

13.5 (i) All progeny from (YY) × (YY) will all be (YY) — none will be affected or carriers, all will be homozygous unaffected.

(ii) Progeny from (YY) × (yy) crosses will be (Yy) — all will be carriers.

(iii) Progeny from (Yy) × (Yy) crosses will be: (YY), (yY), (Yy) and (yy). Thus one quarter of progeny will be homozygous unaffected (YY), one-half will be carriers (yY and Yy), and one-quarter will be affected (yy).

(iv) Progeny from a (YY) × (Yy) cross will be: (YY) and (Yy). Thus one half of the progeny will be carriers, one half will be homozygous unaffected.

14 Biotechnology in food and beverage production

Chapter outline

14.1 Introduction

Although many people imagine that biotechnology is 'new', this is certainly not the case, as biotechnology has been applied by humans, albeit in an empirical way, to food and alcoholic beverage production for many centuries; over 3000 years ago, the ancient Egyptians brewed a type of beer made from wheat and in ancient Greece and Rome, grape wine was produced and widely used in trade. In many countries, brewing and bread baking became the first real 'industrialised' process, being carried on a large scale in a centralised brewery or bakery instead of in the home. These traditional processes have continued to the present day. Throughout the last century, as our understanding of the nature of microorganisms and enzymes increased and process technology developed, these processes have increasingly become based on modern science and technology. In economic terms, alcoholic beverages are by far the most important biotechnological product worldwide (Table 14.1), followed closely by fermented dairy products such as cheese. Such products may be extremely important in a country's economy, for example in Scotland, whisky constitutes a major product of the manufacturing industry.

Table 14.1 Global economic value of microbial food and beverage products.

Product	**Approximate value in million $US per annum (1995 prices)**
Alcoholic beverages	50,000
Cheese	30,000
Glucose/fructose syrups	2,000
Amino acids	1,800
Yeast	1,100
Vitamins	600
Citric acid	450
Amylase enzymes	150

There are many ways in which biotechnology is involved in the food that we eat, particularly in developed countries in which a significant proportion of the diet comprises industrially processed foods. Fermentation processes are used to produce a variety of microorganisms or their products for use in the food industry (Table 8.1). Microorganisms are inoculated into food to bring about particular changes, for example yeast is inoculated into dough to produce leavened bread and lactic acid-producing bacteria are inoculated into milk or milk products to produce a variety of products such as cheese and yoghurt. Enzymes from microorganisms are widely used in the processing of foods such as fruit juices and in cheese production. Vinegar, soya sauce and monosodium glutamate are used widely in food processing and cooking. It is even possible to eat microorganisms themselves as food, not only in the traditional form of mushrooms but also in the form of more recently developed microbial foods derived from particular algae, fungi and bacteria.

In this chapter, we will have an overview of how microorganisms and their products are used in the manufacture of foods and alcoholic beverages. We will also examine some processes in detail to illustrate how developments in gene technology (Chapter 5), protein engineering (Chapter 6), enzyme technology (Chapter 7) and fermentation technology (Chapter 8), are applicable in the food and beverage sectors.

14.2 Food additives

A wide variety of biotechnological products are directly used in the processing of foods as flavourants, natural preservatives and processing aids (Table 14.2).

14.2.1 Vinegar production

Vinegar was probably originally discovered by the natural spoilage of alcoholic beverages by the bacterium *Acetobacter*, and it was realised that vinegar could be used to preserve and flavour food. The modern food industry uses vinegar widely as a preservative, in the pickling of vegetables and as an additive to manufactured foods such as sauces and relishes. Vinegar helps to preserve food by lowering the pH to levels at which spoilage organisms are unable to grow. Vinegar can be made from malt beer (malt vinegar), or wine (wine vinegar).

Table 14.2 Microbial products as food additives.

Product	Production organism	Uses
Vinegar (acetic acid)	*Acetobacter* spp.	Flavourant, acidulant, preservative
Soya sauce	Natural microflora	Flavourant
Yeast extract	*Saccharomyces cerevisiae*	Flavourant
MSG (mono-sodium glutamate)	*Corynebacterium glutamicum*	Flavour enhancer
Nucleotides	*Candida utilis, Saccharomyces* spp.	Flavour enhancer
Citric acid	*Aspergillus niger*	Flavourant, acidulant, preservative
Lactic acid	*Lactobacillus* spp.	Flavourant, acidulant, preservative
Nisin (antibiotic)	*Lactococcus* lactis	Preservative
Xanthan gum	*Xanthomonas* spp.	Thickener
Vitamins and amino acids	Various organisms	Natritional supplement
Aspartame	Derived from amino acids	Low calorie sweetener
Methylketones	*Penicillium* spp.	Blue-cheese flavour
Aceton	*Latococcus* spp.	Yoghurt flavour
Aldehydes and esters	Various organisms	Fruit flavours
Pyrazines	Various organisms	Nutty and roasted flavours
Lactones	*Candida* spp.	Coconut and fruit flavours.

Vinegar production is a two-stage microbiological process. The first stage involves making an alcoholic solution such as malt beer (beer without added hops, Section 14.7.1) or wine. This is an anaerobic process carried out by brewers' yeast. The second stage involves exposing the liquor to the bacterium *Acetobacter*, which obtains energy from the oxidation of the alcohol (ethanol), producing acetic acid as the end product. *Acetobacter* spp. are obligate aerobes, and thus the second stage must be carried out in an aerated bioreactor.

In a traditional process, vinegar is produced in a wooden vat called a generator as shown in Fig. 14.1.

A generator is a large wooden vat with a slatted (perforated) shelf about half way up. Shavings (curls) of a hardwood are placed on this shelf. The batch of liquor is run into the bottom of the vat

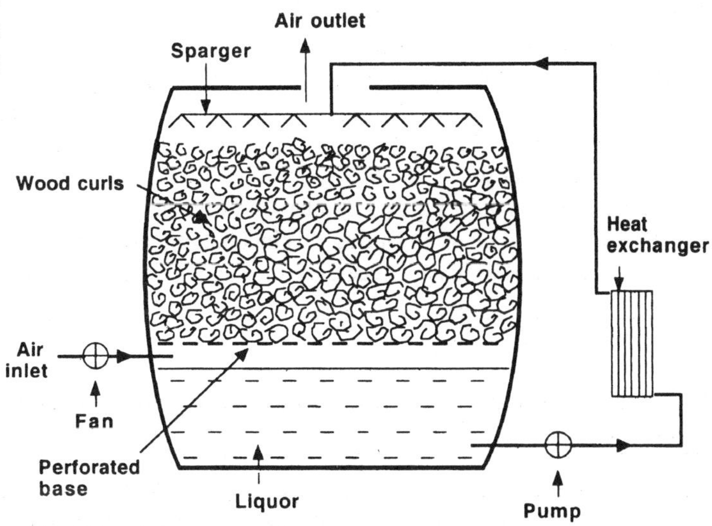

Fig. 14.1 *A vinegar generator.*

and is then circulated around the system; it is pumped up to a rotating sparger (spray) and sprayed over the top of the wood shavings.

When the generator has just been packed with new wood curls, it is inoculated with a culture such as *Acetobacter aceti* var. *xylinum*. As the bacterial cells grow, they secrete a capsule (extracellular layer) composed of cellulose. The capsular material sticks the cells to the surface of the wood curls, and as the organisms grow they form a layer of cells stuck over the surface of the curls. As the liquor trickles down the curls, the bacteria in the layer attached to the surface of the curls oxidise the ethanol to acetic acid. All the ethanol will not be oxidised by one trickle down the generator, so the liquor is recirculated around the system again and again until all the ethanol has been oxidised. The curls are loosely packed in the generator so that air pumped into the vat below the curls circulates around and provides the bacterial cells with the oxygen they require. The process is strongly exothermic (heat-generating), so the liquor is cooled as it circulates around the system by passing it through a heat exchanger.

Once all the ethanol has been converted to acetic acid the vinegar is drained off, filtered to remove any bacteria (cloudiness), diluted to about 4% acetic acid, and packaged.

Assignment 14.1

Would the fact that the bacteria stick themselves to the wood curls be an advantage to the operator of a vinegar generator?

In a modern process, vinegar is produced by placing the alcoholic liquor in a conventional stainless steel fermenter (Section 8.2.1), which may be lined with glass or resin to prevent corrosion. This is inoculated with a culture of *Acetobacter* sp. and the liquid is aerated by blowing in air and by stirring (Section 8.2.4) until all ethanol is converted to acetic acid. In this situation, the bacteria grow as a dispersed cell suspension, and so the fermenter must be reinoculated at the start of each batch.

14.2.2 Soya sauce production

Soya sauce is a widely used flavourant in Oriental cuisine, and, like vinegar, is produced in a two-stage microbiological process. The first stage involves the production of koji, which is a solid substrate fermentation (Section 8.4) designed to enrich the substrate with microbial enzymes. For soya sauce production, rice and roasted wheat are cooked by steaming, cooled and inoculated with spores of the fungi *Aspergillus oryzae* or *Aspergillus sojae*. The inoculated material is then incubated in a humid environment at about 22°C for 3–4 days. As these fungi are strict aerobes, the mixture must be exposed to air; in traditional processes, this is achieved by laying the mixture out in shallow pans or baskets which may be manually agitated at intervals, whereas in a modern process, a forced-aeration reactor (Section 8.4) would be used. During this incubation period, the fungi produces amylase, protease and pectinase enzymes, which are necessary for the next stage of the process.

Assignment 14.2

What functions would be performed by the steaming of the rice-wheat mixture before its inoculation with fungi?

In the start of the next stage of the process, brine (15–20% salt) is mixed with the koji to form what is called a moromi mash. The mash is allowed to stand for a period, during which the enzymes that the fungus produced during the koji process hydrolyse the complex, unavailable nutrients (see Section 8.5) in the rice and wheat forming simple, available forms for the organisms that will carry out the next stage. The mash is sieved or filtered to remove the grains and placed in vats for a period of several months. During this time, a succession of different microorganisms naturally present in the mash carry out a complex series of reactions that eventually result in the formation of soya sauce.

Assignment 14.3

What functions would be performed by the creation of a high salt concentration in the mash before its storage?

As the vats are not aerated, the major reactions are carried out by organisms that are facultatively anaerobic (able to grow in the presence or absence of oxygen) or anaerobic. Initially salt-tolerant organisms such as the bacterium *Pediococcus halophilus* predominate and these organisms ferment sugars to lactic acid (Section 14.5), which reduces the pH of the mixture. Acid tolerant, salt tolerant organisms such as the yeast *Saccharomyces rouxii* then predominate and these organisms ferment sugars to a mixture of aliphatic and aromatic alcohols. Finally, during a long maturation period, a variety of bacteria and yeasts such as *Torulopsis* spp. produce a mixture of aromatic alcohols, esters and phenolic compounds, which give the product its characteristic flavour. Glutaminase enzymes produced by the fungus during the koji phase convert the amino acid glutamine (formed from hydrolysis of proteins in the wheat and rice during the mashing stage) into glutamic acid, which in the brine solution forms mono-sodium glutamate (Table 14.2), a potent flavour enhancer (although it does not have a strong flavour of its own, it increases the natural flavours present in food). A recent technological development has been to inoculate the raw soya sauce at the start of the maturation period with cultures of appropriate organisms, which can reduce maturation time from months to days.

Self assessment exercise 14.1

The flow diagram in the next page represents the production of soya sauce. The individual steps in the process are given in boxes and the inputs and outputs are unboxed. The arrows represent the flow sequence through the process. Some of the process steps have been left out of the boxes on the diagram and are listed below it. Fit these processes into the appropriate boxes and so complete the flow diagram.

14.2.3 Yeast extract production

Yeast extract is produced from the cells of bakers' yeast, *Saccharomyces cerevisiae*, which are produced in fermentation and concentrated into a thick suspension by centrifugation. These processes are described in Section 14.3.1. The yeast cells are then hydrolysed by treatment with acid or by suspending the cells in warm water until they autolyse. The product is then preserved either by partial drying to form a viscous liquid to which salt is added as a preservative, or complete drying to a powder. Alternatively, the yeast extract suspension can be dried by roller drum drying, which involves passing the suspension over steam heated rollers to dry it. In this instance, the heating causes chemical reactions in constituents of the extract which result in the production of 'meaty' flavours, and so the product can be used to impart a meaty flavour into foods, including vegetarian products. In these forms, yeast extract is added to processed foods as a flavourant and flavour enhancer.

Assignment 14.4

How would the hydrolysis of yeast cells develop flavour-enhancing activity?

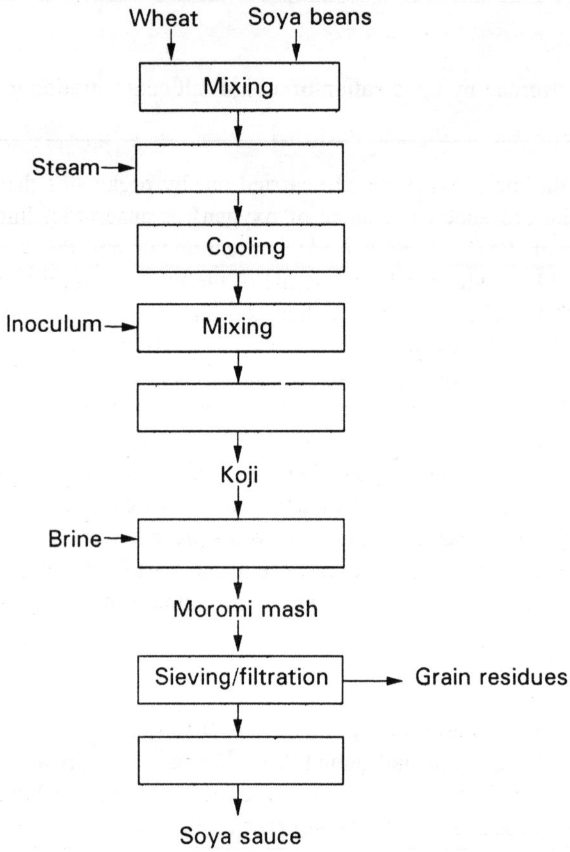

Mixing   Storage/maturation   Incubation   Cooking

14.2.4 Other products

The products listed in Table 14.2 that have not already been considered are produced in conventional fermentation systems (Chapter 8). Glutamic acid and other amino acids, citric acid and vitamins are not normally excreted from cells as they are essential metabolites, so mutant organisms with metabolic blocks which cause accumulation of the product and mutants with 'leaky' cell membranes which promotes excretion of the product are often used in fermentation processes (Section 8.7.2). The production of nucleotides is usually performed by extracting RNA from yeast, which may be a waste product from a brewing process (Section 14.7.1), and hydrolysing the RNA with exonuclease enzyme from the fungus *Penicillium citrinum*, forming a mixture of the constituent nucleotides. The nucleotide adenosine monophosphate is then deaminated by the enzyme adenylic acid deaminase from the fungus *Aspergillus oryzae* to form inosine monophosphate (IMP). The active flavour enhancers, IMP and GMP, can be separated out from the mixture and recovered using ion exchange fractionation.

14.3 Food and beverage inoculants

On an industrial scale some microorganisms are widely used for food processing, usually food conversion in which the nature of the food is altered by the growth and activity of the inoculated microorganism(s). Bakers' yeast and brewers' yeast (strains of *Saccharomyces cerevisiae*) are widely used in bread leavening and alcoholic fermentations, respectively. Lactic acid-producing bacteria (known as lactic acid bacteria, and mainly organisms belonging to the genera *Lactobacillus* and *Lactococcus*,) are widely inoculated into bread to make sour dough products and into milk or milk products to make butter, yoghurt (curds), cheese and other products (Section 14.5). Very often, for industrial scale processes, inocula of the microorganisms are produced centrally in a fermentation unit and distributed out to the production units as required.

14.3.1 Brewers' and bakers' yeast production

The particular yeast is stored as stock cultures, usually in a freeze dried (lyophilised) state. Bakers' yeasts are selected based on their ability to rapidly respire and generate CO_2 in dough. Brewers' yeasts are selected based on their ability to rapidly ferment sugars to ethanol, their ability to flocculate and sediment out from the fermented liquid after fermentation, and for the flavour that they impart into the beverage produced. Distillers' yeasts are selected based on their ability to tolerate high concentrations of sugar and alcohol, as distillation processes are most energy efficient when the alcohol concentration is as high as possible; thus the medium to be fermented is designed to have high initial sugar concentration that in turn should lead to high alcohol concentration (10–12%[v/v]). A stock culture is grown in increasingly large volumes that may finally be centrifuged to concentrate the cells for storage at 4°C for use in several batches. This yeast is inoculated into sterile production fermenters containing a medium based on molasses (Section 8.5), with phosphate and ammonia supplements. Aerated fermenters are used and operated as fed-batch systems (Section 8.6.2) in which nutrient (molasses or ammonia) is fed in at sub-optimal rates.

> ***Self assessment exercise 14. 2***
>
> Explain why nutrients are fed in at sub-optimal rates in the propagation of yeast.

Yeast is recovered from the fermentation medium by centrifugation to form a dense suspension, and the cell-free medium is discarded. The yeast concentrate may be dried further (dewatered) to a cake-like consistency using a rotary vacuum filter and sold in that condition as fresh yeast after mixing it with vegetable oil to hold the cake together, or it can be drum dried to form granular dried yeast. Alternatively, the yeast suspension can be directly spray dried to give small particles of active dried yeast. Fresh yeast remains highly active for just a few days, but its high activity means it can be used for direct inoculation. In contrast, dried yeasts retain viability for months or even years, but need to be reactivated (rehydrated) prior to use. Spray dried yeast retains viability somewhat better then drum dried yeast.

> ***Self assessment exercise 14. 3.***
>
> Draw a flow diagram (one diagram) to show the production of fresh yeast, dried yeast and yeast extract.

14.3.2 Lactic starter production

The production of lactic acid bacteria as food inocula, called lactic starter cultures in the dairy products industry, is shown in Fig. 14.2. Organisms are grown in a non-aerated fermenter in a medium based on milk components, centrifuged to a dense suspension and preserved by deep freezing or freeze drying. Freeze dried cultures have the advantage that they can be stored and transported at ambient temperature. The waste medium produced from such a process contains lactic acid, which can be recovered as a by-product and used as a food acidulant (Table 14.2).

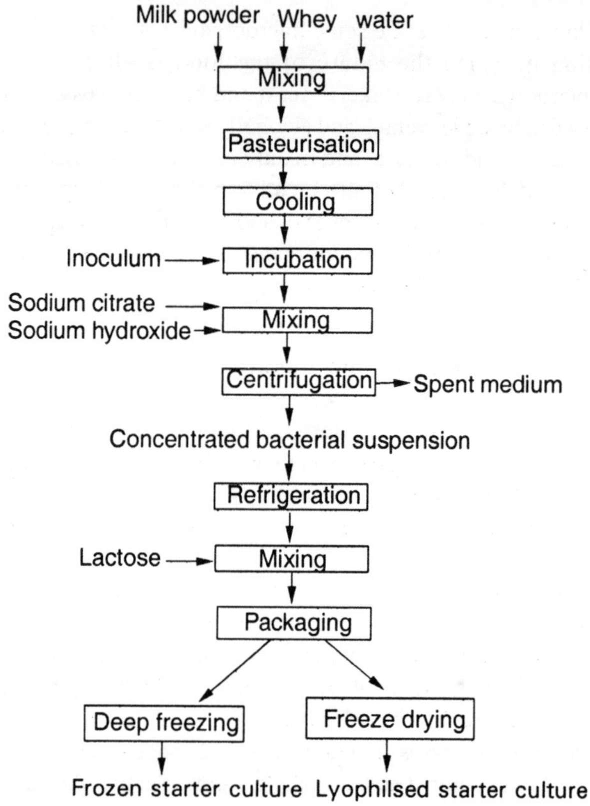

Fig. 14.2 *The production of lactic starter cultures.*

14.4 Enzymes in food and beverage processing

Some uses of enzymes in food are traditional. The juices from papaya (which contains the enzyme papain) and pineapple (which contains the enzyme bromelain) are used as crude sources of proteolytic enzymes for the tenderisation of meat. Extracts from the stomachs of slaughtered suckling farm animals are used as a crude source of enzymes used in cheese manufacture.

> *Assignment 14.5*
>
> In the modern food industry, microbial enzymes tend to be used rather than enzymes derived from plants. Why is this?

A variety of enzymes are used in the processing of foods and beverages (Table 14.3). In the production of enzymes in fermentation systems, the medium is formulated and the system operated such that substances which might repress enzyme synthesis are absent or maintained at appropriately low concentrations (Section 8.6.2), whereas inducers (substances which induce synthesis of inducible enzymes, Section 4.9.4) are maintained at appropriately high concentrations. Techniques of genetic engineering (Section 5.5) are applied to increase the yield of enzymes in fermentations, by creating production organisms with multicopies of the required gene (coding for the desired enzyme) under control of efficient, constitutive promoters (Section 4.9.4).

Table 14.3 Microbial enzymes in food processing.

Enzyme	Produced by	Action	Uses in the food industry
Proteases	*Aspergillus* spp. *Mucor* spp. *Bacillus* spp.	Hydrolysis of proteins	Meat tenderisation Recovery of meat and fish wastes Preparation of soya protein Preparation of biscuit flour Cheese making Haze prevention in beers
Chymosin	Various recombinant organisms	Hydrolysis of κ-casein	Cheese making
Pectinases	*Penicillium* spp. *Aspergillus* spp	Hydrolysis of pectins	Fruit juice extraction Fruit juice clarification Fruit and vegetable maceration
Amylases	*Rhizopus* spp. *Mucor* spp. *Aspergillus* spp. *Bacillus* spp.	Hydrolysis of starches	Maltose production Glucose/fructose production Mashing for alcoholic beverages Preparation of baking flour Liquefaction of starch pastes
Invertase	*Saccharomyces cerevisiae*	Hydrolysis of sugar	Liquefaction of sucrose pastes Glucose/fructose production
Beta-galactosidase	*Kluyveromyces lactis*	Hydrolysis of lactose	Removal of lactose from milk
Beta-glucanase	*Bacillus* spp. *Trichoderma* spp.	Hydrolysis of glucans	Haze prevention in beers
Lipases	Various organisms	Hydrolysis of fats	Emulsification of fats
Glucose	*Aspergillus niger*	Oxidation of glucose	Glucose removal from egg white

14.4.1 Enzymes in cheese production

In the industrial production of cheese (Section 14.5.1), curdling of the milk (coagulation, in which proteins and fats separate out from the other milk components) is brought about by the action of enzymes in rennet, which is added to milk, together with lactic acid produced by starter cultures. Traditionally, rennet has been produced by steeping calf, lamb, kid, piglet or chicken stomachs in brine. In major industrialised countries, calf rennet is the usual choice. However, the supply of rennet worldwide is not sufficient to meet the demand, and rennet is not a product acceptable to vegetarians. Alternative sources of rennet have thus been sought.

The enzymatic activity in rennet which is crucial to milk clotting is protease activity, as this causes breakdown of the structure of colloidal proteins in the milk which in turn causes them to flocculate and coagulate. In calf rennet, the most important enzyme involved in coagulation is chymosin, which specifically hydrolyses the protein kappa casein (κ-casein) between the amino acids at position 105–106 on the κ-casein molecule. Pepsin, another protease found at lower concentrations than chymosin in calf rennet, hydrolyses caseins in a non-specific manner, and although this causes coagulation, it can also result in formation of unacceptable flavours in the cheese that is produced. For this reason, rennet from very young calves is preferred, as the younger the calf, the higher the chymosin levels and lower the pepsin levels.

Although plant proteases such as papain and bromelain have been used in cheese manufacture, their non-specific action on caseins can also lead to the formation of unacceptable flavours in the cheese. Protease enzymes with acceptable coagulation properties have been produced from the fungi *Mucor miehei*, *Mucor pusillus* and *Endothia parasitica* and used industrially in cheese making. The enzymes from these fungi are relatively non-specific in their proteolytic activity, but are acceptable for use in production of a limited number of cheese varieties.

Self assessment exercise 14. 4.

Complete the following statements from the choices given below.

Genetic engineering can best be applied to overcoming the problems associated with the use of calf rennet in cheese manufacture by:

(i) isolation of the protease gene from *Mucor pusillus* and transfer of it to, and expression of it in calves stomachs.
(ii) isolation of the bromelain gene from pineapple and transfer of it to, and expression of it in *Mucor pusillus.*
(iii) isolation of the chymosin gene from calves and transfer of it to, and expression of it in *Mucor pusillus.*
(iv) isolation of the papain gene from calves and transfer of it to, and expression of it in pineapple.

Attempts were thus made to transfer the bovine chymosin gene into a suitable microorganism. The bacterium *E. coli*, the yeast *Kluyveromyces lactis* and the filamentous fungus *Aspergillus niger* were chosen by different companies developing the process. To complicate matters, the bovine gene does not code for active chymosin, but instead codes for an inactive precursor molecule called pre-pro-chymosin. As this molecule is excreted into the stomach, proteolytic enzymes remove a short amino acid sequence from the molecule, forming pro-chymosin (which is also inactive). The low pH of the stomach then produces conformational changes (Section 6.2) in the pro-chymosin molecule, converting it into active chymosin.

For transformation of *Kluyveromyces lactis,* the bovine chymosin gene (or, more correctly, the pre-pro-chymosin gene) was isolated and its sequence modified (shortened) so that it coded for pro-chymosin rather than pre-pro-chymosin. The advantage of producing, in fermentation, pro-chymosin rather than chymosin is that as the former has no proteolytic activity, it can not cause a breakdown of proteins (perhaps even pro-chymosin itself) during fermentation and product recovery. The pro-chymosin gene was then inserted into what is known as an expression cassette (DNA sequences needed for expression of the transgene in the transformed organism, for example, promoter and terminator regions (Section 4.9.1)). In this cassette, the pro-chymosin gene was sandwiched between the promoting and terminating sequences from the beta-galactosidase (β-galactosidase) gene of the organism. In order to promote excretion of the pro-chymosin from the cells of the organism, a leader sequence was attached to the pro-chymosin gene between the promoter and the pro-chymosin gene. A leader sequence codes for a short peptide chain (transit or signal peptide) that assists the transport of the polypeptide to which it attached across the cell membrane (Section 5.9.1); as the polypeptide is transported across the cell membrane, the signal peptide becomes removed from it. Several signal sequences of microbial origin, and the signal sequence of bovine chymosin itself, were found to be effective in *Kluyveromyces lactis*. The expression cassette containing the pro-chymosin gene, promoter, signal and terminator regions was inserted into yeast plasmids, such as pUC 19. Such plasmids have selection markers, usually gene coding for antibiotic resistance, which allows selective isolation of transformed cells.

> ***Assignment 14.6***
>
> Explain how the presence of an antibiotic resistance gene on a plasmid allows the selective isolation of transformed cells.

An added advantage of using the promoting and terminating regions of the β-galactosidase gene from *Kluyveromyces lactis* in an expression cassette is that the β-galactosidase gene in *Klyveromyces lactis* is located on one of the cell's chromosomes. In transformed cells, the expression cassette can become integrated into the β-galactosidase gene by the process of homologous recombination (see Sections 12.7.1 and 12.9.2). The advantage of this in genetic engineering is that while plasmids (and the transgenes they carry) may become deleted from transformed cells (see Section 5.5.6), transgenes incorporated into chromosomes are seldom deleted (being replicated along with the chromosmal genes).

The technique of Southern blotting (Section 5.9.2) demonstrated that the transgene did indeed become incorporated effectively into the host cell chromosome, and SDS-gel analysis (which separates and distinguishes proteins) showed that the recombinant pro-chymosin and chymosin formed from it at low pH were identical to the respective native molecules. The stability and activity of the recombinant chymosin was measured, and the suitability of the recombinant chymosin in cheese making was tested in pilot-scale trials. In order to meet regulatory requirements on genetically engineered, novel food products in force in projected user countries, the producing organisms has to be proven to be non-pathogenic and non-toxic, and both the recombinant chymosin and cheese produced from it are rigorously tested for toxicity and allergenicity in feeding trials with laboratory animals. In order to meet regulatory requirements on large-scale culture of genetically engineered organisms in force in producing countries, the large-scale fermentation system and process operation has to be such that containment of the genetically modified production organism is guaranteed. For example, at the end of the fermentation period, the production organisms are killed by addition of an inorganic acid and benzoic acid (a widely used food preservative). This rules out any possibility of live, genetically engineered organisms being released into the environment during downstream processing. The low pH at this stage also causes conversion of pro-chymosin into active chymosin.

In large-scale production processes, the recombinant yeast is inoculated into a sterile fermenter containing a medium based on milk or milk products (for example, whey). Fed-batch cultures with carbon source (whey) feed give highest levels of enzyme. At the end of the fermentation period, the medium is acidified and benzoic acid is added, as explained above. Once the yeast cells are killed, they are removed by rotary vacuum filtration or centrifugation, and the liquid product stream is concentrated (in essence water is removed from it) by ultrafiltration. The concentrated enzyme solution is then sterilised (to further preserve it from microbial degradation) by filtration. This product is now used in cheese manufacture in some European countries.

Self assessment exercise 14.5
Draw a flow diagram showing the production of chymosin using recombinant *Kluyveromyces lactis.*

14.4.2 Enzymes in the production of sweeteners

Humans tend to have 'a sweet tooth', and prefer sweet dishes and beverages such as tea and coffee sweetened by the addition of sugars. Although in some countries sugars such as maltose are considered preferable, in most countries sucrose is the main sweetening agent used. Sucrose is produced from a variety of plant sources, such as beet and cane.

For industrial food processing, the use of sucrose as a sweetener presents some problems. The worldwide availability of sugar (sucrose) is limited and fluctuates year by year, and likewise so does the cost to the food industry of buying sugar. Sucrose has a relatively low solubility in water, which means that in some processed foods and drinks, the sugar can crystallise out, especially if the food or drink is cooled (which reduces the solubility of the sugar).

Assignment 14.7
What problems would be caused by the crystallisation of sugar in processed foods or drinks?

Sucrose is a disaccharide, composed of a molecule of glucose linked to a molecule of fructose. Both glucose and fructose have higher solubility in water than sucrose, and, although glucose is not as sweet as sucrose to the palate, fructose is sweeter than sucrose and is thus known as a high sweetener. Thus, glucose and especially fructose are suitable alternatives to sucrose for use by the food industry.

Glucose and fructose occur naturally in plants, especially in fruits. However, demand by the food industries in developed countries could not be met by extraction of these sugars from fruits. The enzyme invertase, produced by the yeast *Saccharomyces cerevisiae* and recoverable from autolysed yeast cells, hydrolyses sucrose to glucose and fructose and is used industrially on a limited scale to invert (hydrolyse) sucrose solutions for specialist applications. A plentiful alternative source of glucose is starch, which is composed of a polymer of glucose. Thus, if starch is fully hydrolysed (saccharified), glucose is the product. Although starch can be hydrolysed by acid hydrolysis, bitter flavours are produced as a result, so enzymatic hydrolysis is preferred. Glucose and fructose are isomers of each other (they have the same chemical composition but the atoms in the two molecules have different spatial orientations), and a microbial enzyme that catalyses the conversion of glucose into fructose is available. On this basis, industrial scale processes for enzymatic conversion of starch into glucose and fructose were developed.

Starch is widely available in grain (maize, wheat, rice) and root (potato, cassava) crops. Starch occurs in two forms: amylose and amylopectin. In amylose, glucose molecules are linked in a straight chain by bonds formed between carbon atoms 1 and 4 on adjacent glucose molecules (see Fig 14.3). This type of bonding is known as α-1,4 linkage.

In amylopectin, straight chains of glucose molecules (joined by α-1,4 linkage) are linked to each other in a branched fashion through carbon atoms 1 and 6 on adjacent glucose molecules (see Fig. 14.3). This type of bonding is known as α-1,6 linkage. The proportion of amylose and amylopectin found in starch varies from plant to plant.

(a) Gelatinisation of starch

The first stage in the saccharification process is to solubilise the starch, as starch is present in roots and grains in the form of insoluble and relatively large granules on which enzymes are not highly effective

> ***Assignment 14.8***
>
> How can insoluble starch be converted to a soluble form on which enzymes will be effective?

Thus, the crushed grains or chopped tubers are mixed with water and heated to temperatures of 100°C or more. A heat stable form of the enzyme alpha-amylase (α-amylase), produced using the bacterium *Bacillus licheniformis* or *B. strearo-thermophilus,* is also added at this stage to enhance gelatinisation. Calcium ions (as calcium chloride) are added as they stabilise the enzyme at high temperatures. The mixture is held at high temperatures until gelatinisation (swelling of the starch granules) has occurred.

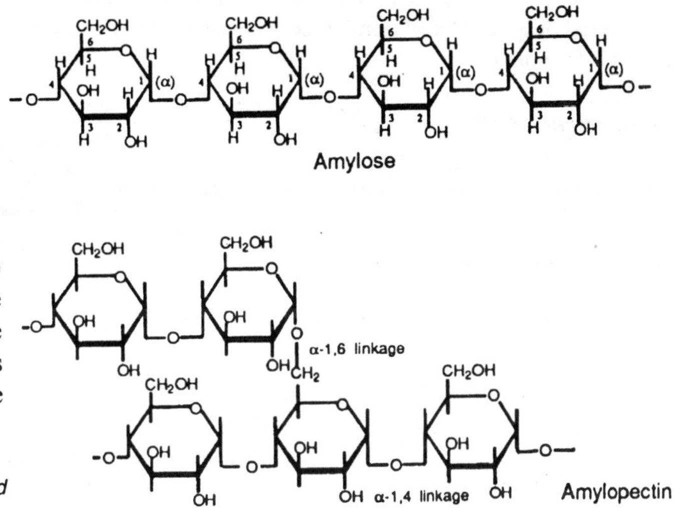

Fig.14.3 *Structures of amylose and amylopection*

(b) Dextrinisation of gelatinised starch

Once the starch granules have gelatinised, the temperature of the solution is lowered to about 90°C, and the mixture is held at this temperature for 60–90 minutes. During this time, the α-amylase enzyme (which remains active despite the high temperatures) randomly hydrolyses α-1,4 links in amylose and amylopectin, producing a mixture of short glucose polymers (2–8 glucose molecules in length) known as dextrins.

The process is thus called dextrinisation. α -1,6 linkages and adjacent α-1,4 linkages are not hydrolysed by α-amylase (Fig. 14.4).

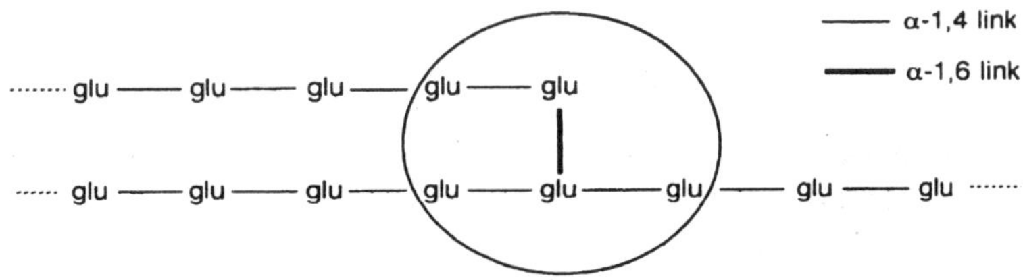

Fig. 14.4 *Dextrinisation of gelatinised starch. Bonds within the encircled area are not hydrolysed by α-amylase*

c) Saccharification of dextrins

The next step is to hydrolyse (saccharify) the dextrins to glucose, and this is achieved by the addition of the enzyme amyloglucosidase (also called glucoamylase) produced using species of the fungi *Aspergillus* or *Rhizopus*. This enzyme rapidly and sequentially hydrolyses the α-1,4 links in dextrins (thus forming glucose) starting with a terminal glucose molecule. It does hydrolyse the α-1,6 links in amylopectin, but at a slow rate. The enzyme does not rapidly hydrolyse dextrins containing more than 5 glucose molecules.

> ***Assignment 14.9***
>
> In the saccharification process, residual α-amylase enzyme as well as the added amyloglucosidase will be active. Predict how α-amylase activity will affect amyloglucosidase activity (will α-amylase activity depress or enhance amyloglucosidase activity, or will there be no effect?).

In practice, saccharification is carried out at pH 4.0–4.5 and at a temperature of about 60°C (as the amyloglucosidase is not as heat stable as α-amylase). It is worth remembering that this process deals with concentrated starch and sugar solutions, which have to be handled at high temperatures in order to reduce their viscosity. Although saccharification is initially rapid, the rate then slows down because as time passes an ever-increasing proportion of the dextrins contain α-1,6 linkages and are thus hydrolysed only slowly. Maximum conversion of dextrins into glucose (in practice about 95%) occurs after 45–50 hours.

Assignment 14.10

How could the rate of saccharification be speeded up?

Self assessment exercise 14.6

Which of the following would you expect to produce glucose from amylopectin a) most rapidly and b) least rapidly?

(i) α-amylase (1 unit per set volume).
(ii) Amyloglucosidase (1 unit per set volume).
(iii) Amyloglucosidase (0.5 units per set volume) and α-amylase (0.5 units per set volume).

This process of saccharification can result in a glucose syrup containing about 30% dry solids, of which 96% is glucose. Such glucose syrup is widely used as a sweetener in drinks, processed sweet foods, and some pickles. Production worldwide is about 7 million tonnes (1 tonne = 1000 kg) of glucose per year.

(d) Isomerisation of glucose

To convert glucose in syrups into fructose, the enzyme glucose isomerase is used. Because of the need (in process efficiency) to use concentrated solutions, glucose solutions need to be maintained at high temperatures to reduce viscosity, so a thermostable glucose isomerase enzyme produced using species of the bacteria *Bacillus, Actinoplanes* or *Streptomyces* has been used. In a batch system (with the enzyme evenly mixed with glucose syrup), glucose is converted at an even rate to fructose over a period of one hour, to give about 36% conversion. Thereafter, the rate of conversion lessens until after about three hours, conversion is maximum at about 44%. In practice, a reaction time of about 2.5 hours is used giving 42% conversion, and the 42% fructose/58% glucose mixture is known as high fructose syrup (HFS 42).

Because of the advantages of using a continuous conversion system with immobilised enzyme, large scale systems using this principle were developed for fructose production.

Self assessment exercise 14.7

Give three advantages of using immobilised glucose isomerase in a continuous system compared to using free enzyme in a batch system.

Systems have been developed with glucose isomerase encapsulated in 1–2mm particles (Section 7.2.4) in a packed bed reactor (Section 7.5.1). Reactors 1.5 m in diameter and 4–5 m in height and containing up to 4000 kg enzyme have been used, with the capability of producing 500,000 kg fructose per day. 42% conversion of glucose syrup to fructose is initially achieved with a column throughput time of less than an hour, but the activity of the immobilised enzyme decreases with time, until after about 4 months throughput times of four hours or more are needed. When this occurs, the process loses its economic advantage and the enzyme is replaced and the whole process repeated.

14.5 Lactic acid fermentations

Throughout the world a wide variety of foods (based on vegetable, dairy and meat products) are produced domestically and industrially by means of lactic acid fermentation. Lactic acid bacteria are widespread in nature, and are usually found in milk and on vegetable material. Domestically, and in some industrial processes such as the production of soya sauce (Section 14.2.2), these naturally occurring organisms are used to carry out the lactic acid fermentation, rather than cultures of the organisms being inoculated into the food. Heating, low moisture content and salt addition are strategies used to suppress competing or spoilage organisms in the foodstuff and thus create conditions in which lactic acid bacteria can predominate. In contrast, most industrial scale processes for conversion of dairy products use starter cultures (Section 14.3.2) as inoculants.

Lactic acid bacteria ferment sugars in the food to lactic acid, which contributes to: preservation (by creating low pH at which spoilage organisms find it difficult or impossible to grow); taste (characteristic sharpness and acidity); texture (dairy products tend to curdle or thicken, and vegetables tend to soften). Certain lactic acid bacteria also produce characteristic flavour and aroma compounds in food, such as diacetyl that has characteristic flavour and aroma. Some lactic acid bacteria secrete proteolytic enzymes, which contribute to flavour production in cheese by creating various amines as a result of protein degradation.

14.5.1 Cheese production

Converting milk into cheese is a means of preserving the valuable milk proteins and fats (as cheese is much less susceptible to microbial spoilage than milk), while at the same time creating a food with taste and texture very different from the original milk.

There are many different variations in the industrial process of cheese manufacture, according to the variety of cheese being produced and to some extent the manufacturer's preference. Milk is received into the production unit, where it is usually pooled and standardised (the fat and protein content may be adjusted to required levels, if necessary, by blending milk batches or by additions of cream or milk powder). The milk may then be homogenised (to evenly disperse the cream in it) and then pasteurised and cooled.

> ***Assignment 14.11***
>
> Give at least one reason why milk is pasteurised before being made into cheese.

The milk, now referred to as cheese milk, is run into the production vessel, called a cheese vat. This is a stainless steel cylindrical vessel rather like a conventional fermenter (Section 8.2.1), but placed horizontally rather then vertically, with temperature control by means of a jacket or internal coil (Section 8.2.2). A rotatable shaft to which blades are attached runs horizontally through the vat. The milk is heated to 30–40°C, rennet or chymosin preparation is added to the milk along with an active starter culture or cultures of more than one organism. The mixture is then incubated for up to 4 hours with gentle mixing by rotation of the blades. During the incubation period, the action of rennet or chymosin (Section 14.4.1) and lactic acid produced by starter culture organisms (Section 14.3.2) leads to coagulation (gelling or curdling) of the milk. Curdling of the milk is further stimulated by a more vigorous chopping of the curdled milk for a period by the rotation of the blades. During this process, the proteins and lipids (fats) in the milk form an insoluble white coagulum (curds) while the remainder of the milk separates out as a straw-coloured solution (whey). The mixture is then removed from the vat and drained to remove the whey. This product may be sold as 'soft curd' cheese. The expulsion of whey from curds may then be stimulated at this stage by heating the curds and/or pressing (squeezing). This process produces 'young cheese'. Salt and perhaps colouring agent may be mixed into the young cheese, which is then formed into blocks. The blocks are then stored for up to several months, usually in a controlled environment, in order for ripening (maturation) to occur.

Ripening of cheese may involve secondary fermentation of residual sugars by starter organisms, and a limited degree of proteolysis by rennet enzymes or chymosin and protease enzymes released by some species of starter organisms. For production of mould ripened cheeses, prior to maturation, blocks are pierced to produce air channels and inoculated with spores of a filamentous fungus, such as *Penicillium roqueforte.* In what is really a form of solid substrate fermentation (Section 8.4), the fungus grows and produces protease and lipase enzymes that enhance degradation of proteins and fats. The breakdown products of this degradation, amines and fatty acids, respectively, impart the characteristic flavour and odour on the cheese.

Self assessment exercise 14.8

The flow diagram on the next page represents the production of mould ripened (blue) cheese. Some of the process steps (boxed) have been left out, and are listed below. Complete the diagram by fitting the process steps into their appropriate places.

Self assessment exercise 14.9

Having examined the cheese making process, explain how the processing of milk to form cheese makes the cheese more stable (less susceptible to microbial spoilage) than the milk from which it was produced.

One major problem that can beset cheese production units is infection of starter organisms by bacteriophages. A bacteriophage (phage for short) is a virus that has a bacterium as its host. A virulent phage kills the host cell in the course of its replication cycle; one phage particle infects the cell, and rapidly (perhaps within 30 minutes or so), the phage is replicated within the cell and the cell bursts and as many as several hundred new infective phage particles are released.

Assignment 14.12

Explain what effect phage infection of starter cultures might have on the cheese production process.

It is vital that starter cultures are free from phage infection, and so during the production of starter cultures (Section 14.3.2), rigorous sterilisation and handling procedures are enforced in order to prevent phage infection of starter organisms, and thorough quality control checks are made on the starter cultures before they are dispatched to production units. An advantage of producing starter cultures in a unit geographically remote from production units is that in such a location it is easier to maintain an environment free from phages (away from areas in which cheese is constantly being produced and phages thus have opportunity to multiply). In production units, phage infection is restricted by thorough cleaning and steam heating of cheese vats and pipe work between batches to eliminate any phages left over from the previous batch of cheese milk. Pasteurisation of milk before inoculation with starter cultures should kill phages already present in the milk. As phage infection is usually strain specific (a particular phage usually is capable of infecting only a certain strain or strains [sub-types] of a particular species of bacterium), starter culture failure can be prevented by inoculating cheese milk with several different strains of the starter organism, or by using different individual strains or combinations of more than one strain in rotation.

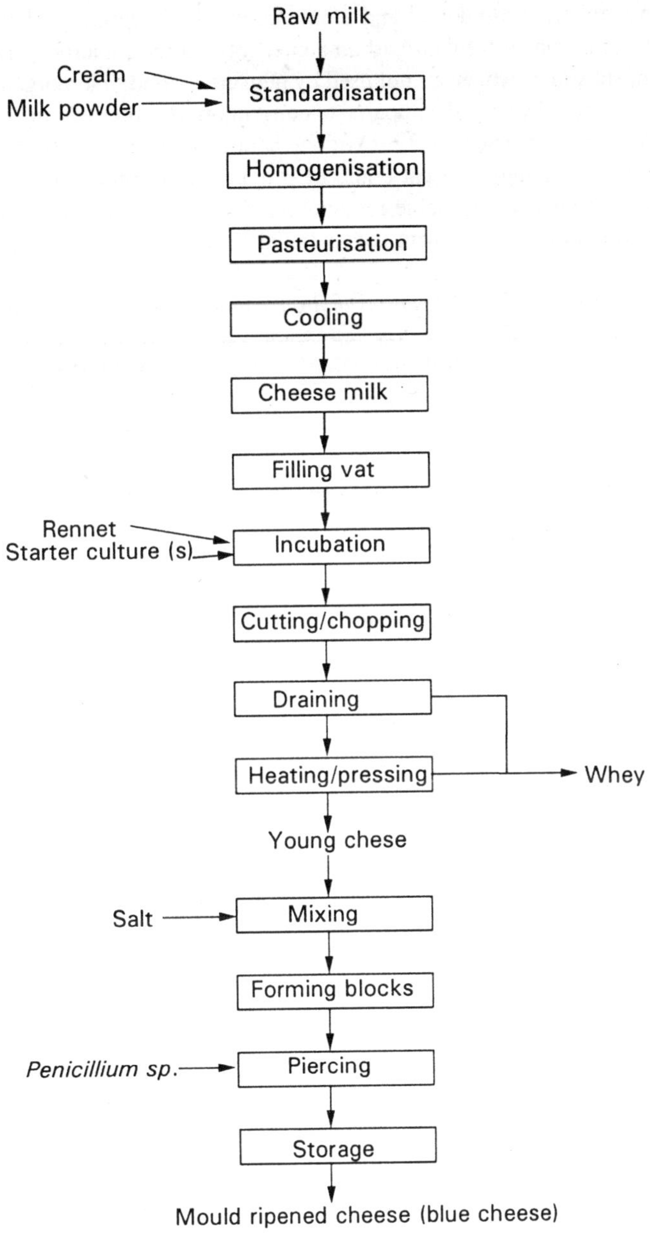

Incubation Mixing Draining Standardisation Heating/Pressing
Piercing Pasteurisation Cutting/Chopping Forming blocks
Homogenisation Storage Cooling Filling vat

14.6 Microbes as a source of food

Using microorganisms as food is not a new concept: the Aztecs in South America harvested filamentous photosynthetic bacteria (called cyanobacteria) from lakes and rivers as food, and for many centuries algae (seaweeds) and fungi (mushrooms) have been collected from the wild and eaten.

Growing microorganisms using industrial scale processes is a more modern concept, which, apart from mushroom cultivation that started earlier, dates back to about 1960. Remember that at this time the world population was increasing rapidly. Europe, Japan and other areas were recovering from the devastation of war, and it was projected that by the end of the century there would be a severe global shortage of protein for (human food) and feed (animal fodder). It seemed logical to turn to microorganisms as an alternative protein source as: on the basis of dry matter, microbial cells contain 40–60% protein; many microorganisms can use ammonia as a nitrogen source, and can thus create protein from ammonia, and ammonia can be manufactured relatively cheaply from abundant atmospheric nitrogen; for their carbon and energy sources, microorganisms are capable of using a wide variety of substrates which are not used by humans as food; microorganisms can grow (and thus produce protein) very rapidly indeed compared to convention food production, for example one cell of a bacterium with a doubling time (Section 8.6.1) of one hour could theoretically produce a mass the size of the Earth in three days if enough nutrient could be provided to it; production units for microbial food processes could be built on land unsuitable for agriculture and can have outputs of protein per unit land area far in excess of anything possible with conventional agriculture (see data for algal ponds in Table 13.1).

Table 14.4 *Use of various organisms in diverse production systems.*

Organism	Substrates	Production	Uses
Mushrooms (various fungi)	Cellulosic waste, dung	Solid substrate fermentation	Food
Various filamentous fungi	Cellulose/starch waste, ammonium salt	Solid substrate fermentation	Feed
Spirulina spp. (blue-green) bacterium)	CO_2 and N_2 from air, sunlight	Open ponds	Food
Chlorella spp. Scenedesmus spp. (algae)	CO_2 from air, nitrate, sunlight	Open ponds	Feed
Chlorella spp.	Molasses, nitrates	Conventional fermenters	Food
Candida spp. *Saccharomyces* spp. (yeasts)	Molasses, ammonia	Conventional fermenters	Food
Candida spp. *Saccharomyces* spp. *Kluyveromyces* spp. (yeasts)	Whey, ammonia	Conventional fermenters	Food
Candida spp. *paecilomyces* sp. (yeasts)	Wood pulp waste,	Conventional fermenters	Feed
Fusarium sp. (filamentous fungus)	Glucose, ammonia	Air-lift fermenter	Food
Candida spp.	Ethanol, ammonia	Conventional Fermenters	Food
Candida spp.	n-Alkanes, ammonia	Conventional Fermenters	Feed
Methylophilus spp. (bacterium)	Methanol, ammonia	Pressure-cycle fermenter	Feed

Several different microbial food and feed processes have been developed, using a variety of organisms in widely different bioreactor systems (Table 14.4). The microbial cell product from such systems is often referred to as single cell protein, or SCP, even though many processes use multicellular organisms. The significance of the distinction between food and feed from a process point of view is that food must conform to more rigid quality standards (in terms of purity of substrate used and purity of the product formed) than feed needs to.

Algal ponds can have high output per area but require sunlight and are open to contamination. Unicellular algae can be difficult and expensive to harvest from the medium in which they are grown. For use as food, the cells need to be disrupted by grinding, as humans are unable to effectively digest the cellulosic cell walls of intact algal cells; this is not a problem if the product is used as feed for ruminant animals, as these beasts can effectively digest cellulose. Algal and cyanobacterial protein products are produced in modest quantities, and have a small market niche as high quality speciality proteins.

Several processes have been developed using oil-based substrates. A great deal of technological development went into these processes, for example, the pressure-cycle fermenter (Section 8.3) was developed expressly for microbial protein production, and was developed up to a scale of 2 million litres working volume. However, despite the successful technological development, oil-based products did not meet with success in the marketplace.

Several processes have been developed using carbohydrate-based substrates. Yeasts have been produced from various substrates, but have found a market niche as a food additive (Section 14.2.3) rather than a protein food as such. The one process that has found success in the marketplace is the production of the filamentous fungus *Fusarium graminearum*. In this process, the fungus is grown in a medium based on food-grade glucose syrup (Section 14.4.2) supplemented with ammonia, mineral salts (phosphate), and choline and biotin that the fungus requires as growth factors. The bioreactor used is a pressure-cycle fermenter, operated as a continuous culture that operates most efficiently (Section 8.6.4) at a dilution rate of 0.2 h^{-1}. The continuous product stream from the fermenter is passed through a cyclone (whirlpool) device to separate the fungal biomass from the medium. The concentrated biomass suspension is then heated to 64°C for about 20 minutes. During this time, heat stable RNase enzyme within the fungal cells degrades the RNA within the cells, reducing levels from 10% to 2%. This reduction in levels of RNA is important, as compared to animal or plant tissue, microbial cells contain relatively high levels of nucleic acids (particularly RNA), and in humans, accumulation of uric acid which results from degradation of nucleic acids may cause health problems. For this reason, it is recommended that the daily intake of nucleic acid in the human diet does not exceed 4 grams. At the same time, the heating process inactivates intracellular protease enzymes, which renders the fungal biomass product less susceptible to autodegradation. The fungal biomass is then dewatered by vacuum filtration forming dryish fibrous material (a filamentous fungus grows in the form of many filaments, called mycelium. Material from egg white is mixed with the mycelial filaments, which are then partially aligned by passing between rollers, and then steam heated to set the egg white. In this way, material with a satisfying meat-like texture is produced. The product is preserved by deep freezing or drying. The fungus itself has a mild flavour, so it is well suited for inclusion as a meat substitute in savoury dishes. The product, trade named MycoProtein or Quorn, is produced in the UK, where it has established a market niche as a vegetarian meat substitute.

Self assessment exercise 14.10

Draw a flow diagram to show the production of MycoProtein

Assignment 14. 13

Give at least two reasons why, in general, microbial protein foods have not met with success in the marketplace.

14.7 Alcoholic beverage production

The production of alcoholic beverages is carried out on an industrial scale in most countries in which alcohol consumption is considered to be acceptable, and worldwide, many different types of beverage are produced. Beers are usually manufactured using grains (particularly barley) and hops, and in most countries, a lager-type beer is produced based on recipes and technology from Central and Eastern Europe. Wines are produced traditionally from grapes, although other fruits or sap from trees are frequently used to produce local varieties of wine. Distillation of beers or wines is used to produce distilled liquors.

14.7.1 Beer production

(a) Malting

Most beers are produced from malt, which is barley which has been germinated and then roasted. For this, barley grains are placed in a tank or bin, covered with water, and soaked (steeped) for about 24 hours. The barley grains absorb water and swell. The grains are then spread out by hand, or some sort of mechanical device, into a shallow layer to germinate. The grain layers are 10–20 cm in depth, so that grains are exposed to oxygen that is necessary for germination. At intervals, usually at least daily, the grains are 'turned' to keep grains evenly exposed to air.

Traditionally, this was done with a malt shovel, a large wooden flat shovel. Modern malting processes are more mechanised and involve blowing air through the bins of grain while the steeping process is taking place. The function of germination is to stimulate the formation of amylase and protease enzymes in the barley grains. The grains are allowed to germinate until the emerging roots are of a certain length, at which stage, sufficient enzymes will have been formed. This process of enzyme induction is akin to the koji process in soya sauce production (Section 14.2.2).

Once barley grains have reached the required stage of germination, they are roasted in a kiln. Normally temperatures of 60–70°C are used. Roasting is done to dry the malt so that it can be stored without suffering from microbial spoilage, and to caramelise sugars (oxidise them upon heating). Caramelised sugars are brown in colour and have characteristic flavours. Thus, when malt is extracted with water during mashing, colour and flavour are given to the beer. If the grains are lightly roasted, a lager-type of beer results, with a light golden colour and soft flavour. If grains are roasted for a longer period of time, a darker type of beer known as ale is produced. If grains are roasted for even longer, then a strongly flavoured, almost black type of beer known as stout results.

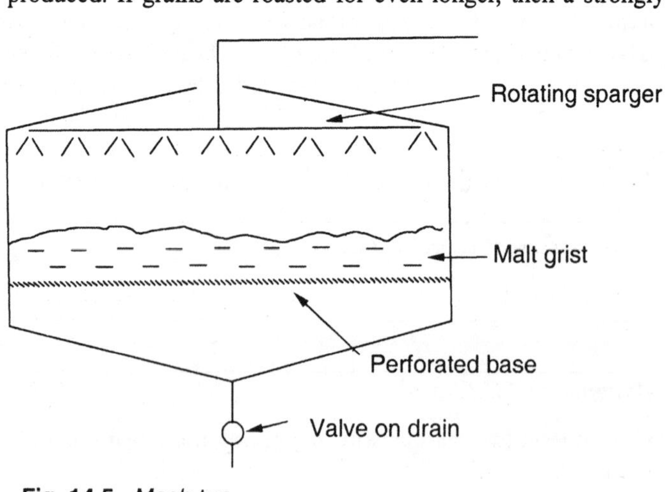

Fig. 14.5 *Mash tun*

(b) Masting

In the next stage of the process a fermentable liquid, called wort (pronounced wert, as in perfect), is produced by a process known as mashing carried out in a vessel known as a mash tun. A mash tun is a low cylindrical vessel with a perforated or slatted base (Fig. 14.5).

Coarsely ground malt grains are layered on the base and are covered in water for 1–2 hours. During this time the grains rehydrate and the amylase and protease enzymes become active. The activity of amylase and protease enzymes during mashing is necessary to convert starch and proteins in malt into forms in which they become available to the yeast used in brewing; brewers' yeasts do not secrete amylases or protease enzymes and thus would not grow well on a mash made from barley. Protease enzymes hydrolyse protein polypeptides into short chain polymers and amino acids. The amylase formed in malt is beta-amylase (β-amylase) which acts on amylopectin (the major component of starch in malt) by hydrolysing the long polymer at alternate a-1,4 bonds (Section 14.4.2). The predominant product of this activity is formation of disaccharide of glucose, maltose. However, as b-amylase does not hydrolyse a 1,6 bonds, a proportion of the short branches of the amylopectin molecule remain unhydrolysed and are known as limit dextrins. Yeast is able to ferment maltose, but is unable to utilise dextrins. The dextrins thus remain in the beer, and they contribute to its flavour.

The temperature at which mashing takes place is important. Protease enzymes from malt work best at around 40°C, whereas amylases from malt work best at around 60°C.

Assignment 14.14
Predict how the nature of the beer produced might be altered by changing the mashing temperature from 40°C to 60°C?

After soaking for 1–2 hours, the liquid (now called liquor or wort) is drained away and collected in a tank. Hot water is then sprayed over the grist by a rotating sparger (arm with nozzles), in order to extract any remaining nutrients from the malt grains. This liquid is added to the tank containing the drained liquor. The liquor produced in this way is referred to as sweet wort, as the maltose in it gives it a sweet taste.

In some processes, a quantity of unmalted grain (such as barley, wheat, rice or other locally available grain) may be added to mash in place of malt, to provide a relatively cheap source of starch and perhaps to produce a beer with characteristic flavour. Commercially available amylase enzymes may also be added in this case to compensate for the reduction in levels of amylase from malt.

(c) Hopping

Hopping involves adding hops to the sweet wort and boiling the wort for about one hour in a vessel traditionally called a kettle. Hops are added to beer primarily to give it flavour; hops contain resinous substances called tannins that have bitter flavours. Hop resins also help to clarify beer after fermentation. They have antibacterial activities that help to prevent microbial spoilage of beer. The boiling process extracts resins (tannins) from the hops, condenses the wort (strengthening the flavour), inactivates proteases and amylases (the activity of which is no longer required — their activity in the finished beer would produce undesirable changes in flavour during maturation and storage) and kills most microorganisms (which would compete with yeast for nutrients in the wort, and could cause spoilage of the finished beer).

After boiling, the hops are removed. This is carried out by straining the wort through a sieving device called a hop press, or passing the wort through a whirlpool or cyclone in which the wort is forced into a whirlpool in such a way that the solid hop material is forced to the bottom of the vessel, from which it is drained off. The wort is now referred to as bitter wort, as the extracted hop tannins give it a bitter taste. The wort is cooled to fermentation temperature by passing it through a heat exchanger and is then run into a fermentation vessel.

(d) Fermentation

Conventionally, beer is produced in a batch fermentation system in a beer fermentation vessel (Section 8.2.1). Once the vessel has been filled to the required level with bitter wort, it is inoculated with beer yeast. This process is traditionally referred to as pitching. For ales and stouts, strains of *Saccharomyces cerevisiae* are used, whereas for lagers, organisms such as *Saccharomyces carlsbergensis* and *Saccharomyces ovarum* are used. Each brewery usually uses its own selected strain of yeast, which is carefully maintained under laboratory conditions. Ales and stouts are fermented at 20–25°C, whereas lagers are fermented at 10°C or less.

The boiling process drives dissolved oxygen from the wort, and although running the wort into the fermenter causes some aeration, the initial oxygen levels in wort are low. Because of this, in some processes, the wort is initially aerated (or roused) until yeast concentration in the wort increases. This can increase the overall yield of ethanol in the process (Section 8.7.1).

Initially, fermentation is very vigorous, with so much carbon dioxide by-product given off that the wort looks as if it boiling, and a frothy foam develops at the surface. After a few days, the rate of fermentation slows down, gas production eventually ceases, and the yeast cells flocculate (clump together) and settle to the bottom of the vessel. When using modern vessels, the flocculation of yeast may be stimulated by cooling the wort. Yeast collects as a creamy dense suspension in the conical base of the fermenter, from which it can be run off.

Assignment 14.15

It is important to remove as much of the yeast as possible from the wort when fermentation has ended. Why is this the case?

The strains of yeast for use in beer production are partly chosen for their ability to flocculate, so that usually the beer becomes clarified by this flocculation. If the beer does not clarify spontaneously, substances (called finings) may be added to stimulate the process. For beer clarification, isinglass (a glycoprotein extracted from the swim bladders of sturgeon fish) is traditionally used as finings, although substitutes of plant origin are now often used.

Continuous fermentation systems for production of beers have been operated on an industrial scale, and continuous systems using immobilised yeast cells at industrial scale are under development. A complication with production of beers in these ways is that flavour and aroma are extremely important to consumers, and flavour development may not be comparable in beers produced in batch and continuous processes.

(e) Maturation

Directly after fermentation, the beer usually does not taste mellow; instead it can taste 'raw' and unpalatable. Maturation involves storing the beer for a period of time so that it mellows and becomes more palatable. This involves both biochemical and chemical processes. Biochemical activity involves secondary fermentation, in which residual yeast in the wort ferments residual sugars. Chemical reactions involve the conversion of unpalatable compounds (such as sulphides) into complexes that taste more palatable. Some beers are matured in barrels (casks), and are thus called cask-conditioned beers. Maturation of other beers is carried out by holding the beer in the fermenter after the yeast has been removed, or transferring it to large stainless steel tanks (maturation tanks). At the end of the maturation period, the beer may be filtered to remove any yeast that remains suspended in the beer.

f) Packaging

Beer not intended for long term storage may be packaged in casks or kegs. Alternatively, for long term storage, beer may be carbonated with carbon dioxide gas and packaged in bottles or cans. The bottles or cans are then pasteurised to kill any spoilage microorganisms they may contain.

14.7.2 Applying genetic engineering to brewing yeasts

> *Assignment 14.16*
>
> From what has been discussed already about the desirable characteristics of brewing yeasts, state what traits of these yeasts, would it be of interest to the brewer to improve by genetic engineering?

Because of the economic importance of alcoholic beverages, much research has been carried out throughout this century on the genetic makeup of brewing yeasts and how the genotype influences characteristics of the organisms that are of importance to the brewer. This knowledge made genetic engineering of yeasts that much easier when the technology to do so was developed. Yeast plasmids occur naturally, and some of these have been modified to make them suitable for genetic engineering, for example by insertion into them of selection marker genes such as genes which code for resistance to copper ions or the antibiotic gentamycin (brewers' yeasts being sensitive to copper and gentamycin). Strong constitutive promoters from yeast genes are known, as well as leader sequences that facilitate excretion of enzyme from yeast cells.

Some yeast plasmids are multi-copy, which means that the plasmid replicates within a host cell and exists there as many copies, which can thus lead to high expression levels of the transgene (the foreign gene inserted in each plasmid may be expressed). However, multi-copy plasmids may be relatively unstable in host cells and the plasmids and transgenes can become deleted (lost from the cell) during cell replication.

Some yeast plasmids have had a sequence from a natural yeast chromosomal gene inserted into them, which means that, once the plasmid is within the yeast cell, by the process of homologous recombination, this sequences on the plasmid may crossover (or swap) with the corresponding sequence on the chromosomal gene (an example of this was discussed in Section 14.4.1). Such plasmids are known as integrative vectors. If the transgene is inserted into such a plasmid so that it is sandwiched within the chromosomal gene sequence, then the transgene can become inserted into the chromosomal gene during recombination, and in this form the transgene is usually very stable, as it is replicated normally as the chromosome is replicated.

(a) Giving yeasts the ability to hydrolyse dextrins

Brewing yeasts are able to uptake and ferment only mono-, di- and tri-hexoses (dextrins with 1–3 glucose molecules). Larger oligosaccharides cannot be utilised (they are said to be 'unavailable' to yeast). Up to 25% of the sugars in wort are composed of oligosaccharide dextrins. They derive from the short glucan branches on amylopectin which are not hydrolysed to sugars (maltose) by amylases from malt (β-amylases) during the mashing process (Section 14.7.1(b))

> *Assignment 14.17*
>
> What would be the advantage to the brewer of using a yeast that is capable of fermenting the dextrins in wort?

Dextrins in wort contribute to the flavour (body) of beer, and so beers lacking dextrins have a mild, light flavour and are referred as 'light' or 'lite' beers. Up to now, such beers have been produced by adding commercially available glucoamylase enzyme (α-1-4 glucan glucohydrolase) to wort during the mashing or fermentation stages in order to bring about hydrolysis of dextrins.

> ***Assignment 14.18***
>
> What would be the advantage to the brewer of using a yeast that produces its own glucoamylase, rather than using commercial enzyme?

Brewing yeasts were first transformed with the *STA2* gene which codes for production of glucoamylase from *Saccharomyces cerevisiae* var. *diastaticus*. Initial experiments inserted the *STA2* gene into a construct containing the *CUP1* gene (which codes for resistance to copper ions) as the selection marker, inserted into a multi-copy bacterial/yeast chimeric (species hybrid) plasmid. This plasmid transformed both ale and lager yeasts, and resulted in the synthesis of extracellular glucoamylase. However, as is common for such multi-copy chimeric (fungal/bacterial) plasmids, they were unstable and growth and fermentation rates were reduced upon transformation. In addition, the glucoamylase lacked α-1-6 debranching activity, which limited dextrin hydrolysis.

An improvement was to construct an all-yeast multi-copy plasmid, which contained the *STA2* gene flanked (upstream) by a constitutive promoter from the yeast *PGK* gene and (downstream) by the *PGK* gene terminator sequence. The *CUP1* gene was also inserted to allow selection of tranformants. A lager yeast transformed by the plasmid produced ethanol at a similar rate to the original yeast and produced a final concentration of ethanol in wort 1% greater.

To improve dextrin hydrolysis even further, genes coding for glucoamylase with α-1-6 hydrolytic activity have been cloned and used to transform brewing strains. A suitable gene from the filamentous fungus *Aspergillus niger* was fused to the coding sequence of a signal peptide of a known gene, and flanked by the constitutive promoter and terminator sequences of a gene coding for a known enzyme in the yeast. Included in the expression cassette were sequences of the *HO* gene of yeast Chromosome IV. This expression cassette was inserted into a plasmid and after introduction into a production strain of yeast it became integrated by homologous recombination into the *HO* gene.

The significance of this was that the *HO* gene is involved in yeast mating (sexual reproduction), which does not occur in production strains of brewers' yeast, so it was anticipated that incorporation of the transgene into the redundant *HO* gene would be unlikely to adversely affect the organism's brewing characteristics. This proved to be the case, and in this organism, the transgene has proved to be stable over a number of years. The advantage of using an *Aspergillus* gene is that *Aspergillus* spp. have traditionally been associated with food products (we have seen that *Aspergillus oryzae* is used in the Koji process used to make soya sauce, and *Aspergillus* spp. are used in the fermentation of citric acid for use in the food industry), and so these organisms are regarded as 'food grade' by regulatory authorities. Approval for use of brewing yeasts transformed by genes from *Aspergillus* spp. is thus made easier.

In an attempt to introduce the glucoamylase gene from *Aspergillus* sp. into yeast so that multiple copies were present in the cell and integrated into the chromosome, the glucoamylase gene from *Aspergillus awamori* was inserted into an expression cassette flanked by a constitutive promoter and a terminator sequence terminator. This cassette was itself flanked by sequences from the yeast gene that codes for rRNA (ribosomal RNA). In yeast, there are 100–150 copies of this rRNA gene, arranged on Chromosome XII as a series of tandem repeats. The expectation was that the transgene would (by homologous crossover recombination events) become integrated into Chromosome XII, and would then be amplified by repeated crossover events. This eventually occurred, and studies with gene probes and Southern blotting showed that transformed cells contained up to 100 copies of the glucoamylase gene. Unfortunately, pilot scale trials showed that, contrary to expectation, the transformed yeast was not capable of complete utilisation of dextrins.

One problem for the brewer with the use of yeasts expressing glucoamylases from *Aspergillus* spp. is that these enzymes are relatively heat stable. Thus, in canned and bottled beers that are designed to have a long shelf life and are pasteurised to prevent spoilage during storage, glucoamylase activity remains after pasteurisation and can lead to unacceptable changes in flavour during long storage periods.

> ***Assignment 14.19***
>
> How could genetic engineering be used to overcome the problem of glucoamylase activity remaining after pasteurisation?

> ***Self assessment exercise 14.11***
>
> Which of the following would you expect to be most effective in producing a beer low in dextrins?
>
> (i) Yeast transformed with a gene for glucoamylase from *Aspergillus* sp. and integrated into the *HO* site on yeast Chromosome IV.
> (ii) Yeast transformed with a gene for glucoamylase from *Aspergillus* sp. incorporated into a multi-copy plasmid
> (iii) Yeast transformed with the α-amylase gene from malt incorporated into the rRNA gene site on the yeast chromosome.

> ***Assignment 14.20***
>
> In what circumstances would it be an advantage to have a yeast that was capable of hydrolysing starches?

(b) Giving yeast the ability to utilise starches

For the complete degradation of starch to available sugars for yeast, both α-amylase and glucoamylase activity is needed. Attempts have been made to engineer yeasts capable of producing a hybrid enzyme having both activities. A gene construct was produced with the α-amylase gene from *Bacillus subtilis* fused with the glucoamylase gene from *Aspergillus awamori*, downstream from a constitutive yeast promoter and a yeast leader sequence. This was expected to result in formation of a fusion protein (hybrid enzyme). Although the recombinant yeast formed in this way was effective in the hydrolysing soluble starch and raw potato starch, it was not effective in the hydrolysis of maize starch (which, like malt starch, is predominantly highly branched amylopectin). This was traced back to defective glucoamylase debranching activity in the fusion protein. Attempts are being made to improve the situation by using genes from other microorganisms and having the two genes expressed independently instead of in the form of a fusion protein.

(c) Giving yeast ability to degrade glucans

β-Glucans from barley are relatively insoluble, and thus can cause hazes, sediments and gels in beer, which may cause processing difficulties (for example the beer may be difficult to filter) or if left in beer may lead to customer rejection of the product. Although malt does contain β-glucanases, these are heat labile and are usually inactivated by the roasting temperature in malt manufacture and during mashing.

Commercially available microbial β-glucanases are sometimes added to mashes or worts, but these suffer from the disadvantages outlined above (see the response to Assignment 14.17). The obvious solution is to engineer beer yeast so that it produces β-glucanases.

The genes coding for β-glucanase from the bacterium *Bacillus cereus,* the filamentous fungus *Trichoderma reesei* and barley have all been cloned. The enzyme from barley shows highest activity at the pH of beer (4.5–4.7) and is more acceptable in food products. The gene from barley has been fused to a yeast signal sequence and placed in an integrative vector downstream from a constitutive yeast promoter. The transformed yeast excreted β-glucanase, reduced the β-glucan content of beer, which in turn improved filterability and reduced haze formation, while flavour and foam stability of the beer were not adversely affected.

(d) Improving the fermentation rates of yeast

Rapid fermentation by yeasts saves time (and hence money) and lessens the chance of spoilage organisms predominating in wort. The major factor that limits the rate of fermentation of maltose (the predominant sugar in malt mashes) by brewing yeasts is the fact that transcription of the genes coding for the maltose permease (*MALT* gene) and for α-glucosidase [which hydrolyses maltose to glucose] (*MALS* gene) are repressed by glucose. Thus, in concentrated worts, which may contain relatively high concentrations of glucose, transcription of the *MALT* and *MALS* genes are repressed and fermentation rates slowed.

> ***Assignment 14.21***
>
> How could genetic engineering be used to overcome repression of synthesis of the *MALT* and *MALS* genes by glucose?

MALT and *MALS* genes have been cloned and inserted into multi-copy plasmids under control of constitutive promoters. Northern blotting and enzyme assays confirmed that *MALT* and *MALS* genes were constitutively expressed in the presence of glucose in transformed yeast. Transformation with the *MALT* gene alone increased the rate of fermentation of maltose, whereas transformation with the *MALS* gene alone did not increase the rate of fermentation of maltose.

> ***Self assessment exercise 14.12***
>
> Answer **true** or **false** to each of the following statements.
>
> (i) A mixed culture containing a yeast transformed with a glucoamylase gene from *Aspergillus* sp. and a yeast transformed with an α-amylase gene from *Bacillus* sp. should be able to utilise starch more effectively than a monoculture of either organism.
> (ii) A mixed culture containing a yeast transformed with the *STA2* gene and a yeast transformed with a glucoamylase gene from *Aspergillus* sp. should be able to utilise dextrins more effectively than a monoculture of either organism.
> (iii) A mixed culture containing a yeast transformed with a α-glucanase gene from barley and a yeast transformed with α-glucanase gene from *Trichoderma* sp. should be able to remove α-glucans from wort more effectively than a monoculture of either organism.
> (iv) A mixed culture containing a yeast transformed with the *MALT* gene and a yeast transformed with the *MALS* gene should be able to ferment maltose in wort more rapidly than a monoculture of either organism.

(e) Improving flocculation properties of brewing yeasts

For modern processes, spontaneous flocculation of yeast at the end of the fermentation phase greatly assists in the production of a clarified beer (if the yeast does not flocculate and sediment out, it would have to be removed by filtration). Flocculation does have a genetic basis, and although several genes involved in the process have been identified and cloned, their exact roles are far from clear. The *FLO1* gene, which codes for a hydrophobic cell wall protein, was shown to be involved in flocculation of flocculent yeasts, and transformation of non-flocculent yeasts with the *FLO1* gene incorporated in multi-copy and integrative vectors under control of constitutive promoters increased flocculation. Unfortunately, such transformed yeasts flocculate during the fermentation period, reducing cell concentrations throughout the wort and thereby reducing fermentation rates; by contrast normal flocculent yeasts express flocculence only at the end of the fermentation period.

> ***Assignment 14.22***
>
> What would be needed in order to transform a non-flocculent yeast into one which flocculated only at the end of the fermentation phase?

> ***Self assessment exercise 14.13***
>
> Would the *MALS* gene regulatory sequence (regulator-promoter-operator) be of use in the control of a *FLO1* transgene introduced into non-flocculent yeast to confer flocculence? Explain your answer.

(f) Improving properties of brewing yeasts to allow reduction in beer maturation periods

After fermentation, the raw beer needs to be matured for a period of days or even weeks, so that off-flavour compounds produced by yeast can be converted by biochemical and chemical conversions into more palatable compounds, and the flavour of the beer thus improved. If the production of off-flavours by yeast during fermentation could be lessened, the maturation stage could be shortened or eliminated, and the overall process time reduced and money saved.

The major off-flavour in raw beer is diacetyl, which is formed during the fermentation phase by the auto-oxidation (chemical oxidation) of α-acetolactate. α-acetolactate is an intermediate in the biosynthetic pathway for valine in yeast (Fig. 14.6).

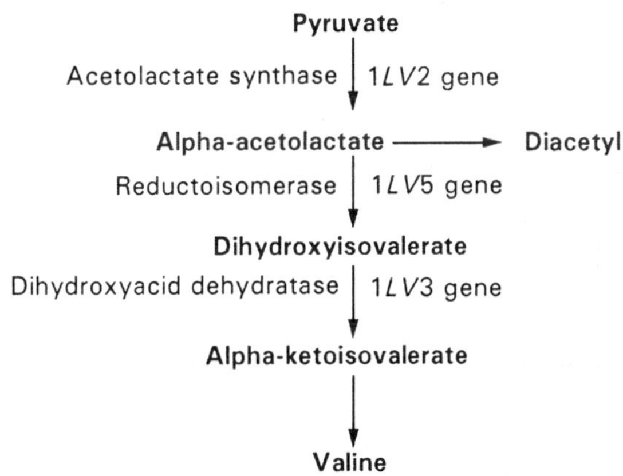

Fig. 14.6 *Valine biosynthetic pathway in Saccharomyces cerevisiae*

Acetolactate synthase activity is inhibited by the herbicide sulphometuron methyl. By isolating brewing yeasts resistant to the herbicide, organisms with an altered (mutant) gene (*ILV3*) for acetolactate synthase were isolated, and were shown to grow slowly on valine-free medium. However, these organisms produced unacceptably slow rates of fermentation. By inserting into brewing yeasts complementary (antisense) sequences to part of the *ILV3* gene, organisms with reduced levels of acetolactate synthase mRNA and with reduced levels of acetolactate synthase activity have been produced. However, these organisms also produced unacceptably slow rates of fermentation

Yeasts have also been transformed with (extra copies of) the gene (*ILV5*) coding for acetolactate reductoisomerase, using multi-copy and integrative vectors. Transformed organisms showed increased levels of reductoisomerase activity and reduced levels of diacetyl formation in worts. In contrast, organisms transformed with (extra copies of) the gene (*ILV3*) coding for dihydroxyacid dehydratase, did not show reduced diacetyl levels in worts, despite having increased levels of dehydratase.

Transformation of yeast using a plasmid carrying both *IVL5* and *IVL3* genes did not result in lowered diacetyl levels in wort, compared to yeast transformed with the *IVL5* gene alone.

Many bacteria possess an enzyme, acetolactate decarboxylase (ALDC), which converts acetolactate to acetoin. Acetoin does not produce off-flavour in beer. The gene for ALDC has been cloned from the bacteria *Enterobacter aerogenes*, *Klebsiella terrigena*, *Lactococcus lactis* and *Acetobacter aceti* var. *xylinum*. Initial trials transformed yeast with the ALDC gene from *Enterobacter aerogenes* under control of a constitutive promoter in a multi-copy plasmid; the transformed yeast produced considerably lower levels of diacetyl in wort. Yeast transformed with the ALDC gene incorporated into the yeast gene for rRNA produced higher levels of ALDC activity (possessing up to 20 copies of the gene) and lower levels of diacetyl. The gene from *Acetobacter aceti* var. *xylinum* has been cloned and inserted into the genome of brewing yeast, and the transformed yeast fermented at normal rates and produced levels of diacetyl in wort below the taste threshold, so that little or no maturation of the beer was required.

Many brewing yeasts produce hydrogen sulphide during fermentation (Fig. 14.7), which, like diacetyl, needs to be removed (converted) during the maturation process.

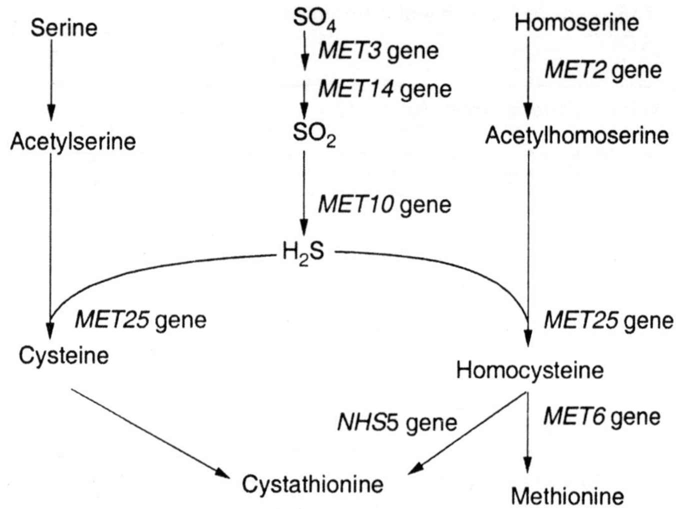

Fig. 14.7. *The sulphate metabolic pathway in brewing yeasts*

Self assessment exercise 14.14

Assume that multiple copies of each of the following genes are inserted into and constitutively expressed in brewing yeast. For each gene, state how its expression might affect H_2S levels in the raw beer produced. Assuming that synthesis of cysteine, cystathionine and methionine are all required for the growth of yeast. Also state whether yeast growth and fermentation rates may be adversely affected.

(i) *MET3* gene.
(ii) Antisense gene to *MET14*.
(iii) *MET2* gene.
(iv) Antisense gene to *MET10*.
(v) Antisense gene to *MET25*.
(vi) *MET25* gene.
(vii) *NHS5* gene
(viii) Antisense gene to *MET6*.

A cloned *NHS5* gene inserted as multiple copies into brewing yeast resulted in reduction of H_2S levels in the raw beer. Similarly, yeast transformed with multiple copies of a *MET25* gene produced raw beer with reduced H_2S levels.

(f) Improving the production of flavour compounds by brewing yeasts

Sulphur dioxide contributes to flavour stability of finished beer by acting as a sequestrating (mopping up) agent for carbonyl compounds which give beer a stale flavour. As we have seen from Fig. 14.7 and Self Assessment Exercise 14.14, inserting (extra) multiple copies of the *MET3* and/or *MET14* genes should lead to an increase in SO_2; this occurred in practice, without appearing to cause any increase in H_2S levels. Antisense genes to *MET10* caused raised levels of SO_2, with lowered levels of H_2S; methionine levels were also lowered, which derepressed the *MET3* and *MET14* genes (methionine causes repression of transcription of the *MET3* and *MET14* genes).

Self assessment exercise 14.15

What effect might expression on an antisense gene to MET2 have on the SO_2 level?

Acetate esters are also important in flavour development of beers. Acetate esters are formed from the corresponding alcohols as:

$$\text{Alcohol } X \xrightarrow[\textit{ATF1} \text{ gene}]{\text{Alcohol acetyl transferase}} \text{Acetate ester } X$$

The *ATF1* gene has been cloned and inserted into laboratory strains of *Saccharomyces cerevisiae*, which produced, increased quantities of acetate esters in culture media. Transformation of brewing strains is underway, and production of beers with improved flavours should be possible using such yeasts.

14.8 Concluding remarks

Using biotechnology in the production of foods and beverages spans many centuries of human history, and in the present day, far exceeds all other biotechnological applications worldwide in economic terms. Microorganisms or their enzymes are used to bring about a wide variety of food and beverage conversions on a domestic scale as well as an industrial scale, and microbial products find many applications in food and beverage processing.

Industrial food and alcoholic beverage producers appreciate the potential of the new biotechnologies, and there is currently great interest in applying genetic engineering, in particular, to improving capabilities of microorganisms used in food and beverage production. These developments are being carried out in parallel with developments in plant genetic engineering (Chapter 12) by which starches, proteins and oils/fats in foods can be modified at source (within the plant) to best suit the needs of the food processors and consumers.

Although in some European countries, genetically engineered yeasts have been used in baking, and recombinant chymosin used in cheese making for several years, the use of genetically engineered organisms and products from them in food remains contentious, and there is a degree of reluctance on the part of some consumers to accept such practices. The argument is particularly concerned with food labelling, and whether or not consumers are being given enough information as to whether or not foods contain products of genetic engineering.

However, in the long-term future, the use of genetically engineered organisms (plants as well as microorganisms) and their products in foods and alcoholic beverages will become more familiar to us all, and is likely to meet broad acceptance. Application of the technology in food and beverage production will do much to meet the ever-increasing demands placed on producers. Although processes to produce microorganisms as food have not in general met with success in the marketplace, a variety of technologies have been successfully developed and are at hand if the need should arise in the future.

14.9 Suggested further reading

1. BIOTOL Series (1992). *Biotechnological innovations in food processing*. Butterworth-Heinemann, Oxford.
2. Crabb, W.D. and Mitchinson, L. (1997). Enzymes involved in the processing of starch to sugars. *Trends in Biotechnology*, **15**, 349.
3. Hammond, J.R.M. (1995). Genetically-modified brewing yeasts for the 21st Century: Progress to date. *Yeast*, **11**, 1613.
4. Roller, S and Harlander, S.K. (1998). *Genetic modification in the food industry*. Blackie, London
5. Tucker, G.A. and Gregory, A. (1995). *Enzymes in food processing*. Blackie, London.

14.10 Responses to assignments

14.1 The fact that the bacterial cells stick themselves to the surface of the wood curls means that they are immobilised on the wood surface. We have seen in Section 7.3 that immobilisation of enzymes on to a solid support means that enzymes can be used in a continuous system without loss of enzyme from the reactor. The situation in a vinegar generator is similar to this; when the next batch of liquor is placed in the reactor, conversion of ethanol to acetic acid will occur rapidly because large numbers of actively metabolising organisms are already present in the reactor, and the vinegar finally produced is easier to clarify by filtration because the organisms remain within the reactor. Also, the fact that the organisms are immobilised on the wood curls means that they are retained within the reactor in the appropriate location, in this case in the aerated (upper) compartment.

14.2 We have seen in Section 8.4 that initial steam heating of the solid substrate used in SSF kills many microorganisms present on the substrate, so that the organism subsequently inoculated onto the substrate can predominate. This is the case with koji. In addition, steaming rehydrates the rice and wheat grains to the required level (the moisture content still remains relatively low, giving conditions suitable for the growth of the fungus but less suitable for many other organisms). Steaming also solubilises starch in the grains making it easier for amylase enzyme to hydrolyse.

14.3 Remember that the purpose of producing koji and mashing it, was to produce a medium in which nutrients are readily available. Thus, if the mash made from the koji was matured without salt addition, the microorganisms present in it could rapidly grow and cause spoilage (putrefaction). The salt adds a selective pressure to the system, allowing only the growth of salt tolerant or salt requiring (halophilic) organisms. As it happens, organisms that produce lactic acid as an end product of their metabolism tend to be relatively salt tolerant, so these organisms tend to predominate initially, and the lactic acid they produce exerts another selective pressure on the system, as only acid tolerant organisms can then survive. Eventually, available nutrients tend to become exhausted (used up by the organisms growing in the maturing sauce), and this lack of nutrients coupled with low pH and high salt concentration make soya sauce a stable foodstuff, not subject to microbial spoilage.

14.4 We have seen in Table 14.2 and Section 14.2.2 that glutamic acid, or at least its sodium salt (MSG), is a flavour enhancer, and glutamic acid will be produced by the hydrolysis of proteins in the yeast cells. We have also seen in Table 14.2 that nucleotides, specifically guanosine monophosphate (GMP), is also a flavour enhancer, and this will be produced by the hydrolysis of nucleic acids. Yeast extract is thus a crude source of flavour enhancers.

14.5 Compared to plants, within the microbial world there is huge variety between different organisms, including the characteristics and activities of the enzymes they produce; it is thus usually possible to isolate an organism that produces a particular enzyme best suited to a particular application in food processing, for example enzymes operative at relatively high temperatures (see Section 14.4.2 for an example of this). The enzyme can then be efficiently and easily produced at low cost in fermentation units. As microbes are relatively easy to genetically modify (by mutation, site directed mutagenesis or genetic engineering), it is relatively easy to modify (improve) the properties of the enzymes produce (for example, see discussion of modification of the enzyme chymosin in Section 6.4.4).

14.6 Suppose the organism to be transformed is sensitive to (will not grow in, or is killed by) antibiotic X. The culture of the organism is transformed by exposure to copies of the plasmid (which carries the transgene and the selection marker gene). However, only some cells, perhaps the vast minority of the cells in the culture, will take up the plasmid and express the genes on it; and it is only these recombinant cells which are of interest. By growing the culture up in/on medium containing the antibiotic, only cells that are expressing the gene coding for antibiotic resistance (and hence have been transformed by the plasmid) will grow. On this basis, transformed cells can be selectively isolated.

14.7 With foods like ice cream, the consumer expects a smooth, even texture. If the ice cream contains crystals of sugar, the product may feel granular, or even 'sandy' in the mouth, and will be unacceptable to consumers. With drinks, the product may again feel sandy when drunk, and if packaged in transparent containers, the crystals may be visible to the consumer, who may then reject the product as unsuitable.

14.8 Your experiences if you cook should give you the answer to this question. If you still cannot answer, reread the response to Assignment 14.2.

14.9 Amyloglucosidase does not act rapidly on dextrins longer than 5 glucose molecules. We have seen that in gelatinisation the α-amylase activity produces dextrins with up to 8 or so glucose molecules, and thus amyloglucosidase will not actively hydrolyse these. Thus, the presence of α-amylase hydrolyses dextrins with 6–8 glucose molecules to shorter chains that can be rapidly hydrolysed by amyloglucosidase. The α-amylase activity thus enhances the amyloglucosidase activity, and in fact, the effect is synergistic (the rate of glucose formation in the presence of both enzymes is greater than the rate of glucose formation with α-amylase and the rate of glucose formation with glucoamylase added together).

14.10 An enzyme which rapidly hydrolysed the α-1,6 links in dextrins would produce α-1,4-linked straight chain dextrins which could then be rapidly hydrolysed by amyloglucosidase. The effect of this combination would thus be the same as the effect of the combination of α-amylase and amyloglucosidase — that is a synergistic one. Such an activity is found in the enzyme pullulanase, and a thermostable pullulanase enzyme can produced using the bacterium *Bacillus acidopullulyticus*.

14.11 If you cannot think of a reason, reread the response to Assignment 14.2; pasteurisation will kill many organisms present in the milk which might otherwise compete with the lactic acid bacteria starter cultures (which are going to be inoculated into the milk), or themselves produce unwanted flavours or aromas in the product. Pasteurisation also kills any pathogenic organisms that may be present in the raw milk brought into the production unit. As it happens, lactic acid bacteria used in cheese production are microaerophilic (they grow best in an environment in which oxygen is present at much lower concentrations than that in the atmosphere). The heating during pasteurisation reduces the solubility of oxygen in the milk and much of the dissolved oxygen is driven out, thus creating ideal conditions for the growth of microaerophiles.

14.12 In phage infection, one virus particle infects one bacterial cell, which may rapidly releases 100 or more phage particles, each of which can then infect another cell to repeat the process. Thus, with each replication cycle, the number of phage particles progressively increases one hundredfold or more: 1 : 100+ : 10,000+ : 1,000,000+ , and so on. In this way, within 4–5 replication cycles of the phage (perhaps 2–3 hours in culture) enough phage particles may be present to infect and lyse all bacterial host cells in the culture. Phage infection of starter cultures can thus result in reduced rates of lactic acid production, or even complete failure to produce lactic acid, in cheese milk, resulting in failure of coagulation. This produces wastage and economic loss.

14.13 The major reason why microbial food protein processes have failed is that the projected dire shortage of food protein has not materialised in most countries of the world, in a large part due to the Green Revolution of the 1970s that boosted agricultural outputs. Thus, conventional protein foods and feeds have remained relatively plentiful and affordable for most people, who have thus not needed to change habits. The prospect of a diet composed of algae or fungi is alien to most people, and even relative small quantities of microbial biomass in the diet may cause intestinal upsets such as flatulence and diarrhoea. Processes based on oil-derived substrates suffered from increased oil price rises of 1973 and overnight became economically uncompetitive. In addition, there were also fears about the safety of products grown using liquid hydrocarbons, as such substrates could be shown to contain trace quantities of carcinogenic (cancer causing) compounds.

14.14 Mashing at 40°C favours protease activity; this leads to a high content of amino acids (proteinaceous material) in wort. These products in beer give stable foam (head). Mashing at 60°C favours amylase activity; this leads to a high concentration of maltose in wort. After fermentation, this results in strong beer (high gravity or alcohol content).

14.15 If yeast is not removed from beer, the yeast may make the beer, cloudy, and beer drinkers do not like cloudy beer. Also, if yeast is left in beer it may autolyse and give the beer a 'yeasty' taste. Autolysed yeast also releases nutrients (such as amino acids and vitamins) into the beer that can stimulate the growth of spoilage microorganisms, particularly *Lactobacillus* spp., which may ferment residual sugars in the beer and turn it sour.

14.16 We have already seen (Section 14.3.1) that brewers require a yeast that rapidly ferments sugars in wort, that strongly flocculates and sediments out once fermentation has ended, and that performs necessary reactions in maturation. We have also seen that yeast cannot break down starches and cannot ferment dextrins. Improvements in these characteristics of yeast are of great interest to the brewer.

14.17 If the yeast used to ferment the wort can utilise dextrins, then the yield of alcohol (the quantity of alcohol produced from a given quantity of substrate [malt]) will increase by up to 33%. Dextrins remaining in wort are digested by amylases in the human intestine to glucose, which is of importance to diabetics and weight watchers; in other words if the yeast can utilise all dextrins, a sugar-free, low-calorie beer can be produced.

14.18 Obviously a heat labile glucoamylase enzyme would be needed. Such an enzyme, having debranching activity, has been isolated from the yeast *Schwanniomyces occidentalis,* cloned and inserted into beer yeast using both multi-copy and integrative vectors. Both types of transformed yeast have been shown to excrete glucoamylase, ferment at normal rates, reduce the concentration of dextrins in beer and produce acceptable canned and bottled beers.

14.19 Commercially available enzyme preparations for addition to beer need to be relatively pure, as beer is classed as a food product, and regulations control what additives can be put into food; such purified enzyme preparations add to the production cost. There is always a danger that too much enzymes may be dosed into wort, resulting in unwanted changes in flavour. Residues of extraneous enzymes in the glucoamylase preparation may also result in unwanted changes in flavour. These drawbacks should be avoided using genetically engineered yeasts.

14.20 We have seen in Section 14.7.1(b) that in the fermentation of some types of beer, starchy grains may be added to the mash during mashing to increase the quantity of starch present and thereby increase the quantity of sugars formed by hydrolysis (by malt β-amylases) and thereby increase the alcohol content of the beer. In these circumstances, hydrolysis of added starch may be incomplete due to low levels of amylases in the limited quantity of malt used. Often in such circumstances, commercially available microbial amylase preparations are added to enhance the hydrolysis of starches. It would make processing easier and save enzyme costs if the yeast itself produced the enzymes that hydrolysed starch.

14.21 If the yeast can be transformed with an expression cassette containing the *MALT* and *MALS* genes downstream from a constitutive promoter (not repressed by the presence glucose), then the transcription of the genes should take place at a given rate irrespective of glucose concentration.

14.22 Obviously the genes involved in flocculation in naturally flocculent yeasts are expressed only at the end of the fermentation phase, perhaps as a result of induction by ethanol (or another metabolite accumulating at the end of the growth phase) or derepression as the effector molecule (such as glucose or maltose) becomes depleted. Thus, the gene(s) used to transform a non-flocculent organism need(s) to be under the control of a regulator and operator from a gene involved in flocculation in naturally flocculent yeasts, or a regulator and operator from some other gene similarly expressed only in the latter stages of fermentation.

14.11 Responses to self-assessment exercises.

14.1 See next page.

14.2 If you cannot answer this exercise, reread Section 8.7.1.

14.3 See next page.

14.4 The correct response is (iii), as what is required is a proteolytic enzyme with the specificity of chymosin. The proteolytic enzyme from *Mucor pusillus* is too non-specific, so its expression in calves' stomachs [(i)] gives no advantage, and anyway the supply would still be limited. Bromelain [(ii)] and papain [(iv)] are also too non-specific. Expression of the chymosin gene in *Mucor pusillus* would allow easy and large-scale production of chymosin in fermentation systems. However, the fungus would also produce its own proteolytic enzyme, which we have seen is too non-specific, so either this enzyme would have to be removed from the chymosin preparation produced by the fungus (which might be technically difficult) or synthesis of the fungus' own proteolytic enzyme would need to be prevented, for example by using antisense technology (see Sections 8.7.2, 12.10.3 and 12.10.6). An easier alternative would be to transform a microorganism other than *Mucor pusillus* and which did not excrete proteolytic enzyme with milk clotting activity.

14.1

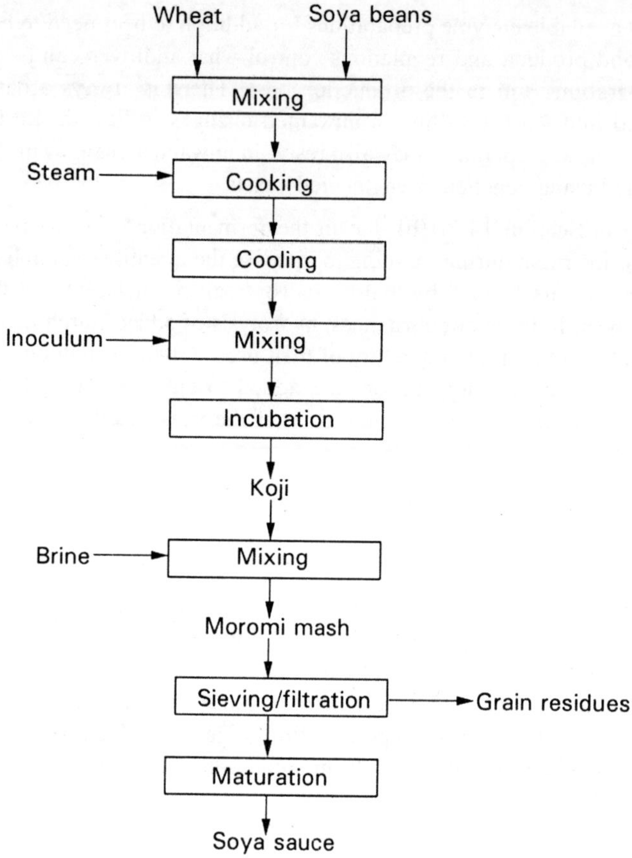
Wheat
Soya beans
Mixing
Steam
Cooking
Cooling
Inoculum
Mixing
Incubation
Koji
Brine
Mixing
Moromi mash
Sieving/filtration
Grain residues
Maturation
Soya sauce

14.3

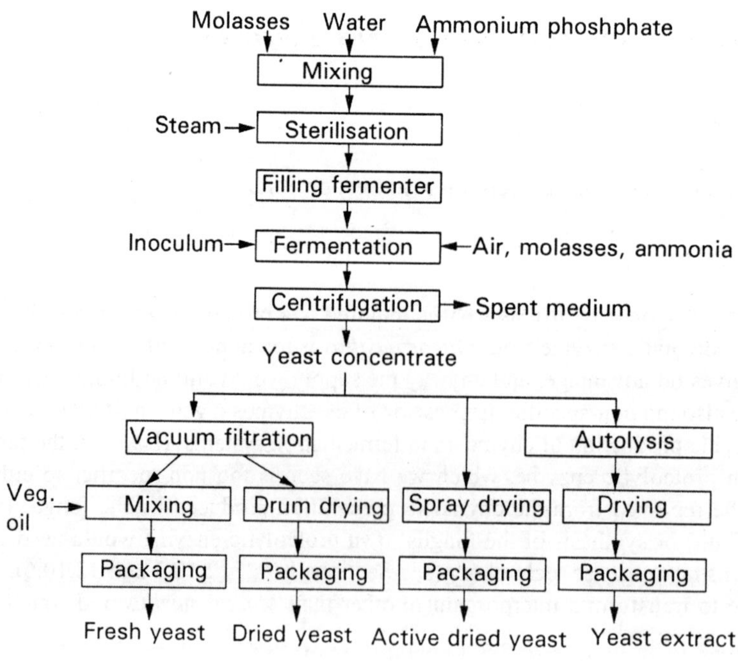
Molasses
Water
Ammonium phoshphate
Mixing
Steam
Sterilisation
Filling fermenter
Inoculum
Fermentation
Air, molasses, ammonia
Centrifugation
Spent medium
Yeast concentrate
Vacuum filtration
Autolysis
Veg. oil
Mixing
Drum drying
Spray drying
Drying
Packaging
Packaging
Packaging
Packaging
Fresh yeast
Dried yeast
Active dried yeast
Yeast extract

14.5

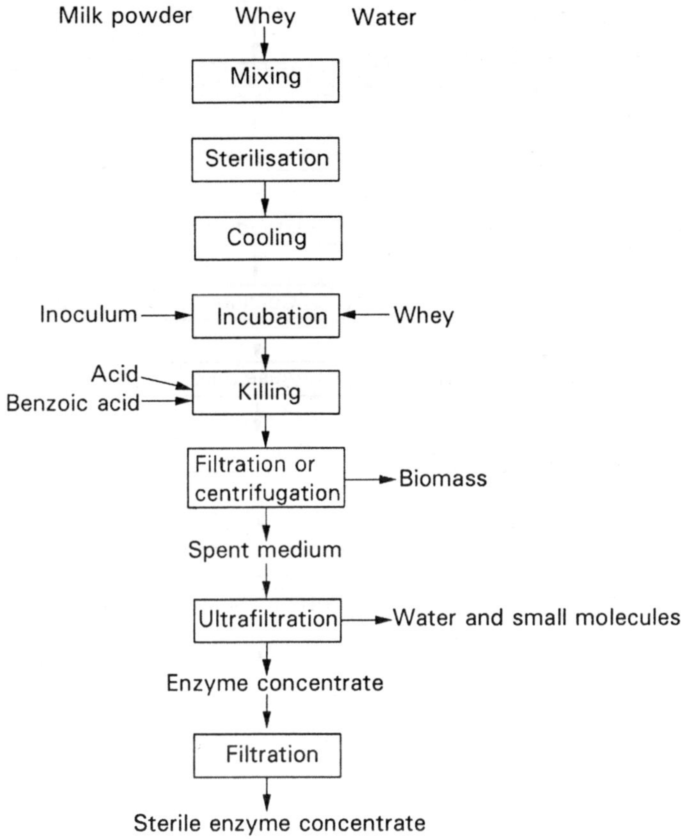

14.6 The correct solution is (iii). α-amylase on its own (i) hydrolyses dextrins in a random fashion, and thus does not rapidly produce glucose. Glucoamylase on its own (ii) produces glucose by sequential hydrolysis of starch polymers or dextrins, starting with a terminal glucose. In combination, (iii), the α-amylase hydrolyses starch polymers to shorter dextrin chains, which increases the number of terminal glucose molecules at which glucoamylase can initiate hydrolysis to glucose.

14.7 If you have difficulty answering this question reread Section 7.4 and 7.5. Immobilising the enzyme can lead to increased enzyme activity and stability. Continuous systems can be operated, which avoid down times and are thus more efficient than batch operation. The enzyme is not lost with each batch, but is conserved within the reactor, which in turn leads to higher purity of the product after enzyme conversion.

14.8 See next page

14.9 We have seen that in the production of cheese that milk is pasteurised and curds heated to expel whey, which kills many spoilage organisms present. The milk and cheese become acidified by lactic acid, and much of the liquid becomes expelled. Salt is added, which results in conditions in which many spoilage organisms cannot easily grow. The expelled whey contains most of the milk sugars, so the only nutrients left in cheese are the complex proteins and fats, which are unavailable to many microorganisms. The interior of a cheese block is anaerobic, so aerobic organisms cannot grow there.

14.8

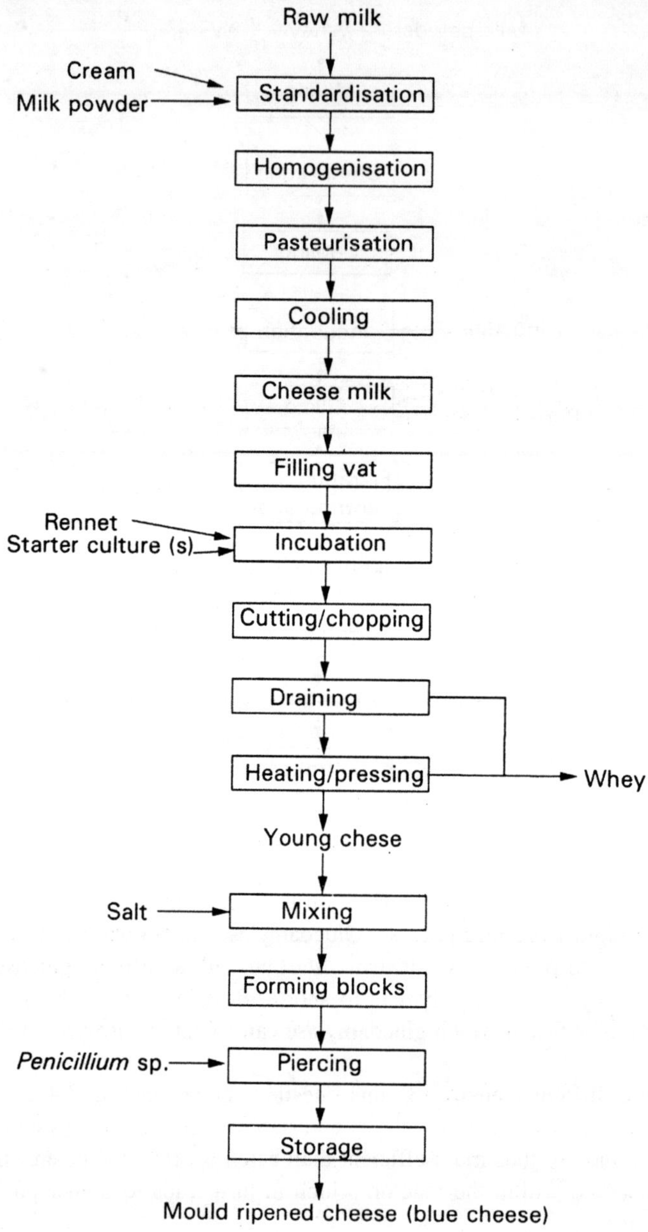
Raw milk
Cream
Milk powder
Standardisation
Homogenisation
Pasteurisation
Cooling
Cheese milk
Filling vat
Rennet
Starter culture (s)
Incubation
Cutting/chopping
Draining
Heating/pressing
Whey
Young chese
Salt
Mixing
Forming blocks
Penicillium sp.
Piercing
Storage
Mould ripened cheese (blue cheese)

14.10

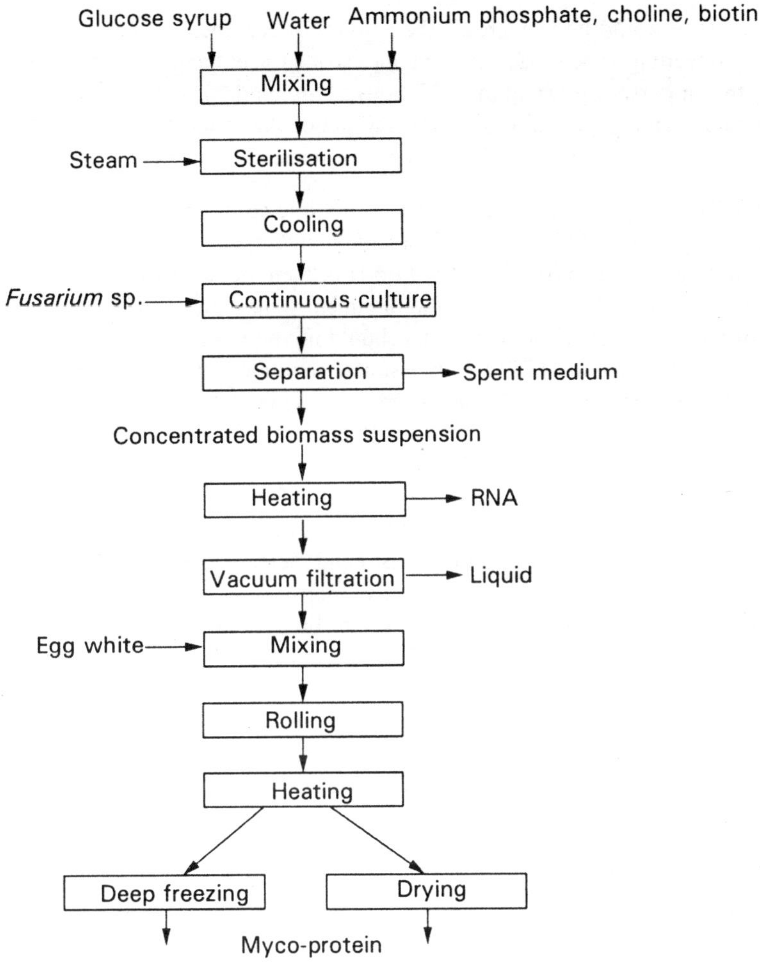

14.11 The correct answer is (ii). The multi-copy plasmid should allow expression of multi-copies of the glucoamylase gene [compared to the single copy in (i)], and thus produce highest levels of glucoamylase. Expressing barley β-amylase [(iii)] confers no advantage, as this enzyme cannot degrade dextrins.

14.12 (i) True; both enzyme activities are necessary for complete utilisation of starch, so the mixed culture would be more effective at starch degradation than either of the monocultures.

(ii) False; the *STA2* gene codes for glucoamylase without α–1–6 debranching activity (less effective at dextrin hydrolysis than *Aspergillus* glucoamylase, which has α–1–6 debranching activity), so a monoculture of the organism transformed with the gene for *Aspergillus* glucoamylase would be as effective as the mixed culture.

(iii) False; assuming the β-glucanases from barley and *Trichoderma* are equally effective, either monoculture should be as effective as the mixed culture.

(iv) False; transformation of yeast with the *MALS* gene does not increase the rate of fermentation, thus the mixed culture would ferment faster than the monoculture of the organism transformed by the *MALS* gene.

14.13 A strategy would be to increase the activity of enzymes (reductoisomerase and dehydratase) involved in converting α-acetolactate to valine, so that acetolactate does not accumulate in the cell (lessening the opportunity for auto-oxidation). This could involve: inserting multiple copies of reductoisomerase and/or dehydratase; inserting genes that code for reductoisomerase and/or dehydratase enzymes that have greater affinity for their respective substrates.

14.14 (i) *MET3* gene. Effective expression of multiple copies of this gene should increase levels of the corresponding enzyme (ATP sulphurylase) and increase the rate of conversion of SO_4 to SO_2. If this is the rate limiting step in conversion of SO_4 to H_2S, then the rate of formation of H_2S would increase and levels in wort be raised. Growth of the organism should not be adversely affected, as the rate of formation of the end products (amino acids) should not be reduced.

(ii) Antisense gene to *MET14*. Expression of multiple copies of this gene should decrease levels of the corresponding enzyme (APS kinase) and decrease the rate of conversion of SO_4 to SO_2. If this is the rate limiting step in conversion of SO_4 to H_2S, then the rate of formation of H_2S would decrease and levels in wort be reduced. Growth of the organism may be adversely affected, as the rate of formation of the end products (amino acids) may be reduced if H_2S formation is rate limiting in their formation.

(iii) *MET2* gene. Expression of multiple copies of this gene should increase levels of the corresponding enzyme (homoserine acetyltransferase) and increase the rate of conversion of homoserine to acetylhomoserine. Acetylhomoserine effectively 'mops up' H_2S by combining with it to form homocysteine. If acetylhomoserine formation is the rate limiting step in the conversion of homoserine to homocysteine, then increasing the rate of formation of acetylhomoserine should increase the rate of 'mopping up' H_2S, and the levels of H_2S in wort be reduced. Growth of the organism should not be adversely affected, as the rate of formation of the end products (amino acids) should not be reduced.

(iv) Antisense gene to *MET10*. Expression of multiple copies of this gene should decrease levels of the corresponding enzyme (sulphite reductase) and decrease the rate of conversion of SO_2 to H_2S. If this is the rate limiting step in conversion of SO_4 to H_2S, then the rate of formation of H_2S would decrease and levels in wort be reduced. Growth of the organism may be adversely affected, as the rate of formation of the end products (amino acids) may be reduced if H_2S formation is rate limiting in their formation.

(v) Antisense gene to *MET25*. Expression of multiple copies of this gene should decrease levels of the corresponding enzyme (acetyl serine sulphhydrase) and decrease the rate of conversion of H_2S to cysteine/homocysteine. Levels of H_2S in wort should be increased. Growth of the organism may be adversely affected, as the rate of formation of the end products (amino acids) may be reduced.

(vi) *MET25* gene. The opposite to (e) would be the case.

(vii) *NHS5* gene. Expression of multiple copies of this gene should increase levels of the corresponding enzyme (cystathionine synthase) and increase the rate of conversion of homocysteine to cystathionine. If this is the rate limiting step in conversion of H_2S to cystathionine, then the rate of conversion of H_2S would increase and levels in wort be reduced. Growth of the organism should not be adversely affected, as the rate of formation of the end products (amino acids) would not be reduced.

(viii) Antisense gene to *MET6*. Expression of multiple copies of this gene should decrease levels of the corresponding enzyme (homocysteine methyl transferase) and decrease the rate of formation of methionine. This may not decrease the conversion rate of H_2S, as homocysteine could be shunted towards cystathionine. Growth of the organism may be adversely affected, as the rate of formation of methionine may be reduced.

14.15 Expression of an antisense gene to *MET2* should decrease the rate of synthesis of acetylhomoserine, decrease the rate of 'mopping up' of H_2S and, unless H_2S is simply shunted into cysteine formation, thus increase the accumulation of SO_2. In practice, an antisense *MET2* gene expressed in brewing yeast did increase levels of SO_2 in wort, although levels of H_2S were increased also.

15 Environmental biotechnology

Chapter outline

15.1 Introduction: What is environmental biotechnology?

A question of this sort is destined to elicit a plethora of answers depending on the perspective. The most immediately tangible environmental biotechnologies are the traditional ones that have had a long successful history of excellence in maintaining the planet so that it is only partially poisoned: those of wastewater treatment and land filling. Without these biotechnologies our planet would undoubtedly be an unpleasant place to be.

These have at their core microbiological activities, although it is easy to argue that the biology has been subsumed into engineering. Indeed, we must admit that, even in activated sludge wastewater treatment, our knowledge of the biology and ecology is still relatively poor. In anaerobic wastewater treatments, the understanding is poorer still. There are still very severe biological problems to be overcome in wastewater treatment. This is where biotechnology will play a pivotal role in future; universal engineering solutions to these problems do not exist.

The current boom being enjoyed by bioremediation techniques applied to contaminated soil and groundwater is perhaps a logical extension of traditional waste treatment. It has evolved as a technology from fundamental biochemical studies performed over the last few decades into a unique and fascinating group of microorganisms—the hydrocarbon oxidisers. Here we have the classical model for the development of a technology—the evolution from a highly specialised scientific activity, accessible to only those few sufficiently skill-loaded, to a technology which can be applied for the benefit of society by companies with less specialist knowledge. The ultimate evolution is to mass application and products for a mass market. This latter extension may not be possible for bioremediation, but there is still plenty of scope for expansion within the market for the products and services.

Allied to the aforementioned environmental biotechnologies are emerging branches, some of which have the function of solving existing waste treatment problems. In response to the finite nature of fossil fuels and the burden being placed on landfill sites globally, biodegradable plastics are beginning to appear in products. Established polymers such as polystyrene foam and polyethylene present specific problems in land filling. Many are extremely durable, and yet are intended for single-use before disposal. Moreover, as they are light, but occupy relatively large volume, they take up great amounts of space in landfill sites, where they remain for many decades without obvious sign of biodegradation. The promise of being able to produce large quantities of plastic materials from renewable resources will extend the lifetime of finite fossil fuel feedstocks.

Biodegradability and ecotoxicity testing are now establishing themselves as instruments of regulatory change in many countries. The spectre of toxicity consents for industries requires that inexpensive, reproducible testing systems be developed. The rapid proliferation of toxicity testing systems based upon microbial activities has obvious ethical advantages over testing on higher animals. Indeed, if the industry is expected to perform its own internal monitoring for its toxicity consents, then even fish kill tests are quite impossible. Biodegradability testing is somewhat behind ecotoxicity testing technologically, and there is a definite need for more sophisticated and accessible techniques in this area. There is no more graphic illustration of this than the testing of biocides for biodegradability. The inherent paradox is both delicious and perplexing: the widespread use of biocides is absolutely essential in many industries, and yet these chemicals are designed to kill target organisms, in many cases microorganisms. Thus would we expect a biocide to be biodegradable?

There is a variety of schisms being researched within environmental science and microbiology; some are destined to become environmental biotechnologies. These are the future developments for our discipline.

Much of the global fossil fuel inventory is burdened by a relatively high sulphur content. Physical and chemical techniques, both pre- and post-combustion, are available for the reduction of the sulphur content, but are often expensive, and are out of step with the world politics of sustainability. Nature has furnished us with microorganisms which are able to specifically remove the sulphur from organosulphur compound in fossil fuels. The research effort which has been deployed may soon lead to the development of bioreactors capable of spearheading desulphurisation biotechnology.

By any criteria, surfactants, e.g., detergents, are very large-scale production chemicals, but unfortunately they also represent large-scale pollutants. Unlike many polluting organic chemicals, the surfactants add an extra dimension to the pollution problem—they are used in large quantities in millions of households. The once-common problem of unsightly foams present in wastewater treatment plants and rivers has been resolved by the production of the so-called 'soft' detergents. But the dual chemical nature of surfactants, which is so important to modern society, is also a contributing factor to their classification as pollutants—their amphiphilic properties can interfere with the functioning of biological membranes. The surfactants also have a nefarious reputation with regard to biodegradability, although much effort has been expended by the industry to improve the situation. The hydrocarbon-oxidising microorganisms face an interesting physiological dilemma in that the substrates for their growth are water-insoluble. Many respond by producing their own surfactants to solubilise or emulsify the hydrocarbons. These are the surfactants that are most often referred to as biosurfactants. The principle advantages that they should offer us in future are lower toxicity and greater biodegradability. They also offer a role in bioremediation.

The 1970s was a period of slow advance in microbial ecology. The arrival of accessible molecular biology changed this situation dramatically. In the future, we can expect genetic engineering to open up new areas of study and application in environmental biotechnology. An area under intensive investigation is the use of reporter gene systems, both in the field and for the development of environmental biosensors. Despite a large potential market for environmental biosensors, it may be argued quite successfully that the only one available commercially is the BOD test, which is itself a test fraught with difficulty and variability in quality of results. If electronic interrogation devices can be developed to detect specific pollutants, based on the high specificity and sensitivity of biological sensing systems, then a market is produced which is in direct competition with analytical chemistry, which is expensive and not field-friendly.

An area in its infancy is the use of microbial systems for the conversion of waste products into useful products. There are still few examples, even in research literature, where this has been accomplished. But this is an intriguing and potentially important future application that should not be omitted in a discussion of environmental biotechnology.

Most of environmental biotechnology is concerned with pollution prevention or amelioration. By necessity this activity is a synergy between scientists and engineers. However, it is often a difficult synergy to achieve. Waste treatment systems have brought together the skills of microbiologists and civil engineers. As the wastes to be treated become more complex, so do the biological and engineering concepts. For the future, the synergy will involve these same disciplines, but will also require the skills of chemical engineering, and an increasing dependence on genetic engineering.

15.2 Aerobic wastewater treatment processes

The over-riding objective of biological treatment of wastewater is the removal of carbonaceous materials that exert an oxygen demand if released to a watercourse without treatment. If released to a river, the microorganisms present in the river water would start to biodegrade this material, resulting in a decrease in the dissolved oxygen concentration (DOC) in the water. If this decrease becomes severe, fish kills will result, along with a general decrease in the water quality. Carbonaceous, oxygen-demanding materials can be broadly subdivided into two categories—those that are readily biodegraded within 5 days, and the total oxidisable material. The former is quantified as the BOD_5 (the biochemical oxygen demand, measured over a five day period, usually expressed in mgl^{-1}) and the latter as the COD (chemical oxygen demand, also usually expressed in mgl^{-1}).

> ***Assignment 15.1***
>
> It is normal for the COD measured on a wastewater sample to be higher than the BOD_5 figure. Why should this be?

Biological wastewater treatment systems started to assume importance on more than a regional scale around the turn of the twentieth century with the invention of the activated sludge process and the percolating filter, which have become the most common forms of treatment. The former is the best example of a dispersed growth system, and the latter a biofilm or fixed film system. The two terms, dispersed growth and fixed film, give us the most fundamental distinction in treatment methods.

15.2.1 *Activated sludge (AS)*

There is a vast array of treatment systems coming under the heading of AS, by far the most common of the dispersed growth processes. Most of systems differ in engineering principles, one system being developed generally to overcome a difficulty or limitation of a previous system. There are many excellent texts that deal with AS in these terms. In this chapter, it is more important to deal with the microbiological principles. The unifying feature of all AS systems is that a proportion of the biomass (sludge) produced is recycled back to the bioreactor (aeration basin) to maintain a high enough concentration of biomass to achieve efficient treatment.

This can be best understood by an illustration (Fig.15.1), and an appreciation of the mass balance considerations in AS.

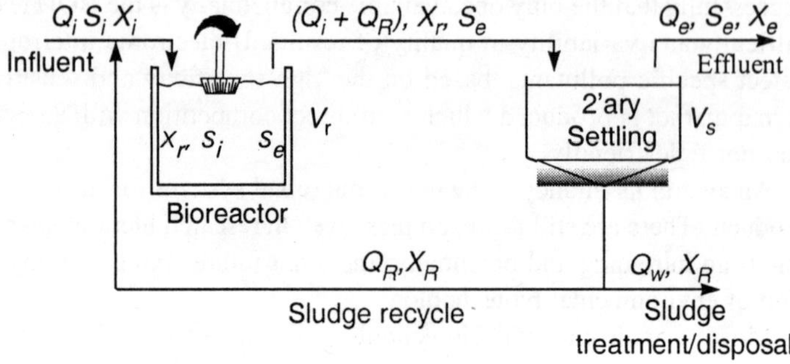

Fig. 15.1 *Mass balance of the major variables in AS treatment.*

Q = flow (m^3d^{-1})
S = substrate concentration i.e., "food" for microorganisms (mg 1^{-1}, gm^{-3} or kg m^{-3})
X = biomass concentration (mgl^{-1} or gm^{-3})
V = volume (m^3)
subscript i = influent
subscript e = effluent
subscript R = recycle
subscript r = reactor
i.e., V_r = reactor volume
subscript w = waste
subscript s = settling i.e., V_s = settler volume

Thus a completely mixed AS system, as deicted above, has some similarities to a laboratory chemostat (Fig. 15.2), which was previously described in Chapter 9.

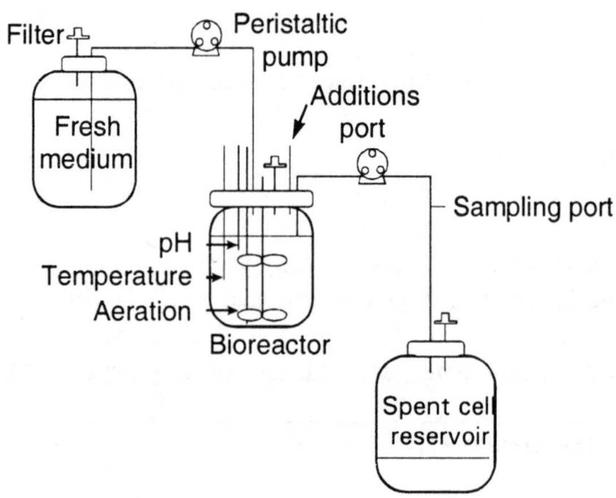

Fig. 15.2 *Laboratory chemostat.*

Some significant differences are:

(a) even a large industrial chemostat tends to be small in comparison to an AS plant;
(b) a chemostat usually produces a relatively high value, relatively low volume product;
(c) the degree of process control in a chemostat is necessarily much greater.

Operational differences between the two systems are:

(a) a chemostat, or particularly a batch fermenter, is usually run with high concentrations of pure substrate(s) to maximise product recovery. An AS plant runs at much lower carbon substrate concentrations (in wastewater engineering terms, organic loadings);
(b) a chemostat is usually operated as an axenic culture of a single organism. An AS plant is open to

the atmosphere, and thus operated with a wide variety of microorganisms (bacteria, fungi, protozoa) and macroinvertebrates i.e., it is not run aseptically;

(c) a chemostat, perhaps as a result of operating with high organic loadings, has the microorganisms dispersed as a uniform suspension. In the AS process, the basic unit of operation is the microbial floc;

(d) only a small proportion of the AS is actually viable and capable of reproduction. Adsorption is an important part of the purification process;

(e) in the AS process, a proportion of the biomass is recycled back to the bioreactor.

In terms of mass transfer characteristics, a chemostat is an example of a S_oB_o (substrate open, biomass open) reactor, whereas an AS plant is an imperfect S_oB_c, in that part of the biomass is recycled. AS plants with total biomass recycle represent perfect S_oB_c systems (substrate open, biomass closed).

Assignment 15.2

It was seen in Chapter 9 that in a chemostat, the biomass is diluted out of the culture vessel when the dilution rate (D) exceeds a critical value (approximately μ_{max}). How would a biomass recycle affect this?

(a) Microbiology of AS

The basic unit of operation in an AS plant is the microbial floc (Fig.15.3), which consists of aggregations of microorganisms along with other organic and inorganic materials. There are various theories purported to explain the formation of these large, macroscopic flocs. An attractive theory is of the filamentous backbone.

Sufficient filaments are required for the initiation of floc formation. The filamentous bacteria form a skeleton upon which the rest of the microbial mass can form. These constitute the floc-forming bacteria (fungi play a much less important role in AS treatment), and the various heterotrophic bacteria present in flocs have the role of BOD removal.

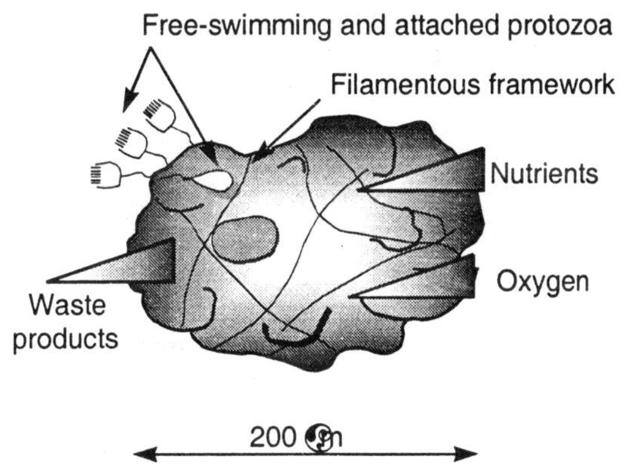

Fig.15.3 *The AS floc (redrawn from Horan, 1989).*

The material, which mediates the attachment of the floc-forming bacteria to the filamentous skeleton, is thought to be the same material responsible for biofilm formation on solid substrata. The primary constituent is bacterial exopolysaccharide (EPS), which manifests itself as bacterial capsules or excreted polymer. There are various genera of bacteria that produce EPS. Bacteria of the genus *Pseudomonas* seem to play a particularly prominent role in this regard. Bacterial division and further production of EPS causes the floc to grow in size. Colonisation by protozoa limits the size of the floc through predation upon bacteria. Eventually equilibrium is reached in which the floc is stable. Oxygen cannot penetrate all the way through the floc, so the centre of the floc is likely to be anoxic. However, this allows for the continuing process of oxygen uptake to the floc. A varying proportion of the biomass in the floc is dead (depending on floc age), but this can contribute to the treatment process through adsorption of organic matter, and residual enzyme activity.

This formation sequence has many features in common with the accepted sequence of events in biofilm formation upon solid surfaces (Sections 15.2.2 and 15.2.3), which can be summarised as: conditioning film formation on the solid surface; transport of microbes to the surface; reversible microbial attachment; EPS excretion; irreversible attachment of microbes; growth of the biofilm though further polymer production and cell division; detachment from the film. The two processes of biofilm and floc formation appear to be remarkably similar. Therefore it should be no surprise that many of the bacteria important in the AS floc are also those important in the fixed film treatment processes, such as the percolating (trickling) filter (Section 15.2.2).

The above sequence is undoubtedly incomplete in detail. An important contribution to flocculation must be made by physical forces, and also the metabolic status of the bacteria. Bacteria that are metabolising rapidly do not form flocs. Flocculation occurs when nutrients are almost exhausted, and this is an important consideration when operating AS plants. At first encounter, it may be imagined that the more rapidly bacteria are dividing, the faster is the treatment process. However, as the cells will not flocculate in this state, the process fails as settlement will be very slow. This can be summarised graphically (Fig.15.4). It can be seen that the preferred condition is to have the microorganisms held at a state close to starvation.

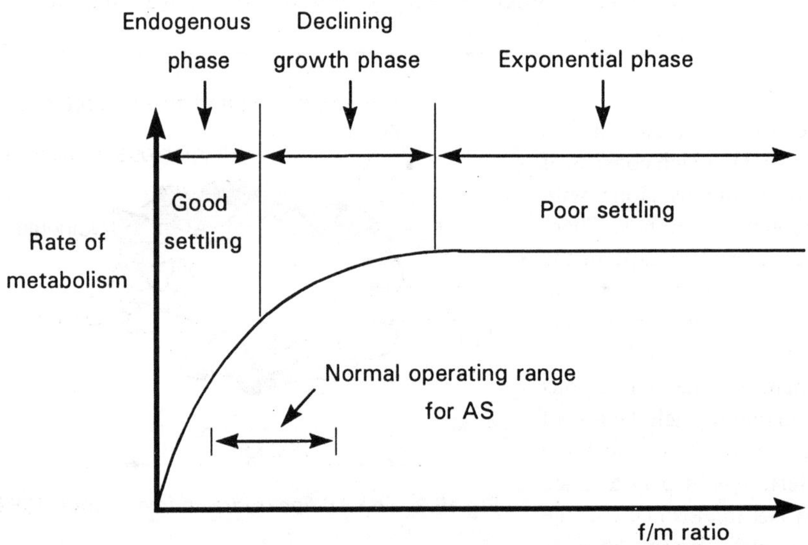

Fig.15.4 *The effect of microbial metabolic rate on settlement in AS.*
[*f/m* = food to microorganism (see later)]

Assignment 15.3

Given a floc size of 200 mm diameter, and assuming an individual spherical bacterium diameter of 1 mm, calculate the settling rate of both using Stokes' law for the terminal settling velocity of a sphere under gravity.

$$V_t = g\frac{(\rho_s - \rho)}{18\mu}d^2$$

where: V_t = terminal settling velocity (ms^{-1}) g = gravitational acceleration (9.81 ms^{-2})
ρ_s = density of bacterial biomass (~ 1050 kgm) ρ = density of water (998 kgm^{-3})
μ = viscosity of water (1.01×10^{-3} $kgm^{-1}s$) d = diameter (mm)

This dramatically illustrates the influence of flocculation upon settling velocity

Although the presence of a number of filamentous bacteria seems to be a requirement for good flocculation, various conditions lead to excessive growth of these forms. In these instances sludge bulking occurs. When the filaments protrude from the floc into the bulk solution, they can then start to form bridging connections between flocs (Plate 15.1), which initiates and catalyses the problem of sludge bulking. Bulking of this nature results in poor settlement of the sludge at secondary sedimentation, and a reduced capacity of the sludge to dewater (to give up water and form a compact sludge).

Plate 15.1 *Photomicrograph of flocs with excessive filamentous growth.* [Note floc bridging.]

Assignment 15.4

A bulking sludge is one with poor settling characteristics. What would be the primary operational affects of a bulking sludge?

Several different filamentous bacteria have been isolated from bulking sludges, and they have widely differing growth requirements. These differences and their interplay in the AS plant often result in the predominance of one type over others. Typical BOD-removing bacteria and filamentous organisms are shown in Table 15.1.

Table 15.1 Some of the bacteria associated with the AS process.

BOD-removing bacteria	Filamentous bacteria
Acinetobacter (also P removal)	*Beggiatoa*
Achromobacter	*Dietzia*
Alcaligenes (Ralstonia)	*Gordonia*
Arthrobacter	*Haliscomenobacter*
Bacillus	*Microthrix*
Cytophaga	*Mycobacterium*
Flavobacterium	*Nocardia*
Pseudomonas	*Rhodococcus*
Zoogloea	*Sphaerotilus*
Nitrifying bacteria	*Thiothrix*
Nitrobacter	*Tsukamurella*
Nitrosomonas	

A related severe operational problem is the production of stable foams (Plate 15.2), thought to be caused principally by the filamentous mycolic acid-containing actinomycetes (the mycolata). The mycolic acids have surface-active properties. A plausible mechanism for the development of these foams involves the hydrophobicity of the cell surface associated with the mycolata (*e.g., Dietzia, Gordonia, Nocardia, Rhodococcus*). The mycolic acids give these organisms a hydrophobic cell exterior. With time this renders the whole of the AS floc hydrophobic, non-wettable, and thus amenable to the attachment of air bubbles characteristic of the AS process. The bubble-floc aggregates are less dense than water and float to the surface where they accumulate as foam. These foams can be stable for long periods and create a range of operational problems. If the problems is left unchecked over long periods, treatment can be lost due to the accumulation of the biomass in the foam, rather than in the bulk aqueous phase.

Plate 15.2 *AS microbial foaming.*

Assignment 15.5
What operational problems might be caused by microbial foaming?

Nitrification

Ammonia is a potent fish poison, and its release to water courses must be strictly controlled. In biological wastewater treatment, ammonia is oxidised to nitrate as a result of the combined action of two genera of aerobic autotrophic bacteria known as nitrifying bacteria. Overall, the process is known as nitrification. No bacterium is known which converts ammonium all the way to nitrate. The ammonium oxidisers (*Nitrosomonas*) are responsible for the oxidation of ammonium to nitrite, and this product is the reactant for the nitrite oxidisers (*Nitrobacter*) which generate nitrate. Both genera assimilate CO_2 as the sole carbon source *via* the Calvin cycle, but this carries with it very high-energy expenditure. As a result, the nitrifying bacteria grow rather slowly (generation times of 10–20 hours), with low growth yield. For example, *E.coli* consumes 2 g of glucose to synthesise 1 g of cells. By comparison, *Nitrobacter* consumes 30 g NH_3 to produce 1 g of cells. This slow growth produces problems in AS treatment when trying to maintain large populations of nitrifiers. Most wastewaters which contain ammonium will also contain organic matter; the nitrifying bacteria must therefore compete with the aerobic heterotrophic bacteria which grow much more quickly. The generally held belief is that the situation is compounded by inhibition of nitrification by a range of physical and chemical variables. However, the belief that the nitrifiers are inherently more susceptible to inhibitory substances than other bacteria in wastewater treatment plants has been called into question recently. But if a wastewater contains a material that is toxic to treatment, the nitrifiers will be washed out of the plant, and it takes much longer for the nitrifying population to recover than it does for the aerobic heterotrophic population.

> ***Assignment 15.6***
>
> Urea (CH_4N_2O) is a common constituent of wastewaters. Calculate the theoretical oxygen demand (ThOD) exerted by 1 g of glucose ($C_6H_{12}O_6$), and 1 g of urea. You should write down the equations for the aerobic mineralisation of both chemicals, and balance them to get the solution to this problem.

(b) Operational considerations: AS plant loadings

The nature of the waste being treated largely determines how a plant is operated. Successful AS treatment has two components: sufficient BOD removal to achieve individual consent levels, and efficient secondary settlement to allow the process to occur within proper time scales. A major factor that influences the eventual success of the treatment is the loading to the plant.

Three loading factors are used in the design and operation of AS. These are: volumetric, organic and sludge loading.

volumetric loading	=	flow of wastewater in relation to aeration tank capacity
organic loading	=	BOD load in relation to the aeration tank capacity
sludge loading	=	BOD load in relation to the total biomass of AS

Volumetric loading

This is described by the hydraulic retention time. Hydraulic retention time (HRT) can simply be expressed as the time that the wastewater spends within the treatment system. Changes effected in the HRT have effects on the microbial populations as the HRT is inextricably linked to the growth rate of microorganisms. However, the major objective of sludge recycle in AS is to increase the mean cell retention time (MCRT). This allows the microbes to stay within the system for much longer than the HRT *i.e.*, an objective is to make the MCRT longer than the HRT. Both these terms are of importance to the understanding of the AS process and have implications for AS as a biotechnological process.

> ***Assignment 15.7***
>
> How is the HRT related to dilution rate (D)?

Organic loading

Wastewaters are highly variable in their organic content, especially if there are variable industrial inputs. The organic loading is expressed as kgBOD$_5$ m^{-3}d^{-1}. For practical reasons, such as oxygen limitation, the greater the organic loading, the higher the BOD$_5$ of the final effluent. In conventional AS, the BOD loading range is 0.5–1.5 kg BOD$_5$ m^{-3}d^{-1}, with a HRT of 5–14 h. The equation for calculation of organic loading is:

$$\text{Organic loading} = \frac{Q_i \times BOD_5}{V_r} = \frac{\text{m}^3 h^{-1} \times \text{gm}^{-3}}{\text{m}^3}$$

$$= \frac{\text{gh}^{-1}}{\text{m}^3}$$

$$= \quad gm^{-3}h^{-1}$$

$$\div 1000 \quad = \quad kgm^{-3}h^{-1}$$

Sludge loading

Both the above loading factors take no account of the biomass present, and there needs to be a factor which relates BOD loading to the amount of AS (biomass) present in the bioreactor. This is the sludge loading and is also known as the food to microorganism ratio (*f/m*).

$$\frac{f}{m} = \frac{\text{organic loading rate}}{\text{volume of sludge}}$$

$$= \frac{Q_i \times BOD_5 \times 24}{V_r X_r}$$

Perhaps a more convenient way to express sludge loading is:

$$\frac{BOD}{X_r} \times \frac{24}{HRT}$$

> **Self assessment exercise 15.1**
>
> Calculate the sludge loading for an AS plant treating domestic waste with an average BOD_5 of 350 mgl^{-1}. The HRT is 5 h, and MLSS (biomass) is maintained at 2800 mgl^{-1}.

If the *f/m* ratio is high, then the microorganisms are not food-limited and therefore they grow exponentially. At first encounter, one may imagine this to be a favourable condition. However, under these conditions the microorganisms are unable to form flocs and remain suspended as individual cells. In this state it is very difficult to settle them in a sludge. Also, as food is in excess, some will not be utilised and will result in a final effluent with high BOD. With low *f/m* ratios, by definition the microorganisms become substrate limited. This encourages floc formation, and there is almost complete substrate utilisation, and thus a settleable, high quality final effluent.

Plant configurations

So far the discussion has been limited to completely mixed AS. A great many configurations exist: usually a new configuration is designed in response to a problem seen with another. Examples are plug flow, step aeration and step feed, contact stabilisation, and the use of pure oxygen plants to perform difficult treatments or to augment overloaded conventional plants.

At the start of plug flow systems there is a high *f/m* ratio, but as treatment progresses, the growth and multiplication of microorganisms depletes the food (i.e., BOD). At the end of the treatment, the *f/m* ratio is low, the condition conducive to good settlement. In contact stabilisation, there is a short period of contact with high levels of biomass to allow adsorption of BOD to the flocs. Note that it is only the secondary sludge that receives a prolonged period of aeration (the stabilisation stage). This means that only a small volume of material requires long aeration, and thus savings are made in space and aeration

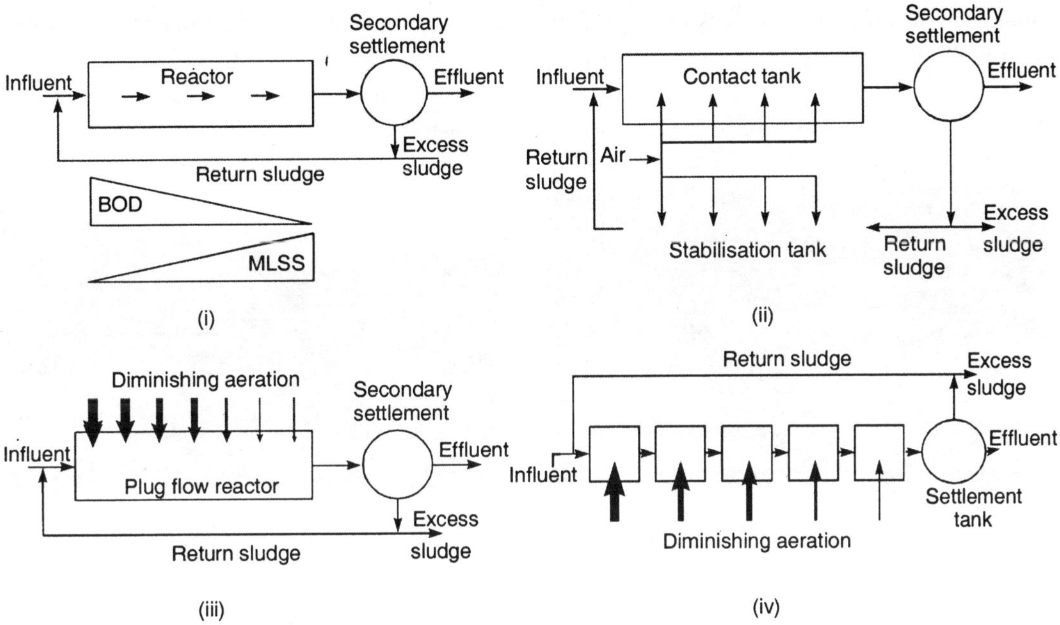

Fig. 15.5 *Some activated sludge plant configurations*
(i) plug flow, (ii) contact stabilisation, (iii) step aeration, (iv) step aeration in a series of tanks (this configuration amount to several complete mix reactors, which in sum equals a plug flow system).

cost at this second stage. Step aeration is a means of balancing the total required aeration over the reactor. At the start of treatment, where the *f/m* ration is high, more oxygen is required. Because it is difficult to maintain plug flow in Fig. 15.5(iii) due to turbulent mixing, an alternative is offered by step aeration in a series of tanks (Fig. 15.5(iv)). Step feed allows the influent to be fed in increments with the same objectives as step aeration. The AS system was originally conceived as a sequencing batch process in which the BOD removal and settlement were performed in a single reactor. With the ability to control plants automatically, this configuration has come back to the forefront of AS treatment.

Aeration

Wastewater treatment plants, because of their size and distribution, have to be energy-efficient. The most expensive part of the AS operation is aeration. It is common practice that the means of aeration is also the means of mixing the contents of the reactor. There are two common methods of aerating AS plants: by surface agitation with impellers (see Plate 15.3) and by diffused air. Surface impellers are immersed at a critical depth within the wastewater. At this depth, wastewater from the bioreactor is thrown out into the atmosphere as an aerosol: the droplets have a large surface area for the absorption of air from the atmosphere. If the impeller is set too deep into the wastewater, there is good mixing of the bioreactor contents, but aeration is inefficient. If the impeller is not in deep enough, good aeration can be achieved, but without good mixing.

Diffused air aeration relies on pumping air through a manifold of pipes set in the bottom of the bioreactor. The pipes are drilled with small holes to allow the air to escape in small bubble streams. The smaller the holes, the smaller the bubbles released, and the greater the surface area for the transfer of air from the bubbles to the bulk liquid. If the holes are too small, they can be clogged by microbial growth; in this way a compromise is reached.

(i)

(ii)

Plate 15.3 *Industrial AS plants in India. (i) distillery effluent treatment plant, Pilkhani, Uttar Pradesh (ii) distillery effluent treatment plant, Shamli, Uttar Pradesh. Both plants are designed to treat spent wash, with BOD of the order of 6000 mgl^{-1}. Courtesy of Lars Enviro Pvt Ltd, Nagpur, India*

15.2.2 Percolating filter

The percolating, or trickling, filter is the most common of the fixed film treatment methods. This is a biofilm treatment in that the biological reactor contains a solid substratum (the medium) to which microorganisms attach. It is this biological film that performs the BOD removal from the wastewater. The treatment *via* biofilms is the method used here to separate the HRT from the MCRT, with the same objectives as recycling biomass in the AS process i.e., to reduce the HRT by increasing the MCRT.

Percolating filters can be rectangular or circular, the latter being the more common. In the circular filter, wastewater is pumped into the bottom and up a vertical riser to a rotary distributor for spreading uniformly over the surface of the medium (Fig. 15.6). The rotary distribution consists of either two or four radial pipes that are pivoted at the centre of the filter on a rotary seal. The wastewater is released from the sparge pipes simply through holes of sufficient size to prevent blockage by solids. They are arranged along the bars to ensure even loading of the medium surface, being more closely spaced towards the periphery. The distributor arms are most often driven by jet action, which helps in the energy-efficiency for which the percolating filter is renown.

The simplicity of the percolating filter means that there are few operating problems. A potential problem is related to overgrowth of the biofilm. With a traditional solid substratum such as crushed rock, the voids volume is relatively low, and excessive biofilm growth can block the voids. Wastewater can accumulate on the surface of the filter, a situation known as 'ponding'. Filters in this state run a very high risk of becoming anaerobic, with the objectionable production of hydrogen sulphide. Also the high density of materials such as rock and blast furnace slag dictate that percolating filters utilising such materials for the medium are limited in height.

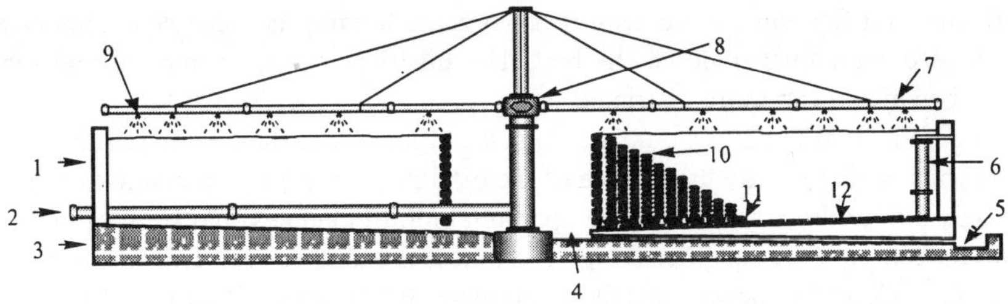

Fig. 15.6 *The main features of the percolating filter. (1) retaining wall, (2) influent pipe, (3) filter base, (4) central well, (5) effluent channel, (6) ventilation pipe, (7) distribution arm, (8) rotary seal, (9) jet, (10) main body medium, (11) larger medium on base, (12) drainage tiles.*

(a) Microbiology of the percolating filter

The ecology of a percolating filter is diverse and ranges from bacteria to higher organisms such as lumbricid worms and dipteran fly larvae. However, the organisms responsible for wastewater purification are the microorganisms that form the biofilm on the support medium. Heterotrophic bacteria play a central role, but, unlike AS, fungi also make a contribution to effective treatment. The biofilms also contain large numbers of protozoa. The EPS-producing and capsulate bacteria are thought to be the first colonisers of the medium. The colonisation by heterotrophic bacteria is rapid, and the genera involved are familiar, as the same organisms are seen in AS e.g., *Pseudomonas, Flavobacterium, Zoogloea*. Within a few days a mature bacterial biofilm can be formed. The fungi take longer to establish: their growth rates and nutrient uptake are slower than those of the heterotrophic bacteria. The comparatively enormous surface area of bacteria greatly aids this latter process. However, over some weeks fungi such as *Fusarium*, *Geotrichum* and *Sepedonium* grow and contribute to BOD removal. The upper surface of percolating filters, by virtue of exposure to sunlight, allows the growth of algae to occur e.g., *Chlorella*, *Euglena*, *Ulothrix*. They may have a limited capacity for BOD removal, but are more likely to be useful in the removal of nitrogen and phosphorus.

The first stage of the purification process is the adsorption of organic matter to the biofilm. Fine particles are flocculated by extracellular polymers and adsorbed. When thus physically trapped the organic matter breakdown process is initiated by bacterial and fungal extracellular enzymes. This solubilises nutrients which can then be directly absorbed and metabolised by the microorganisms of the biofilm.

At the upper surface of the filter, oxygen is abundant due to diffusion from the atmosphere into the liquid and then to the film. The rapidly respiring heterotrophs create a steep oxygen gradient that maintains high rates of oxygen transfer into the film. As the wastewater percolates down through the filter bed, BOD is removed, and the carbon available becomes more dilute due to these rapid heterotrophic activities in the upper regions of the filter. Typically, the majority of the BOD is removed in the top 750 mm of the filter bed. With greater depth into the bed, heterotrophic activity lowers due to the lack of available carbon (BOD).

In the lower reaches of the filter bed, conditions are more favourable for the establishment of nitrifying populations. The autotrophic nitrifiers have lower specific growth rates than the heterotrophs, and they benefit from the restricted growth of the latter in the deeper regions of the bed. Also, as the nitrifiers are strictly aerobic, they are not suited to the lower available oxygen concentrations in the upper regions. And the nitrifiers are sensitive to the presence of toxic metals. Near the upper surface of the filter bed, the heterotrophic biofilm removes many of the toxic materials by adsorption, leaving the lower regions less toxic to the nitrifiers.

Therefore, a stratification of the microbiology of a percolating filter is observed. Evidence for the functioning of this stratification can be seen if the organic loading is increased. This causes heterotrophic activity to extend further down the bed. The nitrifiers cannot compete, and their territory within the filter becomes even more restricted.

The thickness of the biofilm plays a crucial role in the efficiency of the treatment and the maintenance of a stable microbial ecology. Only the surface of the biofilm is efficient in oxidation of organic matter, and therefore only a thin film is required. The optimum biofilm thickness has been estimated to be 150 mm. An important consequence for filter design is that it is the total surface area of filter that is crucial, not the total biomass. Consider some events that occur when the optimum film thickness is exceeded. The defining factor is the low solubility of oxygen in water. As the biofilm becomes thicker, oxygen can only diffuse a certain distance into the film as microbial respiration removes it (Fig. 15.7). An anoxic, and subsequently anaerobic, zone can develop. As the thickness of the film continues to increase, the microorganism at the base of the biofilm (furthest from the aerated surface), become starved of oxygen and soluble nutrients, and are forced into endogenous metabolism i.e., they start to catabolise their structural components and any stored reserves. They can then lose the ability to adhere to the substratum, and become detached from the film, a process known as 'sloughing'. This is a normal operational observation within filters. It allows recruitment of new microbes to the film, and thus the physiology of the biofilm is maintained in optimum condition.

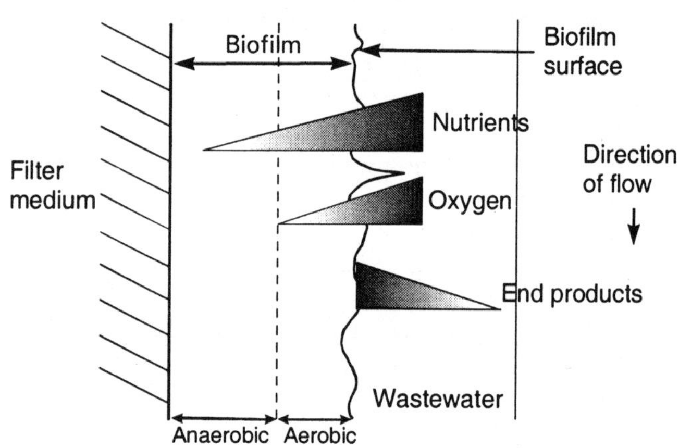

Fig. 15.7 *The percolating filter biofilm.*

(d) Filter media

Given that the surface area of the film is more important than the total biomass concentration, some materials are obvious candidates for use as the support medium. Traditional materials for domestic percolating filters are mineral-based e.g., blast furnace slag, rock, coke. A medium with a very rough surface, such as slag, offers a much larger surface area for colonisation by microorganisms than, say, a smooth rock. It is important that the individual pieces of the filter be of roughly the same size, and also the same size in all directions. This produces the maximum void space to allow percolation of liquid and maximal oxygen diffusion into the filter bed. The materials must also withstand normal weathering process (wetting, drying, freezing, thawing) for long periods—usually they have to be completely stable for at least fifty years.

A most significant development in percolating filter design was the arrival of plastic media. Plastics differ from mineral media in two very important respects: they can be fabricated to have much higher voidage than minerals (around 90% compared to 40–60%) and are much less dense (perhaps 10% of the weight). This makes them an excellent choice for industrial wastewater filters, allied to the fact that plastic media can be designed with very high specific surface area for attachment of microorganisms. As they are very light, they can be built much higher (up to about 12 metres, compared to conventional filters with mineral media in the UK, which are around 1.8 metres deep). This makes them compact and they also have less visual impact. Thus, to treat the same population equivalent, less land area is required as the filter can be built upwards rather than outwards. This is important in countries where land price is high. Also, due to the high voidages, high organic loads can be applied as there is little chance of voids blockage by excessive microbial growth, which overcomes any potential ponding problem.

Self assessment exercise 15.2

Answer the following questions.

1. What are the roles of bacteria that make extracellular polysaccharides ('slime') in AS and in fixed film reactors?
2. What do you imagine to be the principle consequence of protozoan colonisation of biological wastewater treatment plants?
3. Of the three stated methods for determining loading to an AS plant, which is the most useful for design purposes?

15.2.3 Rotating biological contactor (RBC)

The RBC is another fixed film reactor, similar in principle to the percolating filter, but very different in design. Instead of gravity feed of wastewater vertically, wastewater moves horizontally through the bioreactor (Fig. 15.8).

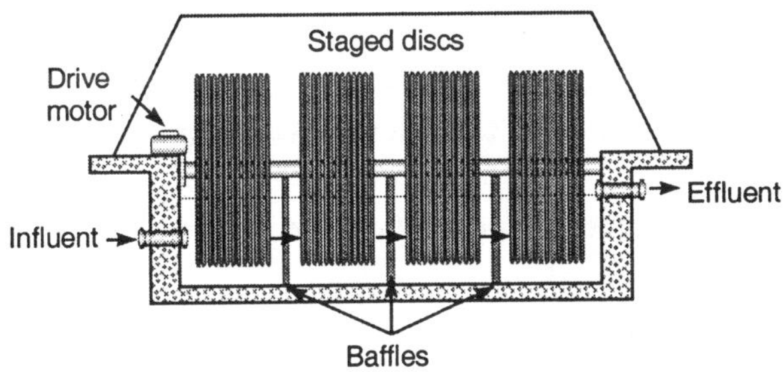

Fig.15.8 *The rotating biological contactor.*

The basic design consists of a series of flat or corrugated discs 2–3m in diameter mounted on a horizontal shaft, driven mechanically so that the discs rotate at right angles to the flow of settled wastewater. The discs are usually plastic—corrugated polythene, PVC or expanded polystyrene. The discs are mounted into a tank such that, when filled with wastewater, the discs are approximately 40% submerged in wastewater. The principles of operation can be summarised as follows: the rotation of the discs through the wastewater means that, with time, a biofilm develops uniformly over the available surface area of the discs; when immersed in the wastewater, the biofilm absorbs nutrients and the microorganisms oxidise them for their metabolism, thus reducing the available oxygen concentration in the biofilm; when the biofilm rotates out of the wastewater and into the atmosphere, the biofilm absorbs oxygen so that it becomes reaerated, and so can continue wastewater purification when resubmerged.

The wet disc rotation in air and the disc speed have the limitations that the discs do not dry out and that the disc peripheral speed does not exceed 0.33 ms^{-1} this is the speed at which microorganisms shear off the disc. This translates to a rotation speed of 1–2 rpm. Thus the rotation of the discs achieves three functions: aeration; mixing of the wastewater; and it helps provide the shear force necessary for sloughing. Discs are arranged in groups (perhaps of 20 discs), separated by baffles to minimise short-circuiting and flow surges. The baffles are not complete bulkheads, and allow slow fluid transfer between compartments. Normally, a minimum of four compartments is used.

Self assessment exercise 15.3

Does compartmentalisation of a RBC produce conditions of complete mixing or plug flow? Of the four configurations shown in Fig.15.5, which most closely resembles the flow regime in such a RBC?

Small plants are designed on the basis of BOD loading per unit area of discs, using both sides of the disc in the calculation. A BOD20/SS30 effluent can be produced at organic loadings of < 6 g BOD $m^{-2}day^{-1}$. For comparison, a conventional low-rate percolating filter can be surface loaded at 1.0–1.2g BOD $m^{-2}d^{-1}$, which is five to six times less than an RBC, to produce a 20/30 effluent. The RBC is an intensification of the percolating filter process and it has many advantages:

(a) Complete wetting of media;
(b) No clogging of media, resulting in reduced retention time;
(c) Unlimited oxygenation;
(d) Land requirement is much lower (around 10% of that required by the equivalent treatment capacity supplied by a low-rate percolating filter);
(e) No distribution problems as can occur with percolating filters;
(f) Low power consumption;
(g) No recirculation of effluent is required;
(h) High BOD removal including substantial nitrification;
(i) Covering with a green plastic cover is environmentally acceptable, and removes wind and frost damage.

The capital cost of RBCs is high compared to other biological unit processes, but their quietness, lack of odour and no fly nuisance makes them ideal where treatment must be done close to houses. The low running costs and lack of required maintenance may offset the capital cost for use in remote locations.

(a) Microbiology of the RBC

In the percolating filter, the growth is rather uniform in thickness and the only realistic way that soluble substrate and dissolved O_2 can be supplied to the biomass close to the support medium is diffusion. With oxygen and substrate limited to the outer portion of the growth, it follows that only this fraction of the total biomass is actively treating the wastewater. By contrast to the percolating filter, the RBC biomass does not form a smooth biofilm, but has many extended projections outwards into the adjacent film of wastewater (Fig.15.9).

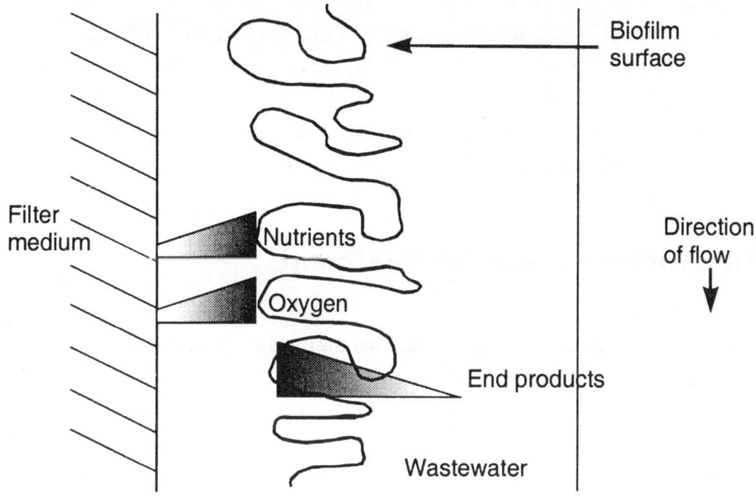

Fig.15.9 *The RBC biofilm*

The straggly nature of the RBC biomass means that the biologically active surface area is much larger than just the surface area of the support medium. The substrate and oxygen can reach a greater proportion of the biomass and render it active in BOD removal.

Rotation of the media also provides turbulence at the interface between biomass and wastewater so that dissolved oxygen and soluble substrate are available to the inner portion of the biomass not just by diffusion, but also by mixing. At a peripheral velocity of $0.33 ms^{-1}$, the Reynolds number at the outer periphery may be as high as 7000, which is well beyond the threshold of laminar-to-turbulent flow (2100) and into the regime of near-turbulent flow. Reynolds number (*Re*) is a dimensionless number which is used to characterise flow as laminar or turbulent. Turbulence within a fluid flow system creates mixing. The exact transition, Re_{tr}, between laminar and turbulent flow varies, depending somewhat on the entrance conditions.

Like the percolating filter, a gradation of microbial species along the reactor path can be seen. The discs in the first two compartments develop a heavy growth of heterotrophic bacteria and fungi of similar genera to those found in other aerobic wastewater treatment systems. As treatment proceeds, the third and fourth compartments receive less strong waste. Here the slow-growing organisms are less likely to be out-competed and this is where significant nitrifying populations develop. Due to the nature of flow in the RBC, there is a tendency for nitrite to accumulate in the compartment where nitrification first occurs. Finally, there is a net loss of nitrogen through the system that approaches 20% due to denitrification (reduction of nitrate to nitrogen gas under anoxic conditions) at the media-biomass interface.

15.2.4 Fluidised bed

These are much less common than the two fixed film processes already examined, and are still in most applications at the experimental stage. They are ideal for treating higher strength wastewater and industrial wastewaters where the loading is variable. The reason for this is that sand can be used as the biomass support medium, and thus the surface area for treatment is massive compared to other media.

The wastewater is pumped upwards at sufficient velocity to fluidise the sand. In addition, fluidisation prevents clogging and excessive film accumulation. The biomass can be maintained at high concentrations without recycling. The expansion of the bed is controlled by the rate or wastewater input, so that treated wastewater can be removed at the top without loss of media. Fluidised beds are compact and therefore land requirement and capital cost is low. However, the high biomass concentrations result in high oxygen demand, which can only be met using pure oxygen, giving an oxygen concentration in the wastewater that is much higher than can be attained with air. This makes treatment very expensive.

15.3 Anaerobic wastewater treatment

Anaerobic treatment of wastewater has found two major niches that would be difficult or impossible to be occupied by aerobic treatments. Firstly, the aerobic treatment systems, particularly the activated sludge process, produce large quantities of excess biomass (sludge) that has to be disposed of by some means as it cannot be released to a watercourse as a result of the high oxygen demand it would exert. This excess sludge can be stabilised by anaerobic treatment. Secondly, many industries produce biodegradable aqueous effluents that are high in BOD. This is particularly true of food processing effluents. The treatment of high BOD wastewaters (above around 1000 mgl^{-1}) by aerobic treatment processes is fraught with hazard and uncertainty. The central issue is one we have already encountered—the low water-solubility of oxygen in water. Naturally, as the BOD increases, so also does the requirement for oxygen to keep the treatment going. But there is a finite limit due to the low solubility of oxygen in water. Aeration contributes a significant proportion to the overall cost of AS treatment, perhaps as much as 80% of the total running cost. An attempt to treat high strength industrial wastes by AS is uneconomical due to aeration cost.

Assignment 15.8

What could be done with high strength wastes to treat them in an aerobic wastewater treatment process?

Many such food-processing wastes are characterised by a great excess of carbon to both nitrogen and phosphorus. This can create conditions which lead to sludge bulking, which is a very severe operational problem in AS. Another important issue relates to the efficiency of oxygen as a terminal electron acceptor. Aerobic microbial metabolism is much more efficient than the anaerobic pathways. As a result, aerobic treatment systems tend to be much more rapid (lower HRT) than anaerobic treatment, but they also produce much higher biomass yields. This results in large quantities of excess sludge for disposal. It may be useful to look at the maximum specific growth rates of some bacteria of importance in wastewater treatment (Table 15.2).

Table 15.2 μ_{max} of various bacteria found in wastewater treatment.

Aerobic:	μ_{max} (h^{-1})
Zoogloea ramigera	5.5
Sphaerotilus natans	6.5
Escherichia coli	25.2
Methanogenic:	
Methanothrix soehngenii.	0.10 (d^{-1})

For many strong wastes, the slow growth (and therefore long HRT) of the methanogenic bacteria is offset by the low biomass yield which overcomes the very serious difficulties posed for high strength waste aerobic treatment. The advantages of anaerobic, methanogenic treatment over aerobic treatment under these circumstances can be summarised:

(a) Very high organic loadings can be achieved e.g., 10 kg BOD $m^{-3}d^{-1}$, compared to 2 for the conventional AS process;

(b) The production of relatively low levels of excess sludge, with a reasonably high level of purification;

(c) Low energy input, especially if the produced biogas, rich in methane, can be recovered in sufficient volumes. Methane is a good end product; it is stable, non-soluble in water (which greatly simplifies recovery), and can be burned to improve the energy budget of the process.

However, methanogenic treatment rarely produces a completely purified effluent from the plant, and some further form of treatment is often required.

15.3.1 Microbiology of methanogenic treatment

The inter-related microbial metabolisms that terminate in biogas production demonstrate some important principles in microbial ecology. On paper it is possible to separate several component parts of the overall metabolism that cannot be done in Nature (Fig.15.10). The entire process depends on community action, which is almost always overlooked in laboratory microbiology. And although several component parts of the overall process can be isolated on paper, each subsequent stage depends on the efficient operation of the previous stage i.e., the end products of one stage are the substrates of the next stage, always bearing in mind that the stages are not temporally separated—all operate at the same time. For example, methanogenesis is totally dependent on the production of acetic acid.

The requirement for several metabolic stages results from the complex nature of the starting materials to be treated. They are, in fact, biopolymers which must be depolymerised to forms available to central microbial metabolism. Also, the polymers are too large to enter bacterial cells. Even the smallest proteins are too large to enter cells by diffusion.

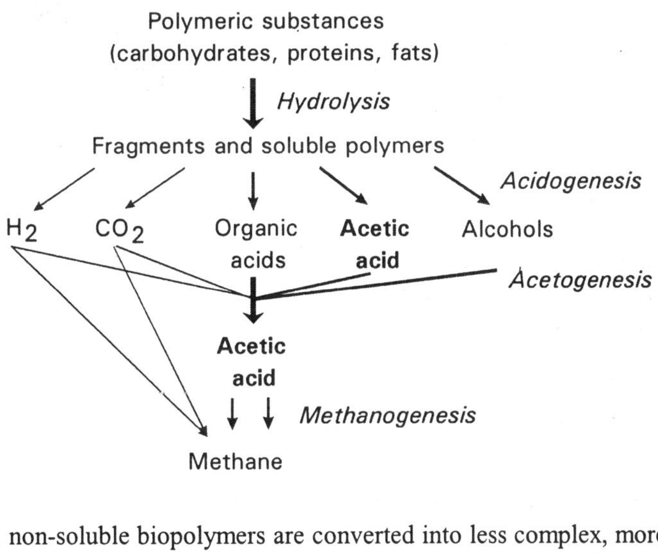

Fig.15.10 *Microbial metabolism ading to biogenic methane during anaerobic wastewater treatment (microbial processes in italics).*

These stages are:

(a) Hydrolysis: at this stage complex, non-soluble biopolymers are converted into less complex, more soluble compounds i.e.,

Carbohydrates (polysaccharides)	→	sugars, oligosaccharides
Lipids	→	fatty acids
Proteins	→	amino acids, oligopeptides

(b) Acidogenesis: the acid forming (fermentation) bacteria convert the soluble polymers into yet simpler compounds. These products are excreted: volatile fatty acids (acetic, butyric and propionic), alcohols, lactic acid, CO_2, H_2, NH_3 and H_2S. New biomass is also formed. The only end products at this stage that can be catabolised directly to methane by methanogenic bacteria are acetic acid, hydrogen and CO_2.
(c) Acetogenesis, in which the products of the above fermentation process (the organic acids and alcohols) are converted to acetic acid by the acetogenic bacteria, with the production of new biomass.
(d) Methanogenesis, the final stage, which in nature is also the final phase of anaerobic decomposition, in which acetic acid, H_2 and CO_2, formic acid and methanol are converted to methane and CO_2, as well as new biomass. In practice, about 70% of the methane produced comes from acetic acid.

It would be right to say that the microbiology of anaerobic digestion is still mysterious in its details. The main reason for this is the difficulty in isolation and handling the microorganisms that come from such a complex environment. Many of the bacteria grow slowly, and in the case of the methanogenic bacteria, vanishingly small concentrations of oxygen in gas mixtures are inhibitory. Hence our knowledge of the biochemistry and molecular genetics of these organisms is much less than our knowledge of many aerobic heterotrophic bacteria.

(a) Hydrolysis and acidogenesis

Proteases are responsible for the hydrolysis of proteins to polypeptides, oligopeptides and amino acids. As smaller entities they can pass through the cell wall for intracellular anabolism/catabolism (Fig. 15.11). Typical proteolytic bacteria are *Bifido-bacterium*, *Clostridium* and *Peptococcus*. It would appear that the most active proteolytic bacteria in anaerobic sludge are the strictly anaerobic clostridia, with numbers of the order of 10^7 ml^{-1} having been recorded.

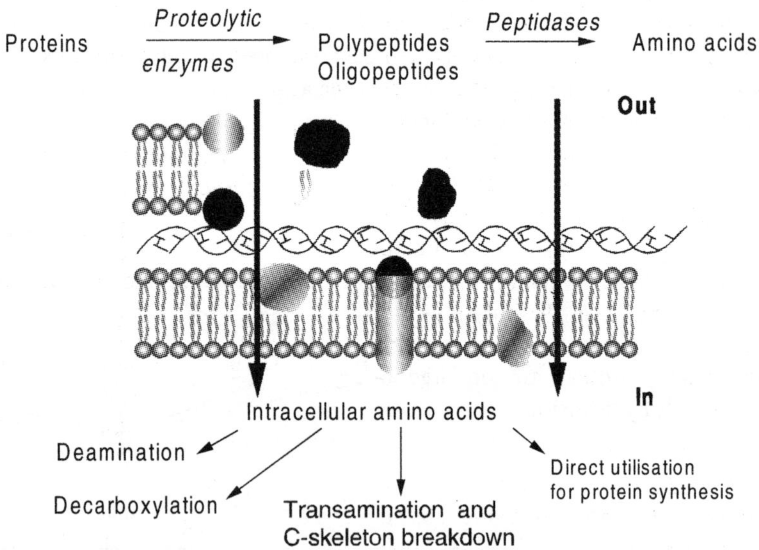

Fig.15.11 *Proteolytic processes (redrawn from Schlegel, 1990).*

Cellulose is the most abundant biopolymer on the planet. It is a basic component of all plant materials. Therefore, many food-processing wastes are likely to contain large quantities of cellulose. Anaerobic cellulolytic bacteria are present in anaerobic sludge in numbers of the order of 10^4–10^5 ml^{-1}. Under anaerobic conditions, cellulose is degraded by mesophilic and thermophilic clostridia. For example, the cellulose fermentation by the thermophile *Clostridium thermocellum* produces ethanol, acetate, lactate, formate, molecular hydrogen and CO_2 as end products. Extracellularly, the cellulose is probably only converted to cellobiose. Intracellularly, cellobiose is converted to glucose and further fermented *via* pyruvic acid to yield volatile fatty acids (VFA), ethanol, hydrogen and CO_2.

The heterogeneous group of organisms bringing about hydrolytic degradations have another important function in anaerobic digestion – they also generate the organic acids required for further treatment by methanogenesis. The most frequently observed end products of acidogenesis are acetic, butyric, lactic and propionic acids. The clostridia appear to be of central importance. Table 15.3 shows some clostridial fermentations.

Table 15.3 Some clostridial fermentations.

Species	Substrates	Fermentation products
C butyricum	glucose, starch, dextrin	butyrate, acetate, CO_2, H_2
C tyrobutyricum	glucose, lactate, glycerol + acetate	butyrate, acetate, CO_2, H_2
C pasteurianum	glucose, starch, mannose	butyrate, acetate, CO_2
C Pectinovorum	pectin, starch, glycogen, dextrin	butyrate, acetate
C acetobutylicum	glucose, glycerol, pyruvate	butyrate, acetate, butanol, acetone, acetoin, ethanol, CO_2, H_2
C Propionicum	alanine, threonine	acetate, propionate, CO_2
C kluyveri	ethanol + acetate + CO_2	caproate, butyrate, H_2
C histolyticum	proteins, amino acids	acetate, lactate, NH_3, H_2

(b) Acetogenesis

The primary function of the acetogenic bacteria in anaerobic digestion is the fermentation of the above primary products to acetate, formate, CO_2 and hydrogen. These are the substrates of the methanogens. Both the acetogens and the methanogens are very ancient organisms in evolutionary terms. The acetogenic bacteria should be regarded as anaerobic, hydrogen-oxidising chemolithotrophs.

Two types of acetogenic bacteria can be recognised:

(a) The hydrogen-producing acetogens, which gain energy by the dissimilation of alcohols higher than methanol, and organic acids larger than acetate, into acetic acid and hydrogen.
(b) The hydrogen-consuming acetogens which catabolise carbohydrates, hydrogen, and CO_2 to acetic acid.

A curiosity of the metabolism of the hydrogen producers is that the accumulation of hydrogen is inhibitory to them. Therefore, they must grow in the presence of organisms that consume hydrogen. In an anaerobic digester, the excreted hydrogen, dissolved in the liquor, is directly assimilated by the methanogens. Here we have a situation that amounts to a mutual symbiosis. The hydrogen-producing acetogens produce the substrates for the methanogens, but the consumption of hydrogen by the methanogens is essential for the survival of the hydrogen-producing acetogens. This phenomenon of inter-species hydrogen transfer is not fully understood.

Assignment 15.9

Explain what is meant by the term 'hydrogen-oxidising chemolithotroph'.

(c) Methanogenesis

The methanogens (Archaea) are the last link in the nutritional chain. This is reflected in their metabolism e.g., the substrate range is very narrow. They contain neither catalase or superoxide dismutase: as a result oxygen is lethal to them. Not only do they require strict anaerobiosis, they also require a redox potential around –300 mV to allow growth to occur. These organisms convert the end products of acid fermentations to gases, mainly methane and CO_2. Roughly equal volumes of both gases are produced in anaerobic digester biogas. Methane is very insoluble in water, but some of the produced CO_2 is not liberated due to its substantially higher water-solubility.

Some of the representative methanogens are shown on Table 15.4. Four groups can be distinguished within the methanogens:

(a) Hydrogen oxidising methanogens: most isolated genera belong to this group;
(b) Hydrogen oxidising acetate-utilising methanogens: these are the most metabolically versatile. Some utilise methanol and methylated amines;
(c) Obligate acetate-utilising methanogens: these are acetate-utilisers unable to utilise hydrogen;
(d) Obligate methylotrophic methanogens: they cannot utilise hydrogen or acetate. They utilise methanol, dimethyl sulphoxide, and methylamines to methane.

Table 15.4 Some representative methanogens and their substrates.

Species	Substrates (H donors)
Methanobacterium arboriphilicum	H_2
M thermoautotropicum	H_2
M formicicum	H_2 or formate
M ruminantium	H_2 or formate
M mobile	H_2 or formate
Methanococcus vannielii	H_2 or formate
Methanosarcina barkeri	H_2, methanol, acetate, mono-, di- or trimethylamine
Methanosarcina mazei	H_2, methanol, acetate
Methanospirillum hungatii	H_2 or formate
Methanothrix soehngenii	acetate

There appears to be a close relationship between gas production and the number of methanogens in an anaerobic digester. They are present in numbers between 10^6 and 10^8 ml^{-1}. It is from acetic acid, and to a lesser extent, propionic acid, that the majority of methane is derived in a digester.

Self assessment exercise 15.4

One stoichiometric equation for the production of methane is given as:

$$C_nH_aO_b + [n - a/4 - b/2]H_2O \rightarrow [n/2 - a/8 + b/6]CO_2 + [n/2 + a/8 - b/4]CH_4$$

Use this equation to calculate the volume (m^3) of CH_4 produced from the digestion of 1 kg of glucose.

15.3.2 *Anaerobic treatment processes*

(a) Flow-through systems

The most fundamental categorisation of anaerobic processes is the distinction between flow-through and contact systems. The most common application of flow-through systems is for the anaerobic digestion of domestic sewage sludge as a means of stabilising the waste sludges from both primary and aerobic secondary treatments. These contain high levels of solids, up to around 60,000 mgl^{-1}. About 70% of the sludge is degradable and up to 80% of this is digested to reduce the solids content substantially before disposal. Anaerobic digesters for this purpose come in three designs, based on the temperature of operation: psychrophilic (5–20°C), mesophilic (20–40°C) and thermophilic (up to 60°C). As mesophilic digesters represent the majority of reactors in use for sludge digestion, these will be examined in more detail. They usually operate at 30–32°C with retention times of 20–40 days. At large wastewater treatment plants, there is sufficient sludge available for continuous operation with gas collected and used to heat the digesters (Fig.15.12).

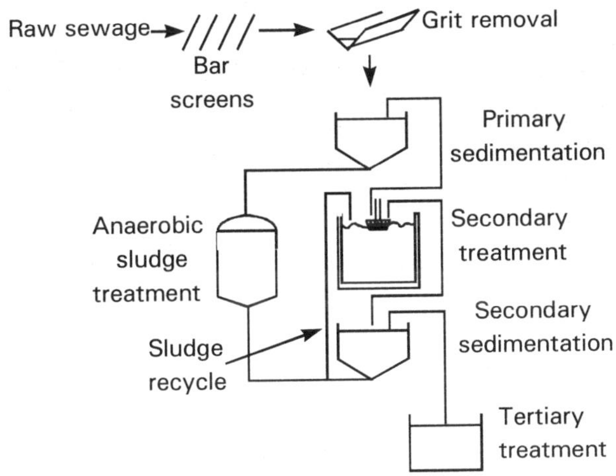

Fig.15.12 *Overall schematic of wastewater treatment.*

In their simplest form, mesophilic digesters are single stage reactors with no recycle and no stirring. In Europe, conventional mesophilic digestion is a two-stage process (Fig. 15.13) in which the primary reactor is both heated and stirred, which optimises acid formation and gas production (by heating) and also decreases the retention time from 30–60 days to <15 days (by improved contact through stirring). The secondary reactor is unheated and unstirred, and thus can be used either to continue digestion in psychrophilic mode, or simply for sludge

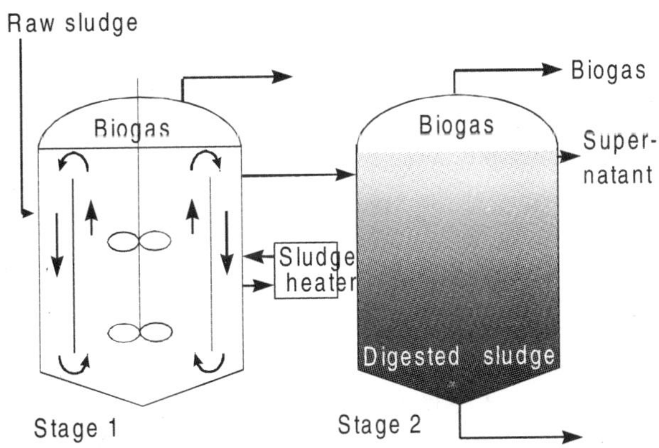

Fig.15.13 *Two-stage mesophilic digester (redrawn from Gray, 1992)*

separation. Temperature fluctuations have relatively little effect on the hydrolytic organisms, but the increasingly more specialist acetogenic and methanogenic bacteria are much more sensitive to change: a drop of just 2–3°C in a mesophilic digester can badly affect biogas production.

(b) Contact systems

There is now great interest in the use of anaerobic treatment of high strength industrial wastes for the reasons outlined earlier. These are the so-called contact systems, in that the MCRT is much higher than the HRT—biomass concentrations are high through biofilm treatment or recycle, and thus there is much more biomass contact with the waste. India has an extensive molasses distillation industry, and the liquid effluent from this industry is often treated anaerobically (Plate 15.4*i*).

(i)

(ii)

Plate 15.4 Industrial anaerobic treatment plants in India.
[i: Petrochemical plant, Hazira ii: Molasses distillery, Visakhapatnam
Courtesy of ADI Systems Inc, Canada.]

Given that high strength biodegradable wastes necessitate anaerobic treatment rather than aerobic, the primary problem has been in maintaining high biomass concentrations within the reactor. In a typical flow-through system, the biomass is diluted out with the treated waste. Thus there have been two strategies developed to maintain high biomass levels:

(a) Recycle a portion of the biomass after separation from the treated wastewater;
(b) Develop treating biofilms within the reactor on inert growth support media.

There are currently four of these contact systems:

(a) Anaerobic AS;
(b) Static media filters;
(c) Sludge blankets;
(d) Fluidised beds.

Anaerobic AS, as a method of treatment, is very similar to aerobic AS (i.e., with sludge recycle), but the reactor is sealed, not open to the atmosphere. Relative to flow-though anaerobic digesters, they operate with short HRT. They can operate with high efficiency (>90% BOD_5 removal), with relatively low reactor volumes. The major application in the UK is to treat sugar beet process water. The remaining systems are more recent developments as high rate systems in which high biomass retention can be achieved in a less complicated way.

The term 'static' media filters refers to reactors containing static support media, which differentiates these processes from fluidised beds, in which the media are mobile. In design they are submerged percolating filters. They are operated with a variety of media from mineral to plastic random packs (Fig. 15.14), the plastics being preferred due to their much greater specific surface area, lower density, and high voidages. If operated in a downflow manner, all the treating biomass is attached, as all the dispersed biomass is washed from the reactor. In upflow anaerobic filters, a significant proportion of the active biomass will be present as dispersed growth in the voids.

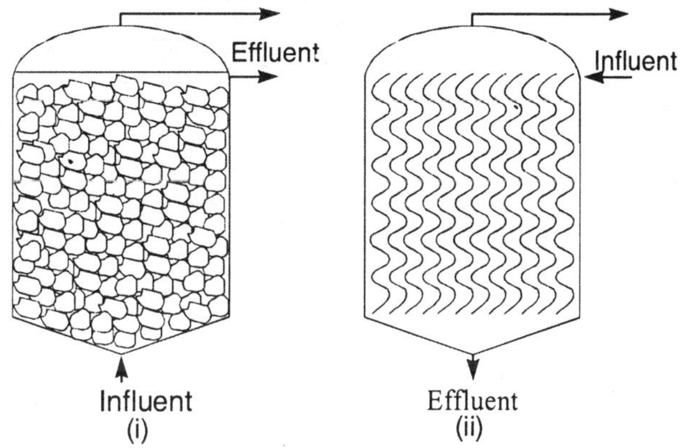

Fig. 15.14 *Static media anaerobic reactors.* (i) *A submerged media reactor* (ii) *A downflow expanded bed reactor (redrawn from Gray, 1992)*

As expected, the MCRT: HRT ratio is extremely high in static media filters. This gives the process more stability than many anaerobic contact systems and they have excellent resistance to toxic compounds. They have been successfully applied to a range of high strength industrial wastes. Typical performance figures are: 80–90% COD removal, and an organic loading of 3.5 $kgm^{-3}d^{-1}$.

Sludge blanket reactors are operated in the upflow mode (Fig.15.15). In this system no support material is used. Instead, a gas–solids separator is installed in the top of the reactor. The reactor provides a very gentle treatment which allows highly stable symbioses of acetogenic and methanogenic bacteria to form, which results in small pellet-like clumps which have very good settling characteristics. Thus a heavily flocculated sludge develops which acts as a separate fluidised bed able to withstand high mixing forces. This is the part known as the sludge blanket. Mixing is achieved by pumping the feed wastewater through the base of the reactor up through the sludge blanket. Thus the main nutrient supply immediately contacts the highest concentration of biomass. Mixing is further improved by the production of biogas (methane is insoluble in water). Overall, the mixing regime is thus mild, and the result is that the sludge flocs or granules do not disintegrate through turbulence. Above the blanket, these stable flocs or granules can easily settle back to the blanket, thereby preventing washout of biomass. The settlement zones slope at an angle of 50° (Fig. 15.15) that promotes the return of sludge from them to the sludge blanket.

The MCRT of the UASB is higher than any contact system, and also the highest organic loadings can be achieved. There have been reports of successful treatment by the UASB on numerous occasions. With a readily degradable carbohydrate waste, 90% removal with a HRT of only four hours has been achieved with very high loadings. With wastewaters with low solids content, an average of 80% removal efficiency is reported. However, it becomes decreasingly successful when the solids content of the waste increases. Compared to other anaerobic systems, operation is difficult, mainly due to problems in maintaining the sludge blanket.

The fluidised bed (Fig.15.15) is a modification of the anaerobic biofilter that is still at the experimental stage. It benefits from the advantages of a fixed bed reactor with regard to cell immobilisation

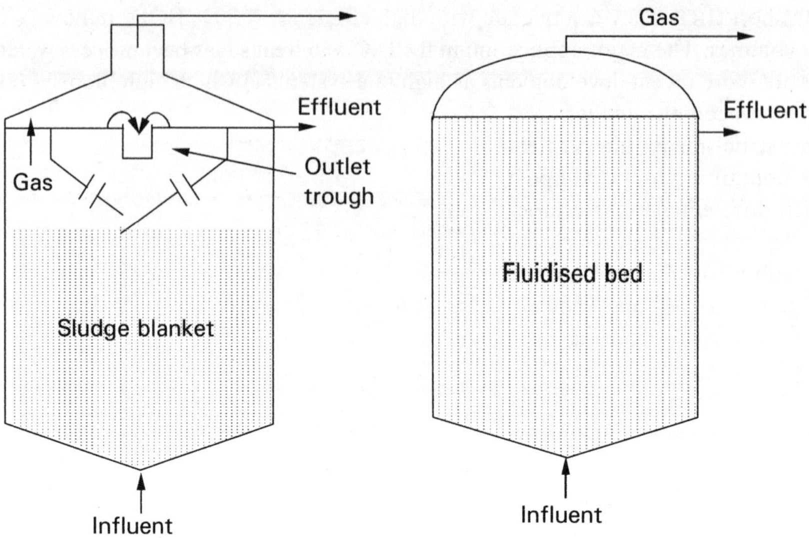

Fig.15.15 *The upflow anaerobic sludge blanket and the fluidised bed*

without the potential problem of blockage caused by excessive growth. The biomass support can be inert (sand) or reactive (activated carbon). As such, the specific surface area is massive. The medium becomes fluidised by applying a high mixing velocity. This requires additional energy input. Stated advantages of fluidised beds are:

(a) High resistance to temperature changes;

(b) Tolerance of toxic compounds;

(c) High organic loadings can be applied;

(d) Tolerance of solids.

Self assessment exercise 15.5

Below is a table into which you must insert a ' +' for a positive response, and a '–' for a negative response at the appropriate places. On occasion, the answer may not be one way or the other, and a +/– response would be appropriate.

Reactor type	**Stationary biomass**	**Fixed film biomass**
Plug flow		
Activated sludge		
Percolating filter		
Rotating biological contactor		
Fluidised bed		
Two-stage mesophilic digester		
Upflow anaerobic biofilter		
Downflow expanded bed		
UASB		

15.4 Bioremediation of contaminated land and water

Bioremediation refers to a treatment technology which uses biological systems to catalyse the destruction or transformation of a large variety of classes of chemical, environmental pollutants. Though biodegradation of wastes is a centuries-old technology, it is only in recent decades that serious attempts have been made to harness natural biodegradative capabilities with the goal of large-scale technological applications for effective and affordable environmental restoration. Over the last decade, bioremediation has become a widely used technology in the US, and has gained acceptance in much of Western Europe.

The chemical structures of organic pollutants have a profound influence on the abilities of microorganisms to metabolise them, especially with respect to the rates and extent of their biodegradation. Biodegradability is essential for the bioremediation of organic pollutants. Some organic compounds are readily biodegraded whereas others are recalcitrant (non-biodegradable). Low- to mid-molecular weight hydrocarbons and alcohols are representative of readily biodegradable chemicals. Xenobiotic (recalcitrant) compounds, especially halogenated hydrocarbons, tend to be resistant to biodegradation and some, such as some dioxins, may be recalcitrant compounds that are more difficult to biodegrade and present the greatest challenges for successful bioremediation. As a rule, branched and polynuclear (consisting of several fused aromatic rings) compounds are more difficult to degrade than straight chain and simple monoaromatic molecules, and increasing the degree of halogenation generally decreases the rates of biodegradation of a series of related compounds.

Trichloroethene (TCE), a widely distributed solvent, and higher molecular weight polychlorinated biphenyls (PCBs), which were widely used in transformers and various chlorinated pesticide preparations, are more difficult to bioremediate because of the requirement to degrade these chemicals cometabolically. Dichloromethane and monochlorobiphenyls, however, are much more easily treated because microorganisms can utilise these compounds as sole carbon and energy sources.

Self assessment exercise 15.6

A

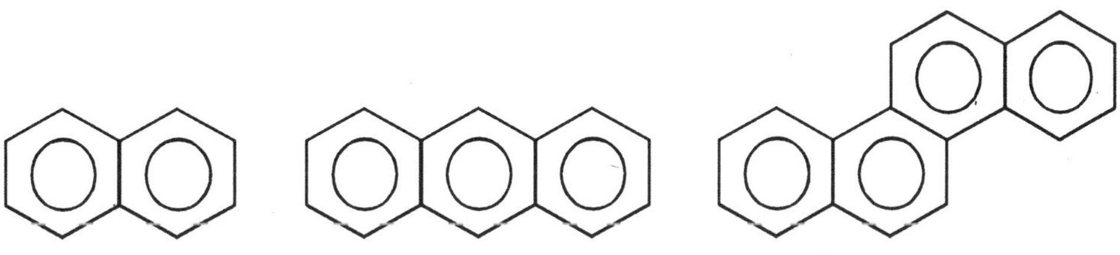

B

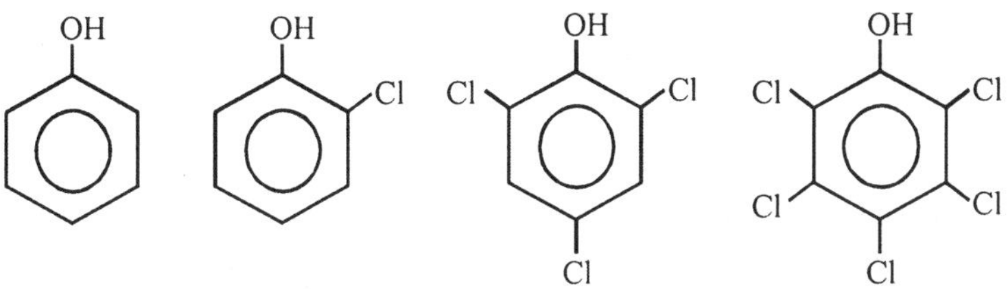

List the compounds in each series (A and B) in order of biodegradability, and state reasons.

15.4.1 Intrinsic bioremediation

Also known as natural attenuation, this is a process whereby the natural microflora and environmental conditions exist for natural attenuation of a pollutant to safe levels within an acceptable time frame. This is generally the first choice for biotreatment because it requires no intervention—just monitoring of the natural process of biodegradation. The lack of specific guidance on appropriate sampling and analytical procedures to ensure that intrinsic bioremediation measurements generate quality data is of concern. The extent to which intrinsic bioremediation is ultimately embraced will depend, to a large degree, on the valid characterisation of site conditions. It should not be seen as a 'do nothing' approach. The key to its success is in proper monitoring of biodegradation with time. If it is clear that the required clean-up will take so long that the cost of monitoring the site will outstrip a more intrusive technique, then intrinsic bioremediation is obviously not the appropriate solution.

15.4.2 In situ bioremediation

As the title implies, *in situ* bioremediation comprises various techniques which minimise intrusion. They do not require impossibly expensive excavations of vast amounts of contaminated vadose-zone (the unsaturated zone above the water table) soils, or unending pump-and-treat schemes for large, deep aquifers. This may save vast sums of money and potentially solve problems that are not approachable by off-the-shelf technologies. Most *in situ* processes involved the stimulation of indigenous microbial populations so that they become metabolically active and degrade the contaminant(s) of concern.

Problems encountered during *in situ* stimulation of microbial populations include the plugging of wells and subsurface formations by the tremendous amounts of biomass that may be generated through microbial growth on hydrocarbons, difficulties in supplying sufficient oxygen to the subsurface, and the inability to move nutrients and electron acceptors to all regions of heterogeneous subsurface environments. Also, it is rarely possible to remove all free product, so reservoirs of slowly released contamination may be present for many years in some situations.

Assignment 15.10

Identify each of the following important organic pollutants, collectively known as BTEX.

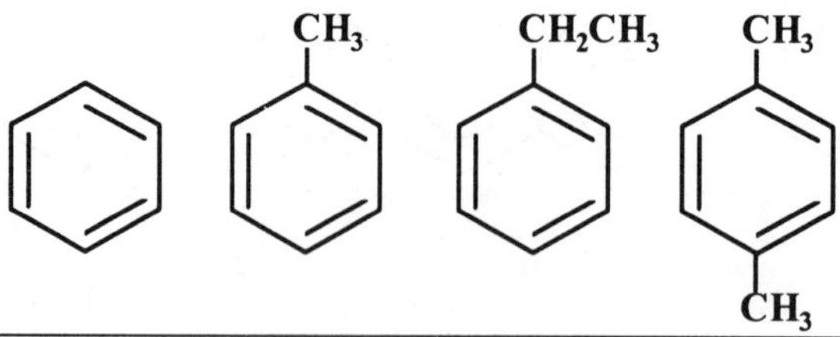

(a) Biostimulation

Biostimulation of indigenous microbial populations to increase the rates of biodegradation so as to remediate the target chemicals at an acceptable rate is a strategy often employed. For sites where biostimulation is to be employed, it has to be demonstrated that a natural degradative population already exists within the contaminated zone and that environmental conditions are insufficient for microbial activity at high rates. Measurements of the physical and chemical properties of field samples can reveal the physicochemical limitations to biodegradative activities, which can then be modelled to indicate the critical limiting factors.

Some common environmental limitations to biodegradation of pollutants are excessively high waste concentrations, lack of oxygen, unfavourable pH, lack of mineral nutrients, lack of moisture, and unfavourable temperature. A variety of methods that modify environmental conditions can be employed to enhance the rates of biodegradative activities by the indigenous microbial populations. Once the limiting environmental conditions are corrected, the ubiquitous distribution of microorganisms allows, in most cases, for a spontaneous enrichment of the appropriate microorganisms.

The availability of molecular oxygen has a profound effect on the biodegradation of a large number of organic pollutants. Oxygen limitation often is the most troublesome problem facing *in situ* bioremediation for hydrocarbons and other pollutants that are biodegraded aerobically. This problem is especially profound when the soil is waterlogged, as waterlogging hinders the circulation of air in soils.

For *in situ* bioremediation of surface soil, oxygen availability is best assured by providing adequate drainage. Air-filled pore space in the soil facilitates diffusion of oxygen to hydrocarbon-utilising microorganisms, while in waterlogged soil, oxygen diffusion is extremely slow and cannot keep up with the demand of heterotrophic decomposition processes. For sites where substantial concentrations of decomposable hydrocarbons create a high oxygen demand in soil, the rate of oxygen diffusion may be inadequate, even in well-drained and light-textured soils. Cultivation (ploughing and rototilling, for example) has been used to turn the soil and assure its maximal access to atmospheric oxygen. Adding dilute solutions of hydrogen peroxide in appropriate and stabilised formulations can also be used to supply oxygen for hydrocarbon biodegradation. The decomposition of hydrogen peroxide releases oxygen which can support aerobic microbial metabolism. To avoid formation of gas pockets and microbial toxicity, the practical concentration of hydrogen peroxide in injected water is kept around 100 ppm.

In situ bioremediation based largely on the injection of air is being extensively used in groundwater treatment. Air sparging is an *in situ* technology which can be utilised either to remove volatile compounds from the subsurface or to induce microbially mediated treatment in aerobic water-saturated soil. During air sparging, air is injected into the saturated zone, usually below the target cleanup zone. Volatile compound(s) dissolved in the groundwater and adsorbed on the soil particles will partition into the air phase and be transported to the vadose (unsaturated) zone. The volatilised compounds can then be collected from the vadose zone by a soil vapour extraction system.

Air sparging can be effective in providing oxygen to aerobic microbes (biosparging, Fig.15.16). These naturally occurring microbes are often capable of oxidizing a wide range of organic compounds, including semivolatile compounds not removed by volatilisation. In addition, aerobic microbes have been shown to be capable of oxidising halogenated compounds such as the halogenated monocyclic aromatics (aryl halides).

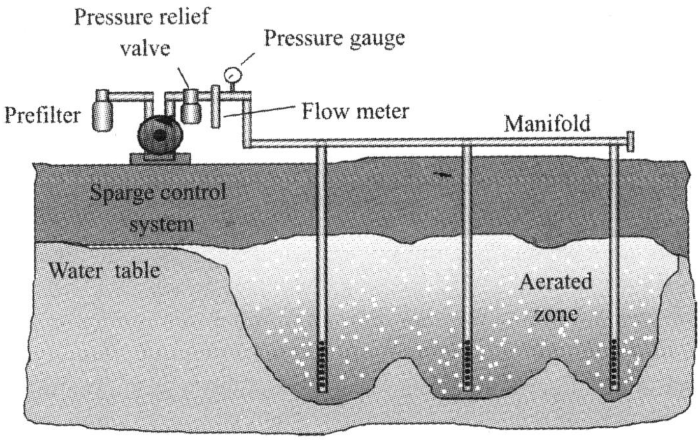

Fig. 15.16 *Biosparging.*

Assignment 15.11

For the compound $C_cH_hCl_{cl}N_nNa_{na}O_oP_pS_s$.

$$\text{The ThOD} = 16\,\frac{[2c + 0.5(h-cl-3n) + 3s + 2.5p + 0.5na-o]}{MW}\ \text{mg mg}^{-1}$$

Write the equation for the aerobic mineralisation of *n*-hexadecane ($C_{16}H_{34}$), and calculate the oxygen demand of 1 g of $C_{16}H_{34}$. Use the above equation to make the same calculation. Do both methods agree?

Bioventing is becoming an attractive option for promoting *in situ* biodegradation of readily biodegradable pollutants like petroleum hydrocarbons. Bioventing (Fig. 15.17) is a process which employs enhanced oxygenation in the vadose zone to accelerate contaminant biodegradation. This technology is also highly effective when paired with bioremediation in the saturated zone (biosparging). When applicable and properly implemented, bioventing often results in a faster and more cost-effective remediation.

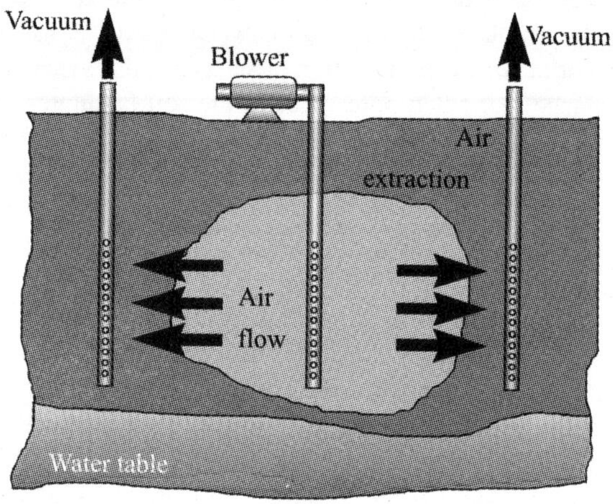

Fig. 15.17 *Bioventing.*

While oxygen availability can severely limit the biodegradation of hydrocarbons, other compounds are more rapidly degraded under anaerobic conditions than when molecular oxygen is available. Thermodynamic considerations, as well as practical experience, indicate that dechlorination of halocarbon compounds is favoured under anaerobic conditions. For the bioremediation of highly chlorinated compounds, sequential anaerobic and aerobic treatment may well be the best practical solution. During an initial anaerobic phase (either intrinsic or stimulated) the higher molecular weight PCB congeners are dehalogenated to form lower molecular weight compounds with fewer chlorines. Subsequently, under aerobic conditions the lower molecular weight congeners are degraded to carbon dioxide, water, and chloride ions. This approach has been demonstrated for the bioremediation of PCB contaminated sediments of the Hudson River and in the Sheboygan River, both in the USA. In these studies, large steel casings were driven into contaminated sediments and nutrients were added following anaerobic dehalogenation of the PCBs. Then agitation of the sediments and forced aeration were used to create aerobic conditions. The lower molecular weight PCB congeners were biodegraded under these conditions. Thus biodegradation of both the higher and lower molecular weight PCB congeners was achieved by sequential anaerobic and aerobic bioremediation.

Biodegradation of oils occurs at the oil/water interface, as microbes degrading the oil require water, and are not metabolically active in an environment lacking water. Biodegradation rates can be limited by the available concentrations of various nutrients. Since microorganisms require nitrogen and phosphorus for incorporation into biomass, the availability of these nutrients within the same area as the hydrocarbons is critical. Under conditions where nutrient deficiencies limit the rate of petroleum biodegradation, the beneficial effect of fertilisation with nitrogen and phosphorus has been conclusively demonstrated and offers great promise as a countermeasure for combating oil spills and other hydrocarbon spills.

> ***Assignment 15.12***
>
> Conventional agricultural fertilisers have low affinity for oil and are usually soluble in water. What problems would arise in using such fertilisers to stimulate biodegradation of oils in highly porous soils or in water?

Atlas and Bartha (1973) developed an oleophilic nitrogen and phosphorus fertiliser that places the nitrogen and phosphorus at the oil-water interface, the site of active oil biodegradation. Because the fertilizer is oleophilic, it remains with the oil and is not rapidly diluted from the site. In the aftermath of the Amoco Cadiz oil spill of 1978 in France, a commercial oleophilic fertilizer was developed by Elf Aquitaine. The product, called Inipol EAP 22, is a microemulsion that contains urea as a nitrogen source, lauryl phosphate as a phosphate source, and oleic acid as a carbon source to boost the populations of hydrocarbon-degrading microorganisms. Inipol EAP22 was also used extensively for the treatment of the 1989 *Exxon Valdez* spill in Prince William Sound, Alaska.

For the bioremediation of pollutants in surface soils it is generally easy to add nutrients as agricultural fertilisers. However, getting nutrients to subsurface soil and groundwater populations is more complex. One method that has been explored is the use of electrical currents to help move the nutrients to the pollutants. This approach to *in situ* remediation utilising low-level direct-current electric potential differences to transport ionic nutrients has been used for enhancing the biodegradation process to clean radioactively contaminated soils.

A remediation technology using hydraulic soil fracturing to form channels, and electric currents to drive nutrients through the soil, called the Lasagna process, has been developed to improve the treatment of contaminated soils and groundwater. The approach, which integrates electrokinetics with desorption and degradation in treatment zones of the contaminated soil, could treat contamination due to organic and inorganic compounds and mixed wastes. Pilot testing results have shown that the process is an effective alternative for treating contamination in low-permeability soils.

A general problem related to the bioremediation of certain xenobiotic pollutants in soils and sediments is bioavailability. It is a common experience that a pollutant is degraded rapidly when it freshly enters the soil, but the biodegradation levels off and stops with some percentage of the pollutant still present in the soil, and conditions still favourable for degradation. Extraction and chemical analysis can detect these pollutants i.e., they had not become bound residues, yet microorganisms no longer appear to have access to them. In the case of 'old spills', remediated years after the pollution event, much of the pollutant may be in non-bioavailable form, rendering bioremediation efforts futile or at best incomplete. The nature of the pollutant, the nature of the soil, other pollutants and the time elapsed since the spill all have an influence on bioavailability.

(b) Bioaugmentation

Seeding with pollutant-degrading bacteria (bioaugmentation) can be used to enhance the biodegradative capabilities of the indigenous microbial populations. Bioaugmentation (Fig.15.18) involves the introduction of microorganisms into the natural environment for the purpose of increasing the rate or extent, or both, of pollutant biodegradation. The rationale for this approach is that indigenous microbial populations may not be capable of degrading xenobiotics or the wide range of potential substrates present in complex pollutant mixtures.

Bioaugmentation is used in cases where intrinsic bioremediation or biostimulation does not work because of insufficient or unacclimated natural bacterial populations. This is generally the case for only very recalcitrant chemicals. Selected strains of bacteria with the desired catalytic capabilities and other characteristics are injected directly into the contaminated zone, along with nutrients if necessary. This technology can be complemented through the use of radio frequency heating and hydrofracturing.

Bioaugmentation can also be used in cases where, although the natural population may be sufficient to achieve a biostimulation remediation, it may be desirable to increase the rate of degradation and shorten the time frame for full-scale remediation. Several commercial mixtures of naturally occurring microorganisms are being marketed for use in degrading oil pollution in contained treatment bioreactors as well as for *in situ* applications. Often, however, the added costs for this approach are not justified.

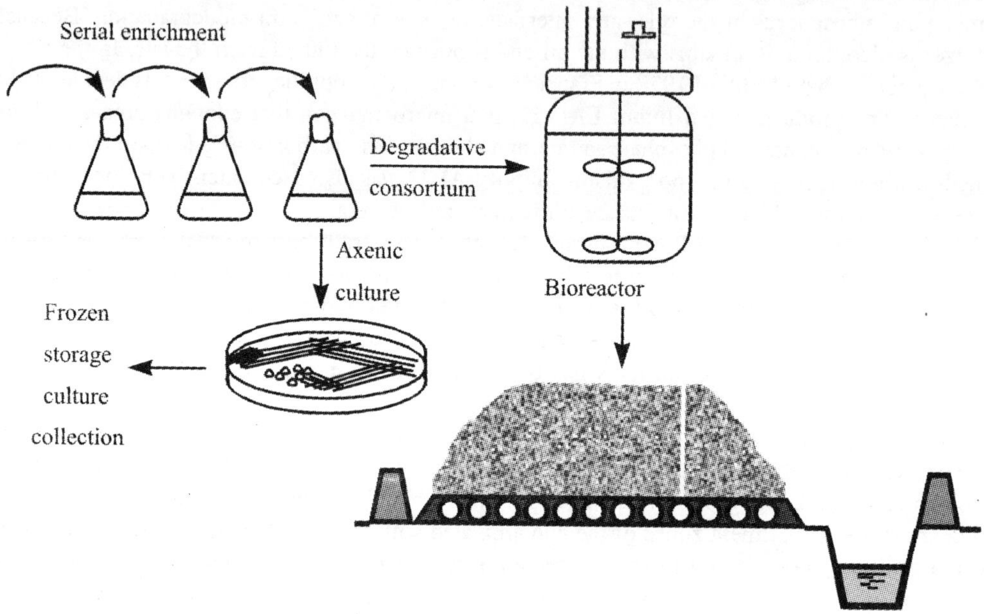

Fig. 15.18 *A scheme for bioaugmentation*
A soil sample from the site to be remediated is taken. It is used to inoculate an appropriate medium containing the target contaminants. Serial enrichment selects for those organisms capable of growing under the imposed conditions. They are grown in mass culture and applied back to the site. Meanwhile the consortium can be broken to a series of axenic cultures, which can be identified, characterised and stored for future use.

One approach for providing the enzymatic capability to degrade diverse pollutants is to use genetic engineering to create microorganisms with the capacity to degrade a wide range of hydrocarbons. Recombinant DNA technology holds great promise for developing microbial strains that can aid in the environmental removal of toxic chemicals. However, considerable controversy surrounds the release of such genetically engineered microorganisms into the environment, and field testing of these organisms has been delayed by the issues of safety, containment, and potential for ecological damage. Given the current regulatory framework for the deliberate release of genetically engineered microorganisms, it is unlikely that any will be utilised for full-scale bioremediation in the near future.

15.4.3 Ex situ bioremediation

These treatment methods are highly intrusive, and therefore the cost and hazards of excavation of potentially large quantities of contaminated soil has to be balanced against the benefits of this approach. They involve removal of contaminated material to another location, either at another site or on the same site. The major benefits that *ex situ* treatments bring are that much greater control over the process can be exercised, and they can treat to completion more quickly. The cost of sampling for laboratory analyses is reduced as there is no need to introduce sampling devices at depth in soil to provide the samples. Also, there is no uncertainty in the overall success of the treatment as all of the soil being treated is available for inspection. With *in situ* treatments, it is expensive in terms of sampling to discover if all pockets of contamination have been treated.

(a) Bioreactor-based treatments

Performing bioremediation in a prefabricated bioreactor gives the ultimate in flexibility of systems with the greatest degree of control over the process. They can be used in various *ex situ* bioremediation treatments. They can be used to treat liquids (e.g., industrial process streams, pumped groundwater), vapours (e.g. solvents vented from contaminated subsurface environments, factory air), or solids in a slurry phase (e.g. excavated soils, sludges, or sediments; plant materials). Bioreactor-based bioremediation technologies can treat soil, sludge, and sediments *ex situ* providing for complete site remediation. Generally for contaminated soils and sediments a bioreactor-based treatment protocol simulates a mixed slurry remediation system that allows for optimal bioremediation process conditions using either biostimulation of indigenous strains or bioaugmentation with superior, selected, naturally occurring strains. They may use suspended microorganisms or adsorbed biofilms, singly or in combination; native microbial populations indigenous to the material being treated; pure microbial cultures isolated from appropriate environments; or genetically engineered microorganisms (GEMs) designed specifically for the problem at hand. They may operate with or without additions of oxygen or other electron acceptors (nitrate, carbon dioxide, sulphate, oxidised metals) and nutrient feeds (nitrogen, phosphorus, trace minerals, co-substrates). Other variables that may be controlled or that must otherwise be accounted for may include ratio of solids to water, biodegradation rates for specific pollutants or mixtures of pollutants, adsorption/absorption of pollutants to matrices such as soil or carriers used to provide surfaces for biofilms, pH, temperature, and redox potential.

Bioreactors, compared to other bioremediation techniques, can be controlled and their processes modelled mathematically with great precision. A general schematic demonstrating the flexibility of bioreactors is shown (Fig.15.19).

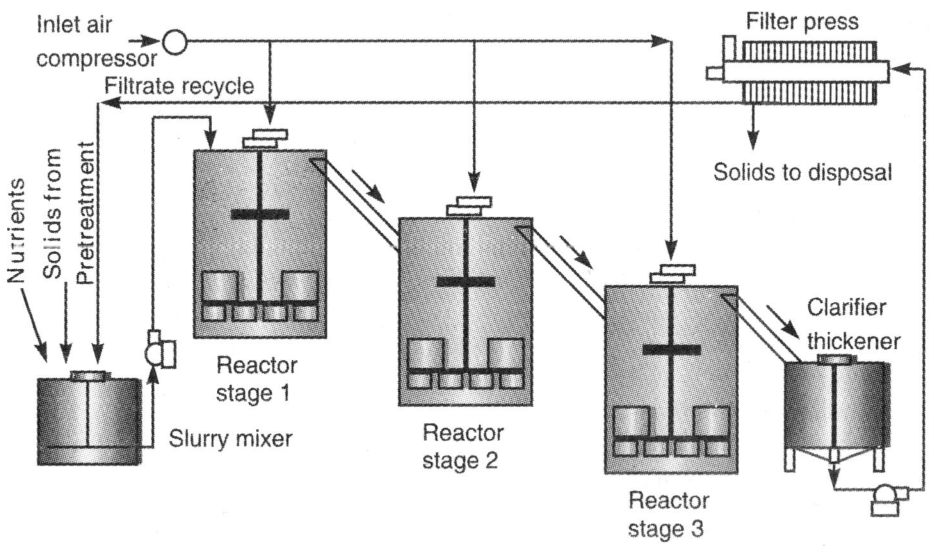

Fig.15.19 *Generalised bioreactor design for bioremediation of soil slurries.*

A variety of bioreactors may be used for treating contaminated soils, sediments, water, and air. Many are similar if not identical to those used in wastewater treatment systems. Some are mobile and can be moved from site to site. Others are at fixed locations, requiring the movement of contaminated materials to the bioreactor.

(b) Composting and landfarming

Landfarming is a biological treatment technology (Fig.15.20) in which waste materials are applied to soil surfaces as solid sludges or aqueous slurries. It has been practiced in the oil industry as a means of treating refinery waste sludges. The soil water content is monitored and adjusted with supplemental inorganic nutrients, added as needed.

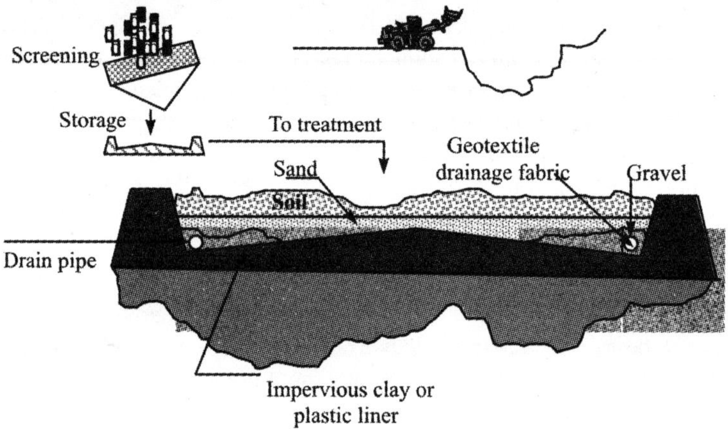

Fig.15.20 *Engineered landfarming.*

Equipment is used to periodically mix the contents to bring the wastes, nutrients and bacteria in contact to enhance biodegradation rates. A potential problem in soil treatment is the residual contaminant concentration that is slowly or not noticeably degraded by soil microorganisms. This non-bioavailable fraction is recognised by its slow transport out of soil micropores. Slow desorption can result from entrapment in intra-particle micropores, especially in the presence of organic matter, which can tightly bind nonionic organic contaminants. Soil leaching experiments can demonstrate whether contaminants are slowly released from the soil matrix, and the addition of surfactants can increase the desorption rate. However, the mixing and surfactant addition that may be necessary to release contaminants increase the cost of landfarming and favour the use of more intensive bioremediation methods such as slurry reactors.

Composting uses windrows or piles constructed on lined areas to encourage biological degradation of contaminants (Fig. 15.21).

Aeration, leachate and runoff control are built into the system design. Blowers are used either to draw or to push air through the soil. Air movement is used to control the concentration of oxygen within the pile as well as the temperature. Depending upon the contaminant type and air emission control requirements, solid-phase peroxide may be used as an oxygen source, thereby reducing the need for engineered air movement.

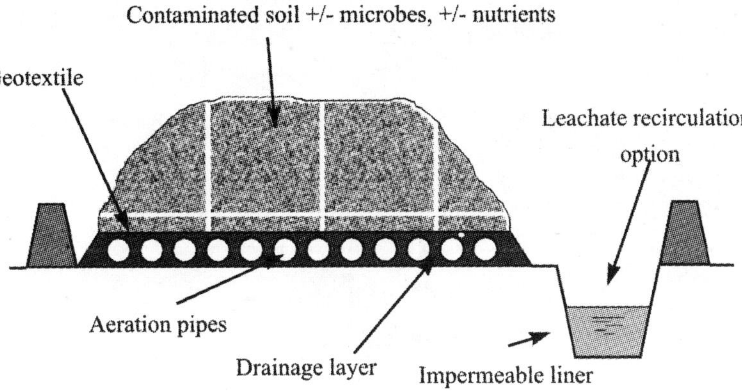

Fig.15.21 *Engineered soil bank.*

Soils are often treated by solid-phase technologies. This usually means placing excavated soils within some type of containment system, e.g ., a lined pit with leachate collection and/or volatile compound entrapment equipment, and then percolating water and nutrients through the pile. Oxygen may or may not be supplied, depending on the bioremediation process being encouraged or supported. Inocula may or may not be added, depending on whether or not an indigenous microbial population can be stimulated to remove the target pollutant(s). Solid-phase treatments are particularly useful for petroleum-contaminated soils.

15.4.4 Heavy metals bioremediation

In most cases, the applicability of bioremediation depends upon the abilities of microorganisms to biodegrade the pollutants to safe levels and to produce nontoxic end products. Toxic metals are a special class of environmental pollutants. Metals cannot be degraded, but only changed from one form (oxidation state) to another. Thus, bioremediation processes for metal-contaminated environments aim at sequestering the metals to make them unavailable to biological components of the ecosystem, or mobilising them in a manner that allows their 'flushing' from the system for collection and disposal. Immobilisation of heavy metals, for example, by sulphide-producing microorganisms (Fig. 15.22) is an effective means of treating some metal-contaminated sites.

The action of sulphate-reducing bacteria produces hydrogen sulphide. In the presence of heavy metals, metal sulphides are produced. Most metal sulphides are extremely insoluble (Table 15.5)

Table 15.5 Metal sulphide solubility products.

Metal sulphide	Solubility product
Ag_2S	1×10^{-51}
Cu_2S	2.5×10^{-50}
CdS	1.4×10^{-28}
ZnS	4.5×10^{-24}
CoS_2	7×10^{-23}
NiS	3×10^{-21}
FeS	1×10^{-19}

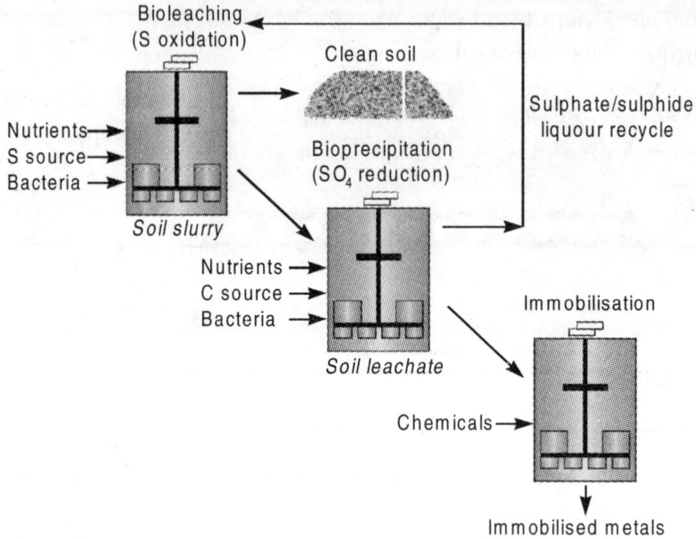

Fig.5.22 *Metals removed by combined sulphur oxidation (leaching) and sulphate reduction (precipitation). [Such a process is being developed by the British Nuclear Fuels Limited.]*

Bioconcentration, that is, uptake without transformation, can also be used to treat heavy metal-contaminated soils and waters. This forms the basis for phytoremediation in which plants are used to extract metals from contaminated soils and waters. Phytoremediation is emerging as a promising method for decontaminating metal-contaminated soils and waters. Various plants concentrate heavy metals, including radionuclides and lead, in the plant biomass. The concentrations of hazardous metals in soils can be reduced and leaching into groundwaters can be prevented by planting Indian mustard (*Brassica juncea*) or ragweed (various *Ambrosia* species) among other plants. Indian mustard can accumulate up to 40% of their biomass as heavy metals—including lead.

Demonstration projects have been conducted at various sites, including Chernobyl in Russia which was heavily contaminated with radionuclides from a nuclear reactor accident. Once the plants accumulate the heavy metals, they can be harvested and the plant biomass incinerated leaving the metals as a residue; metals recovered in this manner can be reused. However, there are great technical difficulties with incineration of radioactive materials.

15.5 Concluding remarks

This chapter has attempted to summarise the important environmental biotechnologies which are currently used on a large scale. The fields of wastewater treatment and bioremediation are very dynamic, with new developments driven by large financial investments. Only some fundamentals, of interest to biologists have been addressed. Many of the techniques appear to suffer from a monetary burden which can only possibly be solved in developed countries. But many local solutions are available using local materials and innovation.

As developing countries become more industrialised, they can learn from the mistakes made by those countries with a long industrial history. Indeed, countries like India, are already trying to learn these lessons, for example the use of wastewater technology in some polluting industries.

Not considered in this chapter is the biological method for treatment of solid municipal wastes – the landfilling technique. Landfill microbiology is assumed to be very similar to the microbiology of anaerobic digestion, but it is even more difficult to investigate. The contents of an anaerobic digester are all available for sampling and inspection; the landfill site is inaccessible to most microbiologists. Landfilling creates two particular problems which the biotechnologist may approach. It is desirable to optimise the production of methane throughout the life of a landfill site to ensure that methane generation can be finished early. This frees up the land for other uses without the fear of continuing hazard from methane.

With time a landfill site creates a hazardous aqueous leachate which varies in composition during the lifetime of the landfill site. These leachates are often amenable to biological treatments by the methods already discussed. However, a great deal remains to be investigated in landfill sites. With increasing pressure to find suitable land for development, and the need to preserve rural resources, there is a movement away from the mass disposal of municipal waste in this way, and this work may never be done.

To the vigilant, it will have become apparent that environmental biotechnologies are different from other biotechnologies in a very important respect. The technologies which constitute waste treatment, be it wastewater, solid waste or contaminated soil or groundwater, are not solely laboratory-based activities. They are very large-scale activities indeed. Also their need in ubiquitous. This has resulted in mass application of these techniques, with a very large industry world-wide, and it could be successfully argued that established environmental biotechnology is the most important branch of a growing science. It also benefits from a great advantage in its acceptability compared to other branches of biotechnology – it has been around for a long time with a very good record in the service of humans.

15.6 Suggested further reading

1. Metcalf and Eddy Inc. (1991).*Wastewater engineering: Treatment, disposal, reuse.* McGraw Hill, New York.
2. van Verseveld, H.W., Arbige, M. and Chesbro, W.R. (1984). Continuous culture of bacteria with biomass retention. *Trends in biotechnology* **2**, 8–12.
3. Horan, N.J. (1989). *Biological wastewater treatment plants: Theory and operation.* John Wiley and Sons, Chichester.
4. Chun, J., Kang, S. O., Hah, Y. and Goodfellow, M. (1996). Phylogeny of mycolic acid-containing actinomycetes. *Journal of industrial microbiology* **17**, 205–213.
5. Gray, N.F. (1992). *Biology of wastewater treatment.* Oxford Science Publications.
6. Hammer, M.J. & Hammer, M.J. (1996). *Water and wastewater technology.* Prentice-Hall, New Jersey.
7. Henze, M., Harremoes, P., Jansen, J.laC., & Arvin, E. (1997). *Wastewater treatment: Biological and chemical processes.* Springer-Verlag, Berlin.
8. Atlas, R.M. & Bartha, R. (1997). *Microbial ecology*, 4th ed. Benjamin/Cummings Science Publishing., Menlo Park.
9. Atlas, R.M. & Bartha, R. (1973). Stimulated biodegradation of oil slicks using oleophilic fertilizers. *Environmental science and technology* **7**, 538–541.

10. Cookson, JT Jr.(1995). *Bioremediation engineering: Design and application.* McGraw-Hill, New York.
11. Schlegel, H.G. (1990). *General microbiology*, 6th ed. pub. Cambridge University Press, Cambridge.

Some of the topics mentioned in the introduction are beyond the scope of this chapter. For the interested reader, some suggested reading on the various topics is included here.

Biodegradable plastics

1. Lee, S.Y. (1996). Review: Bacterial polyhydroxyalkanoates. *Biotechnology & bioengineering* **49**, 1–14.
2. Palmisano, A.C. & Pettigrew, C.A. (1992). Biodegradability of plastics: Consistent methods for testing claims of biodegradability need to be developed. *BioScience* **42**, 680–685.

Biodegradability

1. Grady, C.P.L. (1985). Biodegradation: its measurement and microbiological basis. *Biotechnology & bioengineering* **XXVII**, 660–674.
2. OECD Guideline for the Testing of Chemicals 301: "Ready Biodegradability", October 1990.
3. US EPA712-C-98-075. Fate, Transport and Transformation Test Guidelines. OPPTS 835.3100 Aerobic Aquatic Biodegradation, January 1998.

Biosurfactants

1. Desai, J.D. & Banat, I.M. (1997). Microbial production of surfactants and their commercial potential. *Microbiological and molecular biological reviews* **61**, 47–64.
2. Lang, S. & Philp, J.C. (1998). Surface-active lipids in rhodococci. *Antonie van Leeuwenhoek* , in press.

Ecotoxicity testing

1. Wells, G.G., Lee, K. & Blaise, C. (1998). *Microscale testing in aquatic toxicology: Advances, techniques and practice.* CRC Press, Boca Raton, Florida.

Fossil fuel biodesulphurisation

1. Kilbane, J.J. (1989). Desulfurization of coal: The microbial solution. *Trends in Biotechnology* **7**, 97.
2. McFarland, B.L., Boron, D.J., Deever,W., Meyer, J.A., Johnson, A.R. & Atlas,. R.M. (1998). Biocatalytic sulphur removal from fuels: Applicability for producing low sulphur gasoline. *Critical Reviews in Microbiology* **24**, 99–147.
3. Rhee, S. K., Chang, J.H., Chang, Y.K. & Chang, H.N. (1998). Desulphurisation of dibenzothiophene and diesel oils by a newly isolated *Gordona* strain, CYKS1. *Applied and environmental microbiology* **64**, 2327–2331.

Reporter genes and biosensors

1. Corbisier, P., Ji, G., Nuyts, G., Mergeay, M. & Silver, S. (1993). *luxAB* gene fusions with the arsenic and cadmium resistance operons of *Staphylococcus aureus* plasmid pI258. *FEMS Microbiology letters* **110**, 231–238.
2. Heitzer, A., Malachowsky, B. K., Thonnard, J. E., Bienkowski, P. R., White, D. C. & Sayler, G. S. (1994). Optical biosensor for environmental on-line monitoring of naphthalene and salicylate bioavailability with an immobilised bioluminescent catabolic reporter bacterium. *Applied & environmental microbiology* **60**, 1487–1494.

3. Prosser, J.I. (1994). Molecular marking systems for detection of genetically modified microorganisms in the environment. *microbiology* **140**, 5–17.
4. Willardson, B. M., Wilkins, J. F., Rand, T. A., Schupp, J. M., Hill, K. K., Keim, P. & Jackson, P. J. (1998). Development and testing of a bacterial biosensor for toluene-based environmental contaminants. *Applied & environmental microbiology* **64**, 1006–1012.

Environmental biotransformations

1. Frost, J.W. & Lievense, J. (1994). Prospects for biocatalytic synthesis of aromatics in the 21st century. *New journal of chemistry* **18**, 341–348.

15.7 Responses to assignments

Assignment 15.1

There is a proportion of the biodegradable material that is not degraded within five days. Therefore, the BOD_5 underestimates the true amount of biodegradable material present. This additional material will be oxidised in the COD test. Also, any material that cannot be biodegraded under the conditions of the BOD test will be oxidised in the COD test.

Assignment 15.2

Biomass recycled back to the bioreactor would help maintain a level of biomass within the reactor and prevent washout. The outcome depends on the relationship between biomass concentration in the recycle and dilution rate.

Assignment 15.3

All inputs to the equation are in the form we need except the cell/floc size - they should go into the equation in metres.

Single cell velocity: $$V_t = \frac{9.81}{18} \times \frac{52}{1.011 \times 10^{-3}} \times (10^{-6})^2$$

$$= 2.80 \times 10^{-8}\ \mathrm{ms^{-1}}$$

$$= \mathbf{2.80 \times 10^{-5}\ mms^{-1}}$$

200 μm floc velocity: $$V_t = \frac{9.81}{18} \times \frac{52}{1.01 \times 10^{-3}} \times (2 \times 10^{-4})^2$$

$$= 1.12 \times 10^{-3}\ \mathrm{ms^{-1}}$$

$$= \mathbf{1.12\ mms^{-1}}$$

Assignment 15.4

The operational effects are:

- Protracted periods for secondary sedimentation to the point that the retention time in the sedimentation tank becomes so long as to be operationally impossible;

- Poor settlement also means that sludge (biomass) recycle to the biological aeration tank becomes difficult, and in severe cases, treatment may be lost;
- Increased burden for disposal/sludge treatment.

Assignment 15.5

The problems are similar to those of bulking, with some others:

- The foam may rise and cover walkways, which become slippery and dangerous to personnel;
- Wind-blown foam may contain pathogens;
- Deep foam can prevent gaseous exchange at the air/aqueous interface;
- Deep foam may allow excessive heat development within the bulk fluid.

Assignment 15.6

Urea: $CH_4N_2O + 9/2\ O_2 \rightarrow CO_2 + 2H_2O + 2NO_3$

$\Rightarrow$ 60 g urea require 4.5 × 32 g of O_2

$\Rightarrow$ 1 g urea requires 144/60 = **2.4 g O_2**

Glucose: $C_6H_{12}O_6 + 6O_2 \rightarrow 6CO_2 + 6H_2O$

$\Rightarrow$ 180 g glucose require 6 ×32 g of O_2

$\Rightarrow$ 1 g glucose requires 192/180 = **1.07 g O_2**

It can be seen that, although urea is a much smaller molecule than glucose, it exerts a much higher oxygen demand.

Assignment 15.7

HRT = 1/*D*

Assignment 15.8

There are several things that can be done to try to treat high strength waste by aerobic processes:

- Live with the problems that will occur (particularly high aeration cost and large quantities of excess sludge for disposal);
- Upgrade the process from air to pure oxygen (this increases the dissolved oxygen concentration but also the cost);
- Dilute the wastes with water from another source e.g., process water (this may require a larger reactor);
- Increase the amount of biomass in the bioreactor by increasing the recycle (decreases the *f/m* ratio).

Assignment 15.9

A chemolithotroph is a microorganism that uses inorganic compounds as electron donors. Energy comes from chemical oxidation, and cell biosynthesis from the reduction of carbon dioxide. In this case, hydrogen is oxidised and CO_2 is reduced.

Assignment 15.10

From left to right: benzene, toluene, ethylbenzene, *para*-xylene.

Assignment 15.11

$$C_{16}H_{34} + 49/2O_2 \rightarrow 16CO_2 + 17H_2O$$

$\Rightarrow$ 226 g of $C_{16}H_{34}$ requires 784 g of O_2

$\Rightarrow$ 1 g of $C_{16}H_{34}$ requires **3.47 g of O_2**

From the equation, $\text{ThOD} = 16\dfrac{[2c+0.5(h-\text{cl}-3n)+3s+2.5p+0.5na-o]\text{ mg mg}^{-1}}{\text{MW}}$

$$= 16\frac{[2\times16+0.5(34.0)+0+0+0]}{226}$$

$$= 16\times\frac{[32+17]}{226}$$

$$= 3.47 \text{ mg mg}^{-1}$$

$$= \mathbf{3.47 \text{ g g}^{-1}}$$

Yes, the methods agree.

Assignment 15.12

Oil is immiscible in water. In highly porous soils, the application of a water-soluble fertiliser will lead to it running off the oil and penetrating through the soil. Thus the surface oil would not be fertilised. In water, the fertiliser will dissolve in the water, and the target oil will remain without fertiliser.

15.8 Responses to self-assessment exercises

15.1 Sludge loading $= \dfrac{350}{2800} \times \dfrac{24}{5} = 0.6 \text{ kg}^{-1}\text{kg}^{-1}\text{d}^{-1}$

15.2 Question 1. The EPS in AS plants helps to form flocs by providing slime for the attachment of other organisms and filamentous bacteria. In the fixed film reactors, the slime helps in the initial formation of a biofilm on a solid substratum. As the biofilm develops, more EPS is produced and the microorganisms become irreversibly attached to the biofilm.

Question 2. Protozoa graze the microbial films. This has the effect of controlling biomass levels. It also makes space for new microbial colonisation, which helps to equilibrate the biomass physiology over long periods.

Question 3. The most useful is the *f/m* ratio (sludge loading) as it is the only one of the three which takes the biomass into consideration. Control of the *f/m* ratio is necessary for good settlement.

15.3 The provision of compartments in an RBC produces conditions of complete mixing within each compartment, but over all the four compartments, the effect is plug flow. Not one of the flow regimes in Fig.15.5 exactly describes flow in such an RBC, but the closest is D (step aeration in a series of tanks).

15.4 Glucose is $C_6H_{12}O_6 \Rightarrow n = 6$, $a = 12$ and $b = 6$

$\Rightarrow$ 1 mole glucose $\rightarrow$ 3 + 12/8 – 6/4 moles of CH_4

= 3 moles of CH_4

$\Rightarrow$ 180 g glucose $\rightarrow$ 3 moles of CH_4

$\Rightarrow$ 1000/180 moles glucose $\rightarrow$ 16.7 moles of CH_4

1 mole of a gas occupies 22.4 l at STP

16.7moles of methane = **373.4 litres**

15.5

Reactor type	**Stationary biomass**	**Fixed film biomass**	**Plug flow**
Activated sludge	–	–	+/–
Percolating filter	+	+	+
Rotating biological contactor	–	+	+
Fluidised bed	–	+	–
Two-stage mesophilic digester	–	–	+/–
Upflow anaerobic biofilter	+/–	+/–	+
Downflow expanded bed	+	+	+
UASB	+	–	+/–

The activated sludge process can be operated in various configurations, including completely mixed and plug flow. Percolating filters have to be operated downflow: if operated upflow, they would be anaerobic. Therefore, there is no contribution to treatment from suspended biomass, so the biomass is entirely stationary. In the RBC, the biomass rotates in and out of the wastewater. Fluidisation in the fluidised bed precludes the possibility of plug flow. A two-stage mesophilic digester could be used to operate in a regime approximating plug flow, if the second reactor was set up to operate as a bioreactor also (i.e., by heating). The upflow anaerobic biofilter benefits from having a proportion of the biomass suspended, although the majority is attached. The downflow expanded bed operates entirely as a fixed film as the suspended biomass is washed out of the bottom of the reactor. The best operation of the UASB is when there is minimal disturbance to the anaerobic flocs, so in theory the sludge blanket is stationary, although this may be difficult to achieve. For the same reasons, it should be possible to operate a plug flow regime.

15.6 They are in decreasing order of biodegradability left to right in both series. Increasing the number of rings in the polynuclear aromatic compounds makes them less biodegradable, probably as a result of decreased access of enzymes to sites of attack, and decreasing water solubility as the molecules increase in size. Polycyclic aromatic compounds of five or more rings are very poor substrates for bacterial growth. Increasing halogenation of the aromatic ring leads to a sharp increase in the stability of the molecule, making such compounds less biodegradable.

15.9 Acknowledgements

Several companies and individuals have contributed material to this chapter. Their contributions are greatly appreciated. AEA Technology gave permission for the use of some of their bioremediation diagrams. ADI Systems and Lars Enviro Pvt. have installed wastewater treatment systems in various locations in India, four of which can be seen in photographs in this chapter. Some material relating to the bioremediation of metals was contributed by Dr Harry Eccles of British Nuclear Fuels plc. Their web sites are appended.

ADI Systems Inc - http://www.adi.ca/sys
AEA technology - http://www.aeat-env.com/html/remediation_decontamination.html
BNFL - http://www.bnfl.com/BNFLWebSite.nsf/HTML/ResearchandTechnologyCentre?OpenDocument
Lars Enviro Pvt. Ltd. - http://www.larsenviro.com

16 Bio-informatics and pattern recognition in DNA and protein sequences

Chapter outline

16.1 Introduction

In this chapter we will consider how information may be derived from a given DNA or protein sequence. Such an approach relies very heavily on computer software technology and, though indirect, it is often preferred to the technically demanding and time-consuming experimental approach. This approach has benefitted greatly from recent advances in recombinant DNA technology which facilitate both the formation of cDNA, through the activities of reverse transcriptase and DNA polymerase, and its subsequent cloning into appropriate vectors for sequencing.

Fortunately the process of evolution has been somewhat conservative in that once a process has been found to work in primordial organisms, it continues to function in all species derived thereafter. A striking example is provided by the essential universality of the genetic code and the similarities in transcription among different organisms. Furthermore, evolution seems to have often involved the duplication and subsquent divergence of genes and this, in turn gave rise to multigene families where a certain sequence may indicate a certain function.

For example, the zinc-finger family of proteins (Fig. 16.1) is expected to bind to DNA and to probably regulate transcription, while a member of the protein kinase family is expected to phosphorylate certain amino acids in certain proteins, including itself in some cases. For these reasons, a newly found DNA sequence may well be similar to one or more already known sequences and such similarities are likely to provide strong clues as to the likely identity and function of the sequence in question.

Generally speaking, the distribution of amino acids within a given protein has been found to be non-random and independent of length and/or species. Although the amino acid composition of individual proteins does not appear to be very distinctive, a number of useful features can be gained from such a composition, e.g., the low level of charged amino acid residues in the *lac* repressor molecule indicates poor solubility and high tendency towards aggregation of purified molecules. The amino acid composition, however, does not reveal any information regarding the functional organisation of individual amino acids or specific region of the encoding sequence without utilising state-of-the-art technologies in computing and associated software technologies.

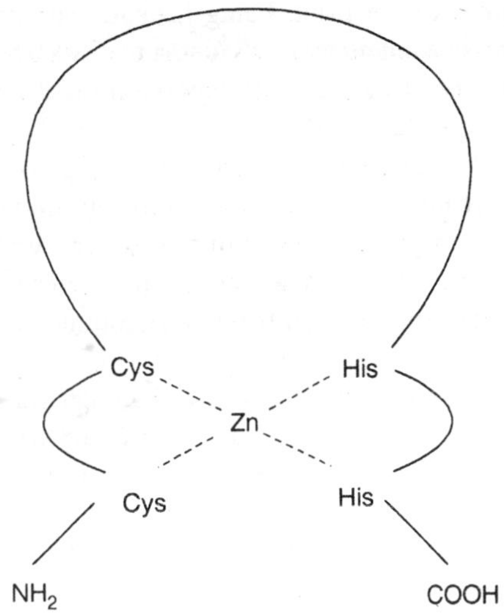

Fig 16.1 *Zinc-finger family of proteins.*
A diagrammatic representation of the C_2H_2 zinc-finger highlighting the region which specifically binds DNA due to the presence of two Cysteine (Cys) and two Histidine (His) residues around the zinc ion.

> ***Assignment 16.1***
>
> Explain how peptide mapping and sequence comparison help in understanding the evolutionary origin of a given enzyme.

Searching for sequence similarities requires the use of computers, sequence databases and appropriate algorithms. In the next sections we will describe the sequence databases and some of the most effective tools available.

16.2 Sequence databases

An international collaboration between GenBank, the European Molecular Biology Laboratory (EMBL) Data Library and the DNA Data Bank of Japan (DDBJ) produces a comprehensive computer database of DNA and RNA sequences, containing at present nearly 1000 million sequenced DNA and RNA base pairs. Accompanying each sequence is information as to where it was published (or if unpublished, by whom submitted), the organism from which it was derived, the function, if known, together with annotation of features of biological relevance such as the sites of promoters, RNA splicing sites, catalytic sites, *etc.*

In the beginning, this database was compiled by scanning journals and then manually entering of the sequences detected therein together with bibliographic and biological annotation. However, in the last five years or so the increasing flood of sequence information has rendered this approach impractical and a new system is currently in operation in which the sequence data and annotation are directly submitted to the database by the authors. In fact, journals that publish the bulk of the sequence data now require that authors must submit their data to the database before it will be considered for publication. On submission, the data are routinely checked for self consistency. With this process, which has been called "electronic data publishing", sequence data appear in the database more or less simultaneously with publication or even earlier if released by the author. An increasing fraction of the data is not appearing in print, being instead submitted to the database in lieu of publication. Indeed, for the massive amounts of DNA sequence data to be generated by the various genome sequencing efforts now underway or planned, deposition of the data in a readily accessible computer database is the only reasonable way to make the data widely available.

All those who determine DNA or RNA sequences are urged to submit their sequences directly to the database, in electronic or computer readable form. Direct submission will ensure the most accurate and timely transmission of their data as well as any benefits from checking of the data for self-consistency.

The data are available on-line and can be obtained on magnetic tape or CD-ROM. Details on access or copies of the database can be obtained from any of the following centres:

GenBank: National Center for Biotechnology Information
National Library of Medicine
Bldg. 38A, Rm. 8N–803
Bethesda, MD 20894, USA
Phone: 1-(301)-496-2475
Fax: 1-(301)-480-9241
E-mail: info@ncbi.nlm.nih.gov

EMBL Data Library: European Bioinformatics Institute
Hinxton Hall, Hinxton, Cambridge CB10 1RQ, UK
Phone: 44-(0)-1223-494400
Fax: 44-(0)-1223-494472
E-mail: datalib@ebi.ac.uk

DDBJ: National Institute of Genetics
Mishima, Shizuoka 411, Japan
Phone: 81 559 75 0771
Fax: 81 559 75 6040
E-mail: ddbj@ddbj.nig.ac.jp

Alternatively many research centres throughout the world subscribe to and maintain copies of the database that can be used by nearby researchers. It is expected that many satellite copies of the database will be established and updated daily.

Also invaluable are protein sequence databases. In many instances these have been derived by translation of the corresponding nucleic acid sequences. As will be seen, it is generally preferable to compare the amino acid sequences of proteins rather than the corresponding DNA sequences, even if the function of a new DNA sequence is completely unknown, as a protein must be translated in each of its six possible codon sequences for the comparison. In any case protein sequences are available as the International Protein Sequence Database or the Protein Information Resource (PIR) from the National Biomedical Research Foundation (NBRF), the Martinsried Institute for Protein Sequences (MIPS) or the Japanese International Protein Information Database (JIPID). Sequences totalling around ten million amino acids are included.

NBRF: Kathryn Seidman
3900 Reservoir Road NW, Washington, DC 20007, USA
Phone: 202 687 2121
Fax: 202 687 1662
E-mail: pirmail@gunbrf.bitnet

MIPS: Hans Werner Mewes
MPI/GEN, Max Planck Institut fur Biochemie, 8033 Martensried, Germany
Phone 49 89 8578 1
Fax: 49 89 8578 2655
E-mail: mewes@dm0mpb51.bitnet

JIPID: Akira Tsugita
Research Institute for Biosciences, Science Univ. of Tokyo, Yamasaki, Noda 278, Japan
Phone: 81 471 23 9777
Fax: 81 471 22 1544
E-mail: tsugita@jpnsut31.bitnet

As for the DNA sequences, it is expected that research centres throughout the world will acquire their own copies of the protein sequence databases for the use of local researchers. Another valuable protein database is PROSITE, in which biologically significant sites and motifs are emphasised and proteins are organised by families; it is available from GenBank or EMBL.

There are many other computer databases that may be valuable to biotechnology researchers. A more complete listing is beyond the scope of this chapter but information can be obtained from two databases, namely the Listing of Molecular Biology Databases (LIMB) and the Directory of Biotechnology Information Resources (DBIR):

LIMB: Christian Burks
Los Alamos National Lab., T-10, MS K710, Los Alamos, NM 87545, USA
Phone: 505 667 6683
Fax: 505 665 3493
E-mail: limb@genome@lanl.gov

DBIR: NLM Specialized Info. Services Div., 8600 Rockville Pike, Bethesda, MD 20894, USA
Phone: 301 496 6531 or 496 1131

16.3 Sequence comparisons

Let us return to our problem of extracting information from a new DNA sequence. At the outset one can scan the sequence for long open reading frames (ORFs), where an open reading frame is defined as a series of successive DNA triplets (codons) uninterrupted by termination codons. If both ends, the initiation and termination codons, of the ORF are present in the sequence, then the first triplet, reading from the 5' end, is the translation initiation signal, ATG (which is, in the RNA transcript, AUG, the codon for methionine) and the last triplet is the first transcription termination codon, namely TAA, TAG or TGA. Since there are three termination codons out of 64 possible codons, we would expect that in a random DNA sequence having equal numbers of each base, the probability for any codon to be a termination codon would be 3/64 = 0.047 and on average an ORF would be about 21 (= 1/0.047) codons long. In prokaryotic DNA, a much longer ORF is commonly found, thus providing strong evidence for a protein encoding sequence.

The identification of the correct reading frame enables one to predict the sequence of the translated protein with some confidence. The same is often true for eukaryote cDNA, but less often for eukaryote genomic DNA, since coding sequences, exons, are almost invariably interrupted by non-coding sequences (introns).

Protein coding regions in eukaryote genomic DNA sequences can often be recognised by using computer programs that search for the non-random patterns of codon utilisation characteristic of coding regions, together with other statistical tests. The power of such searches is enhanced by using the codon biases specific for the species being investigated.

Unlike eukaryotic organisms where AUG is the only codon for initiation, prokaryotic organisms may utilise any one of the following codons: GUG, UUG and AUU. This aspect is the one and only example in which protein synthesis in eukaryotes appears to be simpler than that of prokaryotes. The initiation codon AUG, usually found near (within 10 to 100 base pairs) the 5'-end of the mRNA transcript, is first identified by the 40S ribosomes which are invariably prebound with met tRNA. Following this interaction, translation of the initiation codon to methionine takes place and the rest of the process follows until a termination codon is recognised.

Assignment 16.2
Define what is meant by an open reading frame (ORF) highlighting diagramatically the key features which are required for successful transcription in prokaryotes as well as eukaryotes.

16.3.1 Special features in DNA sequences

The majority of DNA sequences do not appear to have any special features which may make them unusual. However, several sequences with two-fold axis of symmetry have now been shown to be recognition sequences for enzymic activities, other repeated sequences appear to be a common feature in certain regulatory regions and, in some cases, certain sequences may confer special properties such as the formation of a left-handed helical region.

The following features may be seen in a given DNA sequence:

(a) The palindrome

The palindrome is a sequence of inverted repeats round an axis of symmetry (shown below). The sequence to the right of the axis of symmetry on the bottom (3'—to—5') strand matches exactly that to the left of the axis of symmetry on the top strand (5'—to—3') reading the top sequence from right to left.

5'———————ACGGT ACCGT———————3'
3'———————TGCCA TGGCA———————5'

Palindromes vary in length, usually up to 50 base pairs, and when such a palindrome is separated by a spacer as shown below, it is referred to as "inverted repeat". Such regions are generally referred to by crystallographers as region of dyad symmetry.

5'———————ACGGT GGCTTAG ACCGT———————3'
3'———————TGCCA CCGAATC TGGCA———————5'

These palindromes or inverted repeats can, at least in theory, assume different structures which are otherwise known as "cruciforms". Both palindromes and inverted repeats have significant effect on the structure of messenger RNA as well as single-stranded DNA, found in certain phages, due to the formation of intra-strand hydrogen bonding between the complementary bases. Consequently, a palindrome yields an intra-strand double-stranded segment which is usually referred to to as "hair-pin", while inverted repeats produce a double-stranded segment with a terminal single-stranded loop (Fig. 16.2) known as a "stem-and-loop" structure.

(b) Repetitive DNA
Another class of special DNA sequences consists of repetitions of particular sequences. These may be either tandem, i.e., head to tail, or interspersed on either strand throughout a genome, or combinations, i.e., interspersed tandem repeats. Roughly one-half of the DNA in higher eukaryote genomes is repetitive. While mostly of unknown function, the repetitive DNA tells us things about genome evolution and is useful for other reasons noted below.

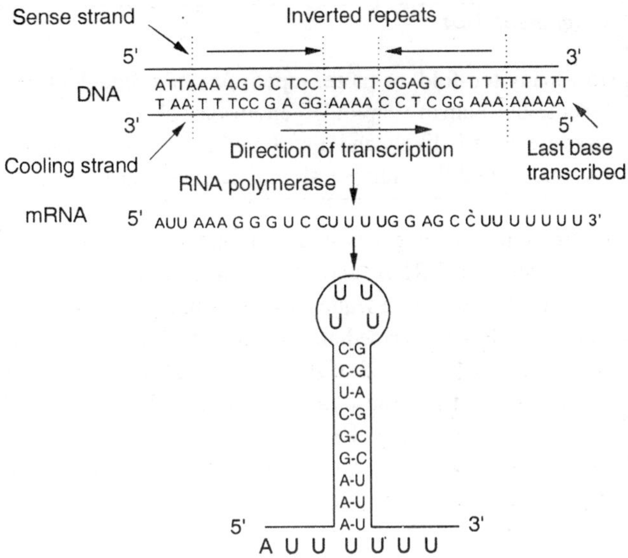

Fig. 16.2 *"Stem-and-loop" structure.*
Formation of "stem-and-loop" structure following transcription due to the presence of inverted repeats and the subsequent formation of intra-strand hydrogen bonding in mRNA.

Among the most important tandemly repeated sequences are:

(i) Satellite DNA, so called because it may appear as a distinct satellite peak when DNA fragments are separated on a density gradient, because of its uniform base composition. Satellite sequences are clustered near the centromeres of each chromosome. In some insects, repetitions of an eight base pair sequence constitute one-quarter of the total genome DNA; in mammals, the repeated sequences are longer, though perhaps descended from short repeats, and comprise a smaller fraction of the genome.

(ii) Minisatellite sequences are typically hundred-fold repetitions of a roughly 30 base sequence. In humans, the number of repetitions at a given location on a chromosome is highly polymorphic, i.e., it is likely to vary from one person to another or between the two homologous chromosomes of a single individual. Therefore minisatellite sequences have become widely used in forensic studies of DNA samples.

> ***Assignment 16.3***
>
> Discuss how polymorphic DNA sequences might be used in a criminal case to establish the guilt or innocence of a suspect. Discuss possible experimental procedures and sources of error.

(iii) Microsatellite sequences are tandem repeats of short sequences such as dinucleotides or trinucleotides. Because of their remarkable abundance in mammalian DNA and polymorphic lengths, they are now being used as landmarks in genome mapping projects. Expansions of trinucleotide (CCG or CAG) repeat numbers have been associated with several human genetic diseases, including Huntington's disease and Fragile X syndrome, the most common genetic cause of mental retardation in males.

> ***Assignment 16.4***
>
> How many ten-fold, or more, repetitions of a dinucleotide would you expect in the human genome? Assume the human (haploid) genome has 3×10^9 base pairs and, for this exercise, that the four bases are equally likely and distributed at random. Experimentally there are about 10^5 copies of such CA repeats and this number should be much larger than your estimate.

(iv) Some functional genes are also found to be tandemly repeated, including those coding for histones, for ribosomal RNA, and for immunoglobulins.

Interspersed repetitive sequences are commonly a result of transposition events in a genome, sometimes involving an RNA intermediate as in a viral infection. In human DNA, the most abundant interspersed repeat family is composed of Alu sequences, so called because most contain the sequence AGCT and are thus cleaved by a restriction enzyme called *Alu* 1. An Alu sequence contains about 300 base pairs and there are around 5×10^5 copies in human DNA; thus Alu sequences comprise about 5% of total human DNA. Most Alu sequences are inactive but a few are still managing to get themselves transposed, *via* an RNA intermediate, into new locations in the genome, sometimes causing genetic disease. Alu sequences are common in introns and it is not unusual for them to comprise 25% of the DNA in a gene. Although they typically differ in sequence by around 15% from any consensus, they are a serious problem in the assembly of long sequences from short sequence fragments.

In general, a DNA sequence that contains a repetitive element, such as an Alu or a microsatellite sequence will be found similar to all the other sequences in a database that contain the same repetitive element. Special considerations may have to be taken to ensure that these similarities do not overwhelm the finding of more subtle similarities between, for example, protein coding regions in the test sequence and the database. One approach is to focus only on protein coding sequences, as explained in Section 14.3.4, but even then it is found that many proteins contain repetitive elements that must be allowed for.

In any event, if we wish to compare a given DNA sequence with those cited in a data base, then our first task is to test the sequence for alignment with other sequences of the database. The alignment of two sequences is achieved by writing one sequence above the other and the difference is highlighted by introducing gaps in either sequence.

For example, an alignment of two sequences may be written as follows

```
AA  GTCCAGGAT
AATCGT  AGGAT
```

In this alignment, gaps of two bases have been introduced into each of the two sequences in such a way that the remaining bases are the same at each position. Alternatively, one could have aligned the two sequences as

```
AAGTCCAGGAT
AATCGTAGGAT
```

such that in this alignment the two sequences differ from each other by four base substitutions, namely T for G in the third position, C for T in the fourth, etc. Which of these two alignments is to be preferred and how does one identify the best way of alignments?

16.3.2 *The iterative, dynamic programming approach*

In general, for two sequences of typical length, say longer than 100 bases, it is not practical to enumerate all possible alignments. Instead, it is possible to find the "best" alignments using an iterative process, called dynamic programming, so long as some very reasonable restrictions are placed upon the scoring scheme which is used to evaluate the alignments. There are two equivalent ways to formulate the problem; in one we seek the similarities between two aligned sequences, in the other we treat the differences or distances between them. We shall consider the similarity formulation.

Let $a=a_1, a_2 \ldots a_n$ and $b=b_1, b_2 \ldots b_m$ denote two sequences where each entry is one of the four bases, A, C, G or T. (The same approach can be used with other choices of the sequence elements, such as the twenty amino acids of protein sequences.) An allowable scoring scheme is one in which for any position of the two aligned sequences a similarity score can be assigned such that the total score for the alignment is the sum of the scores for each position. Let s(a,b) be such a score for a pair of elements (a,b).

Then we must have s(a,a) >0, while for most (a,b) pairs it is necessary that s(a,b)<0. This condition is required in order that two sequences are not said to be more similar simply by being longer. We must also introduce a penalty, – g(a), for introducing a gap opposite element a. The similarity of sequences **a** and **b** is defined to be

$$S(\mathbf{a},\mathbf{b}) = \max \sum_{i=1}^{L} s(a_i,b_j)$$

where the maximum is taken over all alignments. The upper limit to the sum is L = m + n.

The dynamic programming algorithm can be derived as follows. Consider an alignment of $a_1 \ldots a_i$ with $b_1 \ldots b_j$. It can end in one of three possible ways

$$\begin{matrix} \ldots a_1 & \quad \ldots a_i & \quad \ldots a_o \\ \ldots o & \quad \ldots b_1 & \quad \ldots b_j \end{matrix}$$

where the symbol o has been used to indicate that a gap has been introduced in one of the sequences, let $S_{i,j}$ be the similarity of the sequences. If the optimal alignments ends in a_i /o, then the alignment of the sequences $a_1 \ldots a_{i-1}$ and $b_1 \ldots b_j$ was optimal and the best that could be done was to add a_i opposite a gap. In this case $S_{i,j} = S_{i-1,j} - g(a_i)$. Similarly if the optimal alignment ends at o/b_j, $S_{i,j} = S_{i,j-1} - g(b_j)$. In the remaining case, where the optimal alignment ends in a_i /b_j, the sequences $a_1 \ldots a_{i-1}$ and $b_1 \ldots b_{j-1}$ must have been optimally aligned and $S_{i,j} = S_{i-1,j-1} + s(a_i,b_j)$. Thus the similarity $S_{i,j}$ can be calculated as the maximum over the three possibilities enumerated, or

$$S_{i,j} = \max\,(S_{i-1,j} - g(a_i),\ S_{i-1,j-1} + s(a_i,b_j),\ S_{i,j-1} - g(b_j)).$$

This is the basic dynamic programming algorithm; it provides an iterative way of calculating similarities without having to enumerate all possible alignments.

In order to start the process, we set $S_{o,o} = 0$ and

$$S_{o,j} = \sum_{k=1}^{-j} g(b_k) \quad \text{and} \quad S_{i,o} = \sum_{k=1}^{-i} g(a_k).$$

The process of constructing an optimal sequence alignment can be visualised as finding a path in a two-dimensional matrix-like array in which one sequence is placed along the horizontal axis and the other is along the vertical axis. This is explained in Fig. 16.3.

Assignment 16.5

Align the following sequences using the two-dimensional matrix-like array method

(i) GGCTATCTAA
    GG   ATCTAA

(ii) TATC AGTAC
TAT AAAG C
(iii) ACGTAACGTATA
GT CGTATA

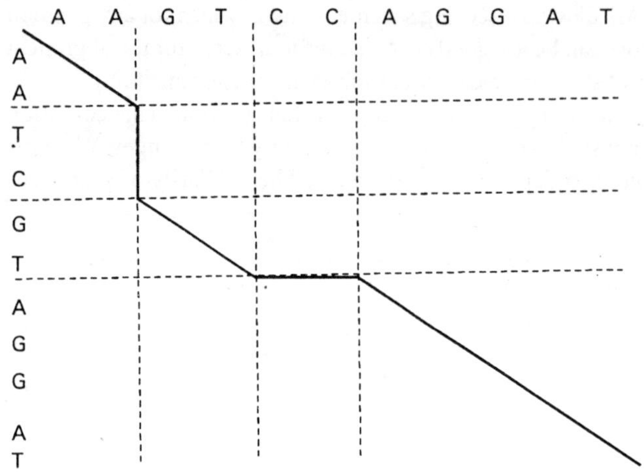

Fig. 16.3 *An alignment of two sequences represented as a path through a matrix.* One sequence is along the horizontal rows, the other along the vertical columns. A diagonal line signifies that the base occupying that row is aligned opposite to the base occupying the column (the bases need not be identical); a vertical line indicates a gap in the horizontal sequence, while a horizontal line indicates a gap in the vertical sequence. Any continuous set of segments running from left to right and top to bottom represents a possible alignment. The path shown represents the following alignment

AA GTCCAGGAT
AATCGT AGGAT.

When one is comparing a new sequence to a database of sequences, a similarity can be calculated of the new sequence to each sequence in the database. For those showing an interesting degree of similarity, the alignment can be reconstructed by the use of pointers at each (i,j) showing whether the optimal path came from (i–1,j), (i–1,j–1), or (i,j–1), or by a trace-back method.

Often two sequences which do not show a high degree of overall similarity will contain sequence segments having striking similarity, indicative perhaps of a common function or evolutionary origin. For such cases, a new algorithm is needed, one that can identify similar segments embedded in otherwise dissimilar sequences.

As before, consider two sequences a and b with a similarity measure s(a,b). The objective now is to find

$$\max S(a_i, a_{i+1} \ldots a_{j-1}\, a_j, b_k, b_{k+1} \ldots b_{l-1}, (b_l)) \qquad 1 < i < j < n \text{ and } 1 < k < l < m$$

and the maximum is to be taken over all values of i, j, k and l in the indicated ranges. Define $H_{i,j}$ to be the maximum similarity of two sequence segments ending at a_i and b_j. Thus

$$H_{i,j} = \max (0, S(a_x, a_{x+1} \ldots a_i, b_y, b_y+1 \ldots b_j)) \quad 1 \le x < i,\ 1 \le y < j.$$

The important feature here is that H is set to 0, if the sequences left of i,j are dissimilar. With this change, the dynamic programming algorithm becomes

$$H_{i,j} = \max (0, H_{i-1,j-1} + s(a_i, b_j), H_{i-1,j} - g(a_i), H_{i,j-1} - g(b_j)).$$

It is also possible to introduce somewhat more general gap penalties without a big computational change. If, in particular the penalty for a gap of k bases is taken to be a linear function of k, i.e., g(k) = c + dk, the algorithm can be readily generalised.

16.3.3 *Faster methods*

The computational time for comparing a sequence of n bases with a database of m bases, using the dynamic programming algorithm is proportional to n × m. As databases have grown, the times to make such comparisons have beome inconveniently long, of the order of a day on powerful scientific workstations or even supercomputers. Therefore scientists have sought more rapid methods of sequence comparison, which do not sacrifice too much accuracy. In one approach, special computer hardware including large scale integrated circuits has been developed explicitly for sequence comparisons; however to date such hardware is expensive and not widely available. Powerful parallel computers have also been used for database comparisons, giving substantial improvements in speed. However, for most laboratories, the use of faster algorithms is the answer.

It can be readily appreciated that sequence comparisons would be much faster if one did not need to include the possibility of gaps. There would then be only n + m possible alignments of two sequences of lengths n and m which could be readily enumerated and evaluated. Moreoever, computer scientists have developed a number of methods for rapid text searching that could be used. In particular, the use of "lookup tables" has been very effective for accelerating sequence comparisons. In this approach, attention is first focused on k letter words, i.e., oligonucleotides of k bases. There are 4^k such possible words. Any region of high local similarity between two sequences will have a significant number of words in common between the sequences in the region of similarity. Therefore one can search for regions of high local similarity by seeking regions having high numbers of words in common.

A start can be made by making an ordered list of the words. For example, let k = 4 and the first word be AAAA, the second be AAAC, etc. Then for each word, one can prepare a list of where the word occurs in the first of the sequences or in the database. Word positions in the second sequence are then compared to this list and regions having a high density of words in common are identified. In a widely used set of programs (FASTA and relatives thereof), a dynamic programming alignment is then done for each region identified and a similarity score is calculated. Inasmuch as the great majority of sequences in a database have no similarity to any comparison sequence, this approach provides a very rapid way of eliminating the unrelated sequences and focusing attention on candidate similarities.

With the FASTA program, it is possible to compare a protein sequence having a few hundred amino acids with a protein database in a matter of minutes on an IBM PC. Use of the program has been clearly illustrated by Pearson, from whom the program is available. In general, it seems that FASTA will detect most of the similarities that would be found by a full scale dynamic programming approach, but occasionally some are missed.

Recently a still faster comparison program, BLAST, has been described, and is becoming available through sequence databases. This program seeks the best local alignments without gaps.

16.3.4 *Choice of scoring weights*

In any alignment method, the scoring weights s(a,b) and gap penalties g(a) must be specified. In the comparison of DNA sequences, such choices are difficult to justify unless one has further information concerning the function of the sequences. In general, the proper choice of weights for DNA sequences is context dependent and non-local. Consider for example a DNA sequence that codes for a protein; many base changes would be irrelevant since they would not change the encoded amino acid, because of the degeneracy of the genetic code. Other base changes might preserve the general character of the amino acid and thus be unimportant. However, identification of which base changes should be treated as neutral depends on identification of the correct reading frame. In a protein coding region, a gap is likely to be very serious since it will shift subsequent reading frames — unless compensated by a nearby corresponding gap in the comparison sequence. In a non-coding region, a gap may well be of little importance.

It is more straightforward to make a reasonable choice of weights for comparing protein sequences and therefore it is usually appropriate to translate a new DNA sequence into each of its six possible reading frames (three in each direction) and to compare the resulting six hypothetical amino acid sequences with a protein database. Moreoever, rather distantly related proteins, having say 20% amino acid identity, can usually be picked out by this process, whereas their corresponding DNA sequences would not show a detectable similarity. Of course, if the test DNA sequence does not code for a protein, then comparison with the DNA database will be more informative. However, a suitable choice of weights may still be difficult. If, for example, the sequence codes for a functional RNA molecule, then compensatory base changes, which preserve the possibility of base pairing in some portion of the molecule, may be neutral. Such effects cannot be included in general sequence comparison algorithms, but require special approaches.

When comparing the amino acid sequences of two proteins, a useful approach is based on an analysis of observed substitutions in homologous proteins, i.e., proteins believed to be descended from a common ancestor and known to have a common function. Dayhoff and colleagues have pioneered the mutation data matrix (MDM) approach as early as 1968. Data for deriving the matrix are obtained from sets of aligned, closely related proteins, for example one set might be globin molecules and another set cytochrome C molecules. From these sequences, one can estimate the frequency with which one amino acid is replaced by another and compare this with the frequency of the two amino acids occurring by chance. An element in the mutation data matrix is the logarithm of this ratio. Note that since logarithms are used, one can add the scores s(a,b) for the individual terms in the sequence comparison as is required for most algorithms.

	Cys	Ser	Thr	Pro	Ala	Gly	Asn	Asp	Glu	Gln	His	Arg	Lys	Met	Ile	Leu	Val	Phe	Tyr	Trp
Cys	12																			
Ser	0	2																		
Thr	−2	1	3																	
Pro	−3	1	0	6																
Ala	−2	1	1	1	2															
Gly	−3	1	0	−1	1	5														
Asn	−4	1	0	−1	0	0	2													
Asp	−5	0	0	−1	0	1	2	4												
Glu	−5	0	0	−1	0	0	1	3	4											
Gln	−5	−1	−1	0	0	−1	1	2	2	4										
His	−3	−1	−1	0	−1	−2	2	1	1	3	6									
Arg	−4	0	−1	0	−2	−3	0	−1	−1	1	2	6								
Lys	−5	0	0	−1	−1	−2	1	0	0	1	0	3	5							
Met	−5	−2	−1	−2	−1	−3	−2	−3	−2	−1	−2	0	0	6						
Ile	−2	−1	0	−2	−1	−3	−2	−2	−2	−2	−2	−2	−2	2	5					
Leu	−6	−3	−2	−3	−2	−4	−3	−4	−3	−2	−2	−3	−3	4	2	5				
Val	−2	−1	0	−1	0	−1	−2	−2	−2	−2	−2	−2	−2	2	4	2	4			
Phe	−4	−3	−3	−5	−4	−5	−4	−6	−5	−5	−2	−4	−5	0	1	2	−1	9		
Tyr	0	−3	−3	−5	−3	−5	−2	−4	−4	−4	0	−4	−4	−2	−1	−1	−2	7	10	
Trp	−8	−2	−5	−6	−6	−7	−4	−7	−7	−5	−3	2	−3	−4	−5	−2	−6	0	0	17

Fig. 16.4 *Mutation data matrix for 250 PAMs (from George, et al., 1990).*
Entries are positive for pairs of amino acids found to occur in the corresponding positions of aligned sequences with higher frequency than at random. Conversely, negative entries are for amino acids less likely to be found at corresponding positions than random. For this presentation, the amino acids were arranged in groups such that positive numbers are almost always found for within-group pairs. Amino acids within a group tend to share physico-chemical properties. Thus six groups are found, (1) cysteine, a group unto itself, (2) serine, threonine, proline, alanine and glycine — small neutral or hydrophilic molecules, (3) asparagine, aspartic acid, glutamine, glutamic acid — the acid and acid amide molecules, (4) histidine, arginine and lysine — basic molecules, (5) methionine, isoleucine, leucine and valine — hydrophobic molecules, and (6) phenylalanine, tyrosine and tryptophan — large aromatic molecules. Interchanges are regarded as symmetric, for example changing Ser to Gly has the same score as changing Gly to Ser.

Various MDMs will differ from one another in the percentage of accepted mutations (PAM); for example, a one PAM matrix will represent one accepted mutation per hundred sites. Simulations showed that for distantly related sequences (between 73% and 86% difference) the MDM for 250 PAMs was optimal for distinguishing between related proteins and those whose observed similarity is due to chance. Hence the matrix for 250 PAMs has often been used as the standard for sequence comparison studies. For ease of computation, i.e., to avoid floating point operations, the logarithm is often multiplied by a factor of ten and rounded to the nearest integer. Such a PAM 250 matrix is shown in Fig. 16.4. Inspection shows many reasonable features. For example, the two highest figures in the matrix are for a tryptophan paired with itself (17) and cysteine paired with itself (12), indicating that the conservation of these unusual amino acids between two sequences is highly significant. On the other hand, the smallest number (– 8) corresponds to the interchange of tryptophan and cysteine.

> ***Assignment 16.6***
>
> Most amino acids are specified by more than one codon. However, only a subset of these codons are actually used. For example, while *E. coli* most frequently utilises CGU and CGC for the encoding of arginine, yeasts preferentially utilise AGA. Discuss the evolutionary significance of this phenomenon.

Replacements which preserve the general character of an amino acid tend to have positive values, thus replacements of one hydrophobic amino acid by another such as valine by leucine or isoleucine have positive contributions, namely 2 and 4 respectively.

More recently it has been argued from an information theory perspective that a PAM 120 matrix is more appropriate for database comparisons that use programs such as BLAST which find local alignments without gaps. Many investigators are now using this scoring matrix with good results. However, it appears that for detecting more distant relationships, i.e., weak homologies with gaps, a higher PAM value, such as 250, is still appropriate.

A large number of other scoring schemes have been proposed and used by various investigators. Most are attempts to capture some of the essential properties of the amino acids, such as size, charge and hydrophobicity, which are believed to be important determinants of protein structure and function.

16.3.5 *How to judge the significance of a similarity*

Suppose that comparison has been made of a new sequence with a sequence database and a number of somewhat similar sequences have been found. How can we judge the significance of the finding? Experience is probably the best guide. That is, comparison of our test sequence will give, for most of the sequences in the database, relatively low similarity scores and any examples that appear to be outside this distribution should be taken very seriously. If the test sequence contains a repetitive element, one will usually wish to discard any similarities with the same repetitive element in the database.

For any particular scoring scheme a rule of thumb can be developed by this approach to suggest a threshold score, below which it is not worth looking for significance.

Other approaches have been tried with less success. In particular, comparison has sometimes been made with a random sequence having the same base composition as the test sequence in order to derive a significance threshold. However, such an approach may be misleading since, for example, DNA sequences coding for proteins are non-random because of the bias in codon usage. Therefore, two DNA sequences coding for protein are likely to be more similar to each other than either of them to a random sequence.

16.4 Multiple sequence alignment — finding and using motifs

Somewhat different pattern recognition problems arise when one has identified a set of sequences which are believed to perform some function in common or to have a common evolutionary origin. Examples of such sets include DNA sequences which act as promoters, transfer RNA molecules, intron–exon splice sites in eukaryotic RNA, as well as families of proteins performing some function such as catalysis of a reaction or binding of a ligand. Important questions are how to identify the critical features (motifs) of a sequence set which enable the members to perform their function and how to make use of these motifs in order to recognise or design new members of the set.

While we can align any two sequences in a set using one of the algorithms and scoring schemes described in the previous section, an alignment of just two sequences is unlikely to reveal a motif. Instead, one would like to simultaneously align all of the set members and to search for any core of invariant features that we may hope are associated with function. However the rigorous alignment of more than three sequences of typical length, say a few hundred base pairs or amino acids, using a dynamic programming approach is computationally intractable. The reason is that just as in the case of comparing two sequences, we were seeking a best path in a two dimensional matrix, so also in aligning N sequences we would be seeking a best path in an N dimensional matrix, which leads to too many possibilities to systematically explore.

Sometimes there is additional information that can be used to guide an alignment. For example, information about the active site of an enzyme may suggest that only a few amino acids are involved in the catalysis, with the remainder serving to provide an appropriate geometry for the active few. This would suggest that the putative catalytic site should play a special role in the multiple alignment and perhaps be evident as a motif.

For these reasons, a variety of *ad hoc* approaches have been developed for multiple sequence alignment and motif detection, often somewhat tailored to the particular sequences being studied, often intuitive rather than systematic. Under these circumstances, we do not present or advocate any particular approach. Instead, a few references are given to recent work that the interested student may consult.

In several approaches, an alignment is first made of the two most similar sequences and features which these had in common are preserved as further alignments are made with additional sequences. In other approaches, the presence of consensus words or consensus combinations of a few amino acids, separated by gaps have been used as a basis for sequence alignment and/or identification of functional motifs. If information is available on the crystal structures of similar proteins, then this information can be taken into account in establishing alignments. Interactive computer programs have also been found useful in permitting the user to intervene in an otherwise automatic alignment process in order to introduce biochemical or structural constraints or other considerations.

Assignment 16.7

Given that a set of sequences has been aligned, how can one detect and report the significant features?

The presence of certain amino acids at certain positions in the aligned sequences indicates consensus sequences which have been conserved. Insofar as gaps may be present in some or all of the aligned sequences, it will be necessary to allow for variable spacing between the invariant amino acids in reporting the consensus or in seeking new examples of related sequences.

Frequently, however, a strict consensus cannot be found. Instead, certain bases (or amino acids) may be preferred at certain positions, but alternatives are also found. In these cases it may be preferable to report results in terms of a selectivity matrix, in which a column refers to the position in the aligned sequences, a row to the possible candidates at any position (bases for DNA, amino acids for proteins), and the matrix elements reflect the probabilities with which each candidate is found at each position. Suppose that we have aligned a number of DNA sequences. Let $f_{b,i}$ be the frequency with which base b is found at position i in the aligned sequences and let p_b be the frequency with which the base is found in the genome.

Then a good choice for the matrix elements is log $f_{b,i}/p_b$ as can be justified by thermodynamic or information theoretic arguments. Moreover, when logarithms are thus used in the matrix elements, a score for alignment of a new sequence can be obtained by adding a contribution from the base found at each position according to the value of the matrix element. An example for *E. coli* promoter sequences is shown in Fig. 16.5.

(a)

A	9	214	63	142	118	8
C	22	7	26	31	52	13
G	18	2	29	38	29	5
T	193	19	124	31	43	216

(b)

A	0.04	0.88	0.26	0.59	0.49	0.03
C	0.09	0.03	0.11	0.13	0.22	0.05
G	0.07	0.01	0.12	0.16	0.12	0.02
T	0.80	0.08	0.51	0.13	0.18	0.89

(c)

A	-2.76	1.82	0.06	1.23	0.96	-2.92
C	-1.46	-3.11	-1.22	-1.00	-0.22	-2.21
G	-1.76	-5.00	-1.06	-.067	-1.06	-3.58
T	1.67	-1.66	1.04	-1.00	-0.49	1.84

Fig.16.5 *Matrices for an E. coli promoter sequence (from Stormo, 1990).*
This is a sequence of six bases located about 10 bases upstream the transcription initiation site. Data from 242 examples were analysed and the outcome is presented above. In (a) the frequency (number of times) with which each base appears at a given position is presented. Thus, for example, 9 times the base A occurred in the first position, 214 times in the second position, etc. In (b) the rate of appearance of each base at each position is given; thus each entry in (B) is that in (a) divided by 242. The selectivity matrix in (c) has entries equal to the logarithm to the base 2 of the frequency rates given in (b) divided by 0.25, the genomic frequency of each base in *E. coli.* The most common bases would suggest a consensus sequence TATAAT, but almost no promoters have this exact sequence and the selectivity matrix contains much more information that can be used in scanning for potential promoters sequences.

It is also possible to include mutation data, as in a PAM 250 matrix, in forming a consensus matrix. Thus, one can take the frequencies with which various amino acids are found at a position, multiply each frequency by the corresponding row and column from a PAM matrix to form a new matrix, which includes not only observed mutations but the effect of all possible substitutions according to the PAM scoring scheme.

In general, when using multiple alignment programs or motifs derived therefrom, one should be aware of their limitations and regard them as heuristic tools rather than providing definitive methods of analysis.

16.5 Concluding remarks

Any new DNA sequence can be compared with a computer database of known sequences. There is a good chance that it will be related to one or more already known sequences and such relationships can provide important clues as to function or origin. There are international computer databases of DNA and protein sequences. A powerful iterative method, dynamic programming, is available for sequence comparison, provided that similarity scores can be computed additively, one residue at a time. For general database comparisons, faster methods search for sequence segments rich in common words. Multiple sequence alignments are more heuristic in nature but can be used to identify motifs, which in turn can be represented sometimes as consensus sequences, but often better as selectivity matrices.

16.6 Suggested further reading

1. Altschul, S.F., Boguski, M.S., Gish, W. and Wootton, J.C. (1994). Issues in searching molecular sequence databases. *Nat. Genet.*, **6**, 119–129.
2. George, D.G., Barker, W.C. and Hunt, L.T. (1990). Mutation data matrix and its uses. *Methods Enzymol.*, **183**, 333–351.
3. Pearson, W.R. (1990). Rapid and sensitive sequence comparisons with FASTP and FASTA. *Methods Enzymol.*, **183**, 63–95.
4. Smith, D.W. (1994). *Biocomputing : informatics and the genome project.* Academic Press, New York.
5. Staden, R (1990). Finding protein coding regions in genomic sequences. *Methods Enzymol.*,**183**, 163–180.
6. States, D.J. and Gish, W. (1994). Combined use of sequence similarity and codon bias for coding region identification. *J. of Computational Biology*, **1**, 39–50.
7. Stormo, G.D. (1990). Consensus patterns in DNA. *Methods Enzymol.*, **183**, 211–221.
8. Vingron, M. and Argos, P. (1990). Determination of reliable regions in protein sequence alignments. *Protein Eng.*, **3**, 565–569.
9. Waterman, M.S. and Jones, R. (1990). Consensus methods for DNA and protein sequence alignment. *Methods Enzymol.*, **183**, 221–237.
10. Wilbur, W.J. and Lipman, D.J. (1983). Rapid similarity searches of nucleic acid and protein data banks. *Proc. Nat. Acad. Sci. U.S.A.*, **80**, 726–730.

16.7 Useful Internet sites

http://www.seqnet.dl.ac.uk
http://www.sanger.ac.uk
http://www.nih.gov
http://www.ebi.ac.uk/hinxton/hinxton.html
http://www.embl-heidelberg.de
http://www.pir.georgetown.edu
http://www.expansy.ch/prosite/
http://www.ncbi.nlm.nih.gov/Genbank/GenbankOver
http://www.bioinf.man.ac.uk/dbbrowser/bioactivity/
http://www.mrc-lmb.cam.ac.uk/pubseq/
http://www.mips.biochem.mpg.de/

16.7 Responses to assignments

16.1 The amino acid composition of a given protein may shed some light on its physical properties, for example the presence of a low level of charged amino acid residues would indicate poor solubility and a high tendency towards aggregation. There is no way, however, of gaining any further information regarding the functional organisation of individual amino acids in this particular protein. On the other hand, peptide mapping and amino acid sequence comparisons of a given protein or a specific region of the coding sequence may shed light on the functional role and evolutionary origin of a given protein. For example, the possibility that lipoamide dehydrogenase (E3 component) of *Escherichia coli* is functionally interchangeable between pyruvate dehydrogenase and 2-oxoglutarate dehydrogenase was first proposed when a good degree of homology betwen the E3 component of both enzymes was observed following peptide mapping. Further search comparisons revealed that lipoamide dehydrogenase shares a common mechanism of catalysis with two other enzymes namely glutathione reductase and thioredoxin reductase. From an evolutionary viewpoint, these observations suggest that lipoamide dehydrogenase and glutathione reductase may have evolved by divergence from a common ancestral gene.

16.2 Open Reading Frame (ORF) is a nucleotide sequence which, on translation, yields a specific protein. Generally speaking, the ORF starts with ATG (start codon) and ends with any one of the three stop codons. The key features which are required for successful transcription are highlighted below.

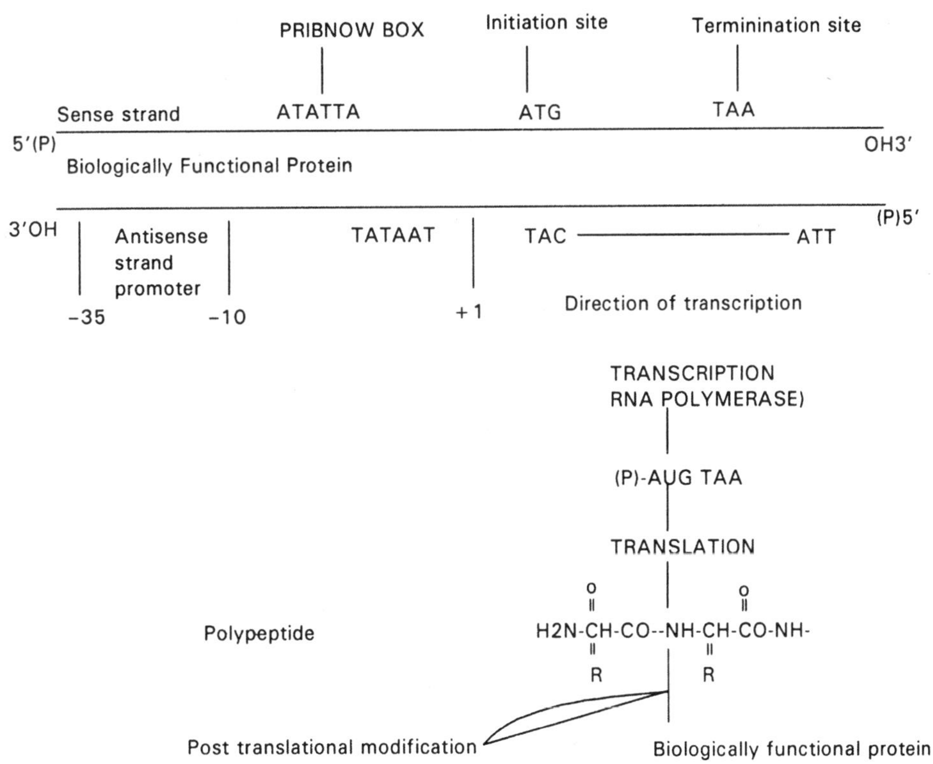

16.3 The number of minisatellite sequences at a given location within the genomic DNA is highly polymorphic, i.e., varies from one person to another, including identical twins. The use of certain restriction endonucleases helps identify those minisatellite sequence which can be identified by genetic fingerprinting technique. It follows, therefore, that the genetic fingerprint of those minisatellites will differ from one person to another and as such this is currently used by forensic scientists to resolve criminal cases. The experimental procedures include: isolation of DNA, restriction digests, agarose gel electrophoresis, Southern hybridisation and finally autoradiography. Sources of errors are essentially human errors and include mislabelling and contamination of DNA samples.

16.4 Although experimental data suggest that CA repeats are in the region of 10^5, the degree of repetition of a particular dinucleotide in this example is 4 + 2 +, i.e., once every sixteen bases. Taking into consideration the assumption that the human (haploid) genome has 3 × 10+9+ base pairs, the repetition can be worked out by dividing 3 × 10 + 9 + by (16 × 10) to give 18.75 + 6 + copies.

16.5 Place the top sequence horizontally and the bottom sequence vertically as shown below for the example (i), then apply the following:

1. A diagonal line indicates alignment.
2. A vertical line indicates a gap in the horizontal (top) sequence.
3. A horizontal line indicates a gap in the vertical (bottom) sequence.

The sequence alignment when you should obtain for examples (i), (ii) and (iii) are as follows:

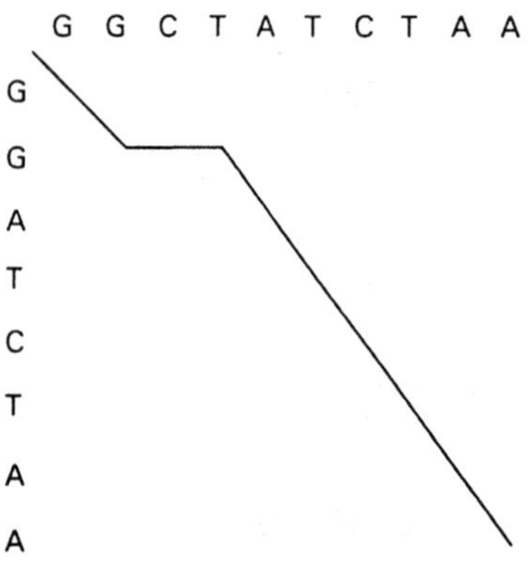

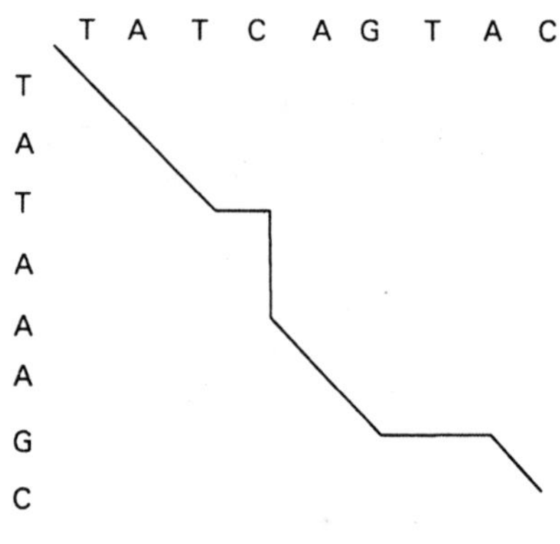

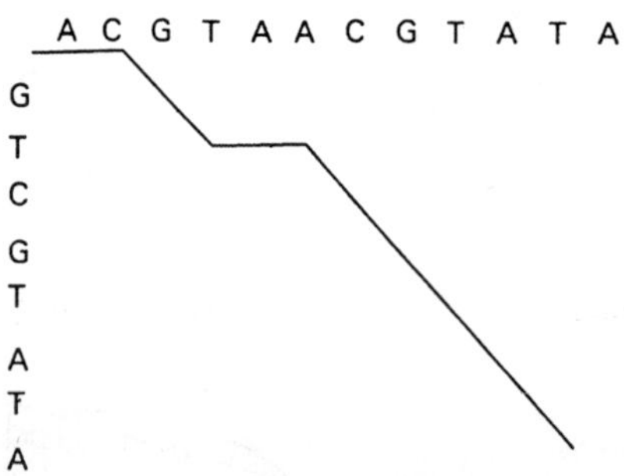

16.6 Although the majority of amino acids can be encoded for by more than one codon, organisms only utilise a subset of these potential codons primarily because the choice is limited by the availability of appropriate tRNAs. Unlike weakly expressed genes, the codons of highly expressed genes frequently contain a guanosine in position 1 and cytidine in position 3.

Furthermore, codon context, i.e., the adjacent codons/nucleotides, appears to be significantly biased, for example, lysine is preferentially encoded by AAA if guanosine is 3' adjacent (AAAG) while AAG is the preferred choice if cytidine is 3' adjacent (AAGC). Furthermore, codon choice and codon context are highly biased insofar as translation initiation was concerned. For example, while prokaryotic organisms utilise AUG, GUG, and UUG as translation initiation codon, eukaryotes only utilise AUG. So clearly the course of evolution has, for one reason or another, favoured the AUG triplet within certain codon context. Codon context could also directly affect the efficiency of transcription by affecting the promoter strength.

16.7 We would like to draw the reader's attention to the fact that codon usage or rather the redundancy of the genetic code means that changes in the third base of each codon may have no effect on the amino acid produced, thus two genes which mismatch every third base might encode identical proteins. Once the alignment is established, construct a homology matrix otherwise known as dot-plot and highlight the possible functional importance of the homologous region. Report the main features in the sequence including the hydropathicity plot which calculates the hydropathic index for each amino acid in the sequence of interest.

17 Marine biotechnology

Chapter outline

17.1 Introduction

The biotechnology revolution has already altered the practice of medicine and the methods of agriculture. A fast rising area of biotechnology comprises the environmental applications of biotechnology, including aquaculture, natural products harvesting and bioremediation.

> ***Assignment 17.1***
>
> Define the terms aquaculture, natural products harvesting and bioremediation.

To provide an insight into the ways in which biotechnology is influencing the marine sciences, a brief summary of the advances being made in a few selected areas of marine biotechnology are provided.

17.2 Aquaculture

Aquaculture, the farming of finfish, shellfish, crustaceans and seaweed, has grown significantly in economic value worldwide during the past decade, reaching *ca.* 14 million metric tons (in 1991). Aquaculture of seaweed and phytoplankton yields high value products as well, expecially algal-derived polysaccharides and chemicals.

The aquaculture industry, however, faces many problems that have to be solved before it can achieve significant growth as an industry. Husbandry of important fish and shellfish species can be advanced by research and many species can be improved by conventional breeding techniques. There are, however, several major contributions that biotechnology can make to enhance aquaculture, that could make the industry far more productive than it is now. Two key factors are genetic manipulation of marine organisms of commercial importance and improved knowledge of molecular and biochemical processes of animal and plant species. Both would allow for significantly improved utilisation of the biological resources of the sea. For example, some progress is being made in regulation of hormonal control of reproduction of finfish and shellfish in fish farming. Another significant contribution that biotechnology can make to aquaculture is in the diagnosis and treatment of diseases of aquacultured species.

Selective breeding of fish, a well established technique, has historically played an important role in the aquaculture industry. What has happened during the past decade is that direct genetic manipulation of fish, using recombinant DNA techniques, has revolutionised the pisciculture industry and made it possible to develop an entirely new approach to fish farming.

Fish are highly amenable to genetic manipulation because fish eggs characteristically are large and, therefore, can be microinjected with DNA constructs, after which external fertilisation and subsequent development can take place.

Assignment 17.2

What traits in fish would it be of interest to manipulate by genetic manipulation?

Enhancement of growth in animals can be induced by introduction of foreign growth hormone (GH) genes and this technique was first demonstrated about a decade ago, using mice. Ova microinjected with rat GH gene constructs yielded larger mice, a principle of growth enhancement that has since been employed in fish species of importance to aquaculture. The first successful growth hormone experiments using fish was the transfer into goldfish of the human GH gene fused to a mouse metallothionein promoter, yielding transgenic offspring significantly larger than the untreated, control fish. Since then, several vertebrate genes have been fused to a variety of promoters and subsequently introduced into fish species. Examples include the expression of a mouse metallothionein-human growth hormone fusion gene in Atlantic salmon and bovine growth hormone (expressed from the Rouse sarcoma virus long terminal repeat and carp β-acting gene promoters) in walleye.

Assignment 17.3

What problems could be caused by promoting growth of farmed fish by insertion of a human growth hormone gene?

Of more practical significance has been the achievement of growth enhancement using fish GH. It was demonstrated that growth hormone in rainbow trout is encoded by two separate genes. A large amount of biologically active rainbow trout GH was prepared by expressing one of these genes in the bacterium *Escherichia coli* and the hormone thus produced administered to rainbow trout. Since exogenous GH application is not very practical for large-scale aquaculture, the next logical step is generation of transgenic fish. Researchers in India have been microinjecting GH into fertilised eggs of catfish and of *Tilapia* spp. in an effort to enhance their sizes. The GH gene has been fused to a Rous sarcoma virus promoter and the cloned constructs were transferred to common carp and channel catfish.

Inheritance and expression of GH genes in transgenic fish are complex genetic and metabolic processes. Important considerations in successful manipulation of these processes include tissue specificity and developmental stage specificity of the transcriptional control elements as well as stable inheritance and expression of the GH genes in the offspring. Fast-growing transgenic fish, however, cannot be used in commercial aquaculture until further progress in research is accomplished to determine physiological, nutritional and environmental factors that maximise performance of individual transgenic fish. Also, resolution of the major issues of safety and environmental impact must be accomplished before large scale, commercial production of transgenic fish can be permitted.

Genetic manipulation of Atlantic salmon has been carried out with the objective of increasing cold resistance of this species. Many marine fish inhabiting cold waters produce proteins which act as "anti-freeze", i.e., protect fish by inhibiting the formation of ice crystals in their serum. These proteins are termed anti-freeze glycoproteins and anti-freeze polypeptides. Atlantic salmon lack genes coding for these proteins and therefore cannot survive in icy waters. The mechanism whereby anti-freeze proteins bind ice crystals and inhibit ice formation has been described. Genes coding for anti-freeze proteins have been transferred, expressed and inherited in Atlantic salmon. Expression of adequate concentrations of anti-freeze proteins in salmon blood can extend the range of culture of these fish.

Shellfish are also amenable to genetic manipulation, especially for enhancing growth, both rate of growth and size of the adult. Bovine GH has been shown to enhance growth rates of California red abalone and similar results have been reported for exogenous application of biosynthetic rainbow trout GH to juvenile oysters.

> ***Assignment 17.4***
>
> What is the disadvantage of enhancing the growth of shellfish by application of exogenous GH, compared to insertion of GH genes into them by genetic engineering?

In contrast to fish and shellfish, lobsters shed their exoskeletons during growth, a molting process that is under hormonal control. Endocrine regulation of molting can improve growth efficiency in the lobster. Endocrine regulation of molting in Crustacea is well understood, but the molecular genetics of marine crustaceans (shrimp, lobster, prawns, etc.) remain to be elucidated before commercial production of these species will be efficient and reliable.

17.3 Marine natural products

Aquaculture of marine macroalgae, i.e., seaweeds, has been practised for several centuries in Asian countries and products from these algae have been widely used as sources of medicine and food. Macro- and microalgae yield a wide range of products, including food additives and supplements, ingredients of culture media, pesticides, plant growth regulators and anti-bacterial, anti-cancer and anti-viral agents.

Microalgae have proven to be useful for large scale production of long chain fatty acids, eicosapentaenoic acid (EPA) and docosahexaenoic acid (DHA). Diets rich in omega-3 oils have been suggested to reduce the risk of coronary vascular disease. The green microalga, *Dunaliella salina,* is grown in large-scale intensive culture in California to produce β-carotene, a vitamin A precursor associated with prevention of cancer. *Dunaliella* cells can accumulate β-carotene to up to ten per cent of their dry weight.

It has even been suggested that "oceanic farming" of marine algae can reduce global carbon dioxide levels.

> ***Assignment 17.5***
>
> Suggest how oceanic farming of marine algae might reduce global CO_2 levels.

Application of biotechnology to cultivation of marine algae presents an opportunity for seaboard countries, especially developing countries with extensive coastal regions. However, to achieve success, understanding of the molecular genetics and application of the techniques of modern molecular biology will be required. Although molecular techniques have not yet been widely applied to achieve strain enhancement or production of transgenic plants and algae of commercial importance, this approach is being taken in several laboratories in the US, Asia and Europe.

A curious twist in the short history of marine biotechnology is that several products of marine macroalgae are essential in many of the techniques used in biotechnology. The contribution of polysaccharides derived from marine macroalgae has been recorded. Notably, agar and its purified derivative, agarose, are mixtures of polysaccharides extracted from red seaweeds, particularly *Gracilaria, Gelidium* and *Pterocladia* spp. Agar forms the basis of solid media used extensively in microbiology, whereas agarose, a thermoreversible, ion-independent gelling agent, is used to prepare gels forelectrophoretic separation of proteins and nucleic acids. Agarose is also widely used in immunological assays and cell culture. Algin (obtained from several species of brown algae), carrageenan (extracted from red algae) and agarose are used for encapsulation and immobilisation of cells of yeast or bacteria. An example of a commercial process based on encapsulation of cells in the case of κ-carrageenan include conversion of glucose to ethanol and production of L-aspartic and L-malic acids. Beads of immobilised cells can also be used for bioconversions, either directly or after further modification.

The marine environment is characterised by physical and chemical properties that are markedly different from those of the terrestrial environment. Furthermore, these properties comprise complex ecosystems which include many sessile (stationary, anchored) organisms. The production of bioactive chemicals is a common means of defence, especially in sessile organisms and in vulnerable soft-bodied organisms. Groups of organisms that have been found to produce bioactive natural products include marine bacteria, dinoflagellates, algae, coelenterates (namely the corals), echinoderms (such as sea cucumbers and starfish), bryozoans, sponges, soft-bodied molluscs (such as sea hares and nudibranchs) and tunicates. Bioactive substances from marine organisms have been studied for several decades and thousands of these chemicals have been described. Many have valuable pharmacological properties.

Sponges have proven to be an important source of bioactive compounds. Dysinin-type sesquiterpenes with anti-helmintic activity, several cytotoxic heterocyclics, and a number of cytotoxic compounds, including anti-tumour and anti-fungal agents, are among the compounds isolated.

Assignment 17.6

A sponge is found to produce an anti-tumour compound. Would this compound best be produced on a large scale by extracting it from sponges grown on a large scale in marine environments?

Self assessment exercise 17.1

Explain the advantages of the following:

(i) insertion of GH gene into carp;
(ii) insertion of anti-freeze glycoprotein gene into Atlantic salmon;
(iii) regulation of molting in lobsters.

Self assessment exercise 17.2

Describe uses of the following:

(i) β-carotene;
(ii) agar;
(iii) agarose;
(iv) carrageenan;
(v) dysinin-type sesquiterpenes.

Self assessment exercise 17.3

Explain the advantages and disadvantages of promoting the growth rate of shellfish by applying exogenous growth hormone compared to insertion of a growth hormone gene by genetic engineering.

17.4 Bioremediation

Bioremediation of pollutants in the environment is a field of biotechnology that is still in its infancy, but one that has great potential for dealing with pollutants that may be extremely difficult or impossible to remove using other approaches. The aspect of bioremediation in the environment that has received most attention is degradation of hydrocarbons, in particular oil products, which enter the environment from oil drilling, loading of tankers, catastrophic oil spills, and by natural seepage from oil-bearing sediments. Environmental parameters which affect microbial degradation of hydrocarbons in the environment, and metabolic and genetic factors that are important in this process, have been extensively reviewed. Biodegradation of petroleum and oils in the environment is a complex process. Physical and chemical factors that affect this process include chemical composition, physical state and concentration of the oil or hydrocarbon, temperature, salinity, oxygen and nutrient availability and water activity. Important biological factors influencing rate of biodegradation include the types of bacteria, fungi and other microorganisms that are present and adaptation of these microorganisms by prior exposure to hydrocarbons.

In view of the complexity of microbial degradation of hydrocarbons, it is perhaps not surprising that the effectiveness of attempts to enhance natural degradation processes is not at all clear. Attempts to treat oil spills by bioremediation have included nutrient addition to enhance growth of the indigenous bacterial community, addition of laboratory-grown inocula of oil-degrading bacteria and the use of biological surfactants derived from bacteria. These methods have worked effectively in estuarine and coastal environments. Open-ocean oil spills, however, have not yet been shown to be treatable by the techniques that have been employed in coastal areas, especially those used to treat beached oil.

Assignment 17.7

Explain why techniques used for bioremediation of beached oil may not be applicable in the open ocean.

17.5 Concluding remarks

In 1983, marine biotechnology, one component of environmental biotechnology, was first recognised as an area of great potential. In the decade that has since passed, rapid progress has been made. More than 1,000 publications of new compounds, natural products, discoveries of the molecular genetics of fish and shellfish growth, metabolism and reproduction, and expansion of biotechnology research on bioremediation, biofouling and related aspects, have appeared in the literature. Where only a few pioneers toiled in their laboratories, there are new major centres of marine biotechnology research and development in many countries.

Promising new antibiotics, anti-cancer therapeutic agents, improved aquaculture stocks, marine polysaccharides and food additives, and potential new energy sources from the sea, driven by molecular genetic control of the marine biological systems, are in the discovery, post-discovery, scale-up, and/or production stages. There has already been and will continue to be major successes, both intellectual and commercial, in marine biotechnology.

Areas of marine biotechnology in an expansion stage are bioremediation and microbial biodiversity inventories and assessments. Yet to be fully exploited are nucleic acid fingerprinting for global biological stock assessments and monitoring, although this application is progressing rapidly. Marine biotechnology requires long-term investment. There are some commercial successes, notably in the seafood industry and bioremediation. The future for environmental biotechnology is very promising and the potential of this aspect of biotechnology will surely be fully recognised.

17.6 Suggested further reading

1. Avron, M. and Ben-Amotz, A. (1992). *Dunaliella: physiology, biochemistry and biotechnology.* CRC Press Inc., Boca Raton, FL.
2. Chen, T.T. and Powers, D.A. (1990). Transgenic fish. *Trends Biotechnol.*, **8**, 209–218.
3. Colwell, R.R.R (1983). Biotechnology in the marine sciences. *Science*, **222**, 19–24.
4. Harvey, W. (1988). Cracking open marine algae's biological treasure chest. *Bio/Technol.*, **6**, 486–492.
5. Kranzfelder, J.A. (1991). (β-carotene production in large scale intensive cultures of *Dunaliella salina. 1991 International Marine Biotechnology Conference*, Baltimore, MD.
6. Kyle, D.J., Gladue, R., Reeb, S. and Boswell, K. (1991). Production and use of omega-3 designer oils from microalgae (1991). *International Marine Biotechnology Conference*, Baltimore, MD.
7. Leahy, J.G. and Colwell, R.R. (1990). Microbial degradation of hydrocarbons in the environment. *Microbial. Rev.*, **54**, 305–315.
8. Moav, B., Liu, Z., Moav, N.L., Gross, M.L., Kapuschinski, A.R., Faras, A.J., Guise, K. and Hackett, P.B. (1992). Expression of heterologous genes in transgenic fish. p. 120–141. In: Hew, C.L. and Fletcher, G.L. (Eds.). *Transgenic fish.* World Scientific Publishing Co., Singapore.
9. North, W.J. (1991). Oceanic farming, a new dimension in cultivating marine macroalgae. *UNESCO International Conference on Global Impacts of Applied Microbiology and Biotechnology*, IS. Valleta, Malta.
10. Paynter, K.T. and Chen, T.T. (1991). Biological activity of biosynthetic rainbow trout growth hormone in the eastern oyster, *Crassostrea virginica. Biol. Bull.*, **181**, 459–462.
11. Renn, D.W. (1990). Seaweeds and biotechnology-inseparable companions. *Hydrobiologia*, **204/205**, 7–13.
12. Scheuer, P.J. (1990). Some marine ecological phenomena: chemical basis and biomedical potential. *Science*, **248**, 173–177.
13. Singleton, F.L. and Kramer, J.G. (1988). Biotechnology of marine algae: opportunities for developing countries. Genet. Eng. Biotechnol. *Monitor*, **25**, 84–90.
14. Zhu, Z. (1992). Generation of fast growing transgenic fish: methods and mechanisms, p. 92–119. In: Hew, C.L. and Fletcher, G.L. (Eds.). *Trangenic fish.* World Scientific Publishing Co., Singapore.

17.7 Responses to assignments

17.1 Aquaculture is the deliberate cultivation by humans of animals or plants in aquatic environments; for the purposes of this chapter it means cultivation in marine environments.

Natural products harvesting implies the collection of certain products from animals or plants, be these cultivated deliberately or growing in the natural state. Bioremediation is the use of biological cells or products to clean up pollution in the environment.

17.2 As with all livestock, the faster the growth rate of the animal the greater the productivity of the farm; fish are no exception, and so enhancing the growth rate of fish would be of great benefit. Increasing resistance to diseases most common in farmed fish would also enhance productivity. The ability to rear warm-water fish in cold waters and *vice versa* would also allow expansion of pisciculture.

17.3 Fish which have been genetically engineered to produce human growth hormone may contain significant residual levels of such growth hormone when they are eaten. This could have an adverse effect upon consumers.

17.4 To promote the growth of shellfish repeated applications of growth hormone may be necessary. In open waters the effect of dilution could mean that large quantities of hormone may need to be added. Residual GH in the environment may have an adverse effect on the natural fauna. The GH would need to be produced by genetically engineered organisms, and may be relatively costly, adding significantly to production costs. These problems do not have to be considered if the GH gene is inserted directly into the shellfish.

17.5 If marine algae could be cultured on a vast scale this would have the effect of removing CO_2 from the atmosphere, as the CO_2 becomes fixed into algal biomass. However, the reduction in CO_2 levels would only occur if the CO_2 (carbon) fixed in algal cells remained fixed. In normal circumstances this may not be the case, as the algae produced may settle out in the oceans and undergo biodegradation (to CO_2 or methane). In order for the carbon to remain fixed it would need to enter the food chain to produce increased animal biomass in the oceans, or would need to become fossilised by geological events.

17.6 It is unlikely that the anti-tumour compound would be produced on a large scale by extraction from cultivated sponges. This would be partly due to the fact that yield of the compound from sponges may be low or many countries requiring the compound are landlocked or do not have climates suitable for the growth of sponges. It is likely that the compound would be produced more effectively and cheaply by fermentation processes using microorganisms with the gene(s) coding for production of the compound inserted into them. It may even be possible to culture sponge cells in large scale fermentation systems under controlled conditions (optimised for production of the compound).

17.7 Beached oil remains fixed in position and is thus amenable to the application of biosurfactants or inocula of degrading microorganisms and any nutrients necessary to promote growth of those organisms. In open-oceans application of inocula/nutrients or biosurfactants may be technically difficult, and the effect of dispersion by current may reduce the effectiveness of inocula/nutrients (by the effect of dilution).

17.8 Responses to self assessment exercises

17.1 (i) Insertion of growth hormone genes into farmed fish can result in increased growth rates and thus greater productivities compared to conventional fish farming. (ii) Insertion of anti-freeze glycoproteins into salmon means that salmon can be reared (farmed) in colder water than would otherwise be possible. (iii) By increasing the molting rate of lobsters increased growth rates are possible, and thus greater productivities.

17.2 (i) β-carotene is used as a food colouring, and dietary supplement as a vitamin A precursor. (ii) Agar is used as a solidifying agent for media in the culture of microbial, plant and animal cells. (iii) Agarose is used in the preparation of gels for molecular separation techniques. (iv) Carrageenan is used in the preparation of gel for immobilisation of cells for immobilised cell production/conversion processes. (v) Sesquiterpenes have use/potential use as anti-helminthic drugs.

17.3 Application of exogenous GH has the following relative advantages: ease of control of dose level and timing of exposure; ability to promote the product to consumers as 'natural'; being technically easier. Application of exogenous GH has the relative disadvantages of possible need for repeated dosage; effect of dilution (in open water); production of GH may be technically difficult and costly; GH introduced into aquatic environments may induce adverse effects in the natural fauna.

Introducing GH genes directly into shellfish may be technically difficult and such shellfish may not be considered acceptable by some consumers. However, the technique is likely to be less expensive than the use of exogenous GH and it is unlikely to contaminate the aquatic environment with GH.

18 Impact of biotechnology on the sustainability of the environment

Chapter outline

"The environmental introduction of any organisms, modified or unmodified, should be undertaken within a framework that maintains appropriate safeguards for the protection of the environment and human health while not discouraging innovation."

(SCOPE/COGENE Statement P)

18.1 Introduction

Today the abuse and over-use of natural resources that result from the pressures of development, population and poverty have reached a point where the problems of how to sustain the human environment has become a major concern of most governments. This chapter, while not ignoring the negative impacts of biotechnology on the sustainability of the environment, is concerned essentially with the positive impacts, both actual and potential. In general, biotechnological impacts on the sustainability of the environment are considered by the scientific community involved in molecular biology research to be positive and by environmentalists to be negative. Three main types of impact are considered: (1) agriculture and increased global food supply; (2) energy use and production; (3)

recycling of waste materials. Attention is drawn to some of the social effects and to problems of environmental safety and security. The need to maintain careful scientific assessments of the effects of the introduction of genetically modified organisms into the environment is stressed.

18.2 Agriculture

> ***Assignment 18.1***
>
> What would be the major positive and negative effects on the environment of increasing food productivity (increasing the yield/output per hectare)?

By increasing food production per hectare, the pressure for converting land with natural ecosystems into agricultural ecosystems diminishes. If productivity can further be increased at requirements that meets the human population growth rates it will be possible to reduce environmental insults and feed the global population. The effects of the ability to increase grain production per hectare on the natural environment is often overlooked. UNEP has estimated that up to 90% of deforestation results from the need for more agricultural land and it has been estimated that if India had continued production at the 1964 level it would have required another 30 million hectares of land in 1990 to produce the 55 million metric tons of wheat actually harvested.

> ***Assignment 18.2***
>
> What would be the major positive effects on the environment of manipulating plants so that they become more efficient in CO_2 uptake and incorporation?

However, enhanced food production requires (using current technology) the use of synthetic pesticides, fungicides, herbicides, etc., and of synthetic nitrogenous fertilisers. The presence of residues of these chemicals has risen at an alarming level in many organisms, including humans, and in rivers, lakes, ground water and also in the atmosphere. Many of these chemicals have been shown to be carcinogenic.

Some of the ways in which biotechnology can help provide a more sustainable environment by decreasing pressures to extend agricultural and horticultural land at the expense of the natural environment are: (i) by protecting plants from pests and diseases; (ii) by providing greater resistance to herbicides in cultivated plants, thus making it possible to reduce the number of applications of herbicides and their concentration and to increase production; (iii) in animal production and (iv) in the conservation of genetic resources. The technology associated with manipulation of useful traits in plants has been discussed in Chapter 12, and in animals in Chapter 13.

18.2.1 Biological plant protection

Two examples will be given: (a) the preparation of a biological insecticide based on a crystal protein from *Bacillus thuringiensis (B.t.)*; (b) the development of increased resistance of plants to viral and bacterial infections.

(a) *Bacillus thuringiensis*

Bacillus thuringiensis is a naturally occurring bacterium which produces on sporulation a protein that has an insecticidal effect. This effect results from the breakdown of the proteins in the spore in the larval gut with the production of a highly specific toxin. This kills

Lepidoptera larvae within 4 to 7 days of ingestion, but it seems to have no effect on vertebrates or invertebrates other than some Lepidoptera, Coleoptera and Diptera. The protein was originally applied directly over insect-infected fields but, although effective, proved to have low-field persistence. The genes for the crystal proteins have been even modified so that it can be expressed in plant cells to confer immunity of the plant. The modified genes have been introduced into cotton, tobacco and tomato plants which produce the bacterial gene product and are toxic to the larvae of specific insect species (Chapter 12).

This type of biological control has many advantages over chemical pesticides. These include: (i) the high specificity of the protein which is unlikely to affect directly non-target organisms, including harmless insects; (ii) the insecticide is contained within the tissues so that there is no risk of unwanted contamination of other plants; (iii) the protein can either be produced more or less uniformly in all the plant tissues or, by linking the gene to tissue-specific regulatory signals, it is possible to produce the protein in tissues, such as leaves, that are normally eaten by the pests; (iv) after more than two decades of field use as a biopesticide there have been no reports of pest resistance to the *B.t.*, toxin; (v) because the toxin is in-built it is probable that the selection pressure will be slower than with repeated sprayings of chemical pesticides; (vi) there will be no unwanted effects of pesticide sprays on beneficial insects and no pesticide residues on plants and in the soil.

Strains of *B. sphaericus* have been isolated that are selectively toxic to certain Diptera including mosquitoes. The use of such strains will be beneficial for the human population but may cause additional pressures on the environment.

(b) The production of virus-resistant plants

When a plant has been infected by one strain of a virus it has increased resistance to infection by another strain. Gene transfer studies using *Agrobacterium tumefaciens* have shown that a condition resembling this natural crossprotection can be induced in cells of certain plants, in particular tomatoes, potatoes and tobacco. The cells subjected to gene transfer are then used to regenerate plants. In such plants, and their progeny, the symptoms of disease either develop more slowly than in control plants or the infection does not develop (Chapter 12).

These studies indicate that it may be possible to provide protection not only against viral but also bacterial and fungal pathogens. Much still remains to be done because most of the studies have been carried out in carefully controlled, "artificial" environments. It remains to be seen if under field conditions, with a higher degree of competition and a wider range of physical and chemical conditions, modified plants are as successful as the unmodified ones.

18.2.2 Herbicide resistance

Assignment 18.3

What are the adverse effects on the environment of using herbicides in agriculture and horticulture?

The development of effective human-manufactured chemical herbicides for weed control in agriculture has greatly increased food production. Weeds cause greater losses in crop production than any other factor. The use of such herbicides has, however, negative effects on non-target species and is causing human concern in relation to the possible effects of toxic residues on humans. The failure to find a perfect herbicide and human concerns about their use has led to studies using genetic engineering to improve the tolerance of the crop plants to specific herbicides.

> ***Assignment 18.4***
>
> How could increasing the tolerance of a crop plant to herbicide be beneficial to the environment?

Efforts have been concentrated on the introduction of genes into several crops, such as cotton, maize, soybean, tobacco, tomato that provide increased resistance to herbicides, including atrazine, glyphosate, imidazolinones and sulphonylureas. Some work, particularly on maize and soybean, has been carried out to make it possible to allow the planting of soybeans after maize without damage to the soybeans because of residual soil levels of herbicide.

> ***Assignment 18.5***
>
> What detrimental effects on the environment might arise from increasing the herbicide resistance of a crop plant?

18.2.3 Biotechnology in animal production

The application of biotechnology to increase animal production (Chapter 13) can also have impacts on the sustainability of the environment. The development of biotechnology-based vaccines, which offer increased safety and greater efficiency, is one example. One recent area of research is to combine vaccine antigens to enhance the immune system so as to increase the effectiveness of the vaccine.

> ***Assignment 18.6***
>
> What are possible adverse effects on the environment of increasing animal production through biotechnological means?

18.2.4 Conservation of genetic resources

Although at present restricted, the potential role of biotechnology in the conservation of genetic resources can be expected to grow. Techniques are already available, for example the use of cryoprotectants, that permit the survival of cell suspension cultures, clusters of cells and embryos (from both plants and animals) at low temperatures and their storage for long periods of time.

> ***Assignment 18. 7***
>
> Why should the preservation of genetic material of plants be important to us?

18.3 Energy

The adverse effects of by-products of fossil fuel use, such as oxides of carbon, nitrogen and sulphur, on the sustainability of the environment are sufficiently well known and need no further elaboration here. Biotechnology can be used in a variety of ways to decrease the quantities of fossil fuel used and to improve the efficiencies in the use of some fossil fuels. Such possibilities include: (i) augmenting the amounts of nitrogen fixed by microorganisms; (ii) the use of fermentation processes to convert biomass to fuels such as methane and various alcohols, and to chemical feedstocks; and (iii) increasing biomass production rates.

18.3.1 Nitrogen fixation

> ***Assignment 18.8***
>
> What would be the positive effects on the environment of replacing chemical nitrogenous fertilisers with biologically fixed nitrogen?

It has been estimated that the genus *Rhizobium* fixes up to 200 million tons of nitrogen globally per year — much more than the total amount fixed industrially (about 40 million tons of nitrogen fertiliser are synthesised annually, mainly by the energy expensive Haber–Bosch process). Although it is unrealistic to think of currently dispensing completely with man-made nitrogen fertilisers, the benefits of nitrogen fixation to a sustainable environment are evident.

Up to the present, the major focus of biotechnological research into increasing biological nitrogen fixation has been on *Rhizobium* strains, including attempts to transfer nitrogen-fixing (*nif*) genes to non-legumes. The transfer of *nif* genes into cereals seems to be achievable, however, the development of a functional nitrogen-fixing system that would allow the cereal to produce its own fixed nitrogen with a minimal reduction in crop yield still seems remote. The possibilities of using plasmid technology to transfer *nif* genes to other soil microorganisms needs further research and field testing, particularly under conditions of stress that impede the effective use of *Rhizobium* strains.

Some work is also being done on the nitrogen-fixing blue–green alga *Anabaena azollae* which lives in association with the water fern, *Azolla caroliniana*, which is traditionally used as a source of nitrogen for rice crops, particularly in South-East Asia. Research is also being carried out on a number of nitrogen fixing bacteria, including *Frankia* spp. and *Azosprillum* spp. which have a symbiotic relationship with trees or crop plants.

18.3.2 Fermentation

Processes for the fermentation of organic wastes to gaseous or liquid fuels developed after the first large increase in oil prices were considered to be unprofitable because of process inefficiencies, uncompetitive costs and the subsequent lowering of the price of fossil fuels. Increasing concern with the state of the environment (and, in a few countries, problems of foreign debts) have led to renewed interest in the process in relation to disposal of wastes and for the fermentation of the biomass obtained from "fuel crops". Research is continuing into the development of biotechnological process for the production of possible substitutes for fossil fuels and chemical feedstocks using wastes and specially collected or cultivated biomass, including some that might be better used for human food. The conversion of lignocellulosic biomass to useful fuels and other by-products is receiving particular attention, as is the need to improve the design of bioreactors in which the fermentations take place. Such research is being directed not only to large-scale industrial fermentations but also to small-scale (one- or multi-family) units, for example biogas generators which produce fuel and natural fertilisers.

18.3.3 Increased biomass production

In addition to the use of biotechnological methods indicated above for the increased production of biomass, research is proceeding in many parts of the world with regard to the rapid propagation of fast growing species that can be multifunctional in relation to energy production and to achieve a sustainable environment. For example, the creation of forest energy farms is appealing for many reasons, not only making available more wood for fuel but also for reclaiming degraded land, providing employment, making available feed for livestock and the natural fauna and making the environment more sustainable. Biotechnology is beginning to play an increasingly important role in the production of seedlings and plantlets designed for special environments and specific uses. These developments have been discussed in Chapter 12.

Self assessment exercise 18.1

Which of the following would, by itself, increase the yield (biomass produced per hectare) of a crop species?

(i) Genetic modification to increase the effectiveness of photosynthesis (CO_2 uptake and utilisation).
(ii) Genetic modification to enable the plant to fix atmospheric nitrogen.
(iii) Genetic modification to increase the plant's resistance to herbicides.
(iv) Genetic modification to increase the plant's resistance to viral disease.

Self assessment exercise 18.2

Answer true or false to each of the following statements:

(i) Environmental improvement could result from genetic modification of a crop species to enable the plants to produce a microbial insecticidal compound, as there is a possibility that the microbial genes could be passed to weedy relatives by natural mechanisms so making them pest resistant also.
(ii) Environmental improvement could result from genetic modification of plants to increase herbicide resistance, as a greater quantity of herbicide could be used in weed control.
(iii) Environmental damage could result from increased animal production, as the loss of natural vegetation could lead to soil erosion.
(iv) Environmental damage could result from the use of nitrogenous fertilisers by the stimulation of non-crop species.

18.4 Treatment of wastes

The effects of inadequately treated or untreated wastes on the environment are all too evident. The accumulation of solid wastes in and near large urban complexes is unsightly, both when left uncollected and when accumulated in dumps. The use of rivers, lakes and coastal regions for the disposal of liquid wastes, especially sewage, is an insult to the environment and a potential menace to the health of humans and to the viability of natural and man-made ecosystems. The negative effects on the environment of both solid and liquid wastes have increased with population growth and, in some countries with prosperity, to the extent that many authorities are now not only expressing concern but also beginning to look for solutions. Biotechnology has been playing an increasingly important role in the treatment of wastes.

Assignment 18.9

What factors lead humans to dump wastes such as sewage into aquatic environments?

Consideration will be given to two main aspects of waste disposal: (1) domestic and agricultural, and (2) industrial and mining.

18.4.1 Treatment of domestic wastes

Domestic and agricultural wastes are usually divided into (a) liquid and (b) solid wastes.

(a) Liquid wastes

It is often overlooked that the purification of water-carried wastes is a biotechnological process involving biological, chemical and mechanical methods. The initial screening in primary treatment to remove wood, grit, cloth metal, plastics, etc., does not concern us here. However, the treatment of suspended solids, dissolved and colloidal matter is partly mechanical, partly chemical and partly biological. That part which needs further development is the biological. The secondary treatment of waste waters involves a fortuitous mixture of aerobic microorganisms on filters or in the activated sludge process. Little work has been done so far to improve the effectiveness either of the micro-organisms in the activated sludge process or in the treatment of the accumulation of solids that settle out, or are precipitated during the treatments. The last stage in the process involving micro-organisms is the digestion, under controlled conditions, of the organic matter in the sludge. This digestion produces mainly methane (CH_4) and carbon dioxide (CO_2). The methane is normally used as an energy source in the treatment plant.

Much work remains to be done to improve the efficiency of the various parts of the treatment process, both the technology and the microbial treatment. Some, but insufficient, work has been done on the optimisation of the aerobic process of sewage sludge treatment and on the improvement of the effectiveness of certain bacteria for the degradation of toxic chemicals in solution. It can be expected that such work will be stimulated by improved legislation and monitoring of environmental pollution and the levy of fines on offenders.

As indicated above, non-existent or weak regulations which encourage polluters to continue to dispose of their wastes into the environment or to treat them inadequately before disposal is a field that is not biotechnological. However, one can foresee, as a result of environmental concern, the introduction of stricter regulations and also of appropriately higher fines — part of which should be used to stimulate research in biotechnology to reduce the human pollution load on the environment.

Up to the present, however, there has been no systematic identification of the microorganisms most suitable for the treatment of wastes, although some work has been done in relation to specific wastes, such as certain persistent, toxic chemicals. Although such a systematic study would involve trials of a large number of bacteria, fungi and yeasts, both known and unknown, similar surveys have been carried out in the search for new antibiotics and much could be gained from the experience already obtained.

(b) Solid wastes

Studies of the possibilities of mixing selected solid wastes with sewage and manure as a means of diposal have, as with the treatment of liquid wastes, been limited by lack of governmental interest in a cleaner environment and to a certain extent by the lack of glamour and insufficient financial incentives.

The amount and composition of solid municipal waste generated per capita per year in a given country depends to a great extent on the level of development. The amount of solid waste per capita disposed into the environment in many developed countries is estimated to be 1–2 kg per day, a large volume of which is paper, plastic and glass. In developing countries the amount is considerably less and comprises mainly vegetable materials which are easier to treat. Much municipal waste is simply hidden in the ground in sites which under good conditions are then covered with a layer of earth. The oxygen initially present is used up by aerobic microbial activity and most of the degradation by microorganisms is anaerobic and methanogenic.

> ***Assignment 18.10***
>
> What environmental problems can be caused by anaerobic degradation of organic wastes?

Advanced biotechnological research on the microorganisms and the processes involved in this anaerobic digestion is being carried out in several countries because of the growing concern about such dumps and the pollution of ground water that results from leaching of toxic materials in the waste and from the by-products of the anaerobic digestion.

Efforts are also continuing with yeasts and a number of filamentous fungal species which produce a wide range of enzymes, which grow on most waste materials, and which in the case of mushrooms can be used directly as human food.

> ***Self assessment exercise 18.3***
>
> Indicate for each of the following processes one way in which it could be expected to produce environmental benefit and one way environmental damage.
> (i) Dumping of municipal waste in rubbish tips.
> (ii) Anaerobic digestion of sewage sludge in sewage plants.
> (iii) Aerobic treatment of sewage effluent.

18.4.2 Treatment of industrial and mining wastes

Although microorganisms have been responsible for the breakdown in the environment of many industrial waste products since the industrial revolution, the use of biotechnology in their treatment is comparatively recent. Genetic manipulation of these microorganisms to enable them to break down synthetic products is even more recent. Unfortunately many small-scale industries and some large, discharge part or all of their wastes into the municipal sewage systems making the treatment of domestic waste difficult and expensive.

> ***Assignment 18.11***
>
> How does industrial waste in municipal sewage make treatment processes more difficult and expensive?

Microorganisms are being used to accelerate the biodegradation of industrial and mining wastes, such as organic halogen compounds, metal- and sulphur-containing compounds, inorganic and organic acids. The extent of these developments is evident from the number of industries concerned uniquely with the treatment of industrial and mining wastes using such methods.

One of the fields in which considerable experience has been gained is in relation to bioleaching, initially for the recovery of metals from low grade ores, particularly copper and uranium but also cobalt, gold, lead and nickel. Bioleaching may also be used for the treatment *in situ* of high-sulphur coal and oil, without which problems of acid rain may be caused when the sulphur is oxidised (on combustion) and discharged into the atmosphere with the formation of sulphuric acid.

18.4.3 Bioremediation

Bioremediation is the use of biotechnological methods to clean up, or detoxify, contaminated land, air or water. Industrial and agricultural activities in particular can lead to pollution of land and water.

Developing technology to clean up contaminated area by microbial detoxification *in situ* is underway in several countries, and may offer a means to reverse many decades of abuse of the natural environment.

18.5 The socio-economic environment

Aspects of biotechnology that require more careful consideration are the unintended social effects and the potential impacts on the socio-economic environment, especially on rural populations, particularly the poor. There are already some indications of the potential impacts, and projections can be made using the analogy of the "Green Revolution", which also resulted from human manipulation of plant genetic material (but at the level of the whole plant rather than at the cellular or gene level as with genetic engineering).

Biotechnology can help to produce effective biocides and to reduce the amount of nitrogenous fertiliser required. This will help to reduce or eliminate the need to use chemical pesticides as well as to reduce the high inputs of synthetic nitrogenous fertilisers. The economic effect on the landless poor is likely to be positive with lower prices for food grains, since it is expected that world food production will increase and biotechnology will also be used to decrease pre- and post-harvest losses. It may possibly also have a negative effect because of reduced needs for labour for weeding and thus of employment, especially for rural women.

In as much as the majority of factories producing chemical biocides and fertilisers are in industrialised countries, the changes introduced by genetic engineering are unlikely to have positive impacts on the labour market in developing countries, but one can expect an improvement in the environments near such factories as production decreases.

The application of biotechnology to increase the productivity and improve the processing of crops grown and consumed in developing countries, such as cassava, millet, sweet potatoes and yams, can be expected to have a positive impact on the sustainability of the environment because of the reduction in the amounts of land required to produce the same outputs, particularly if the land is cultivated at the expense of natural forest.

Similarly if biotechniques are used to decrease dependency on irrigation by adapting food, feed and fibre plants to better resist drought, or to modify them so that they grow in adverse soil such as water-logged, saline and alkaline, some land that has been abandoned could be reclaimed and some of the arid regions used to plant food crops, shrubs and trees. For example, the development of the restriction fragment length polymorphism technique (Chapters 5, 11, 12 and 13) provides a tool which makes possible the selective manipulation of traits related to desirable characteristics, such as drought resistance and greater tolerance of toxic chemicals. In the event of climate change, the advantages of having more resilient plants are obvious.

Finally, tissue-culture techniques can be used to produce identical, disease resistant, high productivity plants by cloning. The use of such plants decreases the amount of land required to produce the same crop and hence pressures on the environment. This may be beneficial for the environment but may have possible socio-economic consequences on the farmer. In general, however, one can expect the use of new developments in biotechnology to have a positive social effect.

18.6 Environmental security and safety

Assignment 18.12

What are the major concerns about the release of Genetically Modified Organisms (GMOs) into the environment?

Such concerns are generally based on ignorance and a fear of the unknown, and the scientists involved and the public authorities are often at fault for failing to adequately inform the general public and to involve appropriate representatives of the general public in environmental release tests. Tests, based on scientific principles, should be organised with appropriate representatives in the fields of interest, such as agronomy, ecology, entomology, microbiology and members of the general public. Misconceptions can then be cleared, as seen with the insecticidal crystal protein of *Bacillus thuringiensis* where the initial opposition was dissipated and authorisation was given for the use of the *B.t.*, toxic protein on a large scale.

The studies that have been undertaken on the introduction of genetically modified organisms into the environment, such as that of the Scientific Committee on Problems of the Environment (SCOPE) and of the Scientific Committee on Genetic Experimentation (COGENE) into the evolutionary and environmental dimensions of the release of "bioengineered organisms", stress that the introduction of any organism, modified or unmodified, into the environment should be undertaken within a framework that maintains appropriate safeguards for the protection of the health of the environment and of humans, but that innovation should not be discouraged.

As indicated above, humans have introduced into the environment many materials such as antibiotics, fungicides, herbicides, pesticides, vaccines and natural compounds released in unnaturally large quantities either deliberately (as with the nitrogenous fertilisers) or "accidentally" or unconsciously (as with domestic and industrial waste products) and these have led and are projected to lead to large changes in the environment. The United Nations Conference on Environment and Development in June 1992 examined some of the problems caused by such introductions.

What is needed is to improve our knowledge of natural mechanisms in order to develop more ecologically sound approaches to releases of both genetically engineered organisms and other man-made materials into the environment. This knowledge will assist us to make projections into the future that will help to avoid the unfavourable consequences of human activities on the environment. It is the responsibility of all scientists to ensure that the results of their professional activities do not have negative impacts on the environment.

One additional positive impact of biotechnology is that the need to be able to monitor effectively the distribution of genetically engineered microorganisms released into the environment has stimulated the development of methods to monitor microorganisms in general. The discovery of the "viable but non-culturable" phenomenon has drawn attention to the deficiencies of some former methods, such as the culture of microorganisms to provide an indication of presence or absence and of abundance.

18.7 Concluding remarks

The potential and actual impacts of biotechnology on the sustainability of the environment are large. The positive impacts outlined above include some of the most important current aspects. If one looks to the future, however, the possible impacts of inter-genetic transfer of genetic material on the sustainability of the environment may be even greater.

Looked at in a dispassionate way, from a purely environmental view point some of the developments in biotechnology will have an impact on the overall longevity of individual human beings. This will lead to increases in the human requirement for natural resources. Unless biotechnological developments lead to an increase in natural resources at a rate that is greater than the human use of the resources the overall impact will be to decrease the sustainability of the environment. On the other hand, it is foreseeable that irresponsible or criminal experimentation with lethal organisms to modify them to resist current treatments could lead to epidemics that could decimate the human population (for instance at one time it was rumoured that the virus [HIV] which causes Acquired Immune Deficiency [AIDS] was a product of genetic engineering.)

The major impacts of biotechnology are likely to be in three main fields — energy, food production and waste treatment (or a reduction in the quantities of waste produced). Reduction in the use of fuel, fossil fuels in particular, and the increased production of organic materials for use as fuels would be the aim of biotechnology. Food production can be increased by better- quality crops and resistance to insects and microbial diseases. This would help in decreasing the pressure to expand the area of land used for agriculture and thus help in saving the natural environment. The impact of biotechnology on waste management are obvious. More research needs to be done to ensure that waste products do not produce environmental insults as are all too obvious in both industrialised and developing countries in urban and rural environments and in bodies of water near to urban complexes.

Up to the present, insufficient use has been made of biotechnology in reclamation of deserts and regeneration of soils. Although these are areas of research that lack somewhat in glamour and financial support, compared to some of the more public eye catching research both the subjects need be taken up more extensively in both industrialised and developing countries. The use of biotechnology has great potential in the production of seedlings of fast growing species for reafforestation and of useful plants with greater resistance to physical and chemical stresses. Reafforestation, desert reclamation and soil regeneration will all increase the sustainability of the environment.

It is appropriate to end this chapter with the words of Virgil, written 2000 years ago:

"The way down to hell is easy. The gates stand open night and day. But to retrace one's steps ... that is toil, that is labour."

Even with the help of biotechnology there will be much toil, much labour before our environment and that of succeeding generations becomes sustainable. One should also remember the words of Lao Tzu:

"As for those who would take the whole world to tinker it as they see fit, I observe they will never succeed: for the world is a sacred vessel not to be altered by man."

18.8 Suggested further reading

1. Anon (1991). p.198 in Baker, F.W.G. and Terry, P.J. (Eds.). *Tropical grassy weeds*,CAB International, Wallingford, 198.
2. Anon (1993). *Bioremediation — special issues of trends in biotechnology*, **11**, 320–367.
3. Colwell, R.R. (1988). New approaches for microbial indices/monitoring in the environment pp. 275–278 in Yasuno, M. and Whitton, B.A. (Eds.). *Biological monitoring of environmental pollution Tokai.* University Press, Tokyo.
4. Ehrlich, H.L.and Brierly, C.L. (1990). *Microbial mineral recovery.* McGraw Hill.
5. Gray, N.F. (1989). *Biology of wastewater treatment.* Oxford University press, Oxford.
6. IPCC (1990). *Intergovernmental panel on climate change: Scientific assessment of climate change.* WMO/UNEP. Cambridge University Press, Cambridge.
7. *Joint SCOPE/COGENE Statement P. xvii.* Mooney, H.A. and Bernardi, G. (Eds.) (1990). Introduction of genetically modified organisms into the environment SCOPE 44. J.Wiley, Chichester.
8. Swaminathan, M.S. (1991) in Swaminathan, M.S. and Hoon V.(Eds.). *Biotechnology: Reaching the unreached.* CSRSARD Publishers, Madras.

18.9 Responses to assignments

18.1 Positive effects would be that enough food could be produced to feed the (increasing) population, thereby reducing pressure to convert natural ecosystems (such as forest) into farmland. Negative effects would be that, in order to achieve this goal using current practices, the use of herbicides, pesticides and chemical fertilisers would be increased, each of which could have detrimental effect on the environment. Adequately feeding the population could itself put pressure on the ecosystem by increasing population growth.

18.2 This would lead to increased biomass yields, and so enable increased food/ biomass production to take place without commensurate increased use of land, reducing pressure on natural ecosystems. If phytoplankton could be manipulated in this way it would, if other nutrients required for growth were available, lead to greater productivity in the oceans, and might thereby reduce atmospheric levels of CO_2 (see Assignment 15.5) .

18.3 Herbicides may damage non-target plant species (which may themselves be food sources for beneficial animals such as insects). Residues may be left in treated plants, soil and water, and become incorporated into the food chain of animals and humans, causing harmful effects.

18.4 Increasing the tolerance of a crop plant to herbicide allows a greater concentration of herbicide to be used on it: this should be more effective in destroying weeds and so repeated applications may not be needed (or fewer applications needed), reducing the overall quantity of herbicide used. Fewer applications of herbicide can lessen adverse effects on non-target species. As fewer applications of herbicide are used, there can be a longer interval after the final application to one crop before another rotation crop is planted, giving more time for biodegradation of herbicide residues in the soil, and so allowing rotational planting of crops with lower tolerance to the herbicide.

18.5 It is possible that increasing the resistance of a plant to a herbicide will allow that plant to accumulate a greater quantity of the herbicide, and that this higher level of residual herbicide could exert toxic effects on animals (including humans) eating the plants. If the crop plant interbreeds with weedy relatives or shares viral pathogens with them (which can lead to gene transfer by the process of transduction), the herbicide-resistance gene(s) may be transferred to weeds themselves, rendering the herbicide useless. In using genetically engineered crops such possibilities must be guarded against.

18.6 Increased animal production, no matter how brought about, can have profound effects on the environment: for example overgrazing of vegetation can lead to soil erosion and desertification. Developing vaccines against trypanosomes would enable cattle to be grazed in areas they cannot be at present, which may lead to such natural habitats being converted to grazing land, to the detriment of the their flora and fauna. Human population growth might also be stimulated, putting even further pressure on the ecosystem.

187 The crop plants used in most countries have been developed over time by selective breeding, so that they may now resemble little their natural progenitors. Potatoes are a good example of this. However, selective breeding can limit the genetic diversity in a crop species. It may become necessary in the future to expand this diversity, for example to introduce drought tolerance or pest resistance into crop species. To do this, the required genes must be sought in their natural progenitors, but in many areas these are being destroyed with the destruction of their natural habitat by human activity. For these reasons plants such as wild potatoes are being preserved, so that the diverse natural genes they contain will be available to us in the future. Modern techniques of plant cell technology mean that this can be conveniently carried out in the laboratory.

18.8 Nitrogenous fertilisers are energy intensive to produce and often mean using fossil fuels with concomitant environmental pollution. Some nitrogenous fertiliser is usually leached from the soil and can accumulate in ground water or surface water: in lakes or rivers this can contribute to eutrophication and lead to growth of unwanted plants, algae or other organisms. Nitrogenous fertiliser can be oxidised or reduced by soil microorganisms in some circumstances to undesirable chemical species (such as ammonia), which at high concentrations can be deleterious to plant growth. Symbiotic nitrogen fixation by *Rhizobium* spp. overcomes these problems: the nitrogen is fixed within the plant itself and rapidly assimilated, so is not leached into the environment. The possibility of introducing bacterial genes for nitrogen fixation into plants is being explored, as is the possibility of expanding symbiotic nitrogen into non-legumes, using genetic engineering.

18.9 To treat sewage there must be effective sewerage systems, which are expensive to construct and maintain. Sewage treatment plants are likewise expensive to construct and maintain, and may have high running costs. Apart from such economic considerations, there are political or sociological ones such as apathy about environmental pollution or lack of effective legislation or monitoring to prevent pollution.

18.10 Organic wastes in municipal dumps or other anaerobic environments can be degraded by anaerobic bacteria with the production of methane. Methane is a greenhouse gas (in fact methane is more potent in producing the greenhouse effect than CO_2). Uncontrolled methane production in the environment from biodegradation of wastes contributes significantly to the greenhouse effect. Other products of anaerobic biodegradation such as hydrogen sulphide or acidic leachates from dumps can also adversely affect local environments.

18.11 Effluents from industrial concerns such as abattoirs, cheese factories and distilleries can contain high concentrations of organic matter, and require extensive treatment in aerobic treatment systems (using energy in many types of process). Such effluents may, however, be suited for anaerobic digestion, which not only reduces their pollution potential but produce energy in the form of methane. Effluents from some industries may contain toxic chemicals, which may reduce the effectiveness of the treatment processes (by exerting toxic effects on the micro-organisms used in them), or they may pass through the system without biodegradative detoxification and enter the environment, and thus require removal from effluent by chemical or physical methods.

18.12 Genetically modified organisms (GMOs) may be able to grow more rapidly than the indigenous microorganisms in soil and water, and so predominate in these environments to the exclusion of other (useful) microorganisms. They may spread to other geographical locations, outside the location of their release. The environment could suffer harm by activities of such organisms in the general ecosystem. Inserted genes in GMOs may become transferred to other micro-organisms by natural gene-transfer mechanisms. We have discussed previously in this chapter how genes from modified plants may become transferred to weeds. Thus, concern about releasing GMOs into the environment centre upon fears that doing so would lead to unexpected adverse effects on the ecosystem and thus to human welfare.

18.10 Responses to self assessment exercises

18.1 The correct response is (ii): plants able to fix atmospheric nitrogen will produce higher yields than unmodified plants — unless nitrogenous fertiliser is applied to the latter; virus resistant plants will produce higher yields than susceptible ones only in the presence of the viral disease.

Increased efficiency of photosynthesis would not in itself produce higher yields — corresponding increases in input on nitrogenous fertiliser would be needed also. Herbicide resistance would not in itself produce higher yields — application of herbicide would be needed also.

18.2 The correct responses are as follows:

(i) False — pests of crops may perform a useful function in controlling crops' weedy relatives: should these become pest resistant, this natural control mechanism would be lost.

(ii) False — using a greater quantity of herbicide would lead to environmental damage: the object of engineering herbicide resistance is to reduce the quantity of herbicide used.

(iii) True.

(iv) True — this is why the use of nitrogenous fertilisers and herbicides are usually combined: without combined treatment with herbicides, nitrogenous fertilisers would stimulate the growth of weeds, and may thus even reduce crop yields: the growth of plants in aquatic environments can lead to environmental problems.

18.3 (i) Benefit — Removing rubbish to dumps leaves a tidy environment. Damage — The CO_2, methane and leachates produced can cause environmental damage.

(ii) Benefit — Digesting sludge reduces its pollution potential, and may render it suitable as fertiliser. Damage — The methane produced (or CO_2 produced from it if it is used as fuel) contributes to the greenhouse effect.

(iii) Benefit — Treatment reduces the pollution potential of sewage. Damage — Most treatments require energy, generation of which may cause environmental pollution.

19 Biotechnology, international competition, and economic, ethical and social implications in developing countries

Chapter outline

19.1 Introduction

Biotechnology has been promoted by many as being essential for human survival, and as a technology that will improve the quality of life in every country. In previous chapters of this book we have seen the great potential of biotechnology, and how some of this has already been realised. This has already had an adverse effect on developing countries, which have been said to be losing US $ 10 billion annually from reduction in their exports due to biotechnology-based product substitutions. International competition implies that there may be some winners and losers in the competition, and it is not yet predictable whether winners will be developing or industrialised countries, the producers or users of techniques, the poor or rich within countries. Although the hope is that biotechnology can improve the life of every person in the world, and allow more sustainable living, the crucial decisions may be dictated by commercial pressure and by the socio-economic goals that society considers to be the most important. These aspects, together with some of the social and ethical implications of biotechnology, are highlighted in this chapter.

19.2 Biotechnology and international trade

Biotechnology can be defined as the use of biological systems (*in vitro* or *in vivo*) to provide, or improve, goods or services. In ancient and modern times people have used biological organisms to provide food, drink, medicine, chemicals, clothing, fuel and to extract metals. The diversity of uses is expanding. Plant and animal breeding and microbial selection have been used to optimise these roles. In addition to the organisms that have been directly selected by human action, other organisms ecologically favoured by the presence of those organisms have been indirectly selected.

Whether countries can use biotechnological techniques to improve life depends on several major factors.

> ***Assignment 19.1***
>
> What factors influence the ability of a country to adopt biotechnological processes?

The barriers that slow the adoption of better techniques and/or varieties should be removed, and conservatively-minded policy makers, scientists and village elders should accommodate biotechnology to boost sustainable local production. These are national issues, but international aid may be required to allow research and to introduce new technology in smaller countries. There are also international questions, such as whether technology can be transferred from countries with a high level of research and capability to countries without it, and how, if at all, intellectual property rights should be protected.

Plant and animal breeding by humans is associated with commerce. International trade for many countries has long been based on biological products. International trade pressures have changed the internal structure of many countries, and they will continue to do so. The need to repay ever increasing loans has had enormous social consequences in many debtor countries, both industrialised and developing. International competition to export products to gain foreign exchange has become intense. It is into this framework that the further use of biotechnology must be viewed, and there could be both positive and negative effects for different countries. Biotechnology will affect every area of countries' economies. Some examples of these effects on developing countries are given in Table 19.1.

The goals of every country may be similar: to improve the quality of life of the citizens, and to maintain living standards at a reasonable level. The perceived method for obtaining the first goal is economic success in international competition, though living standards in many industrialised debtor countries could be considered better than those in some countries successful in international competition. The maintenance of living standards, however, does not require some success in world competition. The maintenance of reasonable lifestyles, and quality of the environment that is consistent with a sustainable way of life in the international community, should be a primary goal of many countries. When more industrialised countries accept this, increased attention may be given to the question of distributive justice in the world. International competition should be adjusted to encourage more sustainable economic policies and protect the environment of the planet. The sooner a steady state is reached the better. Until that time international competition will continue, and some countries will be able to improve their international trade, and others will not.

Developing countries are currently economic losers in international competition, so many would say that the situation for them can only get better.

Table 19.1 Probable effects of biotechnology and international trade of biotechnology products and services on developing countries

Positive effects
- Increased crop production, more food
  — Extension of growing area with drought- and salt-tolerant crops, may increase employment
- Increased livestock health and production
- Increased nutritional qualities
- Increased storage life of foodstuffs, especially useful for the poor who lack good transport and storage facilities
- Decreased dependence on imported fertilisers and chemicals (except herbicides)
- Increased public health, via cheaper multiple disease vaccines, and genetic diagnosis kits
- Faster growing trees, for biomass production, as an energy replacement
- Biodiversity may be preserved, as international companies buy exploitation rights in return for conservation
- Possibility of new products from increased research into genetic resources and from the use of animals or plants as bioreactors.
- Tissue culture allows continuous production and more reliable product quality

Negative effects
- Loss of export markets as products are substituted by production of alternatives in industrialised countries
- Loss of foreign income and employment
- More cash crops may be grown, to produce high value commodities
- Strengthening of large agricultural estates and displacement small-scale landholders and farmers
- Reduced labour needed for cultivation
- Herbicide-tolerant plants may reduce labour market in weeding, and may also increase dependence on foreign imports of chemicals
- Privatisation may increase the portion of seeds that require fees for use and local farmers may lose control of cooperatives
- Greater privatisation increases both legal and financial barriers to use of varieties
- Biodiversity may be reduced as more efficient monocrop systems are introduced and larger size farms are used
- Loss of natural ecosystems in marginal lands as new crops are introduced to those areas

Assignment 19.2

Explain how adoption of biotechnological processes in developed countries can cause economic hardship in developing countries.

However, if commercial forces are left to operate unconstrained by morality, and trade barriers to the import of foodstuffs continue to exist, the situation will clearly get worse for developing countries. This is particularly because of product substitution, and by the increasing ability of industrialised countries to produce enough foodstuffs to become self-sufficient. Products such as sugar, shikonin, coffee, cocoa, vanila, and cotton, are just some potential cases. Exporters of such products are already experiencing difficult times, especially with protectionism. Trade barriers should be removed. The future would be brighter for developing countries if they could produce cheaper foodstuffs and industrial raw materials and products in transgenic plants and animals, especially if at the same time the storage life of foods was increased so that it did not spoil during transport.

The situation in terms of food production and life quality in developing countries may improve nevertheless, because developing countries will become more self-sufficient in food and have better quality foodstuffs and increased energy production from biomass. For example, if a pest-resistance gene saved 1% of the total rice crop in India from disease, it would save that country US $ 300 million a year. However, self-sustainability for most developing countries is several decades away.

As countries become more self-sufficient, international trade may become less important. If developing countries still import high technology products, they would become even more indebted, and the whole international system of trade would be threatened. It would also create more political instability, because as the hopelessness of repayment became evident, countries would withdraw from the international trading system. It is obvious that a potential collapse of the economic system may persuade industrialised countries to attempt to reduce their trade surpluses, or aid the debtor countries so that they remain in the system. It is impossible to predict the outcome. We can hope that trade barriers and protectionism are reduced, but within most countries the protection of small rural farmers is considered socially important; and one must balance the questions of international trade versus national socio-economic structure. Biotechnology could aid the survival of farmers, if more disease-resistant and climate-tolerant crop varieties were introduced. The production of biomass as renewable energy, and production of industrial and pharmaceutical products in crops and livestock, will provide additional need for agricultural production. However, multinational petrochemical and pharmaceutical companies may control the seed needed for such crops, and they could produce hybrid seed rather than open-pollinated varieties so as to maintain the control on seedstocks and hence profits.

Assignment 19.3

Explain how production of hybrid seeds enables seed producers to control seedstocks.

If fees need to be paid for seed, larger farms may succeed better than smaller farms. It is questionable whether in such circumstances biotechnology will support the survival of traditional village structures and small land holders. A free market approach would not do this, unless strong incentives and disincentives were established.

19.3 Privatisation of biotechnology

Biotechnology is, almost by definition, based on applied research rather than basic research. However, the continued development of biotechnology does require much further basic research. The biological features, and even existence, of many organisms, remain unknown. The ecological characteristics of most organisms in the world, especially in the developing world, remain unknown. We need to know the ecology of organisms that we may introduce to new environments. We also need to identify genes associated with useful features that may be applied in biotechnology.

Because of limited research funds, much of the research in developing countries and small industrialised countries is applied research. Applied research is targeted for specific goals, such as the production of particular disease-resistant crops, or specific vaccines. Research facilities include universities, government and private research institutes, and hospitals. The funding in most industrialised countries is from industry and government, but in developing and smaller countries, a lack of industry means that government funding is even more essential. We should all contribute to shared knowledge that people can benefit from, therefore research funding should be shared between countries. Many people believe that the pursuit of knowledge itself is a good thing, though increasingly the pursuit of beneficial knowledge is viewed as more important. To perform research is thought to have a positive effect on the teaching standard at universities, which may be a more significant benefit in small countries than the direct results of research. Another important, but perhaps less worthy, argument people give to promote research is that of national pride. This can be converted to political benefit, if countries are seen to provide much international development assistance.

Research has for many decades also been viewed in terms of business opportunities, both internationally and within nations. As national budgets become more stretched with other needs, many nations are encouraging more research by industry, either independently or in collaboration with government researchers. If research was performed in publicly-funded laboratories, and was published freely, there would be fewer problems with international technology transfer. National governments may transfer technology to other countries as part of development aid. Non-profit making private organisations are also very important in biomedical research in some countries, and they usually allow export of technology. For example, the world's largest gene-mapping laboratory in France, the Genethon, funded by charity, has used automatic DNA sequencing to map the human genome. However, much of the new wave in biotechnology research is being performed by private companies. These companies are being encouraged to perform research in those countries' national interests, including the hope of export earnings from the sale of products and/or technology.

It is obvious that the goals of private companies are not those of whole countries. Many Western biotechnology companies are of very small size, but multinational corporations have been gaining increasing control of technology research. There have even been some takeovers of large biotechnology companies, for example the takeover of Genetech by Hoffman Roche, and the selling of Cetus' PCR technology to Du Pont. Also, many large companies such as Monsanto in the USA realised the commercial potential early on and so invested considerable research money in biotechnology. In Japan, government ministries and large multinationals have been the major investors in biotechnology from the early 1980s, because of greater difficulty in establishing small companies than in the USA. Governments may view competition as being between countries, but it may actually be more between multinational companies.

In some developing countries there has also been major investment in biotechnology, and there is research on numerous applications in China and India. However, there is still a lack of research facilities, qualified personnel, and funding sources to exploit biotechnology. Strategies used depend on the size of the country: small countries can still apply international cooperation agreements to local needs if they lack sufficient resources themselves. No single policy is best.

Large multinational companies may also make development agreements with developing countries. In 1991, the company Merck & Co. made an agreement with Costa Rica, to exclusive rights to new potential "products" it finds in an area of its tropical forests until the year 2000. It is akin to a hunting licence for useful compounds. If successful, a share of the profits will be paid to Costa Rica. This should also encourage other countries to preserve large areas of their forests. It is important to encourage *in situ* conservation and, if no other group will put up the finance, it will be left to large companies who will benefit from the new substances found. This is not such a new phenomenon: industrialised countries have been gathering seeds and genetic resources from other countries for centuries, for the development of new crops and products. In June 1992, at the World Environment and Development Conference in Rio de Janeiro, Brazil, a Biodiversity Treaty was signed, which has important implications for the protection of biodiversity by all countries, and may preserve the intellectual property rights of products derived from the diverse species. However, the USA did not sign this treaty, and it is also too early to say what impact the Biodiversity Treaty will have on industrial investment in biotechnology and technology transfer, though it will have a beneficial long-term effect in the preservation of biodiversity in many countries.

19.4 Patenting of biotechnology discoveries

Many new discoveries are being patented from research performed in national and private laboratories.

Assignment 19.4
What properties must a product or process have in order for it to be patentable?

There is some debate about whether living organisms should be patented, and this is discussed in the following section. First let us consider some general questions concerning patenting.

To qualify for a patent, an invention must be novel, non-obvious and useful. If the claimed invention is the next, most logical, step which is clear to workers in the field, then it cannot be inventive in the patent sense. In the case of natural products, details of a molecule or coding-nucleic acid sequence may have been published, so it may have lost its novelty and non-obviousness. Patents are granted on molecules which have medical uses if the chemical structure, or the useful activity, was novel when the patent was applied for.

Patenting rewards innovation, but the mere sequence of genetic material may not be innovative, even if the technique used to produce it is. There are patents on short oligonucleotide probes used in genetic screening. If someone can demonstrate a use for a larger piece of DNA, then they can theoretically obtain a patent on it. An example of a larger patentable section of genetic material would be a series of genetic markers spread at convenient locations along a chromosome. Another set of genetic markers on the same chromosome could be separately patented if they also met those criteria. To sequence genetic material is not what we normally call an invention. The sequencers of DNA are not charting 'unowned land' but rather they are charting 'uncharacterised land', so the title of mappers is rather suitable in this context. Methods of gene-sequencing, mapping, or expression, can be invented and patented.

Biotechnological process patents can be viewed in a similar way to existing process patents. The information may be used in the study of a particular disease, for example by the introduction of a gene into an animal to make a model of a particular human disease. The process for making "Oncomouse", a mouse that contains activated oncogene sequences that is therefore sensitive to carcinogens, was patented, and it was the first mammal to be patented, in 1988 in the USA. Genetic information can also be used to cure a disease, for example using the technique of gene therapy with a specific gene vector. The direct use of proteins as therapy is well established, and these products may be patented, though we should note, in general, medical procedures have not been patented for ethical and practical reasons.

With the elucidation of the genome sequence of many organisms, including humans, much genetic material will no longer be novel as its sequence will be available in a database. The completion of the genome maps and sequences of many organisms will have many implications for the future of patents in biotechnology. It may be cheaper to buy some genes off companies or to use specialist companies to perform specific gene-splicing experiments, rather than experimenters doing every step themselves: although developing countries have a shortage of research money, the potential for speedier applicable technology when the researchers can concentrate on the special aspects of a project rather than on the more mundane aspects of constructing genetic maps or gene vectors, may eventually convince some governments that buying in genes or expertise is a valid way to use research money. This could also stimulate biotechnology industry in developing countries. If a new technology saves enough money, for example reduces crop losses or environmental degradation, then introducing it as soon as possible may be worth the investment.

A patented product that reaches the commercial market gives the inventors some compensation for the time they spent in research for the development. In the USA the average time required for biotechnological medicines to be approved for commercial sale by the Food and Drug Administration is 21.4 months after results of clinical tests, and it could be ten years after identifying the substance. The period of patent protection varies widely between countries, but in Europe, USA and Japan it has recently been standardised to provide protection for about fifteen years after product approval. Once a product is patented the sales can bring about high income for the company that produces it, and this includes returns for the "inventors". The World market for drugs and medical products made by genetic engineering in 1991was about US $ 3 billion, and by the year 2000, it is expected to be US $ 30 billion. The system is self-sustaining: if patents are awarded companies will invest time in research, and if not there will be less incentive for companies to conduct research. Some processes can be kept industrial secrets, but it can be difficult. Without patents it may be easy for other companies to copy the techniques soon after introduction, and take a share of the commercial market, especially because they do not need to bear the cost of the long period of research for product development. Some system of reward is required to encourage commercial research, which is responsible for a significant number of biotechnological applications. The international recognition of intellectual property rights (patents and variety rights) is thus a basic concern.

We can apply the ethical principle of beneficence here. Does commercialisation of biotechnology lead to more benefits than a ban on it? The benefits should be in terms of general medical or agricultural development, rather than economic prosperity of one company or country over another. Patenting promises useful consequences (e.g., new products/research). If patenting is not permitted, useful information will become trade secrets, or if plant variety rights are not recognised seeds may not be made widely available. However, property rights are not absolutely protected in any society because of the principle of justice, and for the sake of "public interest", "social need", and "public utility", societies can confiscate intellectual property.

People arguing for patenting claim that patent law regulates inventiveness, not commercial uses of inventions. However, there was recent controversy regarding the commercial monopoly held by the company which was able to patent AZT, the first HIV/AIDS treatment, which enabled it to obtain large profits while it held a monopoly. It also meant that the drug was prohibitively expensive for developing countries (as most pharmaceuticals are). There are numerous examples where such commercial monopolies obtained cannot be said to be in the overall public good. Another argument is that if other countries support patents, our country needs to if our biotechnology industry is to compete; however the reverse argument, that some countries do not permit similar patents, is also used to justify exclusions.

19.5 Ethical concerns about patenting of living organisms and genetic material

The patenting of living organisms and genetic material is a contentious issue. In the USA and many other countries, normal patentability criteria apply to any subject matter, that is the invention requires the attributes of novelty, non-obviousness and utility, and the invention should be deposited in a recognised depository. While accepting these criteria, some countries have specifically excluded certain types of invention, for example the European Patent Convention excludes the patenting of varieties of plants or animals. However, in October 1991, the European Patent Office reversed its earlier decision and announced that it intended to grant a patent for "Oncomouse", and transgenic animals containing an activated oncogene. This decision will continue to be debated. There is public rejection of the idea of patenting animals in some countries and in Denmark animal patents are excluded.

In 1961, the Convention on the International Union for the Protection of New Varieties of Plants (UPOV Convention) established international "plant variety rights", and by 1989 there were 19 member countries, which include more than 70% of the world seed market of all countries with a market economy. The requirements for variety rights include stability, homogeneity, novelty, and distinctiveness. The varieties must be generally distributed, and researchers and farmers have exemptions from the payment of royalties on seed that they save from their harvest. A few developing countries have national plant variety rights schemes, but have not joined UPOV. However, no reward is yet given to the farmers who for millenia have established crop varicties which plant breeders use as starting materials, so that Farmers' Breeding Rights have also been talked of. In 1982, the OECD estimated that the contribution of developing countries to the major crops in the USA was several billion dollars annually.

Assignment 19.5
Explain how it is that developing countries can contribute to the major crops of developed countries.

In 1983, at a UN Food and Agricultural Organisation conference, representatives from 156 countries recognised that "plant resources were part of the common heritage of mankind and should be respected without any restriction". Since then an international network of genebanks has been established which will provide genetic material worldwide.

Public attitudes to the patenting of different types of subject matter has been measured in New Zealand in mid 1990 and in Japan in 1991. People were asked if they agreed whether patents should be obtainable for different subject matter. The results for different groups of the population are illustrated in Table 19.2. There was less acceptance of patenting new plant or animal varieties than of inventions in general. Only 51% of the public agreed with patenting of "genetic material extracted from plants and animals" in New Zealand, and only 38% in Japan. There was even lower acceptance of patenting "genetic material extracted from humans": in Japan only 29% agreed. There was more acceptance of patenting of genetic material among those who thought there were benefits to their countries from genetic engineering, and by farmers, and by scientists in Japan. There was less acceptance of patenting among the age group 15–24 year olds and among high school biology teachers in New Zealand.

Table 19.2 Public opinion over patenting
(Results from surveys conducted in New Zealand (NZ) in 1990 and in Japan in 1991)

Occupation of respondents	Public		Scientists		High school biology teachers		Farmers
Country	NZ	Japan	NZ	Japan	NZ	Japan	NZ
Number of respondents	2034	470	258	479	277	227	200
Subject matter: Results expressed as % total supporting patents for							
New inventions	93	91	95	94	88	92	94
Books, information	85	73	81	882	72	77	81
New plant varieties	71	60	66	78	49	61	66
New animal varieties	59	49	63	74	51	60	63
Genetic material from plants/animals	51	37	53	46	34	38	53
Genetic material from humans	–	29	–	35	–	29	–

Arguments against patenting life include metaphysical concerns about promoting a materialistic conception of life: we may begin to view animals the same as we do material goods; living organisms and their genetic material are special and should be considered different to other "materials". Patenting produces excessive financial burdens on farmers and on consumers of medicines (increased costs to consumers, payment of royalties for succeeding generations), and increased burdens on developing countries.

Some critics of ownership go so far as to call those who seek to profit by and to control genetic sequencing, "genomic imperialists". In 1991 and 1992, patent applications for nearly 3,000 human genes based on randomly-sequenced cDNAs were made by the National Institutes of Health (NIH) in the USA. These raised major questions about patenting policy because researchers are reluctant to share genetic sequence data databases while there is the possibility of others abusing it. For instance, the NIH and Medical Research Council in the UK withheld their data from entry into gene databases while preparing patent applications. It would have grave implications for open data exchange if such patents were upheld. A further application for a patent on one thousand human genes by the British Medical Research Council followed. However, there is no demonstrated utility and the technique is obvious to any researcher with appropriate resources, so in addition to ethical or policy issues, patent applications may fail on these grounds. These government bodies may sublicense particular national companies to pursue research on these genes in an attempt to "protect" their national biotechnology industry. Actually, the publication of such sequence markers will make it more difficult for companies to patent those genes and could discourage research, because such genes will no longer be novel, and so the whole policy related to this area is undecided. Public opinion (Table 19.2) could force a policy change regarding the patenting of genetic material, even if it is judged to be legally valid. Judgements should be made considering all the economic, ethical and social implications. In the UK, "patents shall not be granted for an invention, the publication or exploitation of which would be generally expected to encourage offensive, immoral or anti-social behaviour". Similar exclusions are common to European and US patent laws. It is an area of much debate: France and Japan have said that they will not seek such patents, and Britain has said it is opposed to them, despite the MRC application.

19.6 Sharing the benefits from biotechnology — technology transfer

There are major applications and implications of biotechnology. Some of these are presented in Table 19.1. All people should share in the benefits of biotechnology. There are religious backings for this (such as "love thy neighbour as thyself"), the utilitarian ideal that we should try to benefit as many people as possible, and the ethical/legal principle of justice. In the United Nations Declaration of Human Rights, Article 27(1), there is a basic commitment that many countries in the world have agreed to observe (in their regional versions of this declaration). This is that everyone has the right freely to participate in the cultural life of the community, to enjoy the arts and to share in scientific advancement and its benefits. The common claims to share in the benefits of technology should be considered in all aspects of biotechnology, including also the questions of who should make decisions concerning its applications.

Over 100 countries allow some form of patent protection, but developing countries often exclude from patent cover pharmaceuticals and agricultural inputs. Because it is easy to copy production procedures, those countries may reproduce products without paying licensing fees. As a result, technology transfer to countries without patent protection is slow.

A special case is that of the human genome project's results. All people share this genome and thus could be called shared owners of it. Recently efforts to collect DNA samples from all the world's human populations has commenced, to create a general cell bank. It has been argued that the human genome, being common to all people, has shared ownership, is a shared asset, and therefore the maps and sequence should be open to all. Growing numbers of disease-causing and disease-susceptible genes are being sequenced and the mutations characterised. It will be possible to develop DNA probes to diagnose most genetic disorders (Chapter 11).

The number of human proteins that can be made by genetically modified organisms (GMOs) will be increased which would allow conventional symptomatic therapy for many more diseases, and which could be supplemented by somatic cell gene therapy when appropriate. DNA probes have also been used to speed up diagnosis of infectious diseases. Our basic knowledge of human biology will be expanded, which should allow medical treatments to be developed. It is obvious that within the next few decades medicine will undergo a major change in all countries and this represents the beneficial side of the extra knowledge. People in developing countries will indirectly benefit from all the technology developed, which will also be used for the sequencing and manipulation of genes in all other living organisms, as well as humans.

> ***Assignment 19.6***
>
> Outline the techniques of somatic gene therapy.

Technology transfer can be associated with undesirable consequences if diversity and innovation are reduced. Throughout human history about 3,000 plant species have been used for food, but this number has decreased so that there are only about 20 principal food crops used now. This is partly because of external conquest and domination which tends to suppress use of local food crops. Plant breeding used for modern agriculture, including the "Green Revolution", has reduced the number of varieties to those which are the most productive, but at the cost of reduced genetic diversity. It is important to maintain genetic diversity, and because crops may be used in one country, it does not mean they are the best crop for use in another.

> ***Assignment 19.7***
>
> Explain why it is desirable to maintain genetic diversity in crop plants.

Also the yields obtained in experimental plots may be much better than in the farmer's field, due to the different local environment. In a 1991 report, the use of smaller and more efficient animals for food production was recommended for developing countries. This could also have the advantage of avoiding the need for meat storage. It would also maintain genetic diversity. However, in times of food shortage, plants would be more efficient at producing protein.

Another important point is the use of intensive agriculture, with inputs of chemical fertilisers and pesticides, often using multi-application procedures. Although such procedures have beneficial effect on yields, efforts should be made to switch to crop and animal systems less dependent upon intervention. Companies in industrialised countries are continuing much research on applications of biotechnology that require inputs: an example is the development of herbicide-resistant plants, where both seed and herbicide are controlled by the same companies. However, if newly developed pesticides and herbicides have greater biodegradability there may be environmental advantages. There should also be attempts to use biological pest control. Genetic engineering should be used to insert desirable genes directly into openly pollinated crops, which can then be used by farmers in developing countries without dependence upon seed and chemical companies (which are often controlled by the same multinationals). The question is who decides what varieties should be grown in developing countries, and whether this is for local or "international" needs.

Within developing countries, applications of biotechnology should attempt to preserve rural structure, for instance villages could create small-scale biotechnology "factory" supplies to earn income. In developing countries the agricultural sector employs over 80% of the active population, compared to only 5–10% in industrialised countries. Some crops are labour intensive, and others are not: for example, oil palm plantations require about one-third of the labour required in banana plantations. Production of new products such as single cell protein may reduce labour requirement for food production. Weeding is one of the most labour intensive operations, but this will be reduced as herbicide tolerant crops are introduced, leading to loss of work for many people, especially women. However, by introducing new varieties year round crop production may be possible, increasing labour requirements. The effects depend on the country; for example the use of bovine somatotropin (BST) hormone to increase milk production in dairy cows is being opposed by many groups in Western countries because it may favour larger farms, but in some developing countries, such as Mexico or Pakistan, its use would be welcomed because it may reduce imports of milk powder.

Some companies have said that if they do not expect to profit from the sale of biotechnological procedures in developing countries, they will export technology free of charge. This is to be encouraged. Even if such technology is not transferred, the use of genetic engineering techniques allow very simple copying of genes from one organism to another. It would be comparatively easy to isolate the pest resistance gene based on *Bacillus thurengiensis* and its regulatory elements from a transgenic plant and to construct a vector to insert it into a different plant. Some companies have said that they intend to insert specific marker nucleotide sequences into transgenic plants to monitor any such piracy, but it would be very unlikely that they would sue for damage from a developing country.

> ***Assignment 19.8***
>
> Explain how a marker nucleotide sequence would enable a transgenic plant to be identified as originating from a particular company.

Therefore it may be better for such companies to give away new varieties to developing countries, as they will receive public support for such action in the industrialised countries; also the developing countries may not have been able to afford the new varieties anyway. However, this could mean that only varieties of commercial value in industrialised countries will be developed by private companies so that researchers in developing countries would still need to develop local varieties.

> ***Self assessment exercise 19.1***
>
> Explain how the adoption of biotechnological processes in a developed country might (i) increase or (ii) decrease crop production in a developing country.

> ***Self assessment exercise 19.2***
>
> Explain how developments in biotechnological processes might lead to decreased employment in developing countries.

19.7 Changes to international trade and future economic systems

Biotechnology will improve the efficiency of agriculture. This will aid the feeding of the populations of every country, but, as discussed, it may not necessarily mean improved trade. The increased efficiency of agriculture will make many countries self-sufficient in food production. This will mean that agricultural exporters may find difficulty in selling produce, and also food prices may be lowered so that they will lose foreign earnings. The nutritional quality of food is also being improved by genetic engineering, for example, to match the amino acid balance in foodstuffs with human or animal dietary requirements (Chapter 12). Only if countries diversify to new markets and products will they be able to continue to rely on agricultural exports. Countries that do not will find it very difficult to compete in the international economic markets, and one result could be that such countries will increasingly find themselves in debt to countries that export industrial products. In the medium-term, developing countries could switch to quality products such as specific fibre products, or high quality foodstuffs, but in the long-term the solution may be to switch to the production of high value substances via agriculture. Products such as pharmaceuticals or therapeutic proteins could be produced in transgenic plants and animals and exported. However, the research for genetic manipulation of plants and animals to produce such products is concentrated in industrialised countries, and in many cases only small farms would be needed to provide national needs for such compounds. Products for the food industry, such as new sweeteners and oils could be made in plants, which would require larger areas of crops. The concern must be whether these crops can be grown in addition to local food crops.

Biotechnology will certainly continue to change the international trade situation. A benefit for some developing countries may be the development of fast growing biomass, that could be used as a fuel source to reduce the need for oil imports. In addition to the obvious environmental benefits, it would also lessen dependence on imported oil and gas as a source of energy.

We must question the future goals of societies. What is the meaning of life, and does increasing use of technology mean an increased quality of human life, both for individuals and society? If we question this, we may also question the future of international competition in trade. The term competition itself implies that there are winners and losers, and the losers are usually developing countries. If we eliminate competition, then we may allow more global responsibility.

Modern capitalist economic strategy appears to be very successful at increasing the wealth of nations and the general living standards of people in them. However, despite the advantages of market economies, we must really ask whether such commercial competition is in the interests of developing countries. We must also look at the implications for devleoping countries. We have seen the acceptance by some countries that the current economic system is not sustainable because it does not value the environment, and there is debate over how to change the economic system. For all countries, long-term sustainable economic policies are required, different to today's. Genetic resources have been connected with economic prosperity throughout history so the fact that many developing countries do possess such resources is an advantage: they must attempt to minimise the loss of control over them and preserve them. The presence of biological diversity is a great long-term economic asset, more important than the short-sighted policies which destroy them.

Self assessment exercise 19.3

Explain how the adoption of biotechnological processes may make a developing country (i) less dependent on international trade or (ii) more dependent on international trade.

Self assessment exercise 19.4

Explain how developments in plant breeding in developed countries may lead to the success of large farms rather than small farms in developing countries.

Self assessment exercise 19.5

Give the potential advantages and disadvantages for a developing country of leasing product rights for novel plant compounds to companies based in developed countries.

Self assessment exercise 19.6
Explain the advantages to a developing country of using gene banks and genetic engineering expertise of developed countries, rather than develop these within the country.

19.8 The ethical and social impact of biotechnology

It is easy to say that all people should benefit from biotechnology, but will they? The new technology presents both benefits and risks. Scientific risks can be controlled more easily than the risks of unethical applications and adverse effects on societies. We have considered moral responsibility in general questions of biotechnology transfer and use, and it is also important to summarise some of the principles that should be used to ensure more ethical uses of biotechnology, as it affects people as individuals and in society.

The 1990s are the time of a paradigm shift, for scientists to give support to consideration of the social impact of their research. This is already underway in Europe and North America, and it is beginning in Japan. One of the stimuli for consideration of bioethics was the Human Genome Project in these countries. Scientists now refer to ethical, legal and social impact (ELSI) issues of research. A trend for scientists to give some financial support to ELSI research has been initiated in Europe and the USA. In 1992, 5.2% of the Human Genome Project grants from the NIH in the USA were for ELSI research. In the USA in 1992, a total of US $ 9 million was spent on research on the social impact of biotechnology. The European Community has also supported many ELSI projects, and the Canadian genome project is allocating at least 7.5% of its funding to ELSI issues. We can hope that the funding of ELSI research in other countries, such as Japan, and developing countries, will increase, as these ELSI grants are one measure of the seriousness that molecular biologists and geneticists place on considering the social impact of their research. Another measure is the growing willingness of biotechnologists to participate in multi-disciplinary seminars, and the extent to which they start to interact with the general public. To become involved in genuine consideration of ELSI issues requires commitment of time and resources. We need to broaden the understanding across disciplines that have traditionally been closed, and growing consideration must be given to ELSI issues, issues which are not always new, but will become ever more important in the new age of the genetic information explosion. All countries will be affected, and there needs to be research on ELSI issues by people in all societies, for although the science and technology may be similar, social and ethical attitudes of society can vary from country to country.

A recent survey of attitudes to biotechnology and genetic engineering in Japan was compared to surveys in New Zealand (Table 19.2). The most common expectation in both countries of a benefit from genetic manipulation of human cells was in the medical field, and from microbes making useful commodities. Expectation of economic benefits was not cited often, although a higher proportion of respondents in New Zealand expected such benefits, perhaps because the economy is so dependent upon biotechnology in terms of agriculture, and the economic recession has been much harder there. Of the reasons cited for the need for genetic manipulation of animals, a high proportion of New Zealanders cited disease control. In both countries similar proportions cited "new varieties" or "increased production of food" as the main benefits of genetic manipulation of plants and animals, with a trend for more New Zealanders to cite the latter.

The major reasons cited for the unacceptability of genetic manipulation can be grouped into the following groups:

1. unnatural, playing God, unethical;
2. disaster, fear of unknown, ecological and environmental effects;
3. human misuse, insufficient controls, eugenics, cloning, humanity changed;
4. health effects, mutations;
5. reason not stated.

Group 1 concerns may persist with development of the technology, but groups 2 and 4 concerns may be lessened by development of technology. Group 3 concerns can be lessened by regulations. People who did not cite a reason may feel less strongly about the issue, but there is no real indication of what concerns they had. We should also note that many people expressed several of these concerns.

There was also wide diversity in perceived risks from genetic manipulation. These were in general more to do with human misuse and activity, rather than abstract concerns such as "interfering with nature". Specific risks were seen as deformities and mutations. In addition to ecological and environmental concerns, there was also substantial numbers who perceived risk connected with the spread of genes, viruses, and GMOs: what we could call "biohazard". A few people said that science was always associated with danger. There was significant similarity in the reasoning of people in Japan and New Zealand, but data from other countries are not available. Research in all countries, including developing countries, needs to begin to allow society to better prepare for the impact of biotechnology.

The control over human DNA, which is already possible to a limited degree in applications of genetic screening and gene therapy, raises the issue of genetic selection and eugenics. New technologies, such as somatic cell gene therapy, have entered clinical trials in many countries, for a wide-range of diseases and purposes. People in Japan, New Zealand and USA have been shown to support the use of human gene therapy, though they may still have some concerns about it. We need to elevate the importance of individual autonomy, especially in reproduction, while at the same time limiting misuse of new technologies by individuals. We need to maintain a distinction between diagnosis and treatment of disease on the one hand and selection for desirability on the other. Some countries allow the use of sex selection (of foetuses) itself, and in most countries it is condemned but not illegal. With the application of genetic screening we should ask whether the next generation will benefit from being genetically selected. We need data to measure the effects on personal, family and social attitudes. There must be limits on reproductive choice for enhancement, though guaranteeing reproductive freedom is a much more urgent ethical problem in some developing countries.

Universal laws, for example Article 23 of the International Covenant on Civil and Political Rights (which states that "the right of men and women of marriageable age to marry and found a family shall be recognised", and which has been signed by over 75 countries) should guarantee that compulsory eugenics is not introduced. Such statements are based on the ethical principle of respect for human autonomy. While we should not be afraid for society to change, we should be very cautious about possible adverse changes in social attitude because social pressures are very difficult to control. Such a covenant needs to be supported by equal access to social and health services in order to make it effective. In the same covenant there is also supposed recognition of equal access to health care, but what is required is "equal access to equal health care".

Genetic screening for many disease traits and susceptibility to disease will be able to be performed, so confidentiality will be important so that individuals that are found to be carriers of alleles for genetic disease will not be discriminated against. The question of fairness in the use of genetic information with respect to insurance, employment, criminal law, adoptions, the educational system and other areas must be addressed. Education and laws to ensure respect for equality are required. We must also attempt to avoid stigmatisation or ostracism, and labelling in general. We must look at all possible individual psychological responses. There may be a change in attitudes to ourselves and social customs and genetic determination might become popular. This oversimplifies the complex interaction of genetics and environment. In the extreme, determinism eliminates the idea of genuine choice, leaving no room for the belief that we can create, or modify ourselves, or that we can make moral choices.

Most religious approaches support the rationale for obtaining better genetic information, which can be used for human benefit or to alleviate human suffering. It is essential for widespread education to be available in a manner that the public can understand, so they can be involved in decisions that will change the shape of the world. An adequately prepared lay community is the best way to ensure that misuse of genetics does not reoccur. Also, the amount of information obtained may overwhelm existing genetics services and geneticists in industrialised countries. The delay in the introduction of widespread genetic testing in developing countries because of personnel and resource shortages may allow selection of the most appropriate technology, ethically, socially and scientifically.

A common feature of many issues raised by biotechnology is the need to consider the effects of knowledge of technology on future generations. Our traditional views of morality only involve short-term consequences. The ability to genetically engineer all organisms including ourselves changes our moral horizon, as do changes already underway to the global environment. We need to ensure that future generations retain the same power over their destiny as we do, while benfitting from the culture and technology we have developed.

19.9 Ethical criteria for biotechnology

There is a moral imperative to obtain predictive knowledge and data about the wide-ranging consequences of any action. Secondary consequences may be reason enough to prevent the primary action, even when the primary action itself may be good. This imposes a restraint on the use of technology. Researchers may be held accountable for secondary consequences of their research.

We also can ask what criteria are important for scientists to follow in order that we have ethical biotechnology. Although it is possible to develop useful numerical scoring systems, as has been attempted for animal experiments, these are still only guidelines and may only be useful in directing attention to better and more ethical design of experiments. Some important questions useful in assessing the ethicity of an experiment or application are given below.

(i) What is the benefit? To whom? Is it life-saving? Is human benefit greater than monetary benefit?

(ii) Will harm be caused to humans? What it an acceptable level of risk? Are ethical codes followed on free and informed consent for human experimentation?

(iii) Is pain caused? Are animal rights protected as much as possible? Are less sentient animals used for research, and non-animal alternatives developed?

(iv) Is harm caused to the environment? Is the technology used the most environmentally sustainable over the long term?

(v) Is biodiversity protected? Are endangered species protected? Do farmers have affordable or free access to breeding stock? Is planting of diverse crops encouraged?

(vi) Is justice provided to all people, and future generations? Are benefits weighed against risks?

(vii) Is decision-making on safety questions open and independent? Are ethical and social impacts given consideration?

(viii) Are the public and scientists reliably informed and educated about all dimensions of the projects; scientific, social, economic and ethical?

19.10 International cooperation

It would be unethical to withhold information that could save human lives or bring about environmental improvement. International cooperation is required: competition should not sacrifice lives. One thing that should have been learned by people in industrialised countries is that the proportion of prosperous people in society can increase, and this lesson should be applied to the global situation. The improvement in living standard of people in one country is not a threat to the lives of people in other countries, but rather it is respecting equal human rights and justice, which many people claim to recognise. New varieties of crops and animals should be available to all at an affordable price, and without discrimination.

The representatives of the people of the world (legal and political) need to join together to make decisions. Currently, it has been assumed that whoever pays for the research can control the direction of biotechnology, but genetic resources are the shared property of all. Technologies of important international interest and utility should be openly and equally accessible to all people in need of them.

We must also look at the role that individual scientists can and should play in the world. Despite commercial pressures, there is still scope for scientists to promote the introduction of biotechnology in developing countries. In the modern world scientists cannot bury their heads in the sand, or in their individual research, or in the interests of their individual companies and countries. Different countries face different situations, but all can apply biotechnology. In sub-Saharan Africa, especially, the limiting factor is the shortage of trained scientists in biotechnology, in addition to shortage of funds. However, comparatively small investment is required in biotechnology for potentially large returns. International aid organisations would be well advised to also consider this. There are some recent cases of the free introduction of technology from companies who had obtained biological materials from developing countries: this is to be commended. However, much more is required. Public opinion in industrialised countries can influence the "generosity" of multinational companies, and more pressure could be exerted by all peoples of the world to achieve greater cooperation.

The international nature of biotechnological research and its universal applicability mean that it is essential to have international organisations such as the UN taking an active part in the scientific work and considerations of the ethical, legal and social issues of both human genetics and of agricultural uses of biotechnology. Those varieties and organisms favoured by companies in industrialised countries should not be the only ones used. Although many countries are unable to significantly contribute material resources to the scientific advancement of biotechnology, they share in the human relationship to the biosphere, and often the organisms whose genes are being used. Society's interests should transcend proprietory rights, especially given the special nature of living organisms. Humanity does have a chance to build on the supposedly improved international climate in very practical ways.

Self assessment exercise 19.7

Give the potential advantages and disadvantages to a developing country of recognising the patent rights of multinational companies to genetically engineered crop varieties.

Self assessment exercise 19.8

Give the potential advantages and disadvantages to a developing country of technology transfer in biotechnology.

19.11 Concluding remarks

In concluding, not just this chapter but this book, we have seen that the human race stands at the threshold of a biotechnological revolution. Our ability to understand the workings of genes — the fundamental controlling units of all living things — and our ability to move them artificially between unrelated organisms means that humans can control and alter evolution itself, and create novel (transgenic) organisms (plant, animal and microbial). This technology, coupled with improvements in technology associated with the propagation and use of transgenic organisms, promises to improve dramatically areas such as: healthcare; supply of food, fuels, organic chemicals; environmental protection; resource recovery (oil and minerals). The benefits, existing and potential, have been discussed in the pages of this book.

Despite the obvious advantages of the biotechnological revolution, serious issues and concerns are raised by it. These issues are legal, ethical, moral, economic, sociological, environmental and political. Our objective in these last chapters has been to raise these issues. Finding satisfactory solutions to them depends upon education — informing scientists, decision makers and the public in general of the issues. Promoting such education has been our purpose in preparing this book.

19.12 Suggested further reading

1. Anon. (22–27 July, 1990). Genetics, Ethics and Human Values : Human Genome Mapping, Genetic Screening and Therapy. *Proceedings of the 24th CIOMS Round Table Conference*, Tokyo.
2. Anon. (1991a). *Biotechnology in a Global Economy*. U.S. Congress Office of Technology Assessment. U.S.G.P.O., Washington.
3. Anon. (1991b). *Managing Global Genetic Resources: The National Germplasm System*. National Research Council. National Academy Press, Washington.
4. Ahmed, I. (1991). Biotechnology and rural labour absorption. *Biotechnologies in Perspective: Socio-economic Implications for Developoing Countries*. UNESCO, Paris, 57–72.
5. Barton, J.H. (1991). Patenting life. *Scientific American*, **March**, 18–24.
6. Bizley, R.E. (1991). Patenting animal life in Europe. *Biotechnology*, **9**, 619–622.
7. Carey, N.H. and Crawley, P.E. (1990). Commercial exploitation of the human genome: what are the problems? *Human Genetic Information: Science, Law and Ethics*. Ciba Foundation Symposium 149, Elsevier North Holland, Amsterdam, 133–147.
8. Crespi, R.S. (1991). Biotechnology and intellectual property. *Trends in Biotechnology*, **9**, 117–122 and 151–7.
9. Eisenberg, R. (1991). Genes, patents and product development. *Science*, **257**, 903–4.
10. French-Anderson, W. (1992). Human gene therapy. *Science*, **256**, 808–813.
11. Fujiki, N. and Macer, D.R.J. (Eds.) (1992). *Human genome project and society*. Eubios Ethics Institute, Christchurch.
12. Gibbons, A. (1991). Biotech pipeline: bottleneck ahead. *Science*, **254**, 369–70.
13. Hobbelink, H. (1987). Biotechnology and third world agriculture. *Nature*, **330**, 321–322.
14. Juma, C. (1989). *The Gene Hunters: Biotechnology and the Scramble for Seeds*. University Press, Princeton.
15. Kumar, N. (1988). Biotechnology revolution and the third world: an overview. *Biotechnology revolution and the third world*. Research and Information System for the Non-aligned and Other Developing Countries, New Delhi.
16. Lesser, W. (1991). *Equitable patent protection in the developing world: issues and approaches*. Eubios Ethics Institute, Christchurch.

17. Macer, D.R.J. (1990). *Shaping genes: ethics, law and science of using genetic technology in medicine and agriculture.* Eubios Ethics Institute, Christchurch.
18. Macer, D.R.J. (1992a). *Attitudes to genetic engineering: japanese and international comparisons.* Eubios Ethics Institute, Christchurch.
19. Porter, D.G. (1992). Ethical scores for animal experiments. *Nature*, **356**, 101–102.
20. Robertson, I. (1991). Asson, A. and Costarini, V. (Eds.). Will biotechnologies be a threat or an opportunity for the South? *Biotechnologies in perspective: socio-economic implications for developing countries.* UNESCO, Paris.
21. Walgate, R. (1990). *Miracle or menace? biotechnology and the third world.* Panos Institute, London.

19.13 Responses to assignments

19.1 For successful implementation of biotechnological techniques the factors likely to be required are: social acceptance of the technology and a willingness to use it; sufficient resources (natural and financial) being available to allow use of the technology; the technology must be appropriate (to that particular country); necessary infrastructure (political, financial, educational and technical) must be in place.

19.2 Many raw materials, especially biological ones of plant origin, are imported in bulk from developing countries by developed countries in which the plants cannot be grown. Many pharmaceutical products and speciality products for cosmetics and foods fall into this category. Such products are normally given any refinement necessary in developed countries, so that prices paid to producer countries are kept low. Also, with development of biotechnological techniques, such as the ability to produce speciality plant products from large-scale cultures of plant cells *in vitro*, developed countries may be able to replace imports of biological products with home-produced commodities, and so undermine current markets for products from developing countries. In this way biotechnological advancement in developed countries can lead to economic hardship in developing ones.

19.3 Hybrid seeds are produced by crossing selected male parental plants with selected female parental plants. Such seeds give rise to hybrid plants expressing desired characteristics of each parental plant. However, the seeds produced by such hybrid plants will not necessarily be of the same genotype (genetic composition) as the original hybrid seeds, and so may not necessarily give rise to plants having the same phenotype (characteristics) as the parental hybrid plant. Thus, if farmers use seeds from hybrid crops as seedstocks for crops in future years then the desired characteristics of the original hybrid become 'diluted out'. To be certain that all crop plants have the desired genotype then hybrid seeds supplied by the seed producer must always be used. In this way seed producers are able to control seedstocks of hybrid varieties they have developed.

19.4 For a product or process to be patentable it must be: novel (new, not previously developed or used by anyone); non-obvious (not the next logical and obvious progression); utilitarian (must have useful applications). In order that a patent be granted the invention must be deposited in a recognised depository.

19.5 The major staple crops in many developed countries (for example potatoes, wheat, maize and rice) are not indigenous to those countries, but have their origins in developing countries. The same is true for many other crops and for ornamental plants. In almost all cases there has been no financial reward given to the country of origin of the crop species. Today's crop varieties have been highly selected, over many generations, and so the genetic diversity of crop plants is very limited. Thus, if new genetic traits (such as disease resistance) need to be bred into a crop this cannot often be achieved by interbreeding existing crop cultivars; instead desirable genes must be sought in original parental plants still existing in developing countries.

19.6 A genetic disorder arises from a mutation in a gene (or genes) which produces an ineffective enzyme (or enzymes) leading to the disorder. Gene therapy aims to deliver to cells copies of the original (non-mutated) gene, so that expression of this gene produces normal enzymes and so cures the disorder (Chapters 11 and 13). Germ cell therapy involves adding copies of genes to cells (ova) after fertilisation, so that all cells within the individual carry copies of the gene, and so that the gene can be inherited by that individual's offspring. At present this procedure is not regarded as ethical, at least in humans.

Somatic gene therapy involves delivering genes to particular cells of an individual when the disorder becomes apparent; germ cells are not affected and so the gene in question is not inherited by an individual's offspring. This technique is regarded as ethical and is currently being tried out in several countries. One example is in the treatment of cystic fibrosis. In this the defective gene has been characterised and the native gene cloned. Cloned genes are administered to patients as an aerosol (inhaled into the lungs), where it is hoped that the genes will recombine with and be expressed in pulmonary cells.

19.7 The crop varieties in use today have been selectively interbred (inbred) over many generations. While this makes them highly suitable for cultivation under the conditions for which they have been selected, if those conditions change to become unfavourable, then there may not be the natural genetic diversity available to enable development (breeding or propagation) of plants adapted to the new conditions. For example, rice has been bred to thrive in, and has become adapted to, conditions which require copious quantities of water. If, as a result perhaps of global warming, conditions in rice-producing countries were to become drier, there would be a need to produce rice varieties which would thrive in drier conditions. The genes necessary for this may no longer exist in currently used rice varieties; such genes may have been 'bred out' during the selective breeding process. For reasons such as this it is important to preserve the widest possible genetic variation amongst our crop plants and their ancestral wild relatives, so that they can be adapted in the future to meet changing conditions of climate and attack by pests and microorganisms.

19.8 Genetic engineering of plants involves the insertion of a particular gene (or genes) into the plant cells. The inserted gene is normally in a construct containing a promoter (to ensure expression of the inserted gene) and a marker (such as antibiotic resistance; to allow selective isolation of recombinant cells) (Chapter 12). It could be possible to encode in the construct a given nucleotide sequence (fingerprint) unique to that construct, so that the detection of this sequence in the cells of a plant proves that it is derived from the original transgenic plant.

19.14 Responses to self assessment exercises

19.1 Improvements in crop breeding in a developed country may make seeds for improved crop production available in a developing country. However, improved crop production in a developed country might lessen demand for imports from developing countries. Advances in plant cell culture and genetic engineering may allow production of plant compounds or their substitutes by fermentation processes in a developed country, so lessening the need for crop production in a developing country.

19.2 In Response 17.1 above it has been explained that biotechnological developments in developed countries might lead to decreased crop production (and employment in agriculture) in developing countries. Herbicide resistant plants may lead to decrease in requirement for labour for weeding. Other developments in crop manipulation may make crops amenable to mechanised crop management and harvesting, lessening employment. Manipulation of crop plants to make crops less susceptible to pest attack and to improve storage characteristics may allow countries to become self-sufficient in food with decreased employment in agriculture.

19.3 Increased agricultural output in a developing country may allow decreased imports of food/fodder and may provide surpluses which could be used as substrates for fermentation processes to provide fuels, bulk chemicals or fine chemicals (thus further reducing imports). However, such increased crop production may be dependent upon increased importation of seeds, fertilisers and agrochemicals.

19.4 Improved seeds may result in crops which require high inputs of fertilisers, herbicides and pesticides, which are beyond the financial means of small farmers. Likewise the hybrid seeds needed regularly may be too costly for small farmers. Developed countries have highly mechanised farming systems, and crops are generally selected with this in mind: this would favour larger mechanised farming practises if the crops were transferred directly.

19.5 Leasing product rights for novel compounds discovered in the flora of a country has the advantages of financial reward to the country offering such a lease, and the possibility that such agreements offer an incentive to preserve the indigenous flora (with concomitant effects such as preventing deforestation). Employment may be created within the developing country by the activities of the multinational company. However, a feeling may be generated within the developing country that such activities represent economic imperialism and a continuation of exploitation of genetic resources of developing countries by multinational companies. The discovery of novel plant compounds may allow substitution of existing products, and so damage agricultural exports for a developing country (see Response 17.1 above).

19.6 Genetic engineering is based upon fundamental knowledge of the structure and control of particular genes. This knowledge can be prohibitively difficult, expensive and time-consuming for a developing country to acquire. Gene banks may take many years to establish, are relatively costly and are usually based on international ventures. It may thus be much quicker, easier and less expensive overall for a developing country to buy in genetic engineering resources from a developed country, rather than establish its own programmes.

19.7 The potential advantage to a developing country of recognising patent rights to genetically engineered crop varieties is that any particular crop variety will then be made available to that country; failure to recognise such patent rights may mean that seedstocks of any particular crop are withheld. The potential disadvantage of recognising such patent rights is that the company holding the patent right has the ability to control the selling price of the seedstocks:

19.8 The potential advantages to a developing country of biotechnological transfer are that improvements in agriculture, health care, environmental protection and fuel/chemical production may take place more rapidly and at less overall cost than without technology transfer. The potential disadvantages are that technology transfer will: stifle local scientific and technological development; be tied to importation of related commodities (chemical feedstocks); be inappropriate to the situation within that particular country.

Contributors and editors

Ronald M. Atlas is Professor of Biology and Dean of the Graduate School at the University of Louisville. He received his B.S. from the State University of New York at Stony Brook in 1968 and his M.S. and Ph.D. from Rutgers in 1970 and 1972, respectively. Dr. Atlas is President-Elect for the American Society for Microbiology.

He began studying the microbial degradation of petroleum pollutants in 1968 as part of his graduate studies and received his Ph.D. from Rutgers in 1972 with a dissertation on this topic. He then spent a year at the Jet Propulsion Laboratory of Cal Tech in Pasadena, working on Life Detection Systems for Mars before joining the faculty at the University of Louisville in August 1973.

His research at the University of Louisville (U of L) focused on the environmental fate of petroleum and other organic pollutants with projects funded by the U.S. Navy, EPA, DOE, and NOAA. The research was recognized as pioneering and involved both basic research and applied responses to major oil spills. His research also examined biotechnological approaches for upgrading shale oils, biodegrading sulfur and nitrogen-containing compounds, and using molecular biological approaches for tracking microorganisms and containing genetically engineered microbes that are deliberately released into the environment. Studies at U of L resulted in nearly 300 scientific publications. Dr. Atlas is also an active teacher and author of several textbooks and reference volumes, including one on petroleum microbiology. He serves on numerous editorial boards, and as editor of Critical Reviews in Microbiology. He has served as chairperson of the Public Affairs Board Environment Committee of the American Society for Microbiology and on the Recombinant Advisory Committee for the National Institutes of Health.

He chairs the American Society for Microbiology task force on biological weapons. He also serves as the White House representative to the Organization of Economic Co-operation and Development on biotechnology issues.

Dr Frederick William George (Mike) Baker teaches science popularisation at the Institute of Advanced Studies, Caracas, Venezuela. He is interested in Polar Research. During his career, Dr Baker has written books and articles and edited more than 40 scientific books.

Dr Vineeta Bal holds an M.D. in Microbiology and teaches immunology to beginners in the discipline. Dr Bal has worked on the development of a diagnostic test for *T. meningitis* as one of the early problems in immunology, followed by a study on immunology and epidemiology of sporadic viral hepatitis. Her research interests lie in studying the nature of MHC, peptide, T-cell interaction; role of heat shock proteins in autoimmunity, and mechanism of tolerance induction in T-cell clones.

Dr Bal is currently engaged in research on predicting the nature of immune response in terms of activation or tolerance, mechanisms involved in triggering of a Th1-like or Th2-like immune response following immunisation with bacterial antigens and studying the accessory signals necessary to bring about T-cell activation vs. tolerance, etc.

Dr Bal is employed with the National Institute of Immunology, New Delhi, in the capacity of Staff Scientist-I.

Professor D. Balasubramanian was Director, Centre for Cellular and Molecular Biology, Hyderabad, India. Born on 28 August 1939 in Sholavandan, Tamil Nadu, he holds a Ph.D. in Chemistry from Columbia University, New York. He is also a visiting scientist at the National Eye Institute, National Institutes of Health, Bethesda. He is currently the Director of Research at L. V. Prasad Eye Institute, India.

His research interests cover biophysics of eye lens transparency and cataract, biomolecular spectroscopy, micelles, membranes and membrane-active peptides, and protein structure and interactions. He has over 130 research articles published in standard refereed journals of international repute. He has a keen interest in popularising science as demonstrated by his popular articles, science books and educational programmes in the media.

Dr George I. Bell obtained his Ph.D. in Theoretical Physics from Cornell University, New York, in 1957. Dr Bell's research interests laid in studying the various problems in theoretical biology, including theoretical immunology, modelling cell–cell adhesion, repetitive DNA sequences and their evolution. He was a founder of the Center for Human Genome Studies in 1988 and continued to be actively involved in the Human Genome Project until his death.

Professor Charles F.A. Bryce is currently Head of School of Life Sciences at Napier University, Edinburgh.

In the last thirty years he has published in excess of eighty refereed research publications, about eighty non-refereed publications, and designed and produced about thirty teaching packages in a wide variety of media formats from print to multimedia presentations. His research interests cover such topics as protein structural biochemistry, site-directed mutagenesis, enzymology and stopped-flow spectroscopy, educational technology, computer applications, and a review of manpower and training needs in biotechnology. In the course of this work Professor Bryce has undertaken an international survey of the biochemistry curriculum for science and medical students and an education market survey of India and has undertaken with Dr. John Green, a pan-European survey of curriculum content for undergraduate and postgraduate programmes in biotechnology.

For the last ten years Professor Bryce has worked extensively with a number of colleagues in India, Hong Kong, Turkey, USA and a number of countries in Europe on a variety of issues relating to curriculum development, teaching innovations and manpower and training strategies in biotechnology. He has served on a great many national and international committees including UK Deans of Science (Chairman), UK Interest Group on Education in Biotechnology (Chairman), Education Group of the Biochemical Society (Chairman), European Federation of Biotechnology Working Party on Education (Chairman), Specialist Adviser to a number of British Council and COSTED Projects and Executive Editor of the journal *Computer Applications in the Biosciences*.

Nick Christofi is currently Director of the Pollution Researh Unit at Napier Uniersity. He obtained his Ph.D. at the University of Dundee following studies on the role of nitrifying bacteria in eutrophic freshwater lakes. He has worked on the microbiology of deep subsurface environments with respect to the disposal of radionuclides for various organisations in the UK, Switzerland and the UK. He has acted on advisory committees for the UK Department of the Environment, CIRIA (Construction Industry Research and Information Association), BNFL and the US DOE Center for Nuclear Waste Regulatory Analysis.

His recent work is on alkanotrophic *Rhodococcus* species and their role in oil bioremediation and surfactant production, and, with respect to this work, he participates in NATO Advanced Study Institutes (ASI). His work also involves the development of microbial toxicity bioassays and biosensors utilising bioluminescent bacteria. Additional projects deal with the anaerobic degradation of organic pollutants in soils and the use of fluorescent biological and non-biological material to trace pollutant migration in natural systems.

Dr Rita R. Colwell completed her Ph.D. in Marine Microbiology from the University of Washington.

Dr Colwell focused her research on marine biotechnology and the molecular genetics of marine and estuarine bacteria, and on the microbiology of the Chesapeake Bay, coastal recreational waters, and the deep ocean.

Dr Colwell has several top positions to her credit. She has been President of the International Union of Microbiological Societies, President of the American Association for the Advancement of Science and also the President of the American Society for Microbiology. She is currently the Chairman of the Board of Governors of the American Academy of Microbiology and Chairman of the Biological Sciences Section of the American Association for the Advancement of Science. Dr Colwell is also the Presidential Appointee to the National Science Board of The United States. She has also received several honours and awards for her contributions to the subject.

Dr Colwell has authored / co-authored 16 books and has over 450 publications on her name. She headed the University of Maryland Biotechnology Institute at College Park, as its President and is now President of the US National Science Foundation.

Colin J. Cunningham joined the Contaminated Land Assessment and Remediation Research Centre (CLARRC) as research co-ordinator in 1998 as a new graduate. The post has a research role in the bioremediation field and a managerial and administrative role within the wider CLARRC group. He is actively involved in teaching and researching bioremediation and served as a member of the Advisory Group for the Audit of Contaminated Land Research in the UK, DETR. He is co-ordinator of the Department of Trade and Industry Link Bioremediation programme. He won the Conservation Engineering category award of the UK Henry Ford Conservation and Environmental Grants Programme 2000 for bioremediation work done in the UK and Russia.

Dr M.G. Deo is the Director of Cancer Research Institute of the Tata Memorial Centre at Bombay. He has over 22 years of teaching experience at various academic positions including Professorship in Pathology at the All India Institute of Medical Sciences (AIIMS), New Delhi, India. His research interests include cancer biology, growth modulation and oncogenes.

Professor K. Dharmalingam is working at Madurai Kamaraj University, Madurai, as a Professor. He obtained his Ph.D. degree in Biochemistry for his work on "Assembly of Mitochondrial Membranes in Yeast". After his graduation he went to Tufts Medical School, Boston, to work on the restriction - modification systems of *Escherichia coli*. He is continuing his work on restriction and repair in *E.coli* in India and his current research interests also include genetic instability in *Streptomyces* and antibiotic biosynthesis. He is actively involved in teaching various courses primarily for an M.Sc. programme in biotechnology for the past 15 years. He is at present Head and Coordinator of School of Biotechnology, Madurai Kamaraj University, Madurai, India.

Dr E. M. T. El-Mansi is interested in metabolic engineering of micro-organisms, control of carbon flux through degradative, catabolic, anabolic and anaplerotic pathways and control of gene expression in prokaryotic organisms.

Dr El-Mansi teaches microbial biotechnology, microbial biochemistry, molecular microbiology and bioinformatics at the School of Life Sciences, Napier University, Edinburgh, Scotland.

Dr Armin Fiechter holds a Degree in Microbiology from the Swiss Federal Institute of Technology (ETH), Zurich, Switzerland. Dr Fiechter taught at the Department of Biology, ETH, from 1952 till 1992.

During his long career, he also worked as a Guest Researcher in Stockholm, Prague and Wisconsin (Madison, USA) in Molecular Biology. He founded the Institute of Biotechnology and also served as its Chairman. He is currently the Managing Editor of the *Journal of Biotechnology*.

Dr Fiechter's research interests lie in bioregulation (yeast, mammalian cell culture), and on-line analytics, automation and control of advanced process strategies.

Dr N. Gautham has a Ph.D from the Indian Institute of Science at Bangalore and is presently Reader, Department of Crystallography and Biophysics, University of Madras. His research interests cover DNA structure; X-ray crystallography of oligonucleotides; theoretical modelling of DNA and protein structure; protein–nucleic acid interactions; biomolecular structure optimisation; packing of molecules in organic and biomolecular crystals; and molecular recognition. He has published about 16 research articles and 2 general articles. He was a visiting fellow at the Institute de Biologie Moleculaire et Cellulaire at Strasbourg, France, in 1990.

Dr John Green obtained his Ph.D. in Microbiology from the University of Birmingham, U.K., in 1973. During his career, Dr Green worked as Section Head, Department of Bacterial Vaccines at the Lister Institute of Preventive Medicine, Elstree, U.K., for 5 years from 1974, before which he was a Research Fellow at the Department of Microbiology, University of Birmingham.

Dr Green's research interests lie in studying the use of coliphages (viruses of E. coli) as faecal pollution indications and factors that affect coliphage survival in water.

Dr Green is at present teaching at Napier University, Edinburgh, Scotland, on the subjects of fermentation technology, food microbiology/biotechnology and microbial indicators of feacal pollution of water.

Dr Satish Kumar Gupta completed his Ph.D. in Biochemistry from the All India Institute of Medical Sciences (AIIMS), New Delhi, India.

His research interests include immunology, endocrinology, reproductive immunology and immunodiagnostics. He has made significant contributions in the fields of reproductive immunology, sperm-egg interaction, immunocontraception, monoclonal antibodies, B-cell epitope mapping and immunodiagnostics.

He is currently Director of the Institute of Life Sciences, Bhubaneswar.

Dr Russell T. Hill completed his Ph.D. in Microbiology at the University of Cape Town, South Africa, in 1988. Thereafter, he did his post-doctoral research at the same University before shifting to the University of Maryland, USA.

Dr Hill has broad interests in marine microbial ecology and biotechnology, especially in the application of molecular techniques to these areas. His research includes molecular detection of pathogenic bacteria in the marine environment and effects of pollution on microbial ecosystems. Dr Hill is currently working on marine actinomycetes for isolation of novel bioactive compounds, and mechanisms of metal resistance in actinomycetes for bioremediation applications.

Dr Hill is presently working as Research Assistant Professor at the Centre for Marine Biotechnology, University of Maryland Biotechnology Institute, USA.

PROFESSOR KUNTHALA JAYARAMAN was head of the Centre for Biotechnology, Anna University. The major research efforts of her group are in the area of biological pesticides for the control of mosquitoes that transmit several tropical diseases like malaria, filariasis, etc., and also the agricultural insect pests mainly following lepidopteran species. In addition, research efforts for expressing recombinant proteins and purifications are under way. She obtained her Ph.D. at the Indian Institute of Science, Bangalore and spent several years in the United States of America at the Indiana University and Harvard Medical School. Subsequently Prof. Jayaraman has spent a sabbatical year at the Laboratoire d'Enzymology, France, as Maitre d'Recherche CNRS and has been a Fullbright Fellow at NIH, USA. She received the biotechnology prize from the National Academy of Sciences, India and Dr M O T Iyengar Award from the Indian Council of Medical Research. She is currently the Director of Biotechnology Centre at the Vellore Institute of Technology.

DR LEV LVOVITCH KISSELEV obtained his Doctorate of Sciences in Biochemistry from the Bach Institute of Biochemistry, Moscow. Thereafter, he underwent his post-doctoral training at the Engelhardt Institute of Molecular Biology from 1964 to 1975.

Dr Kisselev is a Member of the Scientific Council of the Human Genome Organization (HUGO); Chairman of the Scientific Council for Molecular Biology and Genetics of the Russian Academy of Sciences, and Head of the Scientific Council of the Russian National Human Genome Programme.

Dr Kisselev's research interests are in the fields of protein biosynthesis and the Human Genome Programme. He has also taught courses at the Moscow State University and the College de France, Paris.

Dr Kisselev is currently working at the Engelhardt Institute of Molecular Biology, Moscow.

PROFESSOR JEAN-MARC LAVAL is a Professor of Biochemistry at the Technical University of Compiegne (TUC), France. He obtained his Ph.D. from the Compiegne University in1983. His teaching experience is in the fields of biological physicochemistry, biomembranes and enzymology. His research interests are based on the association between enzymology and electrochemistry, i.e., heterogeneous enzymology, cofactor regeneration and electrosynthesis. Recent research areas include molecular self-assembled structures on electrochemical interfaces that mimic the electron transfer chain of the biological membrane.

PROF. VLADIMIR I. LOZINSKY is the Head of Laboratory for Cryochemistry of (Bio)Polymers at the A.N.Nesmeyanov Institute of Organoelement Compounds of the Russian Academy of Sciences, Moscow, Russia. He obtained his Ph.D. (organic chemistry) at this Institute in 1982 and D.Sc. (Full Professor; Macromolecular Chemistry and Biotechnology) at the same Institute in 1994. His research interests are in the field of cryogenic processes in polymer systems, as well as in the application of cryogenically-processed polymer systems in various practical spheres: cryoprocessed food, biotechnology (the carriers for the preparation of immobilized biocatalysts and special biosorbents), polymeric materials for biomedical purposes, the processing of biopolymer-containing wastes of the leather industry, modification of ice, fixation of thawed soils and grounds, and others. His teaching experience is in biotechnological systems based on immobilized biocatalysts.

Professor Darryl R. J. Macer was born in New Zealand in 1962, and completed a Bachelor of Science with first class honours in Biochemistry from Lincoln College, University of Canterbury. He completed a Ph.D. in Biochemistry at the MRC Laboratory of Molecular Biology, and Trinity College, University of Cambridge, UK.

He has since researched bioethics, as a senior research scholar at Trinity College, University of Cambridge U.K; and as a consultant scientist to the Department of Scientific and Industrial Research, and the Ministry for the Environment, New Zealand. He was the founder, and is the director of Eubios Ethics Institute, Christchurch, New Zealand and Tsukuba Science City, Japan. Since September 1990 he has been a Foreign Professor, and Associate Professor, in the Institute of Biological Sciences, University of Tsukuba, Ibaraki 305, Japan.

He is a member of the UNESCO Bioethics Committee, the coordinator of the International Association of Bioethics Genetics Network, and Editor, *Eubios Journal of Asian and International Bioethics*.

He has authored 6 books (1 in Japanese), edited 2 books (2 in Japanese), and has written about 50 academic papers and articles.

Dr Rita Mulherkar is officer-in-charge of the Laboratory of Genetic Engineering at the Cancer Research Institute, Tata Memorial Centre, Bombay. Her research interests cover cloning of novel growth factors, production of transgenic mice and gene therapy for oral cancer.

Professor Vasantha Pattabhi is a Professor at the Department of Crystallography and Biophysics of the University of Madras. She has a Ph.D from the University of Madras and her research interests include X-ray crystallography and theoretical modelling of peptides, proteins and drugs; and bioinformatics. She has over 80 research articles and 4 general articles published to her credit. She has been a visiting Professor at the University of Pittsburg, University of London and National Research Council, Canada.

Dr Antonio Pena was born in Mexico on 14 February, 1936. He took his M.D. from the University of Mexico in 1960, followed by a Doctorate in Biochemistry from the same University.

During his career, Dr Pena worked as a full-time investigator in biochemistry for 19 years from 1960. He attended the Albert Einstein College of Medicine, New York, as a Post-Doctoral Fellow. He also served as President of three scientific organizations : Mexican Biochemical Society, Mexican Academy of Sciences, and the Pan American Association of Biochemistry and Molecular Biology (PABMB). He was also the Director of the Institute for Cellular Physiology in the National Autonomous University of Mexico from 1979 to 1993.

Dr Pena has over 60 publications to his credit in international journals. He has also authored/edited 10 books. Dr Pena's current interests lie in the study of the interrelation of ion movements in yeast with respect to pH of cells using fluorescent indicators.

Dr Jim Philp obtained his PhD in microbial corrosion in 1987 and subsequently moved to Saudi ARAMCO, where he worked on various projects involving oil-degrading bacteria and microbial corrosion. Since joining Napier University in 1992, he has worked on various research projects related to bioremediation e.g. biosurfactants as oil washing agents. Over the past eight years, he has worked on pilot-scale bioremediation projects in the field in the UK, Russia, and Antigua.

He has worked on the development of immobilised bacteria as toxicity biosensors for the protection of wastewater treatment plants and on other microbial problems associated with wastewater treatment and landfill sites. He is co-ordinator of the Department of Trade and Industry Link Bioremediation programme. He won the Conservation Engineering category award of the UK Henry Ford Conservation and Environmental Grants Programme 2000 for bioremediation work done in the UK and Russia.

Dr Raj Raghupathy has a doctorate degree in Immunolgy from the University of Alberta, Edmonton, Canada. He is presently Associate Professor of Immunology at the Department of Microbiology, Kuwait University. His research interests are in the areas of Vaccinology and Reproductive Immunology.

Dr K.C. Reed is at the Department of Chemistry, University of California, Riverside.

Dr Arthur Charles Robinson holds a Ph.D. in Biochemistry from the University of Aberdeen, Scotland, U.K.

He is currently teaching undergraduate and postgraduate courses in biochemistry and molecular biology/molecular genetics at the Department of Biological Sciences, Napier University, Edinburgh, Scotland. He is also a regular contributor to the journal *Biochemical Education* (Monitor Section).

Dr Robinson's research interests include the study of control of cell division in *Escherichia coli*, organization of eukaryotic repetitive DNA sequence elements and immunological responses to epitotes of the core protein of HIV.

Dr Gregory Alan Smith took his Ph.D. in Biochemistry from The University of Queensland, Australia, in 1992.

He has over fourteen years of research experience in the field of molecular virology and vaccine development. His current research interests include the molecular characterisation of bovine and macropodid herpes viruses, development of recombinant bovine herpes virus vaccine and development of oral vaccine delivery systems for administering immunocontraceptive vaccines to captive and free-ranging kangaroos and wallabies.

Dr Smith is at present working as a Principal Biotechnologist (Animal Biotechnology) at the Queensland Agricultural Biotechnology Centre, St. Lucia, Queensland, Australia.

Dr Manjula Sritharan holds a Ph.D. in Biochemistry from the University of Hull, England, UK and has worked as a postdoctoral research fellow at Harvard School of Public Health, Harvard University, Boston, U.S.A.

Dr Sritharan is currently working as a reader at the Department of Animal Sciences, University of Hyderabad.

Dr Sritharan is currently working on certain aspects of Host Parasite inter-relationship in Lymphatic Filariasis, a tropical disease caused by the filarial parasites Muchereria bancrofti and Brugia malayi.

Professor Gurusaran Prashad Talwar received his Docteur es Sciences from Sorbonne, while working at Institute Pasteur, Paris.

Dr Talwar has made significant research contributions on the mechanism of action of hormones and has pioneered the development of new vaccines, among which a leprosy vaccine and a birth control vaccine are worth highlighting. He received several awards for his outstanding contributions in the field of bio-medical research.

Dr Talwar holds many important positions. He was formerly the Founder Director of the National

Institute of Immunology and Head of the WHO–ICMR Research and Training Centre in Immunology for India and South-East Asia.

He is currently President of two organisations — the Indian Academy of Vaccinology & Applied Immunobiology, and the International Society for Immunology of Reproduction. He is also a Fellow of the Indian National Science Academy, National Academy of Medical Sciences, Indian Academy of Sciences and Fellow ad enudem of the Royal College of Obstetricians and Gynaecologists, London.

Dr Talwar was Professor of Eminence at the International Centre for Genetic Engineering and Biotechnology, New Delhi, India and is now President of the Talwar. Foundation for Research.

Dr Daniel Thomas is presently Director of the Laboratoire de Technologie Enzymatique of the CNRS as well as the Director of the "Zeme cycle" formation in Enzymatic Engineering. He also holds a professorship at the University de Technologie de Compiegne (UTC) and serves as the Vice President of the Science Advisory Board of the University. He is also the Director of Study at the Ecole Pratique des Hantes Etudes in Paris.

He has held several important international responsibilities such as Chairmanship of Biotechnology Committee of OECD, Presidentship of Evaluation Committee of the Programme on "Capital Humain et Mobilite" for the EEC (1993).

Dr K. Veluthambi is Professor and Head of the Dept. of Plant Biotechnology at the School of Biotechnology, Madurai Kamaraj University, India. He is involved in the teaching programmes of post-graduate courses at the University. His research interests include mechanisms of DNA transfer, genetic transformation of rice and pulses, and cloning and analysis of genomes of viruses.

Glossary

Acquired immunity Immunity specific to a foreign antigen resulting from stimulation of the cellular immune system.

Adjuvant A substance incorporated into a vaccine to enhance the immune response to the antigens in the vaccine.

Algorithm A mathematical model for solving a particular problem, especially by the use of a computer.

Allosterism Ability to change conformation (shape).

Antibody (Ab) A protein molecule produced by B-cells which is capable of combining specifically with an antigen. Alternatively termed an immunoglobulin.

Antigen A molecule capable of activating B-cells or T-cells and which induces the formation of antibody.

Antigen-presenting cell (APC) A lymphocyte which engulfs antigen, degrades the antigen into fragments, and presents the fragments on its cell surface in order to stimulate the activities of other lymphocytes of the immune system.

Ascites Pronounced A (as in *apple*) - site (as in *sight*) - ees (as in *freeze*). A suspension of single cells growing in the peritoneal cavity.

Auxins A group of chemical substances which act as plant growth-controlling hormones. The substance 2-4-D (used as a herbicide) is an example.

Auxotrophic Requiring (in a mutant) one or more particular organic nutrients that the wild type (parental) organism does not.

Autoimmunity Immunity directed against components of 'self' instead of against foreign antigens.

B-Cell A type of lymphocyte, originating in bone marrow tissue, which matures into an antibody-secreting cell.

Callus The growth of plant cells producing an irregular disorganised mass of cells. When tissue explants are taken from a plant and placed on a solidified growth medium with a certain concentration of auxin and cytokinin hormones, callus material may develop.

Centromere A point (location) on eukaryotic chromosomes where the two arms (chromatids) of the chromosome are joined together.

Clone A group of cells or organisms having identifical genetic makeup.

Cis (in cis) When applied to control of genetic systems, cis relates to interaction directly with nucleic acid.

Coliphage A bacterial virus having *E. coli* as host.

CT scanning Computed tomography scanning, in which X ray images of a body are taken from various directions, and the results analysed by computer to give an image of the 'slice' through the body at that point.

Cytokines Protein messenger molecules which signal responses between cells of the immune system.

Cytokinins A group of chemical substances which act as plant growth-regulating hormones. Kinetin is an example.

Cytotoxic T-cell A type of lymphocyte which specifically recognises foreign antigen on the surface of a cell which is infected by a virus, for example, and then kills that cell by lysis.

Dermatitis A disorder of the skin.

Diploid Having the normal (double) chromosome content of cells of an adult; one of each pair derived from the paternal line, and the other from the maternal line.

Effector Relating to ability (of cell or molecule) to bring about a given effect or reaction.

Embryogenesis The formation of embryos.

Epitope The smallest part of an antigen which is recognised by and combines with an antibody molecule. A single antigen may exhibit several different epitopes.

Genotype The genetic composition, or genetic component of a cell/organism.

Haematopoietic Of the system which gives rise to cells of the blood.

Haploid Having only one half the chromosome content of an adult. Germ cells (sperm and ova) are characteristically haploid, possessing only one of each of the pairs of chromosomes of the adult cells from which they were formed,

Hepatocyte A liver cell.

Hepatoma Tumour (cancer) of the liver.

Hypo.... Of lower intensity than normal.

Hyper.... Of higher intensity than normal.

HLA The major histocompatibility (MHC) system of humans.

Hypersensitivity An exaggerated immune response to a foreign substance (e.g. pollen, dust) which frequently results in tissue damage/inflammation.

Immune deficiency (Immunodeficiency) A condition, which may be either acquired by infection or genetically inherited, in which one or more immune response is defective.

Immunocyte A cell involved in the immune response.

Immunoglobulin (Ig) A protein molecule (antibody) produced by B-cells which consists of two light and two heavy polypeptide chains. Alternatively termed an antibody.

In situ (From Latin-'in place'). In the location in which the entity is normally found. In its usual place.

Interferons (IFNs) A group of protein molecules (cytokines) which are capable of conferring non-specific immunity to viral infection.

In vitro (From Latin—'in glass') Outwith the living host organism.

In vivo (From Latin—'in the living being') Within the living host organism.

Killer cell (K-cell) A type of lymphocyte which is capable of killing target cells that are coated with specific antibody. The process of killing is known as antibody-dependent cell-mediated cytotoxicity (ADCC).

kDa Kilo-Daltons. A measure of mass of a molecule.

Leukocyte Circulating cell of the blood stream—any type of white blood cell.

Liposome A sac or vesicle formed spontaneously from certain types of lipid molecules when in an aqueous environment.

Locus The site, or location, on a chromosome (within a DNA molecule) of a particular gene.

Lymphocyte A type of white blood cell.

Macrophage A type of cell found throughout the body which engulfs antigen by phagocytosis. The antigen is destroyed following uptake.

Major histocompatibility complex (MHC) A complex genetic locus which specifies the cell surface glycoproteins of lymphocytes that are involved in antigen recognition. The glycoproteins are also responsible for the rejection of transplanted tissue.

Memory (immunological) The phenomenon, mediated by lymphocytes, in which a second or subsequent encounter with antigen gives rise to a more efficient and vigorous immune response compared to that which occurred at the first encounter.

Metastasis The spread of a disease from its site of origin to other parts of the body.

Michaelis-Menten theory Theory relating the rate of an enzyme-catalysed reaction to the concentration of substrate of the reaction.

Mitogen A substance which stimulates cells to divide (multiply).

Monoclonal antibody (Mb or MoAb) An antibody which is produced from a single clone of cells (i.e. all the antibody-secreting cells for a particular MAb are derived from a single ancestral cell).

Myelin A chemical component of nerve cells.

Myeloblast A cell produced by differentiation of a stem cell in the haematopoietic system.

Myeloma Cancer of haematopoietic cells.

Myocardial infarction Damage to the muscles of the heart, resulting from oxygen starvation caused by blood clots during a heart attack.

Natural immunity Immunity/resistance to infection due to physical barriers or chemical secretions. Natural immunity is non-specific and incapable of immunological memory.

Organogenesis The formation of organs.

Phenotype Visible or measurable characteristics of a cell or organism. The features which the genotype determines.

Plasmid Small DNA elements, independent of the chromosome(s), found in some microorganisms.

Pleiotropic Of a gene, having an effect simultaneously on more than one phenotypic characteristic.

Progenitor The one which gave rise to (a cell or organism). An ancestor.

Prophylaxis Procedure or treatment designed to prevent the occurrence of a disease.

Recombination The process of insertion of a gene into a chromosome or a cell, by natural mechanisms or by genetic engineering.

Reporter gene A gene used in genetic engineering experiments, to demonstrate that the gene construct (the foreign gene and its controlling sequences) has been transferred into a host cell/organism.

Somaclonal variation Non-genetic variation in phenotype in plants, resulting from cultural conditions/environment.

Somatic Of cells not involved in reproducton. Of non germ-line cells.

T-cell A type of lymphocyte, originating in bone marrow tissue, which is processed in the thymus in order to become responsive to antigen. T-cells recognise and respond to antigen via specific molecules on their cell surface known as T-cell receptors (TCRs).

T-helper cell A type of lymphocyte which helps to generate cytotoxic T-cells, and which associates with B-cells in order to increase the antibody-mediated response. T-helper cells recognise antigen in association with MHC class II molecules.

Trans (in trans) When applied to control of genetic systems, trans relates to interaction mediated through regulatory molecules such as proteins.

Vaccination The exposure of the body to an antigen with the intention of conferring specific immunity to that antigen.

Xenogenei Of unrelated (foreign) genetic origin.

Index